238
Advances in Polymer Science

Editorial Board:
A. Abe · A.-C. Albertsson · K. Dušek · W.H. de Jeu
H.-H. Kausch · S. Kobayashi · K.-S. Lee · L. Leibler
T.E. Long · I. Manners · M. Möller · E.M. Terentjev
M. Vicent · B. Voit · G.Wegner · U. Wiesner

Advances in Polymer Science

Recently Published and Forthcoming Volumes

Polymer Thermodynamics
Volume Editors: Enders, S., Wolf, B.A.
Vol. 238, 2011

Enzymatic Polymerisation
Volume Editors: Palmans, A.R.A., Heise, A.
Vol. 237, 2010

High Solid Dispersion
Volume Editor: Cloitre, M.
Vol. 236, 2010

Silicon Polymers
Volume Editor: Muzafarov, A.
Vol. 235, 2011

Chemical Design of Responsive Microgels
Volume Editors: Pich, A., Richtering, W.
Vol. 234, 2010

Hybrid Latex Particles – Preparation with Emulsion
Volume Editors: van Herk, A.M.,
Landfester, K.
Vol. 233, 2010

Biopolymers
Volume Editors: Abe, A., Dušek, K.,
Kobayashi, S.
Vol. 232, 2010

Polymer Materials
Volume Editors: Lee, K.-S., Kobayashi, S.
Vol. 231, 2010

Polymer Characterization
Volume Editors: Dušek, K., Joanny, J.-F.
Vol. 230, 2010

Modern Techniques for Nano- and Microreactors/-reactions
Volume Editor: Caruso, F.
Vol. 229, 2010

Complex Macromolecular Systems II
Volume Editors: Müller, A.H.E.,
Schmidt, H.-W.
Vol. 228, 2010

Complex Macromolecular Systems I
Volume Editors: Müller, A.H.E.,
Schmidt, H.-W.
Vol. 227, 2010

Shape-Memory Polymers
Volume Editor: Lendlein, A.
Vol. 226, 2010

Polymer Libraries
Volume Editors: Meier, M.A.R., Webster, D.C.
Vol. 225, 2010

Polymer Membranes/Biomembranes
Volume Editors: Meier, W.P., Knoll, W.
Vol. 224, 2010

Organic Electronics
Volume Editors: Meller, G., Grasser, T.
Vol. 223, 2010

Inclusion Polymers
Volume Editor: Wenz, G.
Vol. 222, 2009

Advanced Computer Simulation Approaches for Soft Matter Sciences III
Volume Editors: Holm, C., Kremer, K.
Vol. 221, 2009

Self-Assembled Nanomaterials II
Nanotubes
Volume Editor: Shimizu, T.
Vol. 220, 2008

Self-Assembled Nanomaterials I
Nanofibers
Volume Editor: Shimizu, T.
Vol. 219, 2008

Interfacial Processe sand Molecular Aggregation of Surfactants
Volume Editor: Narayanan, R.
Vol. 218, 2008

New Frontiers in Polymer Synthesis
Volume Editor: Kobayashi, S.
Vol. 217, 2008

Polymers for Fuel Cells II
Volume Editor: Scherer, G.G.
Vol. 216, 2008

Polymers for Fuel Cells I
Volume Editor: Scherer, G.G.
Vol. 215, 2008

Polymer Thermodynamics

Liquid Polymer-Containing Mixtures

Volume Editors: Sabine Enders
       Bernhard A. Wolf

With contributions by

S.H. Anastasiadis · K. Binder · S.A.E. Boyer · S. Enders ·
J.-P.E. Grolier · S. Lammertz · G. Maurer · B. Mognetti ·
L. Ninni Schäfer · W. Paul · G. Sadowski · P. Virnau ·
B.A. Wolf · L. Yelash

Editors
Dr. Sabine Enders
TU Berlin
Sekr. TK7
Straße des 17. Juni 135
10623 Berlin
Germany
sabine.enders@tu-berlin.de

Dr. Bernhard A. Wolf
Universität Mainz
Inst. Physikalische Chemie
Jakob-Welder-Weg 13
55099 Mainz
Germany
bernhard.wolf@uni-mainz.de

ISSN 0065-3195
ISBN 978-3-642-17681-4
DOI 10.1007/978-3-642-17682-1
Springer Heidelberg Dordrecht London New York

e-ISSN 1436-5030
e-ISBN 978-3-642-17682-1

© Springer-Verlag Berlin Heidelberg 2011
This work is subject to copyright. All rights are reserved, whether the whole or part of the material is concerned, specifically the rights of translation, reprinting, reuse of illustrations, recitation, broadcasting, reproduction on microfilm or in any other way, and storage in data banks. Duplication of this publication or parts thereof is permitted only under the provisions of the German Copyright Law of September 9, 1965, in its current version, and permission for use must always be obtained from Springer. Violations are liable to prosecution under the German Copyright Law.
The use of general descriptive names, registered names, trademarks, etc. in this publication does not imply, even in the absence of a specific statement, that such names are exempt from the relevant protective laws and regulations and therefore free for general use.

Cover design: WMXDesign GmbH, Heidelberg

Printed on acid-free paper

Springer is part of Springer Science+Business Media (www.springer.com)

Volume Editors

Dr. Sabine Enders
TU Berlin
Sekr. TK7
Straße des 17. Juni 135
10623 Berlin
Germany
sabine.enders@tu-berlin.de

Dr. Bernhard A. Wolf
Universität Mainz
Inst. Physikalische Chemie
Jakob-Welder-Weg 13
55099 Mainz
Germany
bernhard.wolf@uni-mainz.de

Editorial Board

Prof. Akihiro Abe

Professor Emeritus
Tokyo Institute of Technology
6-27-12 Hiyoshi-Honcho, Kohoku-ku
Yokohama 223-0062, Japan
aabe34@xc4.so-net.ne.jp

Prof. A.-C. Albertsson

Department of Polymer Technology
The Royal Institute of Technology
10044 Stockholm, Sweden
aila@polymer.kth.se

Prof. Karel Dušek

Institute of Macromolecular Chemistry
Czech Academy of Sciences
of the Czech Republic
Heyrovský Sq. 2
16206 Prague 6, Czech Republic
dusek@imc.cas.cz

Prof. Dr. Wim H. de Jeu

Polymer Science and Engineering
University of Massachusetts
120 Governors Drive
Amherst MA 01003, USA
dejeu@mail.pse.umass.edu

Prof. Hans-Henning Kausch

Ecole Polytechnique Fédérale de Lausanne
Science de Base
Station 6
1015 Lausanne, Switzerland
kausch.cully@bluewin.ch

Prof. Shiro Kobayashi

R & D Center for Bio-based Materials
Kyoto Institute of Technology
Matsugasaki, Sakyo-ku
Kyoto 606-8585, Japan
kobayash@kit.ac.jp

Prof. Kwang-Sup Lee

Department of Advanced Materials
Hannam University
561-6 Jeonmin-Dong
Yuseong-Gu 305-811
Daejeon, South Korea
kslee@hnu.kr

Prof. L. Leibler

Matière Molle et Chimie
Ecole Supérieure de Physique
et Chimie Industrielles (ESPCI)
10 rue Vauquelin
75231 Paris Cedex 05, France
ludwik.leibler@espci.fr

Prof. Timothy E. Long

Department of Chemistry
and Research Institute
Virginia Tech
2110 Hahn Hall (0344)
Blacksburg, VA 24061, USA
telong@vt.edu

Prof. Ian Manners

School of Chemistry
University of Bristol
Cantock's Close
BS8 1TS Bristol, UK
ian.manners@bristol.ac.uk

Prof. Martin Möller

Deutsches Wollforschungsinstitut
an der RWTH Aachen e.V.
Pauwelsstraße 8
52056 Aachen, Germany
moeller@dwi.rwth-aachen.de

Prof. E.M. Terentjev

Cavendish Laboratory
Madingley Road
Cambridge CB 3 OHE, UK
emt1000@cam.ac.uk

Prof. Dr. Maria Jesus Vicent

Centro de Investigacion Principe Felipe
Medicinal Chemistry Unit
Polymer Therapeutics Laboratory
Av. Autopista del Saler, 16
46012 Valencia, Spain
mjvicent@cipf.es

Prof. Brigitte Voit

Institut für Polymerforschung Dresden
Hohe Straße 6
01069 Dresden, Germany
voit@ipfdd.de

Prof. Gerhard Wegner

Max-Planck-Institut
für Polymerforschung
Ackermannweg 10
55128 Mainz, Germany
wegner@mpip-mainz.mpg.de

Prof. Ulrich Wiesner

Materials Science & Engineering
Cornell University
329 Bard Hall
Ithaca, NY 14853, USA
ubw1@cornell.edu

Advances in Polymer Sciences
Also Available Electronically

Advances in Polymer Sciences is included in Springer's eBook package *Chemistry and Materials Science*. If a library does not opt for the whole package the book series may be bought on a subscription basis. Also, all back volumes are available electronically.

For all customers who have a standing order to the print version of *Advances in Polymer Sciences*, we offer free access to the electronic volumes of the Series published in the current year via SpringerLink.

If you do not have access, you can still view the table of contents of each volume and the abstract of each article by going to the SpringerLink homepage, clicking on "Browse by Online Libraries", then "Chemical Sciences", and finally choose *Advances in Polymer Science*.

You will find information about the

– Editorial Board
– Aims and Scope
– Instructions for Authors
– Sample Contribution

at springer.com using the search function by typing in *Advances in Polymer Sciences*.

Color figures are published in full color in the electronic version on SpringerLink.

Aims and Scope

The series *Advances in Polymer Science* presents critical reviews of the present and future trends in polymer and biopolymer science including chemistry, physical chemistry, physics and material science. It is addressed to all scientists at universities and in industry who wish to keep abreast of advances in the topics covered

Review articles for the topical volumes are invited by the volume editors. As a rule, single contributions are also specially commissioned. The editors and publishers will, however, always be pleased to receive suggestions and supplementary information. Papers are accepted for *Advances in Polymer Science* in English.

In references *Advances in Polymer Sciences* is abbreviated as *Adv Polym Sci* and is cited as a journal.

Special volumes are edited by well known guest editors who invite reputed authors for the review articles in their volumes.

Impact Factor in 2009: 4.600; Section "Polymer Science": Rank 4 of 73

Preface

More than half a century has passed since the pioneering books by Flory [1] and by Huggins [2] dealing with some of the most important features concerning the thermodynamics of polymer containing systems. This volume of "*Advances in Polymer Science*" has been composed to update our knowledge in this field. Although most of the experimental observations referring to macromolecular systems could already be rationalized on the basis of the well-known Flory–Huggins theory, quantitative agreement between experiment and theory is normally lacking. The reason for this deficiency lies in several inevitable simplifying assumptions that had to be made during this ground-breaking period of research.

In the meantime, valuable progress could be achieved, thanks to modern computers, improvements of experimental methods, and data handling. This situation has among others provoked a new textbook [3] focusing on polymer phase diagrams. It is the central purpose of this volume to present some further examples for recent developments that were made possible by the above-described improvements. The individual contributions to this issue of the *Advances in Polymer Science* are grouped according to the degree they are connected with the previous text books.

The first part (*B.A. Wolf*) deals with a straightforward extension of the Flory–Huggins theory to account for some aspects of chain connectivity and for the fact that chain molecules may react on changes in their molecular environment by conformational rearrangements. In this manner, several hitherto unconceivable experimental observations (like pronounced composition dependencies of interaction parameters or their variation with chain length) can be understood and modeled quantitatively. This contribution is followed by a chapter devoted to progress in the field of polyelectrolyte solutions (*G. Maurer et al.*); it focuses on the calculation of vapor/liquid equilibria and some related properties (e.g. osmotic pressures) using sophisticated models for the Gibbs energy. Such thermodynamic knowledge is particularly needed for different industrial application of polyelectrolytes, for instance in textile, paper, food, and pharmaceutical industries.

An interesting example for the development and advancement of experimental methods is presented in the third chapter (*J.-P. E. Grolier et al.*), dedicated to the measurement of interactions between gases and polymers based on gas sorption, gravimetric methods, calorimetry, and a "coupled vibrating wire-pVT" technique. Information in this field is of particular interest for polymer foaming and for the self-assembling of nanoscale structures. The fourth section (*S. H. Anastasiadis*) is concerned with interfacial phenomena in the case of polymer blends and reports the current state of the art on measuring and modifying interfacial tensions as well as different possibilities for its modeling. Such information is indispensible for the development and optimization of tailor-made materials based on two-phase polymer blends. The fifth contribution (*S. Enders*) formulates a theory for the simulation of copolymer fractionation in columns with respect to molecular weight and chemical composition. Narrowly distributed polymers are often required for basic research and the removal of harmful components is sometimes essential for special applications.

All previously discussed methods are primarily based on phenomenological considerations, in contrast to chapter six (*K. Binder et al.*), which starts from statistical thermodynamics. This section reviews the state of the art in fields of Monte–Carlo and Molecular Dynamics simulations. These methods are powerful tools for the prediction of macroscopic properties of matter from suitable models for effective interactions between atoms and molecules. The final chapter (*G. Sadowski*) makes use of the results obtained with simulation tools for the establishment of molecular-based equations of state for engineering applications. This approach enables the description and in some cases even the prediction of the phase behavior as a function of pressure, temperature, molecular weight distribution and for copolymers also as a function of chemical composition.

The Editors are well aware of the fact that the above selection is not only far from being complete, but also to some extent subjective. However, in view of the importance of polymer science (worldwide annual production [4] in 2008: $2.8°10^8$ t with a growth rate of approximately 12% per year) and accounting for the significance of thermodynamics in this area, further volumes of the "Advances in Polymer Science" covering missing thermodynamic aspects and presenting further progress in this field are expected.

Berlin   Sabine Enders
Mainz   Bernhard Wolf
Summer 2010

References

1 P. J. Flory, Principles of Polymer Chemistry, Cornell University Press, Ithaca, N.Y. 1953
2 M. L. Huggins, Physical Chemistry of High Polymers, Wiley, N.Y. 1958
3 R. Koningsveld, W. H. Stockmayer, E. Nies, Polymer Phase Diagrams, Oxford University Press, Oxford 2001
4 Statistisches Bundesamt, Fachserie 4, Reihe 3.1, Jahr 2007

Obituary

Prof. Dr. Ronald Koningsveld, for several decades leader in thermodynamics of polymer solutions and blends, was born on April 15, 1925 in Haarlem. In his teen years when he was living in Rotterdam, he was seized by science and music and he started studies of orchestral conducting, piano, and composition at Rotterdam Conservatory. Music remained his love for his whole life. However, following the advice of his father to do something more "practical", he entered the Technical University of Delft to study chemical engineering. After graduation in 1956, Ron joined the Central Research of Dutch State Mines (DSM) in Geleen and in his first years there he was engaged in polymer characterization. In parallel, he started his PhD studies at the University of Leiden under the guidance of A. J. Staverman in the area of phase equilibria in polydisperse polymer solutions with application to polymer fractionation. He obtained the title of Doctor of Mathematics and Natural Sciences in 1967. The papers based on these results rank among the most cited ones of Ron's almost 200 publications cited about 3,000 times (according to WoS). Ron continued working in DSM Research until his retirement in 1985 in various positions including Head of Department of Fundamental Polymer Research (1963–1980) and Managing Director of General Basic Research (1980–1985). In the latter position, Ron also managed external research funded by DSM. He stimulated significantly collaborative fundamental research on polymers in Europe and overseas. The collaboration extended to other countries including Belgium, Czechoslovakia, Germany, United Kingdom, and U.S.A.

Koningsveld is the name well known in the Academia – he was teaching polymer thermodynamics as a guest professor in the University of Essex, University of Massachusetts, Catholic University of Leuven, and ETH Zurich, and for 18 years he was a Professor of Polymer Science in the University of Antwerp.

He received honorary doctorates from the University of Bristol and Technical University of Dresden. Also, he was a consultant to Max-Planck Gesellschaft, Institute of Polymer Research in Mainz. In 2002, Ron's scientific achievements were appreciated by the Paul Flory Research Prize.

It would be difficult to enumerate all Ron's scientific achievements in the field of polymer thermodynamic. One can name the generalizations of the Flory–Huggins Gibbs energy leading to the prediction and experimental verification of coexistence of three phases in pseudobinary system with sufficiently broad distribution; or, the analysis of the functional form of the interaction term leading to the appearance of "off-zero critical concentration", at variance with zero critical concentration associated with theta-temperature. Thanks largely to Ron, polymer scientists realize that the cloud point curve is not the binodal and its maximum or minimum are not identical with the critical temperatures.

Ron had many good friends in the scientific society and some of them (Berghmans, Simha, Stockmayer) are coauthors of his last paper on correlation between two critical polymer concentrations – c^* for the coil overlap and c_s assigned to the maximum/minimum of the spinodal (*J. Phys. Chem. B* 2004, 108, 16168–16173). Unfortunately, Robert and Stocky are no longer with us as well. The scientific community can share Ron's knowledge in phase equilibria in the monograph *Polymer Phase Diagrams*, Oxford (2001) published with coauthors W. H. Stockmayer and E. Nies.

This reminiscence would not be complete without mentioning the second Ron's love – the music. Already in Delft as a student, Ron was engaged in Dutch College Swing Band as a pianist and arranger. During his work for DSM, Ron composed a number of pieces inspired by research of polymers: *Microsymposium Music* performed during Microsymposia on Polymers held every year in the Institute of Macromolecular Chemistry in Prague, *Polymer Music* in six movements for two pianos, *To Science* (inspired by Edgar Allan Poe, *Staudinger March* (commemorating Staudinger's 100th birthday), and *Short Communication*. Some of the readers may remember the "ouverture" to IUPAC Macro in Amherst in 1982, where polymer scientists (Stockmayer, MacKnight, Kennedy, Janeschitz-Kriegel and Ron as pianist) performed *Polymer Music*.

Ron passed away in Sittard on November 26, 2008. We grieve over a famous scientist known all over the world in the thermodynamic community, an outstanding academic teacher and a great personality.

Karel Dušek
Prague

Contents

Making Flory–Huggins Practical: Thermodynamics of Polymer-Containing Mixtures 1
Bernhard A. Wolf

Aqueous Solutions of Polyelectrolytes: Vapor–Liquid Equilibrium and Some Related Properties 67
G. Maurer, S. Lammertz, and L. Ninni Schäfer

Gas–Polymer Interactions: Key Thermodynamic Data and Thermophysical Properties 137
Jean-Pierre E. Grolier and Séverine A.E. Boyer

Interfacial Tension in Binary Polymer Blends and the Effects of Copolymers as Emulsifying Agents 179
Spiros H. Anastasiadis

Theory of Random Copolymer Fractionation in Columns 271
Sabine Enders

Computer Simulations and Coarse-Grained Molecular Models Predicting the Equation of State of Polymer Solutions 329
Kurt Binder, Bortolo Mognetti, Wolfgang Paul,
Peter Virnau, and Leonid Yelash

Modeling of Polymer Phase Equilibria Using Equations of State 389
Gabriele Sadowski

Index 419

Adv Polym Sci (2011) 238: 1–66
DOI: 10.1007/12_2010_84
© Springer-Verlag Berlin Heidelberg 2010
Published online: 13 July 2010

Making Flory–Huggins Practical: Thermodynamics of Polymer-Containing Mixtures

Bernhard A. Wolf

Abstract The theoretical part of this article demonstrates how the original Flory–Huggins theory can be extended to describe the thermodynamic behavior of polymer-containing mixtures quantitatively. This progress is achieved by accounting for two features of macromolecules that the original approach ignores: the effects of chain connectivity in the case of dilute solutions, and the ability of polymer coils to change their spatial extension in response to alterations in their molecular environment. In the general case, this approach leads to composition-dependent interaction parameters, which can for most binary systems be described by means of two physically meaningful parameters; systems involving strongly interacting components, for instance via hydrogen bonds, may require up to four parameters. The general applicability of these equations is illustrated in a comprehensive section dedicated to the modeling of experimental findings. This part encompasses all types of phase equilibria, deals with binary systems (polymer solutions and polymer blends), and includes ternary mixtures; it covers linear and branched homopolymers as well as random and block copolymers. Particular emphasis is placed on the modeling of hitherto incomprehensible experimental observations reported in the literature.

Keywords Modeling · Mixed solvents · Phase diagrams · Polymer blends · Polymer solutions · Ternary mixtures · Thermodynamics

---

B.A. Wolf

Institut für Physikalische Chemie der Johannes Gutenberg-Universität Mainz, 55099 Mainz, Germany
e-mail: bernhard.wolf@uni-mainz.de

Contents

1 Introduction 4
2 Extension of the Flory–Huggins Theory 5
    2.1 Binary Systems 5
    2.2 Ternary Systems 21
3 Measuring Methods 24
    3.1 Vapor Pressure Measurements 24
    3.2 Osmometry and Scattering Methods 25
    3.3 Other Methods 26
4 Experimental Results and Modeling 27
    4.1 Binary Systems 27
    4.2 Ternary Systems 53
5 Conclusions 62
References 63

Symbols

a	Exponent of Kuhn–Mark–Houwink relation (29)
a	Intramolecular interaction parameter (47) for blend component A
A,B,C	Constants of (13)
A_2, A_3	Second and third osmotic virial coefficients
a_i	Activity of component i
b	Intramolecular interaction parameter (47) for blend component B
c	Concentration in moles/volume
E	Constant of interrelating α and $\zeta\lambda$ (34)
G	Gibbs free energy – free enthalpy
g	Integral interaction parameter
H	Enthalpy
K_N	Constant of the Kuhn–Mark–Houwink relation (29)
LCST	Lower critical solution temperature
M	Molar mass
M_n	Number-average molar mass
M_w	Weight-average molar mass
N	Number of segments
n	Number of moles
p	Vapor pressure
R	Ideal gas constant
S	Entropy
s	Molecular surface
T	Absolute temperature
t	Ternary interaction parameter (61)
T_m	Melting point
UCST	Upper critical solution temperature
V	Volume
v	Molecular volume

Thermodynamics of Polymer-Containing Mixtures

w	Weight fraction
x	Mole fraction
Z	Parameter relating the conformational relaxation to β (53)

Greek and Special Characters

$\overline{\omega}$	Parameter quantifying strong intersegmental interactions (42)
$[\eta]$	Intrinsic viscosity
Φ_o	Volume fraction of polymer segments in an isolated coil (27)
Θ	Theta temperature
α	Parameter of (23), first step of dilution
β	Degree of branching (52)
χ	Flory–Huggins interaction parameter
δ	Parameter of (57)
ε	Parameter of (57)
γ	Surface-to-volume ratio of the segments in binary mixtures (24)
φ	Segment fraction, often approximated by volume fraction
κ	Constant of (30)
λ	Intramolecular interaction parameter (23)
μ	Chemical potential
ν	Parameter of (23)
π	Any parameter of (23)
π_{osm}	Osmotic pressure
ρ	Density
τ	Parameter of (44)
ξ	Differential Flory–Huggins interaction parameter for the polymer
ζ	Conformational relaxation (second step of dilution) (23)

Subscripts

$1, 2, 3 \ldots$	Low molecular weight components of a mixture
A to P	High molecular weight components
B	Branched oligomer/polymer
c	Critical state
cr	Conformational relaxation
fc	Fixed conformation
g	Glass
H	Enthalpy part of a parameter
i, j, k	Unspecified components i, j, k
L	Linear polymer
lin	Linear oligomer/polymer
m	Melting
S	Entropy part of a parameter

| s | Saturation |
| o | Quantity referring to a pure component, to an isolated coil, or to high dilution |

Superscripts

–	Molar quantity
=	Segment-molar quantity
E	Excess quantity
Res	Residual quantity (with respect to combinatorial behavior)
∞	Infinite molar mass of the polymer

1 Introduction

The decisive advantage of the original Flory–Huggins theory [1] lies in its simplicity and in its ability to reproduce some central features of polymer-containing mixtures qualitatively, in spite of several unrealistic assumptions. The main drawbacks are in the incapacity of this approach to model reality in a quantitative manner and in the lack of theoretical explanations for some well-established experimental observations. Numerous attempts have therefore been made to extend and to modify the Flory–Huggins theory. Some of the more widely used approaches are the different varieties of the lattice fluid and hole theories [2], the mean field lattice gas model [3], the Sanchez–Lacombe theory [4], the cell theory [5], different perturbation theories [6], the statistical-associating-fluid-theory [7] (SAFT), the perturbed-hard-sphere chain theory [8], the UNIFAC model [9], and the UNIQUAC [10] model. More comprehensive reviews of the past achievements in this area and of the applicability of the different approaches are presented in the literature [11, 12].

This contribution demonstrates how the deficiencies of the original Flory–Huggins theory can be eliminated in a surprisingly simple manner by (1) accounting for hitherto ignored consequences of chain connectivity, and (2) by allowing for the ability of macromolecules to rearrange after mixing to reduce the Gibbs energy of the system. Section 2 recalls the original Flory–Huggins theory and describes the composition dependence of the Flory–Huggins interaction parameters resulting from the incorporation of the hitherto neglected features of polymer/solvent systems into the theoretical treatment. This part collects all the equations required for the interpretation of comprehensive literature reports on experimentally determined thermodynamic properties of polymer-containing binary and ternary mixtures (polymer solutions in mixed solvents and solutions of two polymers in a common solvent). In order to ease the assignment of the different variables and parameters to a certain component, the low molecular weight components are identified by numbers and the polymers by letters. The high molecular weight components comprise linear and branched samples, homopolymers, binary random copolymers,

Thermodynamics of Polymer-Containing Mixtures

and block copolymers of different architecture; the phase equilibria encompass liquid/gas, liquid/liquid and liquid/solid. The only aspects that are excluded are the coexistence of three liquid phases and the demixing of mixed solvent.

This theoretical section is followed (Sect. 3) by a recap of the measuring techniques used for the determination of the thermodynamic properties discussed here. The subsequent main part of the article (Sect. 4) outlines the modeling of experimental observations and investigates the predictive power of the extended Flory–Huggins theory. Throughout this contribution, particular attention is paid to phenomena that cannot be rationalized on the basis of the original Flory–Huggins theory, like anomalous influences of molar mass on thermodynamic properties or the existence of two critical points (liquid/liquid phase separation) for binary systems. In fact, it was the literature reports on such experimental findings that have prompted the present theoretical considerations.

2 Extension of the Flory–Huggins Theory

2.1 Binary Systems

2.1.1 Polymer Solutions

Organic Solvents/Linear Homopolymers

The basis for a better understanding of the particularities of polymer-containing mixtures as compared with mixtures of low molecular weight compounds was laid more than half a century ago [13–17], in the form of the well-known Flory–Huggins interaction equation. By contrast to the form used for low molecular weight mixtures, this relation is usually not stated in terms of the *molar* Gibbs energy $\overline{G}$; for polymer-containing systems one chooses one mole of segments as the basis (in order to keep the amount of matter under consideration of the same order of magnitude) and introduces the *segment molar* Gibbs energy $\overline{\overline{G}}$. For polymer solutions, where the molar volume of the solvent normally defines the size of a segment, this relation reads:

$$\frac{\Delta\overline{\overline{G}}}{RT} = (1 - \varphi)\ln(1 - \varphi) + \frac{\varphi}{N}\ln\varphi + g\varphi(1 - \varphi) \tag{1}$$

$\Delta\overline{\overline{G}}$ stands for the segment molar Gibbs energy of mixing. The number N of segments that form the polymer is calculated by dividing the molar volume of the macromolecule by the molar volume of the solvent. The composition variable φ, representing the segment molar fraction of the polymer, is in most cases approximated by its volume fraction (neglecting nonzero volumes of mixing), and g stands for the integral Flory–Huggins interaction parameter. In the case of polymer solutions, we refrain from using indices whenever possible (i.e., we write g instead of g_{1P}, φ instead of φ_P and N instead of N_P) for the sake of simpler representation. Only if g does not depend on composition does it becomes identical with the experimentally measurable Flory–Huggins interaction parameter χ, introduced in (5).

The total change in the Gibbs energy resulting from the formation of polymer solutions is, according to (1), subdivided into two parts, the first two terms representing the so-called combinatorial behavior, ascribed to entropy changes:

$$\frac{\Delta\overline{\overline{G}}^{\text{com}}}{RT} = (1-\varphi)\ln(1-\varphi) + \frac{\varphi}{N}\ln\varphi \tag{2}$$

All particularities of a certain real system (except for the chain length of the polymer) are incorporated into the third term, the residual Gibbs energy of mixing, and were initially considered to be of enthalpic origin. The essential parameter of this part is g, the integral Flory–Huggins interaction parameter:

$$\frac{\Delta\overline{\overline{G}}^{\text{res}}}{RT} = g\varphi(1-\varphi) \tag{3}$$

In the early days, the Flory–Huggins interaction parameter was considered to depend only on the variables of state, but not on either the composition of the mixture or on the molar mass of the polymer. Under these premises, it is easy to perform model calculations – for instance with respect to phase diagrams – along the usual routes of phenomenological thermodynamics on the sole basis of the parameter g. In this manner, most characteristic features of polymer solutions can already be well rationalized, even though quantitative agreement is lacking. However, as the number of thermodynamic studies increased it was soon realized that (1) is too simple. Above all, it became clear that the assignment of entropy and of enthalpy contributions to the total Gibbs energy of mixing is unrealistic. Maintaining for practical reasons the first term unchanged, as a sort of reference behavior, this means that all particularities of a real system must be incorporated into the parameter g.

This change in strategy has important consequences, the most outstanding being the necessity to distinguish between *integral* interaction parameters g, introduced by (1) and referring to the Gibbs energy of mixing, and *differential* interaction parameters, referring either to the chemical potential of the solvent or of the solute. The partial segment molar Gibbs energies and the corresponding integral quantity are interrelated by the following relation:

$$\overline{\overline{G}}_i = \overline{\overline{G}} - \varphi_k \frac{\partial \overline{\overline{G}}}{\partial \varphi_k} \tag{4}$$

where the subscripts i and k stand for either the solvent or the polymer. The partial molar Gibbs energies $\overline{G}_i$ are customary referred to as chemical potentials μ_i.

Thermodynamics of Polymer-Containing Mixtures

The partial expressions for the solvent (index 1) read:

$$\frac{\Delta\mu_1}{RT} = \frac{\Delta\overline{\overline{G}}_1}{RT} = \ln(1-\varphi) + \left(1 - \frac{1}{N}\right)\varphi + \chi\,\varphi^2 = \ln a_1 \tag{5}$$

and yield the differential parameter χ, the well known original Flory–Huggins interaction parameter, which is related to the activity a_1 of the solvent as formulated above; a_1 can in many cases be approximated (sufficiently low volatility of the solvent) by the relative vapor pressure:

$$a_1 \approx \frac{p_1}{p_{1,\mathrm{o}}} \tag{6}$$

where $p_{1,\mathrm{o}}$ is the vapor pressure of the pure solvent. Otherwise, one needs to correct for the imperfections of the equilibrium vapor.

The Flory–Huggins interaction parameter constitutes a measure for chemical potential of the solvent, as documented by (6) and (5); it is defined in terms of the deviation from combinatorial behavior as:

$$\chi \equiv \frac{\Delta\overline{\overline{G}}_1^{\mathrm{res}}}{RT\varphi^2} \tag{7}$$

In the original theory, χ was meant to have an immediate physical meaning, because of the normalization of the residual segment molar Gibbs energies of dilution to the probability φ^2 of an added solvent molecule to be inserted between two contacting polymer segments. This illustrative interpretation does, however, rarely hold true in reality. Even for simple homopolymer solutions in single solvents, it fails in the region of high dilution because the overall polymer concentration becomes meaningless for the number of intermolecular contacts between polymer segments. Despite this lack of a straightforward interpretation of the Flory–Huggins interaction parameter in molecular terms, the knowledge of $\chi(\varphi)$ is indispensable for the thermodynamic description of polymer-containing mixtures. This information can be converted to integral interaction parameters g [cf. (25)] and gives access to the calculation of macrophase separation (e.g., via a direct minimization of the Gibbs energy of the systems [18–20] and to the chemical potentials of the polymer [cf. (11)].

For practical purposes, the use of volume fractions (instead of the original segment fractions) as composition variable is not straightforward because of the necessity to know the densities of the components and (in the case of variable temperature) their thermal expansivities. For that reason, φ is sometimes consistently replaced by the weight fraction w, and N calculated from the molar masses as M_P/M_1. The χ values obtained in this manner according to (8):

$$_w\chi \equiv \frac{_w\Delta\overline{\overline{G}}_1^{\mathrm{res}}}{RTw^2} \tag{8}$$

are indicated by the subscript w and may differ markedly in their numerical values from χ. The expression for the residual Gibbs energy of dilution is also given an index as a reminder that weight fractions were used to calculate its combinatorial part. Despite the practical advantages of $_w\chi$, we stay with volume fractions for all subsequent considerations, because they account at least partly for the differences in the free volume of the components and because most of the published thermo-dynamic information uses this composition variable.

One of the consequences of composition-dependent interaction parameters lies in the necessity to distinguish between different parameters, depending on the particular method by which they are determined. The Flory–Huggins interaction parameter χ relates to the integral interaction parameter g as:

$$\chi = g - (1 - \varphi)\frac{\partial g}{\partial \varphi} \tag{9}$$

The expression analogous to (5), referring to the solvent, reads for the polymer (index P):

$$\frac{\Delta\mu_P}{RT} = \frac{N\,\Delta\overline{\overline{G}}_P}{RT} = \ln\varphi + (1 - N)(1 - \varphi) + \xi N(1 - \varphi)^2 \tag{10}$$

This relation defines the differential interaction parameter ξ in terms of the chemical potential of the polymer and is calculated from g by means of:

$$\xi = g + \varphi\frac{\partial g}{\partial \varphi} \tag{11}$$

Out of the three types of interaction parameters, it is almost exclusively χ that is of relevance for the thermodynamic description of binary and ternary polymer-containing liquids, as will be described in the section on experimental methods (Sect. 3). The integral interaction g parameter is practically inaccessible, and the parameter ξ, referring to the polymer, suffers from the difficulties associated with the formation of perfect polymer crystals, because it is based on their equilibria with saturated polymer solutions.

Measured Flory–Huggins interaction parameters soon demonstrated the necessity to treat χ as composition-dependent. A simple mathematical description consists of the following series expansion:

$$\chi = \chi_o + \chi_1\varphi + \chi_2\varphi^2 \ldots \tag{12}$$

A more sophisticated approach [21] accounts for the differences in the molecular surfaces of solvent molecules and polymer segments (of equal volume) and formulates $\chi(\varphi)$ as:

Thermodynamics of Polymer-Containing Mixtures

$$\chi = \frac{A}{(1 - B\varphi)^2} + C \tag{13}$$

where these differences are contained in the parameter B. A and C are considered to be further constants for a given system and fixed variables of state.

The thermodynamic relations discussed so far were, above all, formulated for the description of moderately to highly concentrated polymer solutions. The information acquired in the context of the determination of molar masses, on the other hand, refers to dilute solution and is usually expressed in terms of second osmotic virial coefficients A_2 and higher members of a series expansion of the chemical potential of the solvent with respect to the polymer concentration c (mass/volume). For the determination of osmotic pressures, π_{osm}, the corresponding relation reads:

$$-\frac{\Delta \bar{G}_1}{RT\,\bar{V}_1} = \frac{\pi_{osm}}{RT} = \frac{c}{M_n} + A_2 c^2 + A_3 c^3 + \dots \tag{14}$$

Performing a similar series expansion for the logarithm in (5), inserting χ from (12) into this relation, and comparing the result with (14) yields [21]:

$$\chi_o = \frac{1}{2} - \rho_P^2\,\bar{V}_1\,A_2 \tag{15}$$

and:

$$\chi_1 = \frac{1}{3} - \rho_P^3\,\bar{V}_1\,A_3 \tag{16}$$

where χ_o represents the Flory–Huggins interaction parameter in the limit of pair interactions between polymer molecules. $\bar{V}_1$ is the molar volume of the solvent and ρ_P is the density of the polymer.

The need for a different view on the thermodynamics of polymer solutions became, in the first place, obvious from experimental information on dilute systems. According to the original Flory–Huggins theory, the second osmotic virial coefficient should without exception decrease with rising molar mass of the polymer. It is, however, well documented (even in an early work by Flory himself [22]) that the opposite dependence does also occur. Based on this finding and on the fact that the Flory–Huggins theory only accounts for chain connectivity in the course of calculating the combinatorial entropy of mixing and for concentrated solutions, we attacked the problem by starting from the highly dilute side.

The central idea of this approach is the treatment of a swollen isolated polymer coil – surrounded by a sea of pure solvent – as a sort of microphase and applying the usual equilibrium condition to such a system. In a thought experiment, one can insert a single totally collapsed polymer molecule into pure solvent and let it swell until it reaches its equilibrium size. Traditionally, the final state of this process is discussed in terms of chain elasticity. Here, we apply a phenomenological thermodynamic method and equate the chemical potential of the solvent inside the realm of the polymer coil to the chemical potential of the pure solvent surrounding it. In doing so, we "translate" the entropic barrier against an infinite extension of the polymer chain into a virtual semi-permeable membrane. This barrier accounts for chain connectivity and represents a consequence of the obvious inability of the segments of an isolated polymer molecule to spread out over the entire volume of the system. The condition for the establishment of such a microphase equilibrium reads:

$$\ln(1 - \Phi_o) + \left(1 - \frac{1}{N}\right)\Phi_o + \lambda\Phi_o^2 = 0 \tag{17}$$

This relation differs from that for macroscopic phase equilibria [resulting from (5)] only by the meaning of the concentration variable Φ_o and of the interaction parameter λ. Φ_o stands for the average volume fraction of the polymer segments contained in an isolated coil, and λ represents an intramolecular interaction parameter, which raises the chemical potential of the solvent in the mixed phase up to the value of the pure solvent.

By means of the considerations outlined above, we have accounted for chain connectivity. However, there is another aspect that the original Flory–Huggins theory ignores, namely the ability of chain molecules to react to changes in their environment by altering their spatial extension. One outcome of this capability is, for instance, the well-known fact that the unperturbed dimensions of pure polymers in the melt will gradually increase upon the addition of a thermodynamically favorable solvent. These changes are particularly pronounced in the range of high dilution, where there is no competition of different solute molecules for available solvent, due to the practically infinite reservoir of pure solvent.

In order to incorporate both features neglected by the original Flory–Huggins theory into the present approach, we have conceptually subdivided the dilution process into two separate steps as formulated in (18). Such a separation is permissible because the Gibbs energy of dilution represents a function of state.

$$\chi_o = \chi_o^{fc} + \chi_o^{cr} \tag{18}$$

The first term (the superscript fc stands for fixed conformation) quantifies the effect of separating two contacting polymer segments belonging to different macromolecules by inserting a solvent molecule between them without changing their conformation. The second term (the superscript cr stands for conformational relaxation) is required to bring the system into its equilibrium by rearranging the components such that the minimum of Gibbs energy is achieved.

In order to give the second term a more specific meaning, we formulate χ_o^{cr} as the difference in the interaction before and after the conformational relaxation as:

Thermodynamics of Polymer-Containing Mixtures

$$\chi_o^{cr} = \chi^{after} - \chi^{before} \tag{19}$$

Choosing λ, the interaction between polymer segments and solvent molecules in the isolated state, as a clear cut reference point for the contribution of the rearrangement in the second step of dilution, and assuming that the effect will be proportional to λ, we can write:

$$\chi_o^{cr} = -\zeta\lambda \tag{20}$$

where the negative sign in the above expression has been chosen to obtain positive values for this parameter in the great majorit of cases. Denoting:

$$\chi_o^{fc} = \alpha \tag{21}$$

(18) and (20) yield the following simple expression for the Flory–Huggins interaction parameter in the limit of high dilution:

$$\chi_o = \alpha - \zeta\lambda \tag{22}$$

For sufficiently dilute polymer solutions, the only difference between the new approach and the original Flory–Huggins theory is in the second term. According to theoretical considerations and in accord with experimental findings, ζ becomes zero under theta conditions (where the coils assume their unperturbed dimensions) and the conformational relaxation no longer contributes to χ_o.

In order to generalize (22) to arbitrary polymer concentrations, we assume that the composition dependence of its first term can be formulated by analogy to (13). The necessity of a composition dependence for the second term results from the fact that the insertion of a solvent molecule between contacting polymer segments (belonging to different polymer chains) opens only one binary contact within the composition range of pair interactions, whereas there are inevitably more segments affected at higher polymer concentrations. For the second term, we suppose a linear dependence of the integral interaction parameter g on φ. Comparing the coefficients of this ansatz (as they appear in the expression for differential interaction parameter) with (22) for χ_o results in (23):

$$\chi = \frac{\alpha}{(1 - v\varphi)^2} - \zeta(\lambda + 2(1 - \lambda)\varphi) \tag{23}$$

The symbol v instead of B (13) in the above relation indicates that this parameter is related to γ [21], the geometrical differences of solvent molecules and polymer segments as formulated in the next equation, but not identical with γ;

$$\gamma \equiv 1 - \frac{(s/v)_{polymer}}{(s/v)_{solvent}} \tag{24}$$

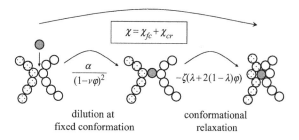

Fig. 1 Assignment of the parameters of (23) to the individual steps of dilution: Two contacting segments belonging to different macromolecules are separated by the insertion of a solvent molecule (*shaded*) between them

The parameters s and v are the molecular surfaces and volumes of the components, respectively. In the limit of $\varphi \to 0$, (23) reduces to (22).

The essentials of the considerations concerning the composition dependence of the Flory–Huggins interaction parameter are visualized in Fig. 1, demonstrating how the dilution is conceptually divided into two separate steps and how these steps contribute to the overall effect. The first step maintains the conformation of the components as they are prior to dilution and does not change the volume of the system; measurable excess volumes are attributed to the conformational rearrangement taking place during the second step of mixing.

By means of the expression:

$$g = -\frac{1}{1-\varphi} \int_{1}^{1-\varphi} \chi \, d\varphi \qquad (25)$$

resulting from phenomenological thermodynamics, the Flory–Huggins interaction parameter χ of (23) yields the following expression for the integral interaction parameter g, required for instance to calculate phase equilibria using the method of the direct minimization of the Gibbs energy [19] of a system:

$$g = \frac{\alpha}{(1-v)(1-v\varphi)} - \zeta(1 + (1-\lambda)\varphi) \qquad (26)$$

This relation contains four adjustable parameters; even if they are molecularly justified these are too many for practical purposes. For this reason, it would be helpful to be able to calculate at least one of them independently. The most obvious candidate for that purpose is λ (17) because it refers to the spatial extension of isolated polymer coils. Radii of gyration would be most qualified for calculation of the required volume fractions of segments, Φ_o, inside the microphase formed by isolated polymer molecules. Unfortunately, however, it is hard to find tabulated values for different polymer/solvent systems in the literature. For this reason, we use information provided by the specific hydrodynamic volume of the polymers at infinite dilution, i.e., to intrinsic viscosities $[\eta]$. The volume of the segments is

Thermodynamics of Polymer-Containing Mixtures

given by M/ρ_P, and $[\eta]M$ yields the hydrodynamic volume of one mole of isolated polymer coils so that Φ_o becomes:

$$\Phi_o = \frac{M/\rho_P}{[\eta]M} = \frac{1}{[\eta]\rho_P} \tag{27}$$

Upon the expansion of the logarithm in (17) up to the second term (which suffices in view of the low Φ_o values typical for the present systems), we obtain the following expression for λ:

$$\lambda = \frac{1}{2} + \frac{[\eta]\rho_P}{N} \tag{28}$$

Relating the intrinsic viscosity to N by means of the Kuhn–Mark–Houwink relation:

$$[\eta] = K_N N^a \tag{29}$$

the intramolecular interaction parameter becomes:

$$\lambda = \frac{1}{2} + \kappa N^{-(1-a)} \tag{30}$$

where $\kappa = K_N \rho_P$.

The insertion of (30) into (22) and employing (15) enables the rationalization of the experimental finding that the A_2 values for the solutions of a given polymer of different chain length do not exclusively decrease with rising M in good solvents, but might also increase. The resulting equation reads:

$$A_2 = A_2^\infty + \frac{\zeta\kappa}{\rho_P^2 \overline{V}} N^{-(1-a)} \tag{31}$$

where A_2^∞ is the limiting value of A_2 for infinite molar mass of the polymer. The reason for an anomalous molecular weight dependence of the second osmotic virial coefficient lies in the sign of ζ, which is positive in most cases, but may also become negative under special conditions. For theta systems, $A_2 = 0$, irrespective of M, and ζ becomes zero. One consequence of the present experimentally verified consideration concerns the way that $A_2(M)$ should be evaluated. Equation (31) requires plots of A_2 as a function of $M^{-(1-a)}$, instead of the usual double logarithmic plots, and does not – in contrast to the traditional evaluation – automatically yield zero second osmotic virial coefficient in the limit of infinitely long chains.

Another helpful consequence of (30) lies in the fact that its second term is almost always negligible (with respect to 1/2) for polymers of sufficient molar mass. This feature allows the merging of the parameters ζ and λ into their product $\zeta\lambda$, and the replacement of the isolated λ by 1/2, as formulated below for the differential interaction parameter χ:

$$\chi \approx \frac{\alpha}{(1 - v\varphi)^2} - \zeta\lambda(1 + 2\varphi) \tag{32}$$

The analogous relation for the integral interaction parameter reads:

$$g \approx \frac{\alpha}{(1 - v)(1 - v\varphi)} - \zeta\lambda(2 + \varphi) \tag{33}$$

By this means, the number of adjustable parameter reduces to three. As will be shown in the section dealing with experimental data (Sect. 4), further simplifications are possible, for instance because of a theoretically expected interrelation between the parameters α (first step of mixing) and $\zeta\lambda$ (second step of mixing) for a given class of polymer solutions. In its general form this equation reads:

$$\zeta\lambda = E(2\alpha - 1) \tag{34}$$

where E is a constant, typically assuming values between 0.6 and 0.95. Equation (34) is in accord with the typical case of theta conditions where $\zeta \to 0$ and $\alpha \to 0.5$. As long as such an interrelation exists, the number of parameters required for the quantitative description of the isothermal behavior of polymer solutions reduces to two. Like with the expression for χ_o (high dilution), the contributions of chain connectivity and conformational relaxation are in (32) (arbitrary polymer concentration) exclusively contained in the second term. Another aspect also deserves mentioning, namely the fact that (32) is not confined to the modeling of polymer-containing systems but can also be successfully applied to mixtures of low molecular weight liquids, as will be shown in Sect. 4.

According to expectation, and in agreement with measurements, all system-specific parameters π (namely α, v, ζ, and λ) vary more or less with temperature (and pressure). The following relation is very versatile to model $\pi(T)$:

$$\pi = \pi_o + \frac{\pi_1}{T} + \pi_2 T \tag{35}$$

where either π_1 or π_2 can be set to zero in most cases.

Up to now, it was the chemical potential of the solvent that constituted the object of prime interest. The last part of this section is dedicated to the modeling of liquid/liquid phase separation by means of the integral Gibbs energy of mixing in the case of polymer solutions. The equations presented in this context can, however, be easily generalized to polymer blends and to multinary systems. Such calculations are made possible by using the minimum Gibbs energy a system can achieve via phase separation as the criterion for equilibria, instead of equality of the chemical potentials of the components in the coexisting phases. The method of a direct minimization of the Gibbs energy [19] works in the following way: The segment molar Gibbs energy of mixing for the (possibly unstable) homogeneous system is calculated by means of (1), where the integral interaction parameter g is in the present case taken from (26). For different overall compositions, it is then checked on a computer by means of test tie lines (connecting arbitrarily chosen data points of the function $\Delta\overline{\overline{G}}$) which values lead to the maximum lowering of the Gibbs energy. In this manner, it is possible to model the binodal curves if $g(T)$ is known. Spinodal curves are also easily accessible by means of these test tie lines, if they are chosen to be very short. In this manner, it is possible to monitor at which concentration the test tie lines change their location with respect to the function $\Delta\overline{\overline{G}}/RT(\varphi)$: Within the unstable range they lie below that function, and within the metastable and stable ranges they are located above it, indicating that homogenization would lead to a further reduction in G. The criterion that (sufficiently short) test tie lines must become parallel to the spinodal line at the critical point gives access to critical data.

Under special conditions, it possible to calculate system-specific parameters from experimentally determined critical concentrations φ_c. The condition for the degeneration of the tie lines to the critical point is that the second and the third derivative of the Gibbs energy with respect to composition must become zero. The application of this requirement to (1) in combination with (26) yields:

$$\frac{1}{1-\varphi_c} + \frac{1}{N\varphi_c} + \frac{2\alpha}{(v\varphi_c - 1)^3} + 2\zeta[\lambda - 3\varphi_c(\lambda - 1)] = 0 \tag{36}$$

and:

$$\frac{1}{(\varphi_c - 1)^2} - \frac{1}{N\varphi_c^2} - \frac{6\alpha v}{(v\varphi_c - 1)^4} + 6\zeta(1 - \lambda) = 0 \tag{37}$$

For the sake of completeness, the coexistence of a pure crystalline polymer with its saturated solution is also considered. Taking the change in the chemical potential of the polymer upon mixing from (10), the equilibrium condition ($T \leq T_m$) reads:

$$\Delta\overline{H}_m - T\frac{\Delta\overline{\overline{H}}_m}{T_m} + RT\left(\ln \varphi_s + (1 - N)\varphi_s + N_P\xi(1 - \varphi_s)^2\right) = 0 \tag{38}$$

where the entropy term of the segment molar Gibbs energy of melting [the second term of (38)] is approximated by $\Delta\overline{\overline{H}}_m$, the segment molar heat of melting, and the melting temperature T_m of the pure crystal; φ_s denotes the saturation volume fraction of the polymer in the solution.

So far, we have not dealt with the question of how the Flory–Huggins interaction parameters are made up of enthalpy and entropy contributions for different systems. This information is accessible by means of (39) and (40) (which neglect the temperature influences on the volume fraction of the polymer, caused by different thermal expansivities of the components). The enthalpy part reads:

$$\chi_H = -T\left(\frac{\partial \chi}{\partial T}\right)_{\text{p}} \tag{39}$$

and the corresponding entropy part is given by:

$$\chi_S = \chi + T\left(\frac{\partial \chi}{\partial T}\right)_{\text{p}} \tag{40}$$

In the above equations, χ can be substituted by any parameter of the present approach to determine its enthalpy and entropy parts, except for the parameter v, which is not a Gibbs energy by its nature.

Organic Solvents/Branched Homopolymers

The different molecular architectures of branched polymers do not require modifications of (32); the particularities of branched polymers only change the values of the system-specific parameters as compared with those for linear analogs in the same solvent [24], as intuitively expected.

Organic Solvents/Linear Random Copolymers

Despite the fact that these solutions represent binary systems, at least three Flory–Huggins interaction parameters are involved in their modeling, like with ternary mixtures. Because of the necessity to account for the interaction of the solvent with monomer A and with monomer B, plus the interaction between the polymers A and B, one should expect the need for a minimum of two additional parameters. Experimental data obtained for solutions of a given copolymer of the type A-ran-B with a constant fraction f of B monomers can be modeled [25] by means of (32), with one set of α, v, and $\zeta\lambda$ parameters. For predictive purposes, it would of course be interesting to find out how these parameters for the copolymer solution (subscripts AB) relate to the parameters for the solutions of the corresponding homopolymers in the same solvent (subscripts A and B, respectively) at the same temperature.

Measured composition-dependent interaction parameters [25] for solutions of the homopolymers poly(methyl methacrylate) (PMMA) and polystyrene (PS) in four solvents on one hand, and for the corresponding solutions of random

Thermodynamics of Polymer-Containing Mixtures

copolymers with different weight fractions f of styrene units on the other hand, are well modeled by the following relation:

$$\pi_{AB} = \pi_A(1 - f) + \pi_B f + \pi^E f(1 - f) \tag{41}$$

in which π stands for the different system-specific parameters α, v, and $\zeta\lambda$. Experimental data indicate [25] that the contribution of the excess term π^E might become negligible for one of the three parameters α, v, or $\zeta\lambda$, but not for the other two.

Polymer Solutions: Special Interactions

The common feature of one group of systems that deviate from normal behavior lies in the solvent, water. The present examples refer to mixtures of polysaccharides and water, which cannot be modeled in the usual manner. Aqueous solutions of poly (vinyl methyl ether) (PVME), exhibiting a second critical concentration, fall into the same category. Solutions of block copolymers in a nonselective solvent represent another instance of the need to extend the approach beyond the state formulated in (32).

Water/Polysaccharides

For the systems characterized by strong interactions between two monomeric units via hydrogen bonds, it is necessary to account for the energy of these very favorable contacts when inserting a solvent molecule between them in the first step of mixing (the parameter α is too unspecific to account for that particularity). This idea has lead to the following extension [26] of (26) for the integral interaction parameter:

$$g = \frac{\alpha}{(1 - v)(1 - v\varphi)} - \zeta(1 + (1 - \lambda)\varphi) + \overline{\omega}\,\varphi^2 \tag{42}$$

where the quadratic term in φ is due to the fact that only two macromolecules are involved in the formation of such energetically preferred intersegmental contacts; $\overline{\omega}$ quantifies the strength of these interactions.

The corresponding expression for χ [obtained according to (9)] reads:

$$\chi = \frac{\alpha}{(1 - v\varphi)^2} - \zeta(\lambda + 2(1 - \lambda)\varphi) + \overline{\omega}\,\varphi(3\varphi - 2) \tag{43}$$

Comparison of this relation with experimental data demonstrates that the parameters ζ and λ can again be merged without loss of accuracy, as shown in (32).

Organic Solvents/Block Copolymers

This is a further kind of system that cannot be modeled by means of the simple (32), referring to typical homopolymer solutions. Like with aqueous solutions of polysaccharides, the reason lies in special interactions between the segments of the different polymer chains. With block copolymers, the interactions are due to the high preference of contacts between like monomeric units over disparate contacts in cases where the homopolymers are incompatible. There is, however, a fundamental difference, namely in the number of segments that are involved in the formation of the energetically preferred structures. Two units are required for the polysaccharides (two segments are involved in the formation of a hydrogen bond), but with block copolymers of this type the interaction of at least three like monomeric units is on the average indispensable to form a microphase. This is another consequence of chain connectivity. For low molecular weight compounds, the number of nearest neighbor molecules is approximately six in the condensed state. The corresponding number of contacting polymer segments on the other hand is only about half this value, because of the chemical bonds connecting these segments to a chain molecule.

Based on these considerations, postulating the simultaneous interaction of three like segments for the establishment of a microphase, we can formulate the following relation for the integral interaction parameter g, by analogy to (42), increasing the power of the composition dependence of the third term from two to three:

$$g = \frac{\alpha}{(1-v)(1-v\varphi)} - \zeta(1+(1-\lambda)\varphi) + \tau \varphi^3 \tag{44}$$

The system-specific parameter τ accounts for the degree of incompatibility of homopolymer A and homopolymer B.

Equation (44) yields, by means of (9), the following expression for the experimentally accessible Flory–Huggins interaction parameter χ:

$$\chi = \frac{\alpha}{(1-v\varphi)^2} - \zeta(\lambda + 2(1-\lambda)\varphi) + \tau \varphi^2(4\varphi - 3) \tag{45}$$

Like with normal polymer solutions, it is also possible to merge ζ and λ for solutions of block copolymers, i.e., to eliminate one adjustable parameter.

2.1.2 Polymer Blends

For mixtures of two types of linear chain molecules, A and B, the segment molar Gibbs energy of mixing is usually formulated as:

$$\frac{\Delta \overline{\overline{G}}_{AB}}{RT} = \frac{(1-\varphi_B)}{N_A} \ln(1-\varphi_B) + \frac{\varphi_B}{N_B} \ln \varphi_B + g_{AB}\varphi_B(1-\varphi_B) \tag{46}$$

Thermodynamics of Polymer-Containing Mixtures

where N_A is the number of segments of component A and N_B is the number of segments of component B. The above equation shows the indices of the variables and parameters to indicate that it refers to a polymer blend. For such systems, the definition of a segment is not as evident as for polymer solutions, where the solvent usually fixes its volume. Sometimes the monomeric unit of one of the components is chosen to specify a segment, but in most cases it is arbitrarily defined as 100 mL per mole of segments, a choice that eases the comparison of the degrees of incompatibility for different polymer pairs.

In the case of polymer solutions, only one component of the binary mixtures suffers from the restrictions of chain connectivity, namely the macromolecules, whereas the solvent can spread out over the entire volume of the system. With polymer blends this limitations of chain connectivity applies to both components. In other words: Polymer A can form isolated coils consisting of one macromolecule A and containing segments of many macromolecules B and vice versa. This means that we need to apply the concept of microphase equilibria twice [27] and require two intramolecular interaction parameters to characterize polymer blends, instead of the one λ in case of polymer solutions.

The conditions for the establishment of microphase equilibria in the case of polymer blends [27], analogous to (17) for polymer solutions, yields two parameters. One, called a, quantifies the restrictions of the segments of a given polymer B to mix with the infinite surplus of A segments surrounding its isolated coil (microphase equilibrium for component A) and an analogous parameter b, referring to the restrictions of the segments of a given polymer A to mix with the infinite surplus of B segments. The following relations hold true for a and b:

$$a = \frac{1}{2N_A} + \frac{1}{N_B \Phi_{o,B}} \tag{47}$$

and:

$$b = \frac{1}{2N_B} + \frac{1}{N_A \Phi_{o,A}} \tag{48}$$

where the Φ values are volume fractions of segments in isolated coils, by analogy to those introduced in (17).

For the calculation of phase diagrams by means of the minimization of the Gibbs energy of the systems [19], we need to translate the information of (47) and (48), based on the chemical potentials of the components, into the effects of chain connectivity as manifested in the integral interaction parameter g. This expression reads [27]:

$$g_{AB} = \frac{\alpha_{AB}}{(1 - \nu_{AB})(1 - \nu_{AB}\varphi_B)} - \zeta_{AB}\left(\frac{2a + b}{3} + \frac{b - a}{3}\varphi_B\right) \tag{49}$$

For the partial Gibbs energies [cf. (4)] of the component i one obtains:

$$\frac{\Delta \overline{\overline{G}}_i}{RT N_i} = \frac{1}{N_i} \ln \varphi_i + \left(\frac{1}{N_i} - \frac{1}{N_k}\right) \varphi_k + \chi_i \varphi_k^2 \tag{50}$$

where i and k are the components A and B. The composition dependencies of the differential interaction parameters, χ_i, can again be calculated from g (49) by analogy to (9) and (11).

Equation (49) formulated for blends of linear macromolecules also provides the facility to model blends of linear polymers (index L) and branched polymers (index B) synthesized from the same monomer [28]. If the end-group effects and dissimilarities of the bi- and trifunctional monomers can be neglected, the parameter α becomes zero. This means that the integral interaction parameter is determined by the parameter ζ_{LB}, i.e., the conformational relaxation, in combination with the intramolecular interaction parameters of the blend components. Because of the low values of Φ_A and Φ_B, the first terms in (47) and (48) can be neglected with respect to the second terms (for molar masses of the polymers that are not too low) so that one obtains the following expression:

$$g_{LB} = -\frac{\zeta_{LB}}{3} \left(\frac{2}{N_B \Phi_B} + \frac{1}{N_L \Phi_L} + \left(\frac{1}{N_L \Phi_L} - \frac{1}{N_B \Phi_B}\right) \varphi_B\right) \tag{51}$$

It is obvious that the conformational relaxation must be proportional to the degree of branching and approach zero upon the transition of the branched polymer to a linear polymer. For the sake of consistency and simplicity, we define the degree of branching, β, again in terms of intrinsic viscosities (cf. (27)) as:

$$\beta = 1 - \frac{\Phi_L^*}{\Phi_B} \tag{52}$$

where Φ_L^* is the volume fraction of segments in an isolated linear coil consisting of the same number of segments as the branched polymer under consideration. We can then write, expanding ζ_{LB} in a series with respect to β and maintaining only the first term for the following calculations, referring to moderately branched polymers:

$$\zeta_{LB} = Z\beta(+H\beta^2 + \cdots) \tag{53}$$

Under the premises formulated above and eliminating the different Φ values by means of (27) and (29) as before, the expression for g becomes:

$$g_{LB} = -\kappa Z \beta \left(\frac{1 + \varphi_B}{\sqrt{N_L}} + \frac{2 - \varphi_B}{\sqrt{N_B}} - \frac{\beta(2 - \varphi_B)}{\sqrt{N_B}}\right) \tag{54}$$

where κ is a constant, which can be calculated from the parameter $K_{L,\Theta}$ of the Kuhn–Mark–Houwink relation (29) of the linear polymer for theta conditions and the density of this polymer as:

Thermodynamics of Polymer-Containing Mixtures

$$\kappa = \frac{K_{L,\Theta}\,\rho_L}{3} \tag{55}$$

The only information required for model calculations concerning the incompatibility of linear and branched polymers on the basis of (54) concerns β, the degree of branching of the nonlinear component, κ (i.e., the viscosity–molecular weight relationship for the linear polymer under theta conditions) and the polymer density, plus $Z = \zeta_{AB}/\beta$, the conformational response of the system normalized to β [cf. (53)].

2.1.3 Mixed Solvents

For the modeling of ternary systems(the topic of the next section), the applicability of (26) to mixtures of low molecular weight liquids would be very helpful, because of the possibility to describe all subsystems by means of the same relation. First experiments [29], presented in Sect. 4, show that this is indeed possible. This means that (26) remains physically meaningful upon the reduction of the number of segments down to values that are typical for low molecular weight compounds. With respect to λ one must, however, keep in mind that this parameter loses its original molecular meaning.

2.2 Ternary Systems

The segment molar Gibbs energy of mixing for three component (indices i, j, and k) with N_i, N_j, and N_k segments, respectively, as formulated on the basis of the Flory–Huggins theory reads in its general form:

$$\frac{\Delta\overline{\overline{G}}}{RT} = \frac{\varphi_i}{N_i}\ln\varphi_i + \frac{\varphi_j}{N_j}\ln\varphi_j + \frac{\varphi_k}{N_k}\ln\varphi_k + g_{ij}\varphi_i\varphi_j + g_{ik}\varphi_i\varphi_k + g_{jk}\varphi_j\varphi_k$$
$$+ t_{ijk}\,\varphi_i\,\varphi_j\,\varphi_k \tag{56}$$

The first three terms stand for the combinatorial part of the Gibbs energy, the next three terms represent the residual contributions stemming from binary interactions, and the last term accounts for ternary contacts.

The double-indexed g parameters are for binary interaction parameters. The first line of the above relation represents the combinatorial part, and the second line the residual part of the reduced segment molar Gibbs energy of mixing. This relation also contains a ternary interaction parameter t_{ijk} that accounts for the expectation that the interaction between two components of the ternary mixture may change in the presence of a third component.

Because of the well-documented composition dependencies of the individual binary interaction parameters, an unmindful use of (56) would lead to totally unrealistic results. This feature requires twofold adaption.

First of all, it is necessary to account for the fact that the contribution of a certain binary contact to the total Gibbs energy of mixing depends on its particular molecular environment, which in the general case also contains the third component. We can allow for that circumstance by multiplying g_{ij} with the factor $(\varphi_j + \varphi_i)$ $= (1 - \varphi_k) \leq 1$.

Secondly, we need to specify whether the composition dependencies of the g_{ij} parameters are formulated in terms of φ_i or of φ_j, because the resulting mathematical expressions are not identical.

In order to enable a straightforward application of the new approach to the most interesting ternary systems (polymer solutions in mixed solvents and solutions of a polymer blend in a common solvent), it is expedient to express the binary interaction parameters for polymer solvent systems (26) and that for polymer blends (49) in the same form. This requirement is met by the relation:

$$g_{ij} = \frac{\alpha_{ij}}{\left(1 - v_{ij}\right)\left(1 - v_{ij}\varphi_j\right)} - \zeta_{ij}\left(\delta_{ij} + \varepsilon_{ij}\,\varphi_j\right) \tag{57}$$

For polymer solutions:

$$\delta = 1 \quad \text{and} \quad \varepsilon = (1 - \lambda) \tag{58}$$

whereas the corresponding equation for a polymer blend (the composition dependence being expressed in terms of φ_B) reads:

$$\delta = \frac{2a + b}{3} \quad \text{and} \quad \varepsilon = \frac{b - a}{3} \tag{59}$$

By means of (57) and the required modification formulated at the beginning of this section, one obtains the following expression for the reduced residual segment molar Gibbs energy of mixing of ternary systems, if one neglects ternary interactions ($t_{ijk} = 0$) for the time being:

$$
\begin{aligned}
\frac{\overline{\overline{\Delta G}}^{\text{res}}}{RT} &= \left[\frac{\alpha_{12}}{(1 - v_{12})(1 - v_{12}\varphi_2(1 - \varphi_3))} - \zeta_{12}(\delta_{12} + \varepsilon_{12}\,\varphi_2(1 - \varphi_3))\right]\varphi_1\varphi_2 \\
&+ \left[\frac{\alpha_{23}}{(1 - v_{23})(1 - v_{23}\varphi_3(1 - \varphi_1))} - \zeta_{23}(\delta_{23} + \varepsilon_{23}\,\varphi_3(1 - \varphi_1))\right]\varphi_2\varphi_3 \\
&+ \left[\frac{\alpha_{31}}{(1 - v_{31})(1 - v_{31}\varphi_1(1 - \varphi_2))} - \zeta_{31}(\delta_{31} + \varepsilon_{31}\,\varphi_1(1 - \varphi_2))\right]\varphi_3\varphi_1
\end{aligned}
\tag{60}
$$

Thermodynamics of Polymer-Containing Mixtures

As one of the three composition variables becomes zero, this relation simplifies to the expression for binary mixtures (57). The extension of (60) to multicomponent systems is unproblematic and enables the calculation of phase diagrams for such mixtures of great practical importance if one calculates the composition of the coexisting phases by a direct minimization of the Gibbs energy [19]. In this manner, it is possible to evade the laborious and sometimes even impossible calculation of the chemical potential for each component.

The implementation of ternary interactions by simply adding the last term of (56) to (60) does not suffice. The reason lies in the fact that the three options to form a contact between all three components out of binary contacts ($1/2 + 3$, $1/3 + 2$, and $2/3 + 1$) might differ in their contribution to the Gibbs energy of the mixture. This supposition results in the necessity to introduce three different ternary interaction parameters. Furthermore, it requires a weighting of these contribution to account for the fact that they must be largest in the limit of the first addition of the third component 3 (highest fraction of 1/2 contacts) and die out as component 3 becomes dominant (vanishing fraction of 1/2 contacts). The simplest possibility to account for $\overline{\overline{\Delta G}}_t^{\,\text{res}}$, the extra contributions of ternary contacts to the residual Gibbs energy, is formulated in (61), where the negative sign was chosen by analogy to the second term of (57):

$$\frac{\overline{\overline{\Delta G}}_t^{\,\text{res}}}{RT} = -[t_1(1 - \varphi_1) + t_2(1 - \varphi_2) + t_3(1 - \varphi_3)]\varphi_1 \varphi_2 \varphi_3 \tag{61}$$

t_1 quantifies the changes associated with the formation of a ternary contact 1/2/3 out of a binary contact 2/3 by adding a segment of component 1. The meaning of t_2 and t_3 is analogous.

Equation (61) makes allowance for differences in the genesis of ternary contacts but it does not yet consider that the number of segments of the third component in the coordination sphere of a certain binary contact might deviate from that expected from the average composition due to very favorable or unfavorable interactions (quasi chemical equilibria). One way to model such effects consists of the introduction of composition-dependent ternary interaction parameters, as formulated in the following equation:

$$\frac{\overline{\overline{\Delta G}}_{t(\varphi)}^{\,\text{res}}}{RT} = -\,[(t_1 + t_{11}\varphi_1)(1 - \varphi_1) + (t_2 + t_{22}\varphi_2)(1 - \varphi_2) \\ + (t_3 + t_{33}\varphi_3)(1 - \varphi_3)]\varphi_1\varphi_2\varphi_3 \tag{62}$$

The relations presented for ternary mixtures open the possibility for investigation of the extent to which their thermodynamic behavior can be forecast (neglecting possible contributions of ternary interaction parameters) if the binary interaction parameters of the three subsystems are known as a function of composition from independent experiments. For such calculations, it is important to make sure that the size of a segment is identical for all subsystems. The fact that most of the experimental information available for polymer solutions uses the molar volume of the particular solvent to fix the size of a segment, requires a conversion in the case of polymer solutions in mixed solvents. If one chooses the molar volume of solvent 1 to define the common segment, this means that the binary interaction parameter for the solvent 2 must be divided by the ratio of the molar volumes $\overline{V}_1/\overline{V}_2$.

3 Measuring Methods

Experimental information concerning the thermodynamic properties of mixtures is primarily accessible via phase equilibria [30]. In the case of polymer solutions, vapor pressure measurements (liquid/gas equilibria) constitute the most important source of data because of the nonvolatility of the solutes and because of the comparatively large composition interval (typically ranging from some 25% to almost pure polymer) over which this method yields reliable data. In order to obtain a complete picture from infinitely dilute solutions up to almost pure polymer melt, these data need to be complemented by further methods. Osmometry (liquid/liquid equilibria) provides χ_0, the Flory–Huggins interaction parameter in the limit of pair interaction between the polymer molecules; this information is also accessible via scattering methods (light or neutrons), which monitor the composition dependences of the chemical potentials. Most published data refer to dilute and moderately concentrated solutions. It is difficult to study the range of vanishing solvent concentration because of the high viscosity of such mixtures. Inverse gas chromatography (IGC) is one of the few sources of information. Thermodynamic information for polymer blends is usually based on small angle neutron scattering. The following sections (Sects. 3.1–3.3) outline how the different methods work and cite some recent relevant publications in this area.

3.1 Vapor Pressure Measurements

The classical method consists of the quantitative removal of air from polymer solutions coexisting with a gas phase and measurement of the equilibrium pressures of the solvent above the solution by means of different devices. Such experiments are very time consuming because the liquid mixtures must be frozen-in and the air that accumulates in the gas phase must be pumped off. In order to obtain reliable data this procedure must be repeated several times to get rid of all gases. By means of this approach it is practically impossible to accumulate comprehensive information for a large number of systems.

For the reasons outlined above, alternative methods were developed that avoid the measurement of absolute vapor pressures. One procedure combines head space sampling with conventional gas chromatography (HS–GC) [31] and yields relative vapor pressures, normalized to the vapor pressure of the pure solvent. A well-defined volume of the equilibrium gas phase is taken out from a thermostated vial

Thermodynamics of Polymer-Containing Mixtures

sealed with a septum by means of a syringe and transferred to a gas chromatograph. The amount of the solvent is registered either in a flame ionization detector (FID) or by means of a cell measuring the thermal conductivity of the gas stream. Such measurements yield the ratio p/p_o, which in many cases can be taken as the activity of the solvent. Whether corrections for the nonideality of the gas are required or not must be checked in each case. The main work required with this method consists of the optimization of the HS–GC, i.e., determination of the best operation procedures for gas sampling, gas chromatography, and data evaluation. However, once these parameters have been determined, HS–GC offers quick access to thermodynamic data because the method is automated.

Another possibility for avoiding the measurement of absolute vapor pressures is provided by sorption methods. In most cases, the polymer is positioned on a quartz balance and the amount of solvent it takes up via the vapor phase is weighted. The so-called "flow-through" variant [32] works with an open system in contrast to the previous method.

Isopiestic [33] experiments also offer access to chemical potentials. This method monitors the conditions under which the vapor pressures above different solutions of nonvolatile solutes (like polymers or salts) in the same solvent become identical, where one of these solutions is a standard for which the thermodynamic data are known. These experiments can be considered to be a special form of differential osmometry (cf. Sect. 3.2) where the semi-permeable membrane, separating two solutions of different composition, consists of the gas phase.

3.2 Osmometry and Scattering Methods

Measurements performed to determine the molar masses of polymers yield – as a valuable byproduct – information on the pair interaction between the macromolecules [30]. The composition dependence of the osmotic pressure π_{osm} observed via membrane osmometry is directly related to the chemical potential of the solvent [cf. (14) of Sect. 2] and provides the second osmotic virial coefficient A_2, from which χ_o, the Flory–Huggins interaction parameter in the limit of high dilution becomes accessible [cf. (15)]. Such data are particularly valuable because they can be measured with higher accuracy than the χ values for concentrated polymer solutions and because they represent a solid starting point for the sometimes very complex function $\chi(\varphi)$. In principle, membrane osmometry can also be operated with polymer solutions of different composition in the two chambers (differential osmometry) to gain data for higher polymer concentrations; however, little use is made of this option.

Scattering methods represent another route to A_2 and χ_o; these experiments do not monitor the chemical potential itself but its composition dependence. Light scattering – like osmosis – can in principle also yield information for polymer solutions beyond the range of pair interaction, but corresponding reports are seldom. In contrast, small angle neutron scattering is an important source of thermodynamic information for polymer blends over the entire composition range.

3.3 Other Methods

In addition to the experiments briefly discussed above, two further equilibrium methods and two nonequilibrium procedures are sometimes employed to obtain thermodynamic information.

The most frequently used additional method is the evaluation of data for liquid/liquid phase separation, i.e., of critical points and of binodal curves [21]. This information is normally obtained by means of cloud point measurements (either visually or turbidimetrically) and the analysis of the composition of coexisting phases. Critical data give access to the system-specific parameters via the critical conditions, as formulated in (36) and (37) for the present approach or by means of equivalent expressions of other theories. If the critical data (temperature, pressure, and composition) are known for a sufficiently large number of polymer samples with different molar mass, and the number of parameters required for a quantitative description of $g(\varphi)$ is not too high, this method yields reliable information. Similar consideration also hold true for the evaluation of binodal curves. In both cases it is very helpful to formulate a theoretically justified temperature dependence of the system-specific parameters.

Liquid/solid equilibria also offer access to thermodynamic information. In this case, it is the differential interaction parameter ξ of the *polymer* that is obtained according to (38) from the known molar mass of the polymer, its melting temperature in the pure state, and the corresponding heat of melting plus the polymer concentration in the solution that is in equilibrium with the pure polymer crystals. Because of the well-known problems in obtaining perfect crystals in the case of macromolecules, special care must be taken with the evaluation of such data.

Vapor pressure osmometry [34–36] constitutes a very helpful nonequilibrium method for obtaining thermodynamic information for solutions of oligomers and polymers of low molar mass, for which osmometry and light scattering experiments do no longer yield reliable data. Such experiments are based on the establishment of stationary states for the transport of solvent via the gas phase from a drop of pure solvent fixed on one thermistor to the drop of oligomer solution positioned on another thermistor. Because of the heats of vaporization and of condensation, respectively, this transport process causes a time-independent temperature difference from which the required information is available after calibrating the equipment.

Inverse gas chromatography (IGC) represents another nonequilibrium method; it yields valuable information on polymer–solvent interactions in the limit of vanishing solvent content [37, 38]. In experiments of this type, a plug of solvent vapor is transported in a column over a stationary phase consisting of the pure polymer melt. The more favorable the solvent interaction with the polymer, the longer it takes until the plug leaves the column. An adequate evaluation of the observed retention times yields access to the chemical potentials of the solvent, i.e., to Flory–Huggins interaction parameters in the limit of $\varphi \to 1$.

4 Experimental Results and Modeling

4.1 Binary Systems

4.1.1 Polymer Solutions

Organic Solvents/Linear Homopolymers

This section gives examples for the typical thermodynamic behavior of polymer solutions. The first part deals with homogeneous mixtures and discusses the molecular weight dependence of second osmotic virial coefficients, the role of glass transition for the determination of interaction parameters, and the reasons for changes in the sign of the heat of dilution with polymer concentration. The second part of this section is dedicated to liquid/liquid phase separation and – among other things – explains in terms of the present approach, why 1,2-polybutadiene is completely miscible with n-butane but 1,4-polybutadiene is not.

One of the major consequences of the thermodynamic approach used here is the postulate that the second osmotic virial coefficients may increase with rising molar mass of the polymer, even for good solvents (better than theta conditions), in contrast to the statements of current theories. Figure 2 shows an example of this behavior, which was already observed by Flory and coworkers [22] in the 1950s and confirmed by independent measurements [39].

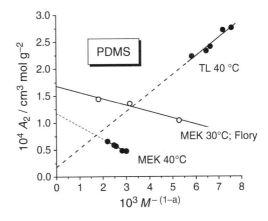

Fig. 2 Molecular weight dependence of the second osmotic virial coefficient for the systems [22, 39] methyl ethyl ketone (MEK)/PDMS and TL/PDMS represented according to (31)

As postulated by (31), the molecular weight dependence of A_2 should not be represented in double logarithmic plots, but as a function of $M^{-(1-a)}$, where a is the exponent of the Kuhn–Mark–Houwink relation. In contrast to the customary evaluation (in double logarithmic plots), this procedure does not in the general case lead to zero A_2^∞ values; in most cases they are very small but, outside experimental errors, different from zero.

The following example of the composition dependence of the Flory–Huggins interaction parameter pertains to the system cyclohexane/poly(vinyl methyl ether) (CH/PVME) [23]. Except for χ_o, obtained via osmometry, all data stem from vapor pressure measurements [40]. This system does not fit into the normal scheme because CH is a good solvent for PVME, despite uncommonly large χ_o values of the order of 0.5. For good solvents, χ_o is usually considerable less than 0.5; for theta solvents, χ_o is typically equal to 0.5 and it increases upon the approach of phase separation. The curves combining the data points in Fig. 3 were calculated by adjusting the parameters of (32). Within the scope of the present approach, the high solvent quality results from fact that the χ values decrease considerably as φ increases so that they are favorable within the range of moderate polymer concentrations, where the system becomes very susceptible to phase separation.

The minima of $\chi(\varphi)$ shown in Fig. 3 represent a consequence of the dissimilar contributions of the dilution in two steps, as demonstrated in Fig. 4. The first term, quantifying the effects of contact formation, is Gibbs energetically very unfavorable and increases with rising polymer concentration because of the parameter v. By contrast, the second term, standing for the contributions of the conformational relaxation, is highly favorable and the more so, the larger φ becomes. The observed minimum in $\chi(\varphi)$ is caused by the fact that the first summand increases more than linearly, whereas the second decreases linearly.

Figure 4 also documents the general observation that the contributions of the two terms of (32) to the measured functions $\chi(\varphi)$ are markedly larger than χ itself; this situation is very similar to the build up of the Gibbs energy from enthalpy and entropy contributions. However, it is not permissible to interpret these terms in this

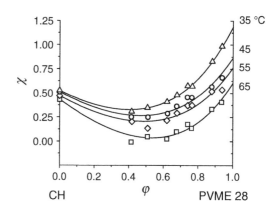

Fig. 3 Composition dependence of Flory–Huggins interaction parameter for the CH/PVME system at the indicated temperatures [40]; the curves are calculated according to (32). *PVME 28* indicates that the M_w of the PVME is 28 kg/mol

Fig. 4 Breakdown of the composition dependence of χ into the contributions resulting from the two steps of dilution (cf. Fig. 1 and (32), $\alpha = 1.599$, $v = 0.398$, and $\zeta\lambda = 1.074$) for the CH/PVME 28 system at 35°C

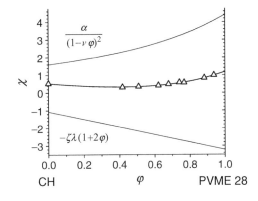

Fig. 5 Interrelation between the leading parameters of (32). *Closed symbols* data from A_2 (M) [39]; *open symbols* data from $\chi(\varphi)$ [23]. Each symbol stands for a different polymer/solvent system, the polymers being PS, PMMA, polyisobutylene, and PDMS

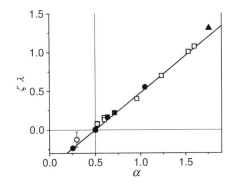

manner: Both terms can be split into their enthalpy and entropy parts, as will be shown later.

Another point of view on the contributions of the two terms of (32) deserves special attention. Namely, the expectation according to the present approach that their leading parameters, α and ζ, should not be independent of each other. The reason for this surmise lies in the fact that contact formation and conformational relaxation share the same thermodynamic background, i.e., the effects of the conformational relaxation of the components should strongly correlate with the effects of contact formation, as discussed in Sect. 2.

The results shown in Fig. 5 demonstrate that there indeed exists such a general interrelation, where each data point represents a certain system and temperature. The results of this graph demonstrate the consistency of the approach because the data [39] obtained from the evaluation of the molecular weight dependence of A_2 (cf. Fig. 2) and from the composition dependence of $\chi(\varphi)$ (an example [40] is shown in Fig. 3) lie on the same line [here $\zeta\lambda = (0.957 \pm 0.00027) \times (\alpha - 0.5)$], despite the fundamentally different experimental methods used for their determination. For the common representation of the data, the ζ values reported in table 2 of [39] were multiplied by -0.5 (i.e., λ was set at 0.5), which is permissible for sufficiently large molar masses of the polymer. The negative sign of this factor results from the fact that ζ of [39], used for the evaluation of $A_2(M)$, was defined with the opposite sign of ζ as compared with the modeling of $\chi(\varphi)$. The reason lies in the interrelation between χ_o and A_2 formulated in (15).

It appears interesting that the interrelation between the leading parameters of the present approach shown in Fig. 5 for simple systems, i.e., for absence of special interactions between the components, is generally valid and holds true for all hitherto studied polymer solutions.

The modeling of homogeneous systems has so far been exemplified by means of solutions of polymers that are liquid at the temperatures of interest. Such systems are, however, the exception rather than the rule, because of the comparatively high glass transition temperatures of most polymers. Typical polymer solutions solidify upon a sufficient augmentation of polymer concentration and the question arises of how this feature is reflected in the thermodynamic data. To study the importance of this loss in the mobility of the polymer chains for the determination of Flory–Huggins interaction parameters, we have studied solutions of PS in different solvents [41] within the temperature range of 10–70°C. These experiments demonstrate that the consequences of the freezing-in of the polymer motion at high polymer concentrations for the measured vapor pressures depend on the thermodynamic quality of the solvent and on the experimental method employed for the measurement.

Figure 6 shows the reduction of the vapor pressures of toluene (TL, a good solvent) and of CH (a marginal solvent) as the concentration of PS rises. As long as the mixtures are liquid these curves display the interaction in the usual manner, i.e., the reduced vapor pressure of TL is considerable lower than that of CH because of the more favorable interaction with the polymer. In the case of CH, this dependence continues smoothly into the glassy range, whereas a discontinuity is observed for TL.

Based on the results shown in Fig. 6, one is tempted to postulate that the solvent quality loses its importance once the solutions become glassy. However, the situation is more complicated under nonequilibrium conditions, as discussed by means of Fig. 7. This graph contains two types of experimental data, one set obtained via

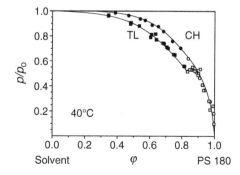

Fig. 6 Dependencies of the reduced vapor pressures of TL (*squares*) and of CH (*circles*) on the volume fraction of PS at 40°C [41]. *Closed symbols* liquid mixtures, *open symbols* glassy mixtures

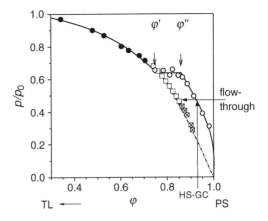

Fig. 7 Reduced vapor pressure of TL as a function of the volume fraction of PS measured at 20°C with either HS–GC (*closed circles* liquid solutions, *open circles* glassy mixtures), or at 30°C, by means of the flow-through method [32] (*squares*). According to [32], the *crossed squares* refer to compositions inside the range of glassy solidification; no information is given for the *open squares*. *Lines* are guides for the eye

HS–GC as usual and another set [32] obtained by means of the so-called "flow through" method, which differs fundamentally.

The vapor pressure data obtained by means of HS–GC measurements for the solutions of PS (below the glass transition temperature of the polymer) in the favorable solvent TL resemble closely the results for the solutions of polyethylene oxide (PEO) in chloroform (below the melting temperature of the polymer), as shown later. The common denominator of these processes lies in the loss of mobility of the macromolecules. The results presented in Fig. 7 can be interpreted in the following manner: The composition range of constant vapor pressure ($\varphi' \leq \varphi \leq \varphi''$) observed with HS–GC measurements reflects the coexistence of two kinds of microphases, one in which the polymer mobility is identical with that in the liquid state at the composition φ', and a glassy microphase of composition φ'', where the segmental mobility is fully frozen-in. The reason why this sort of "tie line" can be observed with HS–GC but not with flow-through experiments lies in the fact that the former method uses a closed system, in contrast to the latter in which additional vapor is always available. Because it is the vapor pressure that is constant in flow-through experiments and not the composition of the mixture, the amount of solvent taken up by the polymer can be constantly replaced. This process comes to an end either as the equilibrium vapor pressure of the liquid mixture is reached at the composition φ' or as kinetic impediments become too large. The two methods under consideration complement each other: HS–GC monitors the upper limit φ'' of the solidification interval, whereas the flow-through method displays its lower limit φ'. Concerning the evaluation of vapor pressures measured via HS–GC above solidified polymer solutions, it is obvious from the present results that such information must not be used to establish $\chi(\varphi)$ dependencies, particularly in the case of thermodynamically favorable solvents.

To conclude the treatment of homogenous solutions of polymers in organic solvents, we deal with the temperature dependencies of the parameters of (32). The knowledge of these changes enables their separation in enthalpy and entropy contributions as formulated in (39) and (40) for χ. In this context, it is of particular interest to check whether the approach helps the rationalization of observed changes in the sign of the heat of dilution upon a variation of the polymer concentration. Solutions of PS in *tert*-butyl acetate (TBA) were chosen for this purpose because of the large temperature interval that was studied for this system. The combination of three different methods was used to obtain interaction parameters in all composition regions of interest: (1) light scattering measurements for dilute solutions in closed cells [42], (2) the determination of absolute vapor pressures (not HS–GC, quantitative removal of air) up to temperatures well above the boiling point of the pure solvent [43], and (3) IGC [37] close to the polymer melt.

The analysis [44] of the thus-obtained temperature dependencies of the system-specific parameters of (32) with respect to the individual enthalpy contributions of the two steps of dilution (cf. Fig. 1), yields $\chi_{H,fc}$ and $\chi_{H,cr}$. How these heat effects depend on polymer concentration is shown for the system TBA/PS at 110°C in Fig. 8.

This graph makes it immediately obvious that the insertion of a solvent molecule between contacting segments at constant conformation of the components constitutes an exothermal process ($\chi_{H,fc} < 0$) at high dilution, whereas the conformational relaxation is endothermal ($\chi_{H,cr} > 0$). In both cases, the absolute values of the heat effects increase with rising polymer concentration. However, the slopes of these two dependencies differ in such a manner that the total heat of dilution is exothermal for low φ values, but endothermal for high polymer concentrations. With the present example, this finding can be rationalized qualitatively in terms of the composition dependence of free volumes and excess volumes. The pure solvent is already highly expanded and the polymer molecules may fill some of the existing voids (this should lead to negative excess volumes and to the evolution of heat, due to the formation of new molecular interfaces). The pure melt, on the other hand, is still densely packed at the same temperature and the addition of a solvent molecule might cause an expansion in volume (resulting in positive excess volumes and in the consumption of heat).

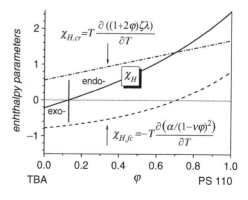

Fig. 8 Composition dependence of the enthalpy part of the Flory–Huggins interaction parameter χ for the system TBA/PS ($M_w = $ 110 kg/mol, narrow molecular weight distribution) at 110°C as compared with the corresponding enthalpy contributions of the two steps of dilution [cf. Fig. 1 and (32)]. The experimental data were taken from [43]

Thermodynamics of Polymer-Containing Mixtures

There exist other examples for inversions in the heats of dilution; in these cases an analogous straightforward molecular interpretation appears difficult. For instance, the system TL/PS shows an inversion [44] from endothermal in the range of moderate polymer concentrations to exothermal at high φ values at 37°C. In this case, the sign of the heat contributions of the two steps of dilution are the same as in the previous case. However, here it is only $\chi_{H,cr}$ which increases (linearly) with rising polymer concentration, whereas $\chi_{H,fc}$ decreases (more than linearly). This combination of the two contributions leads to the opposite inversion, namely from endothermal to exothermal upon an augmentation of φ. Concerning the molecular reasons for this behavior, one may speculate on the basis of the present findings that the insertion of a TL molecule between two contacting PS segments (belonging to different molecules) becomes energetically particularly favorable in the limit of high polymer concentration.

So far, we have dealt exclusively with homogeneous systems; the following considerations concern the possibilities of obtaining the parameters of the present expression for $\chi(\varphi, T)$ from demixing data. The results will demonstrate that the present approach is capable of modeling liquid/liquid equilibria and liquid/gas equilibria with the same set of parameters, in contrast to traditional theories.

The first example refers to solutions of PS in CH. This is probably the system for which the phase separation phenomena are studied in greatest detail [45], namely in the temperature range from ca. 10 to 240°C and for molar masses from 37 to 2700 kg/mol. Figure 9 displays the experimental data [45] together with the modeling, using (32) to describe the Flory–Huggins interaction parameter as a function of composition.

The system-specific parameters used for the modeling of the phase diagrams were calculated from the critical data (T_c and φ_c) measured [45] for PS samples of different molar mass. For this purpose, the critical conditions resulting for the present approach [cf. (36) and (37)] were first simplified: The parameter λ was set at 0.5 (this does not imply a loss of accuracy for the system of interest) and the interrelation between α and $\zeta\lambda$ [cf. (34)] was used to eliminate the parameter α; setting $E = 0.847$ (an average value for solutions of vinyl polymers in organic solvents). This procedure reduces the number of parameters from four to only two (ζ and v) and enables the calculation of their values from the critical temperature by inserting the known numbers of segments N and the critical composition φ_c in the two critical conditions and solving these equations. Because of the large number of different molar masses, yielding different critical data, it is possible to model the temperature dependencies of the parameters ζ and v. The observed maximum in $\zeta(T)$ is expected because of the transition from an upper critical solution temperature (UCST) behavior at low temperatures to a lower critical solution temperature (LCST) behavior at high temperatures, in combination with the fact that $\zeta = 0$ at the theta temperature, irrespective of the sign of the heat of mixing; $v(T)$ also passes a maximum but at a much lower temperature (in the vicinity of the endothermal theta temperature). Within the range of LCSTs, both parameters decrease with rising temperature. The binodal and spinodal curves shown in Fig. 9 for the different PS samples were calculated from the thus-obtained system-specific parameters using

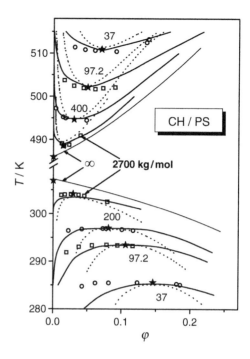

Fig. 9 Phase diagram (demixing into two liquid phases) of the system CH/PS for the indicated molar masses of the polymer (kg/mol). Cloud points (*open symbols*) and critical points (*stars*) are taken from the literature [45]. The data for the high temperatures refer to the equilibrium vapor pressure of the solvent. Binodals (*solid lines*) and spinodals (*dotted lines*) were calculated as described in the text by means of the temperature-dependent parameters [46] ζ and ν

the method of the direct minimization of the Gibbs energy [19], instead of equality of the chemical potentials of the components as the equilibrium condition.

The agreement of information concerning the composition dependence of the Flory–Huggins interaction parameter obtained from different sources is demonstrated by means of Fig. 10. The data points display the results of vapor pressure measurements and the dashed line stems from the evaluation of critical demixing data described above. The interaction parameter is calculated according to (6), (5), and (32) by reading the ζ and ν values from figure 2 of [46] for 308 K, setting $\lambda = 0.5$ and $E = 0.847$. To the author's knowledge, this is the first time that liquid/gas and liquid/liquid phase equilibria have been modeled accurately by the same set of parameters, where only two were adjusted to the experimental data in the present case.

For some technical processes and polymer applications, pressure represents an important variable. For this reason, the extent to which the present approach is suited to describe pressure effects was checked. By means of demixing data as a function of pressure published for the system *trans*-decalin/PS [49] it was shown [46] that (32) is also apt for that purpose.

The systems n-butane/1,4-polybutadiene (98% cis) [n-C$_4$/1,4-PB] and n-butane/1,2-polybutadiene [n-C$_4$/1,2-PB] are the next examples for the modeling of Flory–Huggins interaction parameters [50]. In this case, it appeared particularly interesting to understand why 1,2-PB is totally miscible with n-C$_4$ but 1,4-PB is not. In these experiments we measured the absolute vapor pressures (i.e., not using

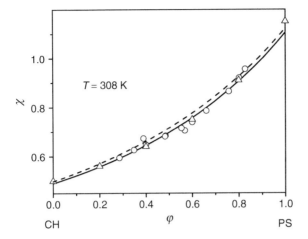

Fig. 10 Comparison of the composition dependence of the Flory–Huggins interaction parameter determined for the CH/PS system at 308 K either from liquid/gas equilibria (*triangles* [47] and *circles* [48]), jointly represented by the solid line, or from liquid/liquid equilibria [46] (*dashed line*)

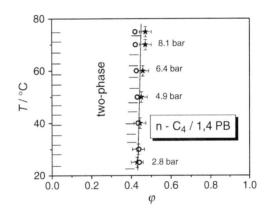

Fig. 11 Isochoric phase diagram for the system n-C$_4$/1,4-PB [50]. The measured data points are drawn as *stars* and those calculated by means of the interaction parameters obtained from the vapor pressure measurements as *circles*

HS–GC) because of the high volatility of n-C$_4$, and studied the segregation of a second liquid phase for the solutions of 1,4-PB under isochoric conditions (instead of the usual isobaric procedure). Figure 11 shows the thus-obtained phase diagram together with the miscibility gap calculated from the measured vapor pressures. Here, it is worth mentioning that the Sanchez–Lacombe theory [4, 51] models the vapor/liquid equilibria for the present systems very well but fails totally when the parameters obtained from such measurements are applied for the calculation of liquid/liquid equilibria.

The good agreement between the prediction of the miscibility gap from liquid/gas equilibria with the actual behavior is a further example of the utility of this approach. The extension of the Flory–Huggins theory by incorporating further contributions of chain connectivity and accounting for the phenomenon of conformational relaxation also enables the rationalization of the fundamentally different solubilities of 1,4-PB and 1,2-PB. The 1,2-isomer interacts favorably with n-C$_4$ because the flexibility of the polymer backbone (pending double bonds) enables the establishment of suitable contacts with the surrounding solvent molecules. With the 1,4-isomer, on the other hand, such a rearrangement is largely impeded because the double bonds are now located in the main chain and make the conformational response much more difficult.

All examples shown so far refer to solutions of noncrystalline polymers. We will now discuss the solutions of a crystalline polymer, namely PEO in chloroform. Figure 12 gives an example of the primary data that can be obtained by measuring the reduced vapor pressure of the solvent by means of HS–GC.

According to the present results, it is possible to distinguish three clearly separable composition ranges, I–III (see Fig. 12). Only for range III do the data not depend on the details of film preparation, i.e., yield equilibrium information. The situation prevailing in the other ranges is discussed in terms of the addition of CHCl$_3$ to solid PEO. Within range I ($1 > w > w''$), the vapor pressure increases steadily up to a characteristic limiting value located well below that of the pure solvent. Within range II ($w'' > w > w'$), p_1 remains constant, despite the addition of further solvent. Finally, within range III ($w' > w > 0$), the vapor pressure rises again and approaches the value of the pure solvent. Range I should be absent for fully crystalline polymers; its existence is due to the amorphous parts of PEO, which can take up solvent until w' is reached. Range II results from the coexistence of the saturated solution with variable amounts of polymer crystals. Finally, no solid material is available in range III and we are back to the normal situation encountered with the solutions of amorphous polymers. According to the present results, it is practically impossible to reach thermodynamic equilibria within range I. Vapor pressures and degrees of crystallinity depend markedly on the details of sample preparation. Measurements within range III do not present particular problems with

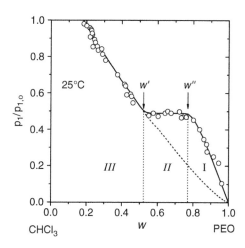

Fig. 12 Reduced vapor pressure of chloroform above solutions of PEO at 25°C as a function of the weight fraction w of the polymer [52]. The three composition ranges are labeled *I*, *II*, and *III*. The *dotted line* extrapolates the behavior of the homogeneous mixtures into the two-phase range

the attainment of equilibria. Range II assumes an intermediate position in this respect.

The observed nonequilibrium behavior at high polymer concentrations can be interpreted in terms of local and temporal equilibria, which are frozen-in during film preparation, i.e., in the course of solvent removal or quenching of the polymer melt. For discussion of these effects it is helpful to compare the fraction of the polymer that does not participate in the liquid/vapor equilibrium with the degree of crystallinity as obtained from DSC measurements. The general findings that the former is always larger than the latter, and that the differences decrease upon dilution, are tentatively interpreted as a trapping of amorphous PEO inside the crystalline material during sample preparation and its gradual release by the addition of solvent. This hypothesis is supported by micrographs showing the existence of such occlusions.

For systems of the present type it is possible to obtain equilibrium information from two sources: in the usual manner via the vapor pressures of the solvent above the solutions within range III (chemical potential of the solvent) and additionally from the saturation composition w' of the polymer (chemical potential of the polymer). The thermodynamic consistency of these data was documented [52] by predicting w' (liquid/solid equilibrium) from the information of liquid/gas equilibria. This match of thermodynamic information from different sources is a further argument for the suitability to the present approach for the modeling of polymer-containing mixtures.

Organic Solvents/Nonlinear Homopolymers

This section deals with the extent to which differences in the molecular architecture of the polymer affects its interaction with a given solvent. In particular, the comparison of linear and branched macromolecules is of interest. In order to obtain a clear-cut answer and for a straightforward theoretical discussion it is important to exclude special end-group effects (i.e., to keep the chemistry of the terminal group as similar as possible to that of the middle groups) and to apply the same criteria to the branching sites. The example [24] chosen refers to solutions of linear and branched polyisoprene (PI) in CH and fulfills the above criteria reasonably well. The number of branching points per molecule of the nonlinear product lies between six and seven. Figure 13 shows the composition dependence of the Flory–Huggins interaction parameter for the two types of systems obtained from HS–GC measurements and from vapor pressure osmometry.

Linear PI interacts with CH considerably more favorably than does the branched analog in the temperature range from 25 to 65°C according to these results, irrespective of polymer concentration. This finding agrees well with the expectation based on the present approach, which states that the first term of (32) (quantifying the first step of dilution, cf. Fig. 1) should only be affected marginally by a transition from a linear to a branched architecture of the polymer, in contrast to

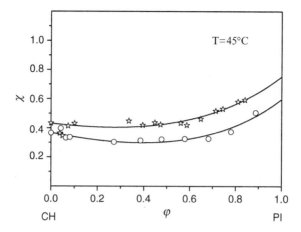

Fig. 13 Flory–Huggins interaction parameters for solutions of linear PI (*circles*, $M_w = 23.2$ kg/mol) and of branched PI (*stars*, $M_w = 21.6$ kg/mol) in CH as a function of polymer concentration

the second term (resulting from the conformational relaxation). This second summand is determined by the parameters λ and ζ, both of which depend on the molecular architecture of the polymer: A higher degree of branching leads to a reduction in the intramolecular interaction parameter λ (accumulation of the segments in a smaller volume) as well as in ζ (diminished possibilities to readjust to a changing molecular environment by conformational relaxation). Because of the negative sign of the second term, these changes lead to larger χ values for the branched polymer. In other words, in the absence of special effects, the thermodynamic quality of a given solvent declines upon an increase in the degree of branching. Another feature worth mentioning is the observation that the interrelation between the parameters $\zeta\lambda$ and α, established for linear macromolecules (cf. Fig. 5), remains valid for branched materials.

Organic Solvents/Linear Random Copolymers

With systems of this type, a new feature comes into play: In spite of the fact that we are dealing with binary systems, we need three different interaction parameters to describe the thermodynamic behavior. This makes the modeling considerably more difficult and is the reason why the present approach requires more adjustable parameters, and the theoretical understanding is far from being satisfactory.

For reasons outlined in the theoretical section (Sect. 2) (8) the study reported here uses weight fractions w instead of the usual volume fractions φ. It was carried out for solutions of poly(styrene-*ran*-methyl methacrylate [P(S-*ran*-MMA)], with different weight fractions f of styrene units, in $CHCl_3$, acetone (AC), methyl acetate (MeAc), and TL at 50°C [25]. Analogous measurement for the solutions of the corresponding homopolymers, PMMA and PS, were also performed for comparison.

For practical purposes, the possibility of predicting the thermodynamic behavior of random copolymers in a given solvent from knowledge of the corresponding homopolymers would be extremely helpful. The present results demonstrate that this is a difficult task and that the choice of the particular solvent plays a decisive role. For all systems under investigation, $_w\chi$ varies considerably with the composition of the mixture. With one exception [CHCl$_3$/P(S-ran-MMA) and $f = 0.5$] the dependencies $_w\chi(w)$ of the copolymers do not fall reasonably between the data obtained for the corresponding homopolymers. In most cases, the incorporation of a small fraction [25] of the monomer that interacts less favorably with a given solvent suffices to reduce the solvent quality for the copolymer, approximately to that for the worse soluble homopolymer. Figure 14 shows an example for which this effect is particularly obvious.

In terms of the $_w\chi$ values measured for a given constant polymer concentration, the polar solvents CHCl$_3$, AC, and MeAc are expectedly more favorable for PMMA than for PS, whereas the nonpolar TL is a better solvent for PS than for PMMA. The shape of the functions $_w\chi(w)$ varies considerably. For AC/PMMA and MeAc/PS, χ increases linearly and for AC/PS more than linearly, whereas it decreases linearly for CHCl$_3$/PS. With three of the systems, one observes minima in $_w\chi(w)$, namely for TL/PMMA, TL/PS, and CHCl$_3$/PS; only MeAc/PMMA exhibits a maximum. On the basis of (32), this diversity of composition influences is easily comprehensible if one keeps in mind that the composition dependence of Flory–Huggins interaction parameters are made up of two separate contributions. The normally nonzero parameter ν of the first term of this relation (which is primarily determined by the differences in the shapes of monomeric units and solvents molecules) leads to a nonlinear composition dependence of $_w\chi$, where the magnitude of this contribution increases as the absolute values of the parameter α rise. The second term of (32) adds a linear dependence, quantified by the parameter $\zeta\lambda$. In agreement with the great diversity of the systems concerning the functions $_w\chi(w)$, all three parameters of the present approach may be positive, negative, or zero.

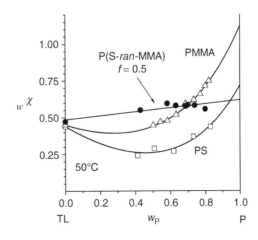

Fig. 14 Composition dependence of the Flory–Huggins interaction parameters [based on weight fractions w_P, cf. (8)] for the solutions of PS, PMMA, and of a random copolymer containing 50 wt% of these monomers in TL at 50°C [25]

The next aspect that deserves discussion concerns the quality of a given solvent for copolymers of different comonomer content, as compared with its quality for the corresponding homopolymers. The naive interpolation of the $_w\chi$ values for the copolymers between the data of the homopolymers according to their composition is at variance with the experimental observations. Only for the system $CHCl_3$/P(S-*ran*-MMA) with $f = 0.5$ does the composition dependence lie reasonably between the $_w\chi(w)$ curves for the homopolymers. All other solvents are approximately as poor for the copolymer ($f = 0.5$) as for the less favorably interacting homopolymer (PS in the case of AC and MeAc; PMMA in the case of TL). For the system TL/P(S-*ran*-MMA), studied in greater detail, the presence of only 10 wt% of styrene units suffice to raise $_w\chi$ to values that are within the range of high polymer concentration, even larger than that of the TL/PMMA system (see Figure 11 of [25]).

The dependencies of the system-specific parameters π on the weight fraction f of the styrene units in the copolymers can be well modeled by (41) for the different solvents. Linear functions, corresponding to $\pi^E = 0$, are exceptions and only observed for the parameters α and $\zeta\lambda$ with $CHCl_3$ and for v with MeAc. For the polar solvents AC and MeAc, $\alpha(f)$ and $\zeta\lambda(f)$ exhibit maxima, whereas minima are observed for TL. The comparison of these excess parameters π^E obtained for the different solvents discloses another interesting feature, namely the fact that all excess parameters π^E exhibit the same sign for the three systems for which the behavior of the copolymer is dominated by the monomeric unit showing the less favorable interaction with the solvent. For α and $\zeta\lambda$, this means that an adverse excess contribution for contact formation is counteracted by a favorable conformational relaxation (AC and MeAc) or conversely, that a favorable excess contact formation goes along with an adverse conformational relaxation (TL).

In conclusion of this section, it is worthwhile noting that the interrelation of the system-specific parameters established for homopolymer solutions (cf. Fig. 5) also holds true for all copolymer solutions studied here (as demonstrated in Figure 15 of [25]).

Aqueous Solutions of Poly(vinyl methyl ether)

This example and the next (cellulose; Sect. 4.1.1.5) concern systems with uncommonly large α values (i.e., very unfavorable contact formation between the solvent and polymer segments) in combination with a similarly favorable conformational relaxation. Literature reports a very uncommon phase behavior [54, 55] for the system H_2O/PVME: The most striking feature is the occurrence of two minima in the cloud point curves instead of one. In addition to the normal critical point at low polymer concentration, the authors report a second critical point at high polymer concentrations for high molar masses of the polymer. Furthermore, they describe a three-phase line occurring at a certain characteristic temperature, even for strictly binary mixtures. The authors used a three-membered series expansion of the integral Flory–Huggins interaction parameter g with respect to φ for the modeling

[55] of their results and confined the temperature influences to the composition-independent term, assuming a linear dependence on $1/T$ to reproduce the observed phase separation upon heating.

The following considerations [53] deal with the question of which criteria (in terms of the system-specific parameters of the present approach) a certain system must fulfill to reproduce the anomalous phase separation phenomena reported in the literature. To that end, the condensed parameter $\zeta\lambda$ is eliminated from the critical conditions calculated on the basis of (33) to yield expressions analogous to (36) and (37). This procedure provides the following relation, containing the parameters α and v, the number of segments N of the polymer, and the critical composition φ_c of the system:

$$\alpha = \frac{[6\varphi_c^3(N-1) + \varphi_c^2(11-2N) - 4\varphi_c - 1](1-v\varphi_c)^4}{6\varphi_c^2 N(1-\varphi_c^2)(4v\varphi_c + v - 1)} \qquad (63)$$

Plotting α according to (63) as a function the critical composition φ_c for a given polymer (i.e., constant value of N) with v as independent variable gives access to the combination of α and v values, yielding more than one solution for φ_c. Figure 15 shows the results for two v values; this graph merely specifies which parameter combinations result in critical conditions, it does not yet refer to a certain temperature. The horizontal lines indicate the first appearance of an additional critical point upon an augmentation of α. Under these special circumstances, a stable and an unstable critical point [53] coincide and form a double critical point. In the general case, the three solutions for the critical conditions correspond to different temperatures and one of them is an unstable critical point.

The minimum α value required for the occurrence of a double critical point is considerably higher for $v = 0.4$ than for $v = 0.5$. A more detailed mathematical analysis [53] of (63) yields a border line for the combination of parameters, which separates the normal from anomalous behavior. For ordinary systems, the combination of α and v values required to produce multiple critical points has so far not been observed. However, for water/PVME systems, such data may well be realistic

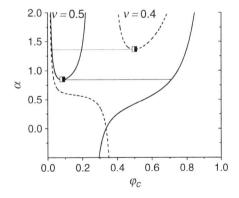

Fig. 15 Modeling of polymer solutions with anomalous phase behavior. Example of plots [53] of α as a function of φ_c according to (63) at the constant v values indicated in the graph for 1. *Solid curves* $v = 0.5$, *dashed curves* $v = 0.4$. The *horizontal lines* mark the minimum value that α must exceed for a given v to generate an additional critical point. *Squares* mark anomalous double critical points

because of the large surface of water as compared with that of the polymer segment, making large γ values and hence also large ν values plausible [cf. (24)]. In view of the pronounced chemical dissimilarities of water and PVME, this should lead to large α values.

The phase diagram shown in Fig. 16 was calculated [53] choosing a combination of α and ν values inside the range of multiple critical points. For this modeling it was (unrealistically but for the sake of simplicity) assumed that only α depends on temperature and that this dependence can be formulated as:

$$\alpha = \alpha_1 + \alpha_2(T - T_s) \tag{64}$$

where α_1, α_2, and T_s are constants. From this graph it is clear that the central features of the phase diagram observed for the water/PVME system can be adequately modeled by the present approach.

An interesting result of the present modeling is an uncommon option to realize theta conditions. Maintaining its definition in terms of $A_2 = 0$, leading to $\chi_o = 0.5$:

$$\chi_{o,\theta} = \frac{1}{2} = \alpha_\theta - (\zeta\lambda)_\theta \tag{65}$$

it is obvious that this relation cannot only be fulfilled in the normal way, with $\zeta_\theta = 0$ and $\alpha_\theta = 1/2$, but also via an adequate combination of α and $\zeta\lambda \neq 0.5$. For such exceptional systems, the unperturbed state results from an exact compensation of an uncommonly unfavorable contact formation between the components ($\alpha > 0.5$) by an extraordinarily advantageous conformational response ($\zeta\lambda \gg 0$). In the case of H_2O/PVME, the plausibility of large α values has already been mentioned. From reports [56] on the formation of a complex between water and PVME and the fact that the system exhibits LCST behavior, one can infer that large ζ values are caused by the very favorable heat effects associated with that process.

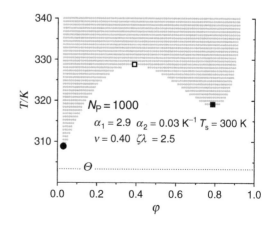

Fig. 16 Spinodal area calculated [53] according to (33) and (64) for an exothermal model system by means of the parameters listed. The *horizontal line* indicates the theta temperature ($\chi_o = 0.5$); *circle* normal critical point, *closed square* stable anomalous critical point, *open square* unstable anomalous critical point

Swelling of Cellulose in Water

The water/cellulose system is unique in several ways. First of all it is so far the only one for which we have observed that the vapor pressures above homogeneous mixtures depend on the particular manner in which the samples were prepared. The results reported here [57] were obtained by means of thin cellulose films (approximately 20–25 μm thick) cast from cellulose solutions in the mixed solvent LiCl + dimethylacetamide. After careful removal of the components of the mixed solvent, these films were kept in a surplus of water at 80°C until the weight of the swollen cellulose film no longer changed. The solvent was then removed stepwise by vacuum treatment and the resulting samples were kept in the measuring cell of the HS–GC until the vapor pressure no longer changed, which was typically after 1 day. The experimental data are highly reproducible but not identical with the results of measurements (of equally reproducibility; not yet published) with cellulose films that were cast from a different solvent. From these findings, one is forced to conclude that at least one set of data does not refer to the macroscopic equilibrium of the system. It looks as if the final arrangement of the polymer chains after total removal of the solvent (e.g., with respect to the degree of crystallinity) could depend on the chemical nature of the solvent employed for film preparation. Under this assumption, and in view of the high viscosity of swollen cellulose, one can then speculate that the molecular environment established upon the removal of a particular solvent is more or less preserved in the swollen state and permits only the establishment of local equilibria.

Figure 17 shows the results for a cellulose sample with 2940 segments (defined by the molecular volume of water) prepared from a solution in LiCl plus dimethylacetamide [57]. The most striking feature is the enormously large range that the Flory–Huggins interaction parameter spans as a function of composition. It falls from $\chi_o = 6$ (for worse than theta conditions, the typical χ_o values are in the order of 0.6) to a minimum of approximately -3.6 (much less than the lowest values observed so far) for φ values around 0.6, and increases again up to -1.7 in the limit of the pure polymer.

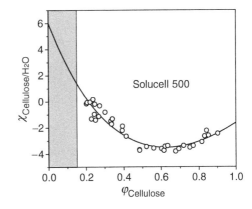

Fig. 17 Composition dependence of the interaction parameter for water/cellulose (Solucell 500) at 80°C, obtained from vapor pressure measurements [57]. The two-phase area is *shaded*. The *curve* was calculated by means of (26) and the following system-specific parameters: $\alpha = 56.8$, $\nu = -0.56$, $\zeta = 37.9$, and $\lambda = 1.34$

For the modeling of the function $\chi(\varphi)$ of Fig. 17, it is essential to use (26) and not (33) because the intramolecular interaction parameter λ deviates strongly from the usual value of 0.5. This observation is conceivable considering the fact that cellulose is not noticeably soluble in water under the prevailing conditions, which means that isolated polymer coils should be widely collapsed. Under these conditions, the average volume fraction Φ_o of the segments within the realm of such a macromolecule will become very high and so consequently will λ [cf. (27) and (17)]. The evaluation of the present data yields $\lambda = 1.34$. All other parameters required for the modeling of the measured Flory–Huggins interaction parameter also lie well outside the normal range. The leading parameter α of the first term of (26) is positive and very large – like with the example of multiple critical points discussed in the previous section (Sect. 4.1.1.4). However, this time the large value is not only due to the chemical dissimilarity of the components, but is also caused by very favorable intersegmental contacts (H-bonds) that must be broken upon the insertion of a solvent. In agreement with the general interrelation of the parameters $\zeta\lambda$ and α, this adverse contribution via α is counteracted by a comparable advantageous conformational relaxation via ζ. The unique behavior of the water/cellulose system is also demonstrated by the value of v, which is negative, in contrast to almost all other polymer solutions studied so far. The only negative value of similar magnitude was observed for the butane/1,4-polybutadiene system [50], which also exhibits a large solubility gap. One might therefore speculate that the pronounced self-association tendencies of the components (due to the unfavorable mutual interaction) causes effective surface-to-volume ratios [cf. (24)] that differ considerably from those expected on the basis of the molecular shapes of the components.

A further, immediately obvious particularity of the present system is the anomalous swelling behavior of cellulose in water, as shown in Fig. 18. To the author's knowledge it is the only case where a high molecular weight polymer takes up more of the pure coexisting liquid than does a sample of lower molar mass.

The results shown in Fig. 18 demonstrate that the miscibility gap of cellulose and water, predicted from the vapor pressures measured above the homogeneous mixture

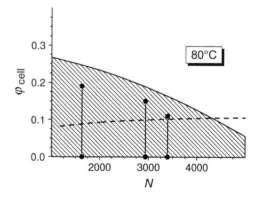

Fig. 18 Composition of the phases coexisting for the system water/cellulose at 80°C as a function of the number of segments N of the polymer determined in swelling experiments [57]. *Symbols* indicate experimental data; the *solid line* was calculated by means of (26) as described in the text; the two phase area is hatched. The *dashed line* (normal behavior) was calculated by means of the original Flory–Huggins theory, setting the interaction parameter equal to 0.54

Thermodynamics of Polymer-Containing Mixtures

(liquid/gas equilibrium), matches the observed swelling behavior (liquid/liquid equilibrium) reasonably well. Above all, it correctly models the observed *diminution* of the two-phase region with rising molar mass of the cellulose. The lack of quantitative agreement should not be overestimated because of the sensitivity of the calculated swelling with respect to the exact value of the central parameter α; a reduction of α by less than 3% would suffice for quantitative matching.

In an attempt to rationalize this unique behavior, we recall that the χ_0 values of the present system are about ten times larger than normal, which means that the tendency to form dilute solutions is practically nil. When adding increasing amounts of water to pure cellulose, the extent of chain overlap (stabilizing the homogenous state) will surpass a critical value below which a cellulose molecule can no longer evade the formation of extremely adverse contacts between its segments and water. At this point, the segregation of a second phase consisting of practically pure water sets in. From simple considerations concerning the chain-length dependence of the size of polymer coils, one can conclude that this critical overlap will be reached at higher dilution by larger molecular weight samples than by smaller molecular weight samples, thus explaining the anomalous swelling behavior of cellulose in water.

The last two examples have dealt with systems for which the first step is uncommonly unfavorable and goes along with a favorable second step. For the mixtures described in the next section, the opposite is the case: here a very favorable first step is followed by a correspondingly adverse second step.

Aqueous Solutions of Pullulan and Dextran

These systems exhibit a common feature, which becomes noticeable in the primary data, i.e., in the composition dependence of the vapor pressures. Unlike normal polymer solutions, $p(\varphi)$ shows a point of inflection in the region of high polymer contents, as demonstrated in Fig. 19. This peculiarity and the necessity to introduce an additional term in the expression for the integral interaction parameter g [cf. (42)] is interpreted in terms of hydrogen bonds between the monomer units of the polymer, on one hand, and between water and the monomers, on the other hand.

The opening of intersegmental contacts – a prerequisite for the dilution of the mixture – is Gibbs energetically adverse and modeled in terms of positive $\overline{\omega}$ parameters. The subsequent insertion of solvent molecules between these polymer segments is, in contrast, very favorable and quantified by negative α values. The reason why the total contribution of the first step of dilution cannot be modeled by a single common parameter lies in the different composition dependencies of the effects of opening and of insertion.

According to the details of the dilution process discussed above, the point of inflection in the vapor pressure curve shown in Fig. 19 can be given an illustrative meaning: In the region of low polymer concentration it is practically only "bulk" water that it transferred into the vapor phase. This situation changes, however, as φ approaches unity; under these conditions the vapor is increasingly made up of solvent molecules taken from the "bound" water (located between two polymer

Fig. 19 Comparison of the vapor pressures above aqueous solutions of two types of polysaccharides [26] at 37.5°C calculated by means of (43) plus (5) and (6). The *dotted line* is the tangent at the point of inflection in the case of dextran

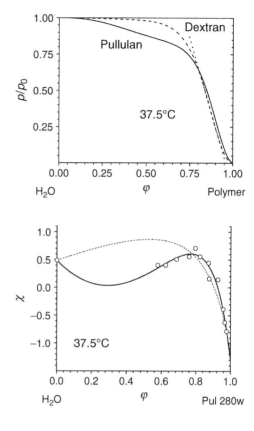

Fig. 20 Composition dependence of the Flory–Huggins interaction parameter for the system [26] water/pullulan at 37.5°C. The *solid line* shows the best fit by means of (43) and the *dotted line* the best fit according to (32)

segments). The intersection of the tangent at the point of inflection with the abscissa can be taken as an estimate of the amount of "bound" water.

Another feature that is immediately visible from Fig. 19 is the higher solvent quality of water for pullulan as compared with dextran. Within the composition range $0.25 > \varphi > 0.75$, the reduced vapor pressure is considerable lower in the former than in the latter case. This situation leads to rather complicated composition dependencies of the Flory–Huggins interaction parameter for the solutions of pullulan, as shown in Fig. 20. From the dotted line of this graph it becomes obvious that a modeling is impossible without an additional term in the relation for the integral interaction parameter (42). The uncommonly low χ values of the system for large volume fraction of the polymer are another outcome of a very stable "intercalation" of a solvent molecule between two segments of the polysaccharide.

Nonselective Solvent/Block Copolymers

The modeling of block copolymers solutions is necessarily much more difficult than the modeling of solutions of random copolymers. Again, the binary system

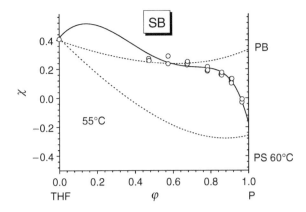

Fig. 21 Composition dependence of the Flory–Huggins interaction parameter for the solutions of a diblock copolymer of styrene and butadiene (*solid line*) in THF at 55°C. The information for the corresponding homopolymer solutions (*dotted lines*) refers to 55°C for PB, and to 60°C for PS [58]

requires three different interaction parameters for its adequate description, but this time a possible incompatibility of the homopolymer blocks is much more consequential. The examples discussed for block copolymers are a diblock copolymer of styrene and butadiene (SB), the corresponding triblock copolymer formed by joining two SB blocks at their butadiene ends (SB)2, and a four-arm block copolymer (SB)4 in which the inner blocks consist of polybutadiene. The investigations reported in [58] use the nonselective tetrahydrofuran (THF) as solvent in all cases. Figure 21 presents – as an example – the composition dependence of the Flory–Huggins interaction parameter measured for the diblock copolymer at 55°C. The results for the other two types of block copolymers and different temperatures look qualitatively very similar.

Like with the aqueous solutions of the polysaccharides discussed in Sect. 4.1.1.6, the present systems require an extra term in the integral interaction parameter g to account for the effect of the first step of dilution, where a solvent molecule is inserted between two polymer segments. With the block copolymers of present interest, the situation is different from that encountered with the polysaccharide solutions because of the microphase separation induced by the incompatibility of the blocks. In this case, the number of segments required for special interactions is larger than two. Geometrical considerations suggest that contacts between more than three segments belonging to different polymer chains are very unlikely, even in the pure melt. This means that the insertion of a solvent molecule will typically destroy advantageous ternary contacts between segments. By analogy to the reasoning in the context of the aqueous solutions of pullulan or dextran, this implies that the extra contribution to g should depend on the third power of φ in the case of block copolymers, as formulated in (44).

Despite these dissimilarities in the molecular details, the α parameters required for the modeling of the experimental findings are in both cases negative. In the case of the block copolymer solutions, it is not the formation of favorable contacts resulting from the addition of solvent that makes $\alpha < 0$, but the destruction of very unfavorable contacts between the two types of monomeric units. Figure 21 shows that the Flory–Huggins interaction parameter is smallest in the limit of $\varphi \to 1$, where the solvent is practically exclusively incorporated into the interphase separating the coexisting microphases. In this concentration range χ can in some cases even fall below the χ value of the THF/PS system. With progressive dilution, the interaction parameters for the block copolymer increase because the solvent is now more and more incorporated into the microphases until they pass a maximum in the range of semidilute solutions. The reason for this thermodynamically worst situation can be rationalized by the fact that the polymer concentration is no longer high enough to enable microphase separation and not yet low enough for intramolecular clustering of the segments of the different blocks. Maxima in $\chi(\varphi)$ of the type shown in Fig. 21 might cause (macro)phase separation. Calculation for the present copolymer solutions and temperatures under investigation with respect to liquid/liquid demixing by means of (36) and (37) and the interaction parameters obtained from liquid/gas equilibria did not, however, result in miscibility gaps, in agreement with the direct experimental observation of the mixtures. According to these calculations, the thermodynamic quality of THF for the block copolymers is already marginal so that one can expect the occurrence of macrophase separation in addition to microphase separation at low enough temperatures.

Before leaving the area of polymer solutions to deal with polymer blends and mixtures of low molecular weight compounds, it appears worthwhile to document once more an experimental finding that is very helpful for the modeling of new systems. This is the existence of a very general interrelation between the leading parameters α and $\zeta\lambda$ of the present approach. Even for systems that behave in a very anomalous manner at higher polymer concentrations, the parameters α and $\zeta\lambda$ suffice for the description of the dilute state of pair interaction between the solutes and interrelate in the usual way [cf. (34)]. Figure 22 shows the data for the studied polymer solutions with specific interactions, together with some typical data for ordinary polymer solutions that do not need an extension of (32) for the integral interaction parameter. The general validity of the function $\zeta\lambda$ (α) reduces the number of adjustable parameters by one and eases the modeling and qualitative predictions considerably.

4.1.2 Polymer Blends

Poly(vinyl methyl ether)/Polystyrene

Out of the many polymer blends investigated so far, PVME/PS is probably the one for which the molecular weight dependence of the critical conditions has been studied in most detail (cf. citations in [59]). The critical temperatures span more than 60°C, and the critical volume fractions of PS lie between 0.13 and 0.68. The comprehensive experimental information that is available makes this system

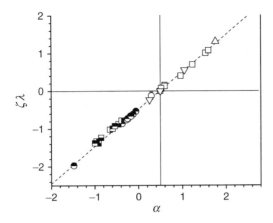

Fig. 22 Interrelation between the parameters $\zeta\lambda$ and α [the *vertical line* at $\alpha = 0.5$ is drawn according to (34)]. The data for solutions of several typical polymers in organic solvents are almost exclusively located in the first quadrant (*open symbols*); they only extend into the third quadrant for systems close to their demixing. Data for anomalous systems (*half-closed symbols*), where the first step of dilution represents the main driving force to homogeneity, are entirely located in the third quadrant

particularly suited for modeling of interaction parameters on the basis of critical conditions.

The modeling presented here still uses the expression for the integral interaction parameter as formulated for polymer *solutions* (26), which leads to the critical conditions specified in (36) and (37). According to the extension of the approach to polymer *blends* (which had not yet been carried out when this study was performed), (49) should have been used for that purpose because it accounts for the fact that the polymer coils A are accessible to the segments of polymer B and vice versa. Both expressions employ a linear dependence of the parameter ζ on the composition of the mixture; the differences between polymer solution and polymer blends only lie in the numerical values of the constants.

Figure 23 shows how the molar masses of the blend components influence the experimentally obtained critical compositions of the mixture. The two curves shown in this graph were obtained by adjusting five parameters, namely ν and λ (which were considered to be temperature independent), plus two parameters for the temperature dependence of α. The fifth parameter concerns ζ, which was either kept constant (variant 1) or set proportional to α (variant 2). Both assumptions model the experimental data with comparable accuracy. In view of the expectation that all system-specific parameters should depend on temperature, the quality of the description with only five parameter is surprising. It must, however, be kept in mind that a naive molecular interpretation of the system-specific parameters is not permissible in the present case.

Despite the similarity of the two variants of modeling presented in Fig. 23, the detailed phase diagrams calculated from the two sets of parameters differ

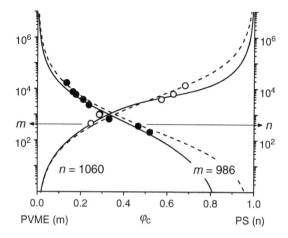

Fig. 23 Combination of numbers m of PVME segments with numbers n of PS segments leading to the critical compositions φ_c. The experimental data, taken from the literature [60–62], were obtained by mixing one PS sample ($n = 1060$, *open circles*) with PVME of different molar mass or, vice versa, one PVME sample ($m = 986$, *closed circles*) with different samples of PS. The *curves* are calculated using two modeling variants as described in the text [59]. *Solid lines*: the conformational relaxation does not depend on temperature; *dashed lines*: it varies linearly with T

fundamentally if both components become high in molar mass. In both variants, α was considered to depend on temperature but variant 1 keeps ζ independent of T, whereas variant 2 applies the proportionality between α and ζ, i.e., treats ζ as a function of T. Variant 1 yields two stable and one unstable critical points [59] (as for the system water/PVME), whereas the demixing behavior remains normal for variant 2. Defining theta conditions for polymer blends by analogy to the usual definition for polymer solutions in terms of critical temperature for infinite molar mass of the polymer according to:

$$\lim_{m,\,n \to \infty} T_c \equiv \Theta \tag{66}$$

one obtains two different theta temperatures, where the corresponding critical concentration is either zero or unity. Conversely, ζ proportional to α yields only one theta temperature, and the corresponding critical composition remains indefinite, like in the original Flory–Huggins theory. The question of which of the predictions comes closer to reality can only be answered by directed experiments.

Shape-Induced Polymer Incompatibility

Demixing of polymer blends consisting of macromolecules synthesized from the same monomers and differing practically only in their molecular architecture plays an important role in the polyolefin industry [63–65]. Numerous experimental and theoretical studies have therefore been performed to investigate this phenomenon; for pertinent literature see [66] and [67]. The present approach offers a particularly simple theoretical access because the first term of the expression for the integral interaction parameters [(49), corresponding to the first step of mixing] can be set to zero.

For the special case of linear and branched polymers of the same chemistry, one obtains a very simple relation if the degree of branching β is introduced in terms of the intrinsic viscosity of the branched polymer as compared with that of the linear analog [(27) and (52)] and the conformational relaxation is set proportional to β (53), which means that it approaches zero as the degree of branching becomes vanishingly small. Under these conditions, one single parameter suffices to model the phase behavior (54); the parameter κ of (54) can either be estimated from the Kuhn–Mark–Houwink relation for the linear polymer and theta conditions [(29) and (55)] or it can be merged with the parameter Z (54). Figure 24 shows an example of the critical conditions calculated for blends of chemically identical linear and branched polymers with $\kappa = 0.27$, which is the typical value for vinyl polymers.

Model calculations [28] along the described lines indicate that the sensitivity to phase separation is particularly pronounced for blend partners of comparable numbers of segments. In Fig. 24 this can, for instance, be seen from the frontmost curve ($\beta = 0.1$) passing a maximum in this range of N_L. Each of the data points on the critical surface of this graph corresponds to a different phase diagram, which can be represented in terms of $Z(\varphi)$, by analogy to the more customary theoretical diagrams $\chi(\varphi)$ or $g(\varphi)$. In order to transform such general phase diagrams into the directly measurable phase diagrams $T(\varphi)$ (phase separation temperatures as function of composition), it is necessary to know how the parameters Z, χ, or g depend on T. There are literature reports [65] on phase separation upon heating as well as

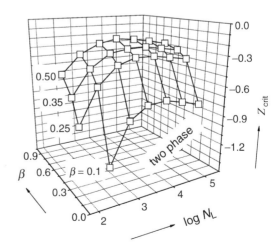

Fig. 24 Three-dimensional representation of the critical surface calculated for blends of a branched polymer consisting of 1000 segments but differing in its degrees of branching β, with its linear analogs varying in the number N_L of their segments. Phase separation may set in (depending on the composition of the blend) if Z falls below its critical value Z_{crit}. The area of possible demixing is located below the critical surface [28]

upon cooling. In view of the fact that the first term of (49) is for the present calculations set at zero ($\alpha = 0$), one might think that shape-induced demixing should be entirely due to unfavorable entropies of mixing and should always be of the LCST type. This interpretation is, however, not permissible because both steps of dilution contribute to the residual Gibbs energy via enthalpy and entropy, as discussed earlier.

4.1.3 Mixtures of Low Molecular Weight Liquids

For the modeling of systems containing more than one low molecular weight component, like polymer solutions in mixed solvents, it would be very advantageous to be able to use the same mathematical expressions for the mixtures of the low molecular weight liquids. Experiments performed to investigate these possibilities have demonstrated that (32) can indeed describe the thermodynamic behavior quantitatively [29], as demonstrated in Fig. 25 for the system water/N-methyl morpholin N-oxide monohydrate [NNMO*H$_2$O]. This graph shows the measured reduced vapor pressures of water as a function of composition, and the curves calculated by means of (32) and the adjusted parameters α, ν, and $\zeta\lambda$.

It is obvious that the parameter λ of (32) (introduced via considerations concerning the establishment of microphase equilibria with polymer-containing systems) loses its physical meaning for the low molecular weight mixtures because the segments of the components are geometrically strictly separated. This is unlike the situation with polymer solutions, where the solvent enters the polymer coil, or

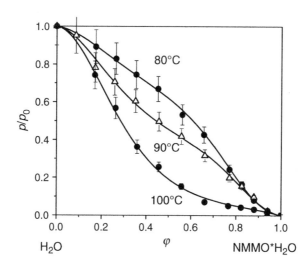

Fig. 25 Equilibrium vapor pressures, p, of water above mixtures of NMMO*H$_2$O normalized to p_o, the equilibrium vapor pressure of pure water, as a function of the volume fraction of NMMO hydrate for the indicated temperatures [29]. The *curves* are calculated according to (32), where $N_{\text{NMMO*H}_2\text{O}} = 6.35$

Thermodynamics of Polymer-Containing Mixtures

with polymer blends, where both components are accessible for segments of the other polymer. The meaning of the parameter ζ, on the other hand, remains unchanged because molecular rearrangements, similar to those occurring with polymer solutions and polymer blends, will also take place in low molecular weight mixtures, due to preferentially interacting sites of the components. According to the present results, a linear composition dependence of the conformational part of the interaction parameter should suffice to describe reality.

4.2 Ternary Systems

The material presented so far has demonstrated the ability to model the thermodynamic behavior of binary systems accurately by means of the present approach. For the description of polymer solutions, it is normally possible to eliminate one of the three parameters of (32) thanks to a general interrelation between α and $\zeta\lambda$ (34). For polymer blends and mixtures of low molecular weight components, a similar general simplification is presently not known. Notwithstanding this situation, it is possible to model the principle features [27] of all types of phase diagrams observed for ternary systems using only two parameters for each binary subsystem.

This section deals with the phase-separation behavior of ternary systems, where a distinction is made between polymer solutions in mixed solvents (Sect. 4.2.1) and solutions of two polymers in a single solvent (Sect. 4.2.2). Furthermore, the systems are classified according to the way the thermodynamic properties of the ternary systems are made up from the properties of the corresponding binary subsystems: *Simplicity* denotes "smooth" changes in the phase behavior of the binary subsystems upon the addition of the third component in its pure form or in mixtures (see later). *Cosolvency* means that the thermodynamic quality of mixture of two components is higher with respect to the third component than expected by simple additivity, i.e., cosolvency reduces the extension of the two-phase region with respect to that expected from additivity. *Cononsolvency*, finally, denotes the opposite behavior, i.e., an extension of the two-phase region beyond expectation.

4.2.1 Mixed Solvents

The use of mixed solvents is widespread, because it offers the possibility to tailor desirable thermodynamic conditions by mixing two liquids with sufficiently different qualities in adequate ratios, instead of the often inconvenient or even impossible variation of temperature. The combination of good solvents with precipitants is the basis of many industrial processes, like membrane production or fiber spinning. In order not to go beyond the scope of the present contribution, the following considerations are limited to complete miscibility of the components of the mixed solvent. There is, however, no particular difficulty to extend the treatment to incompletely miscible components of mixed solvents.

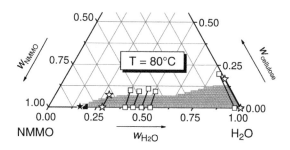

Fig. 26 Phase diagram of the NMMO/H$_2$O/Solucell 400 system at 80°C [68]. *Shaded area* shows the calculated unstable composition range; *open squares* calculated tie lines, *open stars* experimental tie lines, *closed square* calculated critical point, *closed star* experimental critical point

Simplicity

The following example for simplicity refers to a technically important ternary system, namely cellulose solutions in mixtures of the favorable solvent NMMO with the precipitant water. Fibers are formed as thin threads as homogeneous cellulose solutions are spun into water. Figure 26 shows how experimental data for this ternary mixture compares with the modeling [68] on the basis of (60). On the theoretical side, it is important to take care of the fact that the information concerning the binary subsystems usually differs by a diverging definition of the size of a segment.

The unstable area, the critical point, and the tie lines shown in Fig. 26 were calculated by means of the independently determined parameters for the binary subsystems NMMO/water [29] and cellulose/water [57]. The corresponding information for NMMO/cellulose is inaccessible along the present routes, because the vapor pressure of both components is negligibly small. For that reason, it was necessary to adjust the parameters α and ζ for this binary subsystem to the experimentally observed ternary phase diagram; ν was equated to γ (obtained from group contributions) and λ was set at 0.5, the typical value for polymer solutions. This procedure enables the modeling of the phase diagram for the ternary system, which matches the measurements within experimental error. Even if this procedure is not predictive, it helps the discrimination of metastable and unstable compositions and enables assessment of the effects of different molar masses of cellulose on demixing [68].

In another, very abundant form of simplicity the miscibility gaps existing for the polymer solution in either of the two solvents transform smoothly into each other as the composition of the mixed solvent changes.

Cosolvency

A much higher quality of mixed solvents as compared with either of its components is not uncommon; since the first report [69] it has been described in the literature many times. This phenomenon can be easily modeled [27] by means of (60) using physically meaningful combinations of parameters. The example shown in Fig. 27

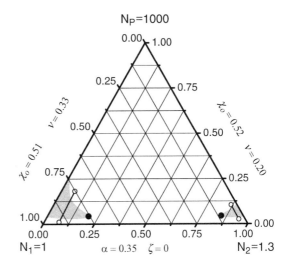

Fig. 27 Cosolvency as a result of unfavorable interactions between components 1 and 2. The numbers of segments of the different components are given at the *corners* of the phase diagram and the characteristic parameters for the binary subsystems are indicated on its *edges*. *Open symbols* composition of coexisting phases, *closed symbols* critical points, *shaded areas* unstable regions [27]

applies to sufficiently unfavorable 1/2 interactions; in their absence the miscibility gap would extend from one binary subsystem to the other throughout the ternary system, i.e., this would be an example of simplicity.

The reason for the complete miscibility of the polymer with mixed solvents containing comparable fractions of their components shown in Fig. 27 lies in the adverse interactions between them. Within a certain range of compositions, the ternary system can avoid these unfavorable contacts between components 1 and 2 by inserting a polymer segment between them and forming homogeneous mixtures.

Cononsolvency

The creation of a miscibility gap by mixing two favorable solvents was reported a long time ago [70] and many examples have been described since. Figure 28 shows a typical modeling of this behavior. For that purpose, we assume that the components 1 and 2 are markedly better solvents for the polymer P than in the case of cosolvency, and that they mix in a combinatorial manner ($g_{12} = \chi_{12} =; 0$).

The reason why the present combination of parameters for the binaries leads to a miscibility gap for the ternary system lies in the particularly favorable interactions 1/P and 2/P as compared with the more or less "neutral" interactions 1/2. Under these conditions, the Gibbs energy of the total system can be lowered by phase separation such that the polymer-lean phase contains practically low molecular weight components only and that many favorable 1/P and 2/P contacts can be formed in the polymer-rich phase.

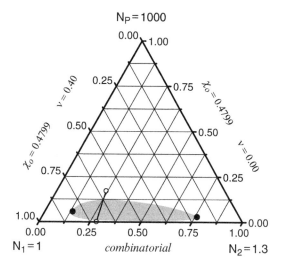

Fig. 28 Formation of an island of immiscibility in the ternary system, caused by favorable 1/P and 2/P interactions [27]. For details, see legend to Fig. 27

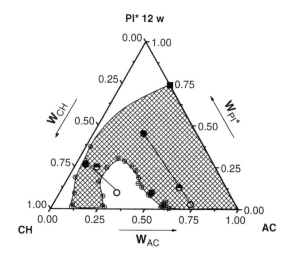

Fig. 29 Phase diagram of the CH/AC/PI* system at 25°C. The polymer sample PI* (M_n = 5 kg/mol, M_w = 12 kg/mol) consists of a mixture of branched and linear chains. *Crossed circles* cloud points, *half-closed circles* overall composition of the coexistence experiments, *open circles* compositions of the polymer-lean phases, *closed circles* compositions of the polymer-rich phases, *closed square* swelling point of PI* in AC. The composition area of possible demixing is *hatched* [71]

Complex Behavior

Phase diagrams for polymer solutions in mixed solvents can look much more complicated than shown so far. Figure 29 gives an example observed in the course of a study concerning differences in the thermodynamic behavior of branched as compared with linear polymers [71].

The reason for uncommon phase diagram often lies in the polydispersity of the polymer sample, which means that we are strictly speaking no longer dealing with ternary but with multinary systems, for which the representation of phase diagrams requires a projection into a plane. In the present case, the polydispersity is due to the

Thermodynamics of Polymer-Containing Mixtures

presence of linear and branched PI in addition to the usual nonuniformity of molar masses. The consequences of broad molecular weight distributions of linear polymers for the shape of phase diagrams are not negligible, but are usually considerably less pronounced than nonuniformities with respect to the molecular architecture of the macromolecules. The reason is that polymers of different chain length are usually completely miscible, whereas this needs not be the case for linear and branched macromolecules, as exemplified when dealing with their solutions in a common solvent.

The strange peninsular of the miscibility gap shown in Fig. 29 is caused by the fact that the PI sample contains both linear and branched material; neither the solution of the linear product nor that of the branched polymer in the same mixed solvent show this particularity [71]. It is, however, very probable that particular interactions between the components of the mixed solvent also play a role in the occurrence of the anomalous peninsula of the phase diagram. This consideration rests on the fact that the CH/AC system exhibits an upper critical solution temperature [72] at $-29°C$. The low mixing tendency of these components might increase the possibilities of the quaternary system to reduce its Gibbs energy via demixing.

4.2.2 Blend Solutions

Solutions of chemically dissimilar polymers in a common solvent play an important role in the processing of polymer mixtures, where this is particularly true for incompatible polymer pairs but also for the production of homogeneous films consisting of two compatible polymers. Like with polymer solutions in mixed solvents, one can observe all the deviations from additive behavior discussed earlier.

Simplicity

The modeled example given in Fig. 30 for this behavior shows the gradual disappearance of a miscibility gap existing between two moderately incompatible polymers upon the addition of a solvent of comparatively low thermodynamic quality.

The phase diagram of Fig. 30 looks very similar to the one measured for the solutions of linear and branched PI in CH and shown in Fig. 31. For these experiments, the originally synthesized branched material (PI* of Fig. 29) was to a large extent freed from the linear components by means of the large-scale method of spin fractionation [73]. Despite the fact that the boundary between the homogeneous and the two-phase area was only mapped, instead of the usual cloud point measurements, the results of Fig. 31 testify to the existence of shape-induced incompatibility of polymers. It is remarkable that this phenomenon can be observed for comparatively low molar masses of the components.

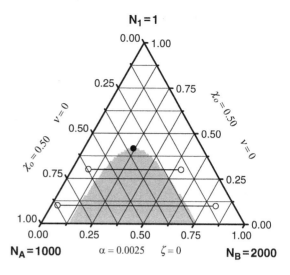

Fig. 30 Phase diagram for a moderately incompatible polymer pair and a solvent of moderate quality that dissolves polymer A and polymer B equally well. *Open symbols* composition of coexisting phases, *closed symbol* critical point, *shaded area* unstable region [27]

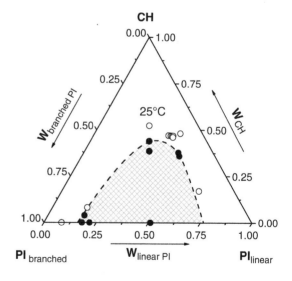

Fig. 31 Phase diagram of the CH/branched PI/linear PI system at 25°C obtained by mapping homogeneous (*open symbols*) and inhomogeneous (*closed symbols*) mixtures. The M_w of the linear polymer is 21.6 kg/mol and that of the branched material is 18 kg/mol. The two-phase region is *hatched* [71]

Cosolvency

According to model calculations, the phenomenon of cosolvency should also occur for solutions of polymer blends in a common solvent. For the example shown in Fig. 32, the components of the blend were chosen to be highly incompatible, and the solvent to be bad for polymer B but favorable for polymer A.

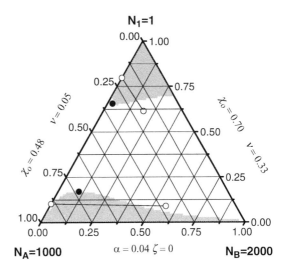

Fig. 32 Phase diagram calculated according to (60) by means of the parameters indicated on the *edges* of the triangle for the ternary mixture solvent 1/polymer A/polymer B. *Open symbols* composition of coexisting phases, *closed symbols* critical points, *shaded areas* unstable regions [27]

As with the example presented for cosolvency in the case of polymer solutions in mixed solvents (Fig. 27), the origin of cosolvency for polymer blends in a common solvent can be interpreted as a dissection of a miscibility gap that would normally bridge the Gibbs phase triangle from one binary subsystem to the other binary system (here from 1/B to A/B) by special interactions between the completely miscible components (here 1/A). With the example of Fig. 32, the thermodynamic quality of the solvent for polymer A is almost marginal; in this manner polymer B becomes completely miscible with certain solutions of polymer A in solvent 1.

Cononsolvency

This phenomenon is generally characterized by the existence of islands of immiscibility inside the Gibbs phase triangle, i.e., phase separation is absent for all binary mixtures. According to model calculations along the present lines, closed miscibility gaps should be comparatively abundant for solutions of two favorably interacting polymers in a common solvent that is sufficiently favorable for both polymers; Fig. 33 shows an example of the outcome of such calculations. A slight modification of the binary interaction parameters for the polymer solutions changes the size of the miscibility gap and its location inside the Gibbs phase triangle considerably. This is, for instance, made evident by the fact that the island disappears by increasing both χ_o values from 0.482 to 0.483, i.e., a slight reduction in the thermodynamic quality of the solvent brings the polymer solutions closer to phase separation.

The explanation for the occurrence of islands of immiscibility under the conditions specified in Fig. 33 lies in the high preference of 1/A and 1/B contacts over A/B contacts (even if A and B interact favorably), as demonstrated by the

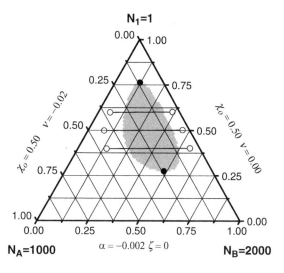

Fig. 33 Phase diagram calculated under the assumption that the polymers A and B are compatible and that the solvent is moderately favorable for both A and B [27]

position of the tie lines. Under these circumstances, the Gibbs energy of the ternary system can be reduced as compared with the homogeneous mixture by forming two liquid phases, one preferentially containing polymer A and the other polymer B. This phase separation leads to a reduction in the number of A/B contacts (associated with lower entropies of mixing than the corresponding 1/A and 1/B contacts) and in a corresponding increase in number of the more favorable polymer/solvent contacts.

The predictions of model calculations of the type shown in Fig. 33 were checked [74] by means of the systems THF/PS/PVME and CH/PS/PVME. This choice was made because of the availability of the thermodynamic information for all binary subsystems. One of the questions to be answered by this comparison between theory and experiment concerns the extent to which the phase behavior of the ternary system can be predicted if the corresponding information for the binary subsystems is available.

Figure 34 shows how experiment and the prediction by means of (60) compare in the case of THF; the data for THF/PS and THF/PVME were taken from [75] and that for the polymer blend from [59].

It is obvious from Fig. 34 that the modeling predicts the phenomenon of cononsolvency but fails to capture the details of demixing. The extension of the calculated island is considerably larger than experimentally observed. If the solvent THF is replaced by CH (which is less favorable for both polymers), the extension of the measured island is considerably increased. Again, the modeling does predict an island, but its size and location in the phase triangle are at variance with reality.

From these results, it must be concluded that the interaction between two chemically different segments is influenced by the vicinity to a segment of the third component. In other words, it is necessary to account for ternary interaction

Fig. 34 Measured and calculated phase diagram for the THF/PS/PVME system at 20°C. *Circles* show measured cloud points. The modeling was performed by means of (60) and the binary parameters are shown on the *edges* of the triangle. The calculated spinodal area is *shaded*. *Open stars* composition of the coexisting phases, *closed stars* critical points [74]

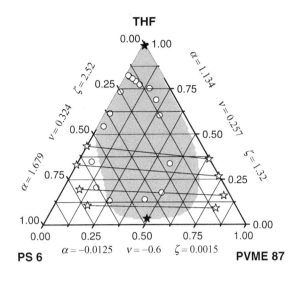

Fig. 35 Phase diagram of the PS/PVME/THF system at 20°C. *Circles* measured cloud points, *closed stars* calculated critical points, *open stars* calculated tie lines. The values of the specific ternary interaction parameters (61) are indicated on the *edges*; the binary interaction parameters are the same as in Fig. 34 [74]

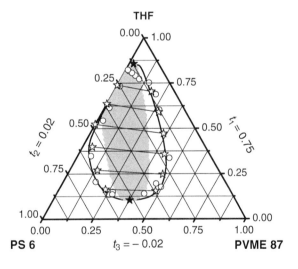

parameters. Figure 35 shows the experimentally determined phase behavior of the system THF/PS/PVME at 20°C, again along with the results of model calculations on the basis of (61) and (60) by means of the ternary interaction parameters stated at the edges of the triangle.

The agreement between the actually measured demixing behavior and that modeled on the basis of binary interaction parameters plus composition-independent ternary interaction parameters is surprisingly good. However, the results also demonstrate how sensitive the calculated phase diagrams can be with respect to the details of some interaction parameters. For instance, the analogous experiments performed with CH (less favorable solvent) instead of THF using a molecularly disperse PS sample require at least one composition-dependent ternary interaction parameter for their modeling. Indications exist that this complication is due to the presence of PS molecules differing markedly in their molar mass.

One important consequence of the results presented for solutions of compatible polymers in a common solvent is this: The suggested idea to prepare homogeneous polymers films containing both types of macromolecules from joint solutions by solvent evaporation will probably not work. The reason is that solutions containing comparable amounts of polymers A and B need to pass the unstable area of the phase diagram upon the removal of solvent, which means that they inevitably demix into two phases: one rich in polymer A and the other in polymer B. Despite the fact that the system enters the one-phase region again as the solvent content falls below a certain value, the high viscosity of the coexisting liquids will normally prevent homogenization.

5 Conclusions

The theoretical concepts presented in this chapter and the experimental examples given for their validity demonstrate how the Flory–Huggins theory can be made practical with reasonable effort. The central features of the approach are the provision for chain connectivity in dilute polymer-containing systems (by means of microphase equilibria) and the variability of macromolecules with respect to their spatial extension (expressed in terms of conformational relaxation after mixing). Both particularities contribute to the Flory–Huggins interaction parameters and are quantified in a second, additive term, which becomes zero for most of the theta systems. In contrast to the original Flory–Huggins theory, the interaction parameters are no longer independent of concentration; complicated functions $\chi(\varphi)$ are sometimes necessary to model experimental data, including minima and maxima in this dependence. It is therefore no wonder that several parameters are needed to gather the particularities of a certain system. In many cases, two parameters suffice for the quantitative description because of some possible simplifications and interrelations, as described in Sect. 2. With complex systems (like water/cellulose) up to four parameters might, however, be required.

There is one finding that speaks strongly for the validity of the present approach, namely the fact that several types of phase equilibria can be described quantitatively by means of the same set of parameters (cf. the systems n- C_4/1,4-PB and $CHCl_3$/PEO). Another eminent advantage of the present approach is its general applicability to very different classes of polymers (including branched macromolecules and copolymers of different architecture); furthermore, there is no obvious reason why it should fail for multicomponent systems.

So far, the extension of the Flory–Huggins theory has enabled the modeling of several hitherto unexplainable anomalous phenomena, like uncommon molecular weight dependencies of second osmotic virial coefficients, the existence of multiple critical points for binary systems, or the odd swelling behavior of cellulose in water.

Thermodynamics of Polymer-Containing Mixtures 63

Furthermore, it has helped a better understanding of ternary mixtures with respect to the conditions that the subsystems must fulfill for the occurrence of cosolvency or cononsolvency, as well as concerning the necessity for the use of ternary interaction parameters. Suggested further investigations concern mixtures containing charged macromolecules and a more detailed analysis of the predictive power of the present approach.

Acknowledgments The author is grateful to Dr. John Eckelt (WEE-Solve AG, Mainz, Germany) and to Prof. Spiros Anastasiadis (University of Crete, Greece) for their constructive criticism, which has certainly improved the readability of this contribution.

References

1. Flory PJ (1953) Principles of polymer chemistry. Cornell Univ. Press, Ithaca
2. Simha R, Somcynsky T (1969) On statistical thermodynamics of spherical and chain molecule fluids. Macromolecules 2(4):342–350
3. Trappeniers NJ, Schouten JA, Ten Seldam CA (1970) Gas–gas equilibrium and the two-component lattice-gas model. Chem Phys Lett 5(9):541–545
4. Sanchez IC, Lacombe RH (1976) An elementary molecular theory of classical fluids. Pure fluids. J Phys Chem 80:2352
5. Dee GT, Walsh DJ (1988) A modified cell model equation of state for polymer liquids. Macromolecules 21(3):815–817
6. Beret S, Prausnitz JM (1975) Perturbed hard-chain theory – equation of state for fluids containing small or large molecules. AIChE J 21(6):1123–1132
7. Chapman WG, Gubbins KE, Jackson G, Radosz M (1989) SAFT – equation-of-state solution model for associating fluids. Fluid Phase Equilib 52:31–38
8. Hino T, Song YH, Prausnitz JM (1994) Liquid–liquid equilibria for copolymer mixtures from a perturbed hard-sphere-chain equation of state. Macromolecules 27(20):5681–5690
9. Fredenslund A, Jones RL, Prausnitz JM (1975) Group-contribution estimation of activity-coefficients in nonideal liquid-mixtures. AIChE J 21(6):1086–1099
10. Heil JF, Prausnitz JM (1966) Phase equilibria in polymer solutions. AIChE J 12(4):678–685
11. Vimalchand P, Donohue MD (1989) Comparison of equations of state for chain molecules. J Phys Chem 93(10):4355–4360
12. De Sousa HC, Rebelo LPN (2000) A continuous polydisperse thermodynamic algorithm for a modified Flory–Huggins model: the (polystyrene plus nitroethane) example. J Polym Sci B Polym Phys 38(4):632–651
13. Chang TS (1939) The number of configurations in an assembly and cooperative phenomena. Proc Cambridge Philos Soc 35:265
14. Flory PJ (1941) Thermodynamics of high polymer solutions. J Chem Phys 9:660
15. Flory PJ (1942) Thermodynamics of high polymer solutions. J Chem Phys 10:51
16. Huggins ML (1941) Solutions of long chain compounds. J Chem Phys 9:440–440
17. Huggins ML (1942) Theory of solutions of high polymers. J Chem Phys 46:151
18. Horst R (1995) Calculation of phase diagrams not requiring the derivatives of the Gibbs energy demonstrated for a mixture of two homopolymers with the corresponding copolymer. Macromol Theory Simulat 4:449
19. Horst R (1996) Calculation of phase diagrams not requiring the derivatives of the Gibbs energy for multinary mixtures. Macromol Theory and Simulat 5:789
20. Horst R (1998) Computation of unstable binodals not requiring concentration derivatives of the Gibbs energy. J Phys Chem B 102(17):3243

21. Koningsveld R, Stockmayer WH, Nies E (2001) Polymer phase diagrams. Oxford University Press, Oxford
22. Flory PJ, Mandelkern L, Kinsinger JB, Shultz WB (1952) Molecular dimensions of polydimethylsiloxanes. J Am Chem Soc 74:3364
23. Wolf BA (2003) Chain connectivity and conformational variability of polymers: clues to an adequate thermodynamic description of their solutions II: Composition dependence of Flory–Huggins interaction parameters. Macromol Chem Phys 204:1381
24. Eckelt J, Samadi F, Wurm F, Frey H, Wolf BA (2009) Branched versus linear polyisoprene: Flory–Huggins interaction parameters for their solutions in cyclohexane. Macromol Chem Phys 210:1433
25. Bercea M, Eckelt J, Wolf BA (2008) Random copolymers: their solution thermodynamics as compared with that of the corresponding homopolymers. Ind Eng Chem Res 47:2434
26. Eckelt J, Sugaya R, Wolf BA (2008) Pullulan and dextran: uncommon composition dependent Flory–Huggins interaction parameter of their aqueous solutions. Biomacromolecules 9:1691
27. Wolf BA (2009) Binary interaction parameters, ternary systems: realistic modeling of liquid/liquid phase separation. Macromol Theory Simulat 18:30
28. Wolf BA (2010) Polymer incompatibility caused by different molecular architectures: modeling via chain connectivity and conformational relaxation. Macromol Theory Simulat 19:36
29. Eckelt J, Wolf BA (2008) Thermodynamic interaction parameters for the system water/NMMO hydrate. J Phys Chem B 112:3397
30. Tanaka TE (2000) Experimental methods in polymer science. Academic, New York
31. Petri H-M, Wolf BA (1994) Concentration dependent thermodynamic interaction parameters for polymer solutions: quick and reliable determination via normal gas chromatography. Macromolecules 27:2714
32. Krüger K-M, Sadowski G (2005) Fickian and non-Fickian sorption kinetics of toluene in glassy polystyrene. Macromolecules 38:8408
33. Ninni L, Meirelles AJA, Maurer G (2005) Thermodynamic properties of aqueous solutions of maltodextrins from laser-light scattering, calorimetry and isopiestic investigations. Carbohydr Polym 59(3):289
34. Karimi M, Albrecht W, Heuchel M, Weigel T, Lendlein A (2008) Determination of solvent/polymer interaction parameters of moderately concentrated polymer solutions by vapor pressure osmometry. Polymer 49(10):2587–2594
35. Gao ZN, Li JF, Wen XL (2002) Vapor pressure osmometry and its applications in the osmotic coefficients determination of the aqueous monomer glycol and polymer polyethylene glycol solutions at various temperature. Chin J Chem 20(4):310–316
36. Eliassi A, Modarress H, Nekoomanesh M (2002) Thermodynamic studies of binary and ternary aqueous polymer solutions. Iran J Sci Technol 26(B2):285–290
37. Schreiber HP (2003) Probe selection and description in Igc: a nontrivial factor. J Appl Polym Sci 89(9):2323–2330
38. Voelkel A, Strzemiecka B, Adamska K, Milczewska K (2009) Inverse gas chromatography as a source of physiochemical data. J Chromatogr A 1216(10):1551–1566
39. Bercea M, Cazacu M, Wolf BA (2003) Chain connectivity and conformational variability of polymers: clues to an adequate thermodynamic description of their solutions I: Dilute solutions. Macromol Chem Phys 204:1371
40. Petri H-M, Schuld N, Wolf BA (1995) Hitherto ignored influences of chain length on the Flory–Huggins interaction parameter in highly concentrated polymer solutions. Macromolecules 28:4975
41. Bercea M, Wolf BA (2006) Vitrification of polymer solutions as a function of solvent quality, analyzed via vapor pressures. J Chem Phys 124(17):174902–174907
42. Wolf BA, Adam HJ (1981) Second osmotic virial coefficient revisited – variation with molecular weight and temperature from endo- to exothermal conditions. J Chem Phys 75:4121
43. Schotsch K, Wolf BA, Jeberien H-E, Klein J (1984) Concentration dependence of the Flory–Huggins parameter at different thermodynamic conditions. Makromol Chem 185:2169

44. Bercea M, Wolf BA (2006) Enthalpy and entropy contributions to solvent quality and inversions of heat effects with polymer concentration. Macromol Chem Phys 207(18):1661–1673
45. Saeki S, Kuwahara S, Konno S, Kaneko M (1973) Upper and lower critical solution temperatures in polystyrene solutions. Macromolecules 6(2):246
46. Stryuk S, Wolf BA (2003) Chain connectivity and conformational variability of polymers: clues to an adequate thermodynamic description of their solutions III: Modeling of phase diagrams. Macromol Chem Phys 204:1948
47. Schuld N, Wolf BA (1999) Polymer–solvent interaction parameters In: Brandrup J, Immergut EH, Grulke EA (eds) Polymer handbook, 4th edn. Wiley, New York
48. Krigbaum WR, Geymer DO (1959) Thermodynamics of polymer solutions. The polystyrene–cyclohexane system near the Flory theta temperature. J Am Chem Soc 81:1859
49. Wolf BA, Jend R (1977) Über die Möglichkeiten zur Bestimmung von Mischungsenthalpien und -volumina aus der Molekulargewichtsabhängigkeit der kritsichen Entmischungstemperaturen und -drucke. Makromol Chem 178:1811
50. Stryuk S, Wolf BA (2005) Liquid/gas and liquid/liquid phase behavior of n-butane/1,4-polybutadiene versus n-butane/1,2-polybutadiene. Macromolecules 38(3):812
51. Sanchez IC, Balazs AC (1989) Generalization of the lattice-fluid model for specific interactions. Macromolecules 22:2325
52. Khassanova A, Wolf BA (2003) PEO/CHCl: crystallinity of the polymer and vapor pressure of the solvent – equilibrium and non-equilibrium phenomena. Macromolecules 36(17):6645
53. Wolf BA (2005) On the reasons for an anomalous demixing behavior of polymer solutions. Macromolecules 38(4):1378
54. Šolc K, Dušek K, Koningsveld R, Berghmans H (1995) "Zero" and "off-zero" critical concentrations in solutions of polydisperse polymers with very high molar masses. Collect Czech Chem Commun 60(10):1661–1688
55. Schäfer-Soenen H, Moerkerke R, Berghmans H, Koningsveld R, Dušek K, Šolc K (1997) Zero and off-zero critical concentrations in systems containing polydisperse polymers with very high molar masses. 2. The system water-poly(vinyl methyl ether). Macromolecules 30(3):410–416
56. Meeussen F, Bauwens Y, Moerkerke R, Nies E, Berghmans H (2000) Molecular complex formation in the system poly(vinyl methyl ether)/water. Polymer 41(10):3737–3743
57. Eckelt J, Wolf BA (2007) Cellulose/water: l/g and l/l phase equilibria and their consistent modeling. Biomacromolecules 8:1865
58. Xiong X, Eckelt J, Zhang L, Wolf BA (2009) Thermodynamics of block copolymer solutions as compared with the corresponding homopolymer solutions: experiment and theory. Macromolecules 42:8398
59. Wolf BA (2006) Polymer–polymer interaction: consistent modeling in terms of chain connectivity and conformational response. Macromol Chem Phys 207(1):65
60. Halary JL, Ubrich JM, Monnerie L, Yang H, Stein RS (1985) Polym Commun 26:73
61. Ben Cheikh Larbi F, Lelong S, Halary JL, Monnerie L, Yang H (1986) Polym Commun 22:23
62. Rätsch MT, Kehlen H, Wohlfarth C (1988) Polydispersity effects on the demixing behavior of poly(vinyl methyl ether)/polystyrene blends. J Macromol Sci Chem A25:1055
63. Greenberg CC, Foster MD, Turner CM, Corona-Galvan S, Cloutet E, Butler PD, Hammouda B, Quirk RP (1999) Effective interaction parameter between topologically distinct polymers. Polymer 40(16):4713
64. Hill MJ, Barham PJ, Keller A, Rosney CCA (1991) Phase segregation in melts of blends of linear and branched polyethylene. Polymer 32(8):1384–1393
65. Hill MJ, Barham PJ, Keller A (1991) Phase segregation in blends of linear with branched polyethylene – the effect of varying the molecular-weight of the linear polymer. Polymer 33(12):2530
66. Stephens CH, Hiltner A, Baer E (2003) Phase behavior of partially miscible blends of linear and branched polyethylenes. Macromolecules 36(8):2733–2741

67. Singh C, Schweizer KS (1995) Correlation-effects and entropy-driven phase-separation in athermal polymer blends. J Chem Phys 103(13):5814–5832
68. Eckelt J, Eich T, Röder T, Rüf H, Sixta H, Wolf BA (2009) Phase diagram of the ternary system NMMO/water/cellulose. Cellulose 16:373
69. Cowie JMG, McEwen IJ (1983) Polymer co-solvent systems.6. Phase-behavior of polystyrene in binary mixed-solvents of acetone with N-alkanes – examples of classic cosolvency. Polymer 24(11):1449–1452
70. Wolf BA, Willms MM (1978) Measured and calculated solubility of polymers in mixed solvents: co-nonsolvency. Makromol Chem 179:2265
71. Samadi F, Eckelt J, Wolf BA, López-Villanueva F-J, Frey H (2007) Branched versus linear polyisoprene: fractionation and phase behavior. Eur Polym J 43:4236–4243
72. Francis AW (1961) Critical solution temperatures. Advances in Chemistry Series, vol 31. ACS, Washington, D.C.
73. Eckelt J, Haase T, Loske S, Wolf BA (2004) Large scale fractionation of macromolecules. Macromol Mater Eng 289(5):393
74. Bercea M, Eckelt J, Morariu S, Wolf BA (2009) Islands of immiscibility for solutions of compatible polymers in a common solvent: experiment and theory. Macromolecules 42:3620
75. Bercea M, Eckelt J, Wolf BA (2009) Vapor pressures of polymer solutions and the modeling of their composition dependence. Ind Eng Chem Res 48:4603

Adv Polym Sci (2011) 238: 67–136
DOI: 10.1007/12_2010_93
© Springer-Verlag Berlin Heidelberg 2010
Published online: 21 August 2010

Aqueous Solutions of Polyelectrolytes: Vapor–Liquid Equilibrium and Some Related Properties

G. Maurer, S. Lammertz, and L. Ninni Schäfer

Abstract This chapter reviews the thermodynamic properties of aqueous solutions of polyelectrolytes, concentrating on properties that are related to phase equilibrium phenomena. The most essential phenomena as well as methods to describe such phenomena are discussed from an applied thermodynamics point of view. Therefore, the experimental findings concentrate on the vapor–liquid phase equilibrium phenomena, and the thermodynamic models are restricted to expressions for the Gibbs energy of aqueous solutions of polyelectrolytes.

Keywords Aqueous solutions · Counterion condensation · Excess Gibbs energy · Osmotic coefficient · Polyelectrolytes · Salt effects · Thermodynamics · Vapor–liquid equilibrium

Contents

1 Introduction... 74
2 Structure and Characterization of Polyelectrolytes 76
3 Experimental Data for the Vapor–Liquid Equilibrium of Aqueous
   Polyelectrolyte Solutions... 80
   3.1 Aqueous Solutions of a Single Polyelectrolyte 81
   3.2 Aqueous Solutions of a Single Polyelectrolyte and a Low Molecular
       Weight Strong Electrolyte.. 85
4 Gibbs Energy of Aqueous Solutions of Polyelectrolytes 91
5 Thermodynamic Models... 94
   5.1 Cell Model of Lifson and Katchalsky 95
   5.2 Counterion Condensation Theory of Manning...................................... 101
       5.2.1 Contribution from the Polymer.. 103
       5.2.2 Contribution from Condensed Counterions................................ 104

G. Maurer (✉), S. Lammertz, and L. Ninni Schäfer

Thermodynamics, Department of Mechanical and Process Engineering, University of Kaiserslautern, 67653 Kaiserslautern, Germany

e-mail: gerd.maurer@mv.uni-kl.de

5.2.3	Contribution from Free Counterions	106
5.2.4	Contribution from Coions	107
5.2.5	Contribution from Water	107
5.3	Modifications of Manning's Theory	108
5.4	NRTL Model of Nagvekar and Danner	109
5.5	Pessoa's Modification of the Pitzer Model	113
5.6	VERS-PE Model	117
6	Summary	131
References		131

Abbreviations

A	Inverse length (in model of Lifson and Katchalsky)
A	Repeating unit (in model of Lammertz et al.)
A_ϕ	Debye-Hückel Parameter
$A_{s,s}$	Second osmotic virial coefficient for interactions between solutes S in water
$A_{s,s,s}$	Third osmotic virial coefficient for interactions between solutes S in water
a	Anion
a	Radius
$a_i^{(k)}$	Activity of species i normalized according to composition scale k
a_w	Activity of water
$a_{i,L}^{(0)}$	Binary interaction parameter between species (groups) i and L
$a_{i,L}^{(1)}$	Binary interaction parameter between species (groups) i and L
$a_{MX}^{(0)}$	Binary interaction parameter between cations M and anions X
$a_{MX}^{(1)}$	Binary interaction parameter between cations M and anions X
$a_{p,p}$	Binary interaction parameter between repeating units
$a_{p,\text{Cl}}$	Binary interaction parameter between repeating units and the chloride ion
b	Distance between two electrolyte groups in a polyelectrolyte backbone
b	Numerical value in Pitzer's model ($b = 1.2$)
b^*	Configurational parameter
$b_{i,L,k}$	Ternary interaction parameter between groups i, L and k
BaPSS	Poly(barium styrene sulfonate)
C	Repeating unit that will never dissociate (in model of Lammertz et al.)
c	Cation
c_i	Concentration of species i
c_i	Molarity of species i

Aqueous Solutions of Polyelectrolytes

CI	Counterion
$c_{k,b}$	Molarity of monomeric groups saturated with counterion k
c_m	Molarity of repeating units
c_P	Concentration of polyelectrolyte P
c_s	Molarity of salt S
$\tilde{c}_i$	Mass density of solute i
CaPAM	Calcium salt of copolymer of acrylic acid and acrylamide
D	Repeating unit undergoing a chemical reaction (in model of Lammertz et al.)
DMO	Differential membrane osmometry
DS	Degree of substitution
e	Proton charge
EMF	Electromotive force measurement
EQDIA	Equilibrium dialysis
F	Dissociated repeating unit (in model of Lammertz et al.)
F	Free energy
f	Short-range parameter
f^{el}	Function in the theory of Lifson and Katchalsky
f_i	Functions (in model of Lammertz et al.); $i = 1, 2$
FPD	Freezing point depression
$f(M)$	Molecular mass distribution function
G	Gibbs energy
G_{ji}	Binary interaction parameter (in model of Nagvekar and Danner)
$G_{ji,ki}$	Interaction parameter (in model of Nagvekar and Danner)
g_{ji}	Energy parameter (in model of Nagvekar and Danner)
GDM	Gel deswelling method
h	Length of a polyion
HPAA	Poly(acrylic acid)
HPAMS	Poly(2-acrylamido-2-methyl-1-propane sulfonic acid)
HPAS	Poly(anethole sulfonic acid)
HPES	Poly(ethylene sulfonic acid)
HPMAA	Poly(methacrylic acid)
HPMSS	Poly(methyl styrene sulfonic acid)
HPP	Poly(phosphoric acid)
HPVB	Poly(vinyl benzoic acid)
HPVS	Poly(vinyl sulfuric acid)
HPVSA	Poly(vinyl sulfonic acid)
HPSS	Poly(styrene sulfonic acid)
I	Ionic strength
I_m	Ionic strength (on molality scale)
$I_{m,MX}$	Ionic strength (on molality scale) of an aqueous solution of MX
I_s	Ionic strength (on molarity scale)
ISO	Isopiestic experiments
j	Abbreviation
j	Component
K	Chemical reaction constant (in model of Lammertz et al.)

k	Boltzmann's constant
k	Component
k	Concentration scale
k	Degree of counterion condensation at infinite dilution (in model of Lammertz et al.)
KPA	Poly(potassium acrylate)
KPAM	Potassium salt of copolymer of acrylic acid and acrylamide
l_B	Bjerrum length
LiCMC	Lithium carboxymethylcellulose
M	Cation
M	Molecular mass
M_r	rth moment of distribution function for molecular mass
M_n	Number-averaged molecular mass
M_w	Mass-averaged molecular mass
M_w^*	Relative molecular mass of water divided by 1,000
m_i	Molality of species i
m°	Unit of molality $m^\circ = 1 \text{ mol}/(\text{kg water})$
m_j^*	Modified molality of species j (in model of Pessoa and Maurer)
MgPAM	Magnesium salt of copolymer of acrylic acid and acrylamide
MO	Membrane osmometry
MX	Salt (cations M and anions X)
n	Mole number
N_A	Avogadro's number
$n_{p,\text{diss}}$	Number of moles of dissociated repeating units
$n_{\text{freeCI}}^{(p)}$	Number of moles of counterions originating from P (in Manning's theory)
$n_{\text{freeCI}}^{(s)}$	Number of moles of counterions originating from S (in Manning's theory)
n_T	Total mole number
NaCMC	Sodium carboxymethylcellulose
NaDS	Sodium dextran sulfate
NaPA	Poly(sodium acrylate)
NaPAM	Sodium salt of copolymer of acrylic acid and acrylamide
NaPAMA	poly(sodium acrylamido-co-trimethyl ammonium methyl methacrylate)
NaPAMS	Sodium salt of HPAMS
NaPES	Poly(sodium ethylene sulfate)
NaPMAA	Poly(sodium methacrylate)
NaPP	Poly(sodium phosphate)
NaPSS	Poly(sodium styrene sulfonate)
NaPVAS	Poly(sodium vinyl sulfate)
NH_4PA	Poly(ammonium acrylate)
NMR	Nuclear magnetic resonance
P	Polyelectrolyte
P	Polydispersity (M_w/M_n)

Aqueous Solutions of Polyelectrolytes

p	Pressure
p_w	Vapor pressure of water
PAAm	Poly(allylamino hydrochloride)
PDADMAC	Poly(diallyldimethyl ammonium chloride)
PEI	Poly(ethyleneimine)
PMETAC	Poly(2-(methacryloyloxy) ethyl trimethyl ammonium chloride)
PTMAC	Poly(trimethyl ammonium methyl methacrylate)
PVA	Poly(vinyl alcohol)
PVAm	Poly(vinyl amine)
PVBTMAC	Poly(vinyl benzene trimethyl ammonium chloride)
q	Number of charges
q_{gl}	Surface parameter of the globular form of the polyelectrolyte
q_i	Surface parameter of species i (in model of Lammertz et al.)
q_{max}	Maximum number of charges
q_{st}	Surface parameter of the stretched polyelectrolyte
R	Universal gas constant
R	Radius of a cylindrical cell around a polyion
r	Exponent
r	Distance
r_i	Volume parameter of species i (in model of Lammertz et al.)
T	Temperature
UV/VIS	Ultraviolet/visible light
V	Volume
V_p	Volume in Manning's theory
VO	Vapor pressure osmometry
x_i	Mole fraction of species i
X	Anion or counterion
X_j	Modified mole fraction of component j
X-DP	Salt (with counterion X) of dextran phosphate
X-DS	Salt (with counterion X) of dextran sulfate
z_{Cl}	Absolute valency of counterion
z_j	Absolute valency of ion j
z_j^*	Modified absolute valency of ions j (in model of Pessoa and Maurer)
z_M	Absolute valency of cation M
z_X	Absolute valency of anion X
z_p	Absolute valency of a repeating unit of polyelectrolyte P

Greek Symbols

α	Constant ($\alpha = 2$) in Pitzer's model
α	Total degree of dissociation of the repeating units (in model of Lammertz et al.)

α_{ji}	Nonrandomness parameter (in model of Nagvekar and Danner)
$\alpha_{ji,ki}$	Nonrandomness parameter (in model of Nagvekar and Danner)
β	Dimensionless parameter
Γ	Salt-exclusion parameter
Γ_i	Activity coefficient (on molality scale) of species i (in model of Lammertz et al.)
$\Gamma_{l,Z}$	Activity coefficient (on molality scale) of species l in state Z (in model of Lammertz et al.)
γ	Activity coefficient
γ_{LK}	Dimensionless parameter in the theory of Lifson and Katchalsky
Δ	Difference
ΔT_{FP}	Freezing point depression
ε	Relative permittivity of pure water
ε_0	Permittivity of vacuum
Φ	Osmotic coefficient
Φ_p	Osmotic coefficient (for pressure)
Φ_p^0	Osmotic coefficient (for pressure) at infinite dilution
Φ_T	Osmotic coefficient (for temperature)
$\Phi_s^{(c)}$	Osmotic coefficient (for pressure) on molarity scale due to salt S
$\Phi_{p+s}^{(c)}$	Osmotic coefficient (for pressure) on molarity scale due to salt S and polyion P
φ_p	Volume fraction of the polyelectrolyte
$\phi(r)$	Electrostatic potential that depends on radius r
θ_k	Degree of condensation of a counterion k
θ_z	Ratio in Manning's theory
$\theta_z^{(0)}$	Limit for θ_z in Manning's theory
κ	Inverse radius of the ionic cloud (Debye–Hückel theory)
λ	Charge density parameter
λ_{ij}	Binary interaction parameter (in model of Pessoa and Maurer)
$\lambda_{ij}^{(0)}$	Binary interaction parameter (in model of Pessoa and Maurer)
$\lambda_{ji}^{(1)}$	Binary interaction parameter (in model of Pessoa and Maurer)
μ_i	Chemical potential of component i
Ξ	Volume fraction of polyelectrolyte (in model of Lammertz et al.)
v	Number of repeating units of a polyelectrolyte molecule
v^*	Number of dissociated repeating units of a polyelectrolyte molecule
v_M	Stoichiometric coefficient for cation M in salt MX
v_X	Stoichiometric coefficient for anion X in salt MX
π	Osmotic pressure
Θ_{gl}	Surface fraction of the polyelectrolyte in its globular shape
Θ_L	Surface fraction of group L
Θ_{st}	Surface fraction of the polyelectrolyte in its stretched shape
ρ_i^*	Specific density of pure solvent i
$\bar{\rho}_i^*$	Molar density of pure solvent i

Aqueous Solutions of Polyelectrolytes

σ_i	Parameter of species i (in model of Pessoa and Maurer)
τ_{ji}	Binary interaction parameter (in model of Nagvekar and Danner)
$\tau_{ji,ki}$	Interaction parameter (in model of Nagvekar and Danner)
v_p^*	Molar volume in Manning's theory
$\bar{v}_{s,\text{pure}}$	Molar volume of pure solvent s
$\varpi^{(0)}$	Configurational parameter
$\varpi^{(1)}$	Configurational parameter

Subscripts

A	Repeating unit (in model of Lammertz et al.)
a	Anionic component
c	Cationic component
C	Repeating unit that will never dissociate (in model of Lammertz et al.)
CI	Counterion
Cl	Chloride ion
COI	Coion
cond.	CI contribution due to condensed counterions
D	Repeating unit undergoing a chemical reaction (in model of Lammertz et al.)
F	Dissociated repeating unit (in model of Lammertz et al.)
Free CI	Free counterions
Free COI	Free coions
id.liq.mix.	Ideal liquid mixture
id.mix.	Ideal mixture
H	Hydrogenium ions
K	Potassium ion
k	Contribution
local	Local
LK	Lifson and Katchalsky
M	Cations
m	Solvent component
Ma	Manning's theory
Mg	Magnesium ion
MX	Salt (cations M and anions X)
Na	Sodium ions
p	Polyelectrolyte
Pb	Lead ions
pure liquid	Pure liquid component
pure water	Pure water
rp	Repeating unit of polyelectrolyte

sym	Symmetrical convention
w	Water
$(w + s)$	In an aqueous solution of the salt
$(w + s + p)$	In an aqueous solution of (salt + polyelectrolyte)
X	Ion X

Superscripts

(c)	On molarity scale
Comb.	Combinatorial
E	Excess
el	Contribution from electrostatics
fv	Free volume
id.mix.	Ideal mixture
(k)	Characterizes the concentration scale
LR	Long-range
(m)	On molality scale
ref	Reference state
SLE	Solid–liquid equilibrium
SR	Short-range
vdW	Van der Waals
(x)	On mole fraction scale
∞	Infinite dilution
Δconf	Caused by a difference in the configuration (in model of Lammertz et al.)

1 Introduction

Polyelectrolytes are polymers of a single repeating unit (monomer) that is an electrolyte or of several repeating units (monomers), where at least one of the repeating units is an electrolyte. That electrolyte can dissociate in water and in aqueous solutions resulting in negative or positive charges on the polymer backbone. Polyelectrolytes are very soluble in water, particularly when, in addition to the ionic monomers, the other monomers are also hydrophilic. The large variety of monomers means that there is a huge variety of polyelectrolytes. The number of different repeating units and the number of each of those repeating units determines the primary structure of a polyelectrolyte, i.e., the chemical nature and the molecular mass. However, that information is not sufficient to characterize a polyelectrolyte. As typical of polymers, polyelectrolyte samples reveal a molecular mass distribution (polydispersity). Furthermore, when a polymer consists of more than a single repeating unit, the secondary structure is important for its properties: the different repeating units might be statistically distributed or arranged in, more or less uniform, blocks. The polymer can be a linear structure or a branching one. It might have a certain shape in space (tertiary structure) that can depend on the surrounding solution. All these parameters influence the properties of an aqueous solution of polymer, but there are more parameters when the polymer has electrolyte groups. These electrolytes can be weak or strong electrolytes, resulting in different degrees of dissociation/protonation. The electrolyte groups of the backbone might be all cationic or all anionic, but they might also be partially cationic and partially anionic. Such polyelectrolytes are called polyampholytes. There is another parameter that has an important influence on the properties of polyelectrolytes in aqueous solutions: the distance between the electrolyte groups in the polymer backbone. When that distance is small, the attractive electrostatic forces between the ionic groups in the backbone and their counterions in the aqueous solutions become so strong that, even if the repeating unit is a strong electrolyte, one observes an ion pairing, i.e., some of the counterions condensate (at least partially) with the ions of the backbone. Therefore, even at high dilution in water such polyelectrolytes are not completely dissociated and the degree of dissociation might depend on the composition of the surrounding aqueous phase. The large number of parameters that influence the properties of aqueous solutions of polyelectrolytes is reflected in the variety of areas where such solutions are found and applied. Table 1 gives some typical examples of applications. These applications take advantage of the particular thermodynamic properties of aqueous solutions of polyelectrolytes. Therefore, there is a need for methods to describe such properties. In applied thermodynamics, the properties of solutions are described by expressions for the Gibbs energy as a function of temperature, pressure, and composition. From such equations all other thermodynamic state functions can be derived.

There are many well-established models for the Gibbs energy of nonelectrolyte solutions and also several methods to describe conventional polymer solutions. However, the state of the art for modeling thermodynamic properties of aqueous solutions of polyelectrolytes is far less elaborated. This is partly due to the particular features of such solutions but is also caused by insufficiencies in the knowledge of the parameters that characterize a polyelectrolyte, for example, the polydispersity and the different structures (primary, secondary etc.) of the polyelectrolytes. The development and testing of thermodynamic models has always been based on reliable experimental data for solutions for which all components are well characterized. Such characterization is particularly scarce for biopolymers and biopolyelectrolytes. Furthermore, such polymers are generally more complex than synthetic polymers. Therefore, the present contribution is restricted to a discussion of the thermodynamic properties of aqueous solutions of synthetic polyelectrolytes that consist of only two different repeating units that are statistically distributed. Furthermore, it is restricted to systems where sufficient information on the polyelectrolyte's polydispersity is available.

Table 1 Applications of polyelectrolytes

Application	Product	References
Stabilization of colloid systems as dispersing agents	Poly(acrylic acid), gelatin, sodium carboxymethylcellulose	[1–6]
Sludge dewatering, flocculating agents	Acrylamidecopolymers, Poly(diallyldimethyl ammonium chloride)	[7, 8]
Retentions aids in paper industry	Poly(ethyleneimine), cationic starches, poly (diallyldimethyl ammonium chloride)	[9, 10]
Thickeners	Gelatin, Sodium carboxymethylcellulose, pectin, arab gum, carrageenan	[11]
Gelling agents	Gelatin, pectin, carrageenan	[12]
Temporary surface coatings for:		
Textile industry	Poly(acrylic acid) sodium salt, Sodium carboxymethylcellulose, poly(acrylic acid) ammonium salt	[12]
Capsules in pharmaceutical applications	Gelatin, Sodium carboxymethylcellulose, cellulose acetate phthalate, copolymers of methacrylic acid	[12]
Corrosion-protecting coatings	Poly(styrene sulfonic acid), poly(acrylic acid)	[12]
Cosmetic industry	Copolymers of acrylic acid	
Antistatic coatings	Copolymers with styrene sulfonate units, cationic polyelectrolytes	
Adhesives for:		
Food industry	Gelatin	[13]
Paper industry	Sodium carboxymethylcellulose	[14]
Dental material/dental composites	Zinc polycarboxylate, polyacrylic acid-glass cements, poly(methyl methacrylate)	[15, 16]
Controlled release of drugs and responsive delivery systems	Cellulose acetate phthalate, poly(dimethylamino ethyl methacrylate-co-tetraethyleneglycol dimethacrylate) gels	[15, 16]
Polymeric drugs	Poly(N-vinyl pyrollidone-co-maleic acid), sulfonated polysaccharides	[15, 16]

2 Structure and Characterization of Polyelectrolytes

Polyelectrolytes are primarily characterized by the backbone monomers and the electrolyte and/or proton-accepting groups attached to those monomers. Table 2 gives an overview of some of the most important anionic and cationic synthetic polyelectrolytes. Styrene, the vinyl group, cellulose, and dextran are the most important backbone monomers for hydrocarbonic polyelectrolytes. The most important dissociating groups in synthetic, organic polyelectrolytes are sulfonic, acrylic, benzoic, phosphoric, and sulfuric acid. By dissociation, such polymers become electrically charged species, carrying negative charges. Therefore, such polymers are also called "anionic polyelectrolytes," whereas "cationic polyelectrolytes" have proton-accepting groups. By protonation, such formerly neutral groups can be positively charged. The most important proton-accepting groups are NR_3^+ and NH_2^+. Short nomenclatures are often used to abbreviate the chemical

Aqueous Solutions of Polyelectrolytes

Table 2 Important anionic and cationic polyelectrolytes (cf. Scranton et al. [16])

Polyelectrolyte (Abbreviation)	Repeating unit	Polyelectrolyte (Abbreviation)	Repeating unit
Poly(acrylic acid) (HPAA)	$-(CH_2-CH)_n-$ COOH	Poly(vinyl sulfuric acid) (HPVAS)	$-(CH_2-CH)_n-$ OSO$_3$H
Poly(methacrylic acid) (HPMAA)	CH$_3$ $-(CH_2-C)_n-$ COOH	Poly(vinyl sulfonic acid) (HPVS)	$-(CH_2-CH)_n-$ SO$_3$H
Poly(styrene carboxylic acid)	$-(CH_2-CH)_n-$ (phenyl) COOH	Poly(2-acrylamido-2-methyl-1-propane sulfonic acid) (HPAMS)	CH$_3$ $-(CH_2-C)_n-$ O=C–NH H–C—CH$_3$ CH2 SO$_3$H
Poly(styrene sulfonic acid) (HPSS)	$-(CH_2-CH)_n-$ (phenyl) SO$_3$H	Poly(phosphoric acid) (HPP)	O ‖ $-(O-P)_n-$ OH
Poly(vinyl benzoic acid) (HPVB)	$-(CH_2-CH)_n-$ (phenyl) CH$_2$COOH	Sodium dextran sulfate (NaDS)	$-OCH_2$ O HO– –O– HO OSO$_3$Na
Sodium carboxymethyl-cellulose (NaCMC)	ROCH$_2$ OR OR O RO– –O– –O–R O OR OR CH$_2$OR R=H or CH$_2$CO$_2$Na		

Table 2 (continued)

Polyelectrolyte (Abbreviation)	Repeating unit	Polyelectrolyte (Abbreviation)	Repeating unit
Poly(ethylene-imine) (PEI)	$\left[NH-CH_2-CH_2\right]_n$	Poly(vinyl amine) (PVAm)	$\left[CH_2-CH\right]_n$ with NH_2
Poly(trimethyl ammonium methyl methacrylate) (PTMAC)	$\left[CH_2-CH\right]_n$ with CH_3, $C=O$, $O-CH_2-CH_2-N^+-CH$ with Cl^-, CH, CH		
Poly(allylamino hydrochloride) (PAAm)	$\left[CH_2-CH\right]_n$ with CH_2, $NH_2\ HCl$	Poly(diallyldimethyl ammonium chloride) (PDADMAC)	$\left[CH_2 \quad CH_2\right]_n$ with $CH-CH$, $H_2C \quad CH_2$, N^+ Cl^-, $H_3C \quad CH_3$
Poly(vinylbenzene trimethyl ammonium chloride) (PVBTMAC)	$\left[CH_2-CH\right]_n$ phenyl, CH_2, $H_3C-N^+-CH_3\ Cl^-$, CH_3	Poly(2-(meth-acryloyloxy) ethyl trimethyl ammonium chloride) (PMETAC)	$\left[CH_2-CH\right]_n$ with CH_3, $C=O$, O, H_2C, CH_2, $H_3C-N^+-CH_3\ Cl^-$, CH_3

formula. The abbreviation usually consists of two parts: one part stands for the backbone and electrolyte group (cf. Table 2), the other for the (approximate) molecular mass. A polyelectrolyte material rarely consists of one single type of molecule, but of a variety of molecules of different molecular masses.

In principle, a distribution function $f(M)$ has to be used to characterize that material: $f(M)\ dM$ is the fraction of polymers with a molecular mass between $M - dM/2$ and $M + dM/2$, with the normalization:

$$\int_0^\infty f(M)\ dM = 1. \tag{1}$$

However, as such distributions are difficult to determine, it is common practice to characterize a polymer sample by the number-average (M_n) and the mass-average (M_w) molecular masses, which are the first members in a series of moments:

$$M_r = \frac{\int_{M=0}^{M_i=\infty} f(M) M^r \, dM}{\int_{M=0}^{M=\infty} f(M) M^{(r-1)} \, dM}, \tag{2}$$

with $M_1 = M_n$ and $M_2 = M_w$, or by using one of those average molecular masses and the polydispersity P, which is the ratio of M_w to M_n:

$$P = \frac{M_w}{M_n}. \tag{3}$$

There might be a variety of different counterions in a polyelectrolyte, therefore, the quota of different counterions can be used to further characterize a polyelectrolyte. That quota can undergo some changes, e.g., when a polyelectrolyte is dissolved in an aqueous solution of electrolytes or of other polyelectrolytes. The degree of dissociation of a polyelectrolyte is also often used for characterization. However, from the view of thermodynamics, that property depends on the surroundings and therefore it is more suited for characterizing the state of a polyelectrolyte instead of characterizing the polyelectrolyte itself.

Various experimental methods such as potentiometric titration, conductometry, polarography, electrophoresis, spectroscopy (NMR, UV/VIS), osmometry, light scattering (static and dynamic laser light scattering, X-ray scattering, and neutron scattering), viscometry, sedimentation, and chromatography (e.g., size exclusion chromatography and gel electrophoresis) have been used to characterize polyelectrolytes in aqueous solutions (for a recent review cf. Dautzenberg et al. [12]). Experimental information on the average molecular mass of a polyelectrolyte is mostly derived from laser light scattering, osmometry or viscometry, i.e., from methods that are also used to determine the thermodynamic properties of polyelectrolyte solutions, e.g., the activity of water. The polydispersity of polyelectrolytes is usually determined by size-exclusion chromatography. Potentiometric titration is often used to determine the degree of functionalization and the chemical reaction equilibrium constants for the dissociation/protonation reactions, i.e., properties characterizing the number of anionic groups saturated by hydrogen ions (in an anionic polyelectrolyte) or the number of protonated groups (in a cationic polyelectrolyte). The number of ionic groups in an anionic polyelectrolyte is sometimes determined by atomic absorption spectroscopy. X-ray structural analysis and neutron scattering are typical methods for investigating the structure of polyelectrolytes. From the viewpoint of thermodynamics, a polyelectrolyte should be characterized by all single polymers comprising the polyelectrolyte sample, the number of functional groups (ionic as well as neutral groups), the state of the ionic groups (e.g., number and nature of dissociable counterions of anionic groups as well as the number of protonated cationic groups), the secondary structure, and the concentration of any single polyelectrolyte in the sample. However, that information is almost never available. In most cases, the chemical nature of such polyelectrolyte samples is only characterized by the backbone monomers and the kind of ionic groups, as well as an estimate of the ratio of electrolyte groups to backbone monomers. The accessible information on the counterions is often limited to the chemical nature and a more or less rough estimate for the ratio of different counterions, e.g., the ratio of hydrogen counterions to sodium counterions of an anionic polyelectrolyte. Very often, the degree of polymerization is given only as an estimate of either the number-averaged or the mass-averaged molecular mass, but detailed information on the polydispersity is missing. Thus, the characterization of the polyelectrolyte is often far from satisfactory (at least from the viewpoint of thermodynamics) and, consequently, reported thermodynamic data are often of very limited use, e.g., for testing and developing of models for describing and predicting the thermodynamic properties of such solutions.

3 Experimental Data for the Vapor–Liquid Equilibrium of Aqueous Polyelectrolyte Solutions

Because polyelectrolytes are nonvolatile, the most important thermodynamic property for vapor + liquid phase equilibrium considerations is the vapor pressure of water p_w above the aqueous solution. Instead of the vapor pressure, some directly related other properties are used, e.g., the activity of water a_w, the osmotic pressure π, and the osmotic coefficient Φ. These properties are defined and discussed in Sect. 4. Membrane osmometry, vapor pressure osmometry, and isopiestic experiments are common methods for measuring the osmotic pressure and/or the osmotic coefficient. A few authors also reported experimental results for the activity coefficient γ_{CI} of the counterions (usually determined using ion-selective electrodes) and for the freezing-point depression of water ΔT_{FP}. The activity coefficient is the ratio of activity to concentration:

$$\gamma_{CI}^{(k)} = \frac{a_{CI}^{(k)}}{k},$$

(4)

where k in the denominator is used to express a certain concentration scale (e.g., mole fraction x, molarity c or molality m). Superscript (k) indicates that the activity coefficient and the activity are defined using a certain reference state, which depends on the selection of the scale used to express the composition of the solution. Some authors report experimental data for the freezing point depression of an aqueous solution:

$$\Delta T = T^{SLE} - T^{SLE}_{purewater},$$

(5)

and convert that data to an osmotic coefficient $\Phi_T^{(k)}$ by:

$$\Phi_T^{(k)} = \frac{\Delta T}{\Delta T_{id.liq.mix.}^{(k)}},$$

(6)

Aqueous Solutions of Polyelectrolytes   81

where $\Delta T_{\text{id.liq.mix.}}^{(k)}$ is the freezing point depression of an ideal aqueous solution of the polyelectrolyte. Subscript T to the osmotic coefficient Φ indicates that the osmotic coefficient is here defined with the freezing point depression. Superscript (k) is again used to indicate that the definition of *the ideal mixture* depends on the chosen concentration scale. However, it also depends on an assumption about the dissociation of the polyelectrolyte. It is common practice to assume that in an ideal mixture the polyion is completely dissociated.

3.1   Aqueous Solutions of a Single Polyelectrolyte

Tables 3–6 give a survey of literature data for the vapor–liquid equilibrium of aqueous solutions of a single polyelectrolyte with various counterions. Abbreviations (shown in Table 2) are used to characterize the polyelectrolyte and the experimental procedures (MO membrane osmometry; DMO differential membrane osmometry; VO vapor pressure osmometry; ISO isopiestic experiments; EMF electromotive force measurements including also measurements with ion-selective electrodes as well as titration; FPD freezing point depression; GDM gel deswelling investigations). Table 3 gives a survey for aqueous solutions of poly(styrene sulfonic acid).

Table 3 Survey of literature data for thermodynamic properties of aqueous solutions of polyelectrolytes with styrenesulfonic acid as the backbone monomer (without any other salt)

Molecular mass $(\times 10^{-5})$	Counterion	Counterion molality	Method	Exp. prop.	References
0.4 and 5	Na^+; H^+	0.01–1.4	VO	Φ_p	[17]
		0.8–7.54	ISO	Φ_p	
5	$H^+, Li^+, Na^+, K^+, Cs^+, NH_4^+,$ N^+R_1; $[R_1 \equiv (CH_3)_4,$ $(C_2H_5)_4, (C_4H_4)_4]$	0.05–1.1	DMO	Φ_p	[18]
0.4–5.2	$Na^+, H^+, Ca^{++}, Cu^{++}, Cd^{++}$	10^{-3}–10^{-2}	MO	Φ_p	[19]
0.4–5.2	Cu^{++}, Na^+	10^{-3}–10^{-2}	MO	Φ_p	[20]
0.2–1.0	$Na^+, Tl^+, Cd^{++}, Ca^{++}$	5×10^{-4}–10^{-2}	MO, EMF	Φ_p; γ_{Cl}	[21]
0.4	Li^+, Na^+, K^+, Cs^+	6×10^{-3}–0.3	FPD	Φ_T	[22]
0.4	Cd^{++}, Mg^{++}	6×10^{-3}–0.15	FPD	Φ_T	[23]
4.6	$H^+, Li^+, Na^+, K^+, Ca^{++},$ Ba^{++}, NH_4^+, N^+R_1; $[R_1 \equiv C_3H_7, (C_2H_5)_4,$ $(CH_3)_4, CH_2C_6H_5]$	0.04–3.3	ISO	Φ_p	[24]
5	Na^+	0.7–1.44	MO, VO	π	[25]
4.3	Na^+	4×10^{-4}–0.37	MO	π	[26]
5	Na^+	0.4–2.7	ISO	π	[27]
1.3	Na^+	5×10^{-4}– 4×10^{-3}	ISO	a_w	[28]

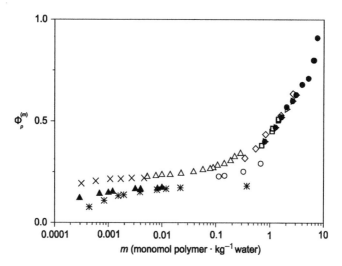

Fig. 1 Osmotic coefficient of aqueous solutions of NaPSS of varying molecular mass (*M*) at 25°C (unless otherwise indicated): *open triangles* 4 × 10^4 (0°C) [22]; *times* 4 × 10^4 [19]; *right-pointing triangle* 4 × 10^4 [17]; *star* 4.3 × 10^5 [26]; *closed circles* 5 × 10^5 [17]; *open diamonds* 5 × 10^5 [27]; *open squares* 5 × 10^5 (35°C) [25]; *open circles* 5 × 10^5 (no temperature given) [18]; *closed triangles* 5 × 10^5 [19]

Figure 1 shows some typical experimental results for the osmotic coefficient of aqueous solutions of poly(sodium styrene sulfonate) (NaPSS). The osmotic coefficient $\Phi_p^{(m)}$ is plotted versus the concentration of the polyelectrolyte (expressed as the molality of the "repeating units" or "monomer groups"). The figure reveals that the osmotic coefficient of a diluted aqueous solution of a polyelectrolyte is well below unity even at very small concentrations, e.g., at monomer-group molalities below about 0.001 mol/kg. It also reveals that, at low polymer concentrations, the influence of the concentration of the polyelectrolyte on the osmotic coefficient is rather small (e.g., when the monomer-group molality is increased from about 0.0002 to about 0.1 mol/kg, the osmotic coefficient of an aqueous solution of high molecular weight NaPSS increases only from about 0.2 to about 0.25), whereas at higher concentrations the osmotic coefficient increases strongly with increasing polymer concentration (e.g., from about 0.4 to about 0.8, when the monomer-group molality is increased from about 1 to 10 mol/kg). Furthermore, experimental results from different sources often do not agree with each other, but it seems that most experimental results confirm that there is only a very small influence of the molecular mass (*M*) of the polyelectrolyte on the osmotic coefficient.

Table 4 gives a similar survey for other polyelectrolytes. Figure 2 shows the influence of the backbone monomer of the polyelectrolyte on the osmotic coefficient. At constant polyelectrolyte concentration (again expressed as the molality of the repeating units), the osmotic coefficients might differ by, for example, a factor of five. For example, at 25°C, the osmotic coefficient of a 1 mol/kg aqueous solution of monomer groups of poly(sodium ethylene sulfate) (NaPES) is about 0.22, whereas

Aqueous Solutions of Polyelectrolytes

Table 4 Survey of literature data for thermodynamic properties of aqueous solutions of a single polyelectrolyte at around 300 K (without any other salt)

Polymer	Molecular mass/ ($\times 10^{-5}$)	Counterion	Counterion molality	Method	Exp. prop.	References
HPMSS	1.5	Na^+, H^+, Ag^+, Tl^+, Zn^{++}, Cd^{++}, Pb^{++}	4×10^{-4}–0.1	EMF	γ_{CI}	[29] [30]
HPAA	2.6	Na^+/H^+	0.28–0.77	MO, VO	π	[25]
HPAA	0.012	Na^+	0.1–3.1	ISO	Φ_p	[31]
HPAA	0.46	Na^+/H^+, N^+ (n-C_4H_9)$_4$/ H^+, Li^+, K^+, N^+R_1; [$R_1 \equiv (C_2H_5)_4$, $(CH_3)_4$, (n-C_3H_7)$_4$]	0.2–4.5	ISO	Φ_p	[32]
HPAA	1.2	Na^+	4×10^{-3}– 0.23	EMF	γ_{CI}	[33]
HPAA	1.2	Na^+	1.7×10^{-3}– 0.25	EMF	γ_{CI}	[34]
CMC	2.5	Na^+, Li^+, K^+, N^+R_1; [$R_1 \equiv (C_2H_5)_4$, $(CH_3)_4$, (n-C_4H_9)$_4$]	0.16–2	ISO	Φ_p	[32]
PMETAC	1.7	Cl^-	0.38–1.1	MO, VO	π, Φ_p	[25]
PAAm	0.5	Cl^-	0.44–1.9	MO, VO	π, Φ_p	[25]
HPVB		Na^+	0.32–0.77	MO, VO	π, Φ_p	[25]
HPVAS	2.5	Na^+, Li^+, K^+, Ca^{++}, Ba^{++}	0.11–2.5	ISO	Φ_p	[35]
HPP	0.61	Na^+	0.13–2.3	ISO	Φ_p	[36]
HPES	1	H^+, Li^+, Na^+, K^+, NH_4^+, N^+R_1; [$R_1 \equiv (C_2H_5)_4$, $(CH_3)_4$, (n-C_3H_7)$_4$, (n-C_4H_9)$_4$, $(CH_3)_3CH_2C_6H_5$]	0.19–6.1	ISO	Φ_p	[37]
PVA/ HPVAS	1.7	Co, Ni, Cu^{++}	5×10^{-4} –0.12	GDM	a_w	[38]
PVA/ HPVAS	0.7–0.9	Na^+, Cu^{++}, Li^+, La, Cs^+, Mg^{++}	5×10^{-4}–0.2	GDM	a_w	[39]
HPAS	0.1	Na^+, Li^+, Cs^+	1×10^{-3}–0.3	VO/ MO	Φ_p	[40]
HPAA	0.03; 0.07	Na^+, NH_4^+	7×10^{-4} –4×10^{-3}	ISO	a_w	[41]
HPMAA	0.06; 0.14	Na^+	7×10^{-4}– 3×10^{-3}	ISO	a_w	[41]
HPES	0.02; 0.07	Na^+	4×10^{-4}– 5×10^{-3}	ISO	a_w	[41]

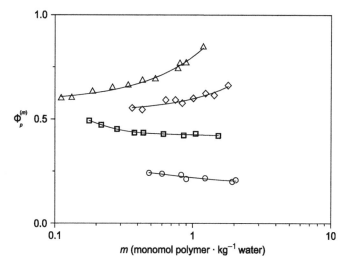

Fig. 2 Osmotic coefficient of aqueous solutions of a polyelectrolyte from isopiestic measurements at 25°C [32, 35–37]: *open triangles* NaPVAS, $M = 2.5 \times 10^5$; *open diamonds* NaCMC, $M = 2.5 \times 10^5$, DS = 0.95; *open squares* NaPP, $M = 6.1 \times 10^4$; *open circles* NaPES, $M = 1 \times 10^5$. *DS* degree of substitution (carboxymethyl groups per glucose unit)

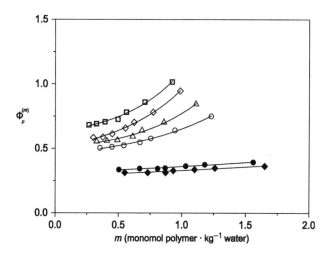

Fig. 3 Influence of counterion on the osmotic coefficient of aqueous solutions of poly(acrylates) at 25°C from isopiestic measurements (Asai et al. [32]): *open squares* $-N(n\text{-}C_4H_9)_4$; *open diamonds* $-N(n\text{-}C_3H_7)_4$; *open triangles* $-N(n\text{-}C_2H_5)_4$; *open circles* $-N(n\text{-}CH_3)_4$; *closed circles* $-$Li; *closed diamonds* $-$K

it is about 0.44 when ethylene sulfate is replaced by phosphate (i.e., for NaPP), and is 0.6 and 0.8 for sodium carboxymethylcellulose (NaCMC) and poly(sodium vinyl sulfate) (NAPVAS), respectively.

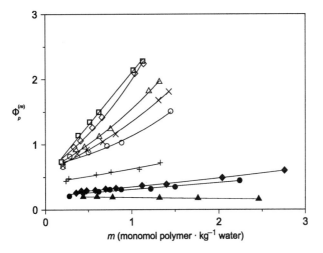

Fig. 4 Influence of counterion on the osmotic coefficient of aqueous solutions of poly(ethylene sulfonates) at 25°C from isopiestic measurements by Ise and Asai [37]: *open squares* $-N(n\text{-}C_4H_9)_4$; *open diamonds* $-N(n\text{-}C_3H_7)_4$; *open triangles* $-N(n\text{-}C_2H_5)_4$; *times* $-N(n\text{-}CH_3)_4$; *open circles* $N(CH_3)_3CH_2C_6H_5$; +, $-NH_4$; *filled diamonds* $-H$; *closed circles* $-Li$; *closed triangles* $-K$

Figures 3 and 4 show some typical examples of the influence of the nature of the counterion of a polyelectrolyte on the osmotic coefficient. The osmotic coefficient is typically very small for inorganic counterions, but it can be increased by a factor of about 10 by organic counterions, for the same temperature and polyelectrolyte monomer-group molality. Figure 4 shows that the osmotic coefficient of an aqueous solution of a poly (ethylene sulfonate) increases in the counterion series K^+, Li^+, H^+, NH_4^+, $N^+(CH_3)_3CH_2C_6H_5$, $N^+(CH_3)_4$, $N^+(C_2H_5)_4$, $N^+(n\text{-}C_3H_7)_4$, and $N^+(n\text{-}C_4H_9)_4$.

3.2 Aqueous Solutions of a Single Polyelectrolyte and a Low Molecular Weight Strong Electrolyte

There have been many investigations on the influence of a low molecular weight strong electrolyte on the thermodynamic properties of an aqueous solution of a polyelectrolyte. A survey on literature data is given in Table 5. The experimental methods already mentioned above are also common for investigating aqueous solutions of both a polyelectrolyte and a salt. However, also equilibrium dialysis (EQDIA) and EMF-measurements with ion-selective electrodes have been used in such experimental investigations. In EQDIA, an aqueous polyelectrolyte solution and an aqueous solution of a low molecular weight salt are separated by a membrane that is permeable to water as well as to the ions of the salt and the counterions of the polyelectrolyte. In phase equilibrium, the concentration of the free ions in the coexisting phases are

Table 5 Survey of literature data for thermodynamic properties of aqueous solutions of a single polyelectrolyte at around 300 K (with an added salt)

Polymer	Polymer conc. (g/dm^3)	Salt	Salt conc. (mol/dm^3)	Method	Experimental properties	References
NaPSS	2.5–31	NaCl	0.01–0.1	MO	π	[42]
NaPSS	0.5–7.5	NaCl	0.005–0.5	MO	π	[26]
NaPSS	68–556	NaCl	0.056–0.91	ISO	Φ_p	[27]
BaPSS	2–52	LiCl, NaCl, KCl	0.001–1		Solubility	[43]
NaPSS	0.2–21	NaCl, Na$_2$SO$_4$, NaCNS	1.7×10^{-4}–0.1	EMF, EQDIA	Γ	[44, 45]
NaPSS	0.08	Pb(NO$_3$)$_2$, HNO$_3$, NaNO$_3$	0.01–0.02	EQDIA	$\theta_H, \theta_{Na}, \theta_{Pb}$	[46]
NaPSS	<0.8	NaNO$_3$	0.005–3.7	EQDIA	θ_{Na}, θ_H	[47]
NaPA	3.7–623	NaCl	0.046–3	ISO	Φ_p	[48]
NaPA	0.09–9.1	NaCl, Na$_2$SO$_4$	1.7×10^{-4}–0.1	EMF, EQDIA	Γ, Φ_p	[45]
KPA	5	KCl	0–0.1	EMF, VO	θ_K, γ_{Cl}	[49]
NaPMAA	0.08–49	NaBr	0.001–0.3	DMO	Φ_p	[50]
NaCMC LiCMC	1–6	NaCl, LiCl, NH$_4$Cl, CaCl$_2$, Mg Cl$_2$	0.005–1	MO	π, Φ_p	[51]
NaCMC	0.01; 0.025	NaCl	0.1–5.7	EMF	γ_{Cl}	[52]
CaPAM	0.87	CaCl$_2$	0.002–0.02	EMF	γ_{Cl}	[53]
NaPAM	0.19–1.3	NaBr	4×10^{-4}–4×10^{-2}	EMF	γ_{Cl}	[54]
NaPAM, KPAM, MgPAM	0.22–1.67	NaNO$_3$, KNO$_3$, Mg(NO$_3$)$_2$	3×10^{-4}–0.05	EQDIA, EMF	$\gamma_{Cl}, \theta_{Na}, \theta_K, \theta_{Mg}$	[55]
X-DP,X-DS; X = (Li$^+$, Na$^+$, K$^+$, Cs$^+$, Rb$^+$, Mg^{++}, Ca^{++}, Cd^{++})	0.5–1.0	LiCl, NaCl, KCl, CsCl, RbCl, MgCl$_2$, CaCl$_2$, Cd(NO$_3$)$_2$	$c_p/c_s = 0$–12	EMF	γ_{Cl}, θ_X	[56]
PEI	5.2	KCl	0.001–0.1	EMF, VO	θ_{Cl}, γ_{Cl}	[49]
PVBTMAC	90–670	NaCl	0.05–0.85	ISO	Φ_p	[27]
PAAm	0.09–9.4	NaCl, Na$_2$SO$_4$	1.7×10^{-4}–0.1	EMF, EQDIA	Γ	[44]
PMETAC	0.2–21	NaCl, Na$_2$SO$_4$	1.7×10^{-4}–0.1	EMF, EQDIA	Γ	[44]
NaPA, NH$_4$PA, NaPMAA, NaPES, NaPSS	0–40	NaCl	0–0.3	ISO	a_w	[28]

Aqueous Solutions of Polyelectrolytes

determined, e.g., by titration or by ion chromatography. The results are often reported as the "degree of condensation" θ_k of a counterion k or the "salt-exclusion parameter" Γ. θ_k is the degree of electrolyte groups in the polymer that are neutralized by ionic species k:

$$\theta_k = \frac{c_{k,b}}{vc_p}, \tag{7}$$

where $c_{k,b}$ is the molarity of monomeric electrolyte groups saturated with counterion k, v is the number of repeating units, and c_p is the molarity of the polyelectrolyte. The salt-exclusion parameter Γ is the ratio of the difference in the molarity c_s of the counterion on both sides of the membrane to the molarity of (monomer) electrolyte groups of the polyelectrolyte c_p in the aqueous phase:

$$\Gamma = \frac{c_{s,(w+s)} - c_{s,(w+s+p)}}{c_p}. \tag{8}$$

When ion-selective electrodes have been used, the activity coefficient of the counterions is sometimes presented as a function of the "charge density parameter" λ (from the theory of Lifson and Katchalsky):

$$\lambda = \frac{e^2}{4\pi\varepsilon\varepsilon_0 kTb}, \tag{9}$$

where e, ε_0, ε, k and b are the proton charge, permittivity of vacuum, relative permittivity of pure water, Boltzmann's constant, and the distance between two electrolyte groups in a polyelectrolyte backbone, respectively.

Figure 5 shows some typical results for the osmotic pressure π of aqueous solutions of NaPSS and NaCl. At high ionic strength, the slope of the ratio of osmotic pressure to the (monomer) molarity c_p does not depend on the concentration of the polyelectrolyte. That slope increases with decreasing ionic strength and – at constant, but lower ionic strength – with increasing polymer concentration. In such experiments, the ionic strength is adjusted by the amount of dissolved NaCl; a high ionic strength causes a condensation of sodium ions to the polyelectrolyte backbone. The osmotic pressure is primarily caused by the added salt, and small amounts of the polyelectrolyte cause a change in the osmotic pressure very similar to that observed in an ideal solution. The strong increase of the osmotic pressure with decreasing ionic strength, but constant polymer concentration, is at least partially due to the increasing degree of dissociation of electrolyte groups of the polymer.

When a low molecular weight salt MX is dissolved in an aqueous solution of an anionic polyelectrolyte of counterions CI, both cations (CI and M) compete for the anionic groups in the polymer. Such competition could result in a change in the degree of dissociation of the ionic groups, i.e., the ratio of charged to neutral repeating units in the backbone. Some examples are shown in Fig. 6. When the lead ion concentration is increased in an aqueous solution of 0.001 mol/dm^3

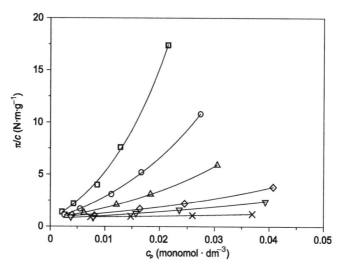

Fig. 5 Reduced osmotic pressure vs. polymer concentration for NaPSS at 25°C in aqueous NaCl solutions of various ionic strength I: *open squares* $I = 0.005$ mol/dm^3; *open circles* $I = 0.01$ mol/dm^3; *open triangles* $I = 0.02$ mol/dm^3; *open diamonds* $I = 0.05$ mol/dm^3; *open inverted triangles* $I = 0.1$ mol/dm^3; *crosses* 0.5 mol/dm^3 [26]

sulfonic acid groups at constant ionic strength [fixed by a mixture of NaNO$_3$ + Pb(NO$_3$)$_2$], the relative amount of sulfonic acid groups neutralized by lead ions (θ_{Pb}) also increases and, consequently, the relative amount of sulfonic acid groups neutralized by sodium ions (θ_{Na}) decreases. However, the decrease of θ_{Na} is not completely compensated by the increase of θ_{Pb} and, therefore, the relative amount of dissociated sulfonic groups increases. θ_{Pb} decreases and θ_{Na} increases when the ionic strength is increased at constant lead concentration. The sum ($\theta_{Pb} + \theta_{Na}$) also increases because at the higher ionic strength more ionic species compete for the charged repeating units of the backbone. When, at constant ionic strength, sodium nitrate is replaced by nitric acid, θ_{Pb} increases and θ_{Na} decreases and the sum ($\theta_{Pb} + \theta_{Na}$) reveals a small change.

There are also many reports on the application of low angle static light scattering, particularly laser light scattering, in investigations of aqueous polyelectrolyte solutions. Light scattering experiments are common for determining the mass-averaged molecular mass of a polymer, but the technique has also been applied to the determination of osmotic virial coefficients in aqueous solutions.

Osmotic virial coefficients are commonly used to express the osmotic pressure π as a function of solute concentrations. For an aqueous solution of a single solute the osmotic virial equation is:

$$\frac{\pi}{RT} = \frac{\tilde{c}_s}{M_n} + \tilde{c}_s^2 A_{s,s} + \tilde{c}_s^3 A_{s,s,s} + \ldots, \qquad (10)$$

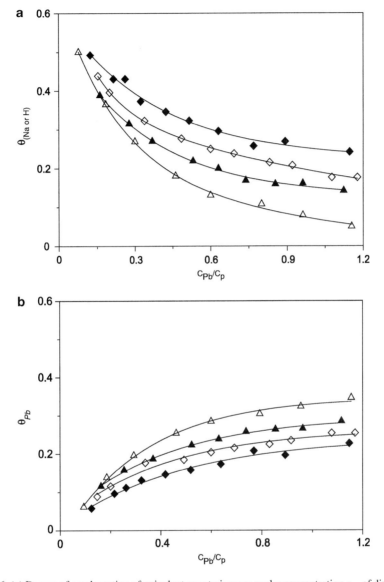

Fig. 6 (a) Degree of condensation of univalent counterions vs. molar concentration c_{Pb} of divalent Pb ions in NaPSS at 25°C, $M = 3.54 \times 10^5$, $c_{\text{sulfonated groups}} = 0.001$ mol/dm^3. *Closed diamonds* PSS/Pb/Na, $I = 0.02$ mol/dm^3; *open diamonds* PSS/Pb/H, $I = 0.02$ mol/dm^3; *closed triangles* PSS/Pb/Na, $I = 0.01$ mol/dm^3; *open triangles* PSS/Pb/H, $I = 0.01$ mol/dm^3 [46]. (**b**) Degree of condensation of divalent counterions vs. molar concentration c_{Pb} of divalent Pb-ions in NaPSS at 25°C, $M = 3.54 \times 10^5$, $c_{\text{sulfonated groups}} = 0.001$ mol/dm^3. *Closed diamonds* PSS/Pb/Na, $I = 0.02$ mol/dm^3; *open diamonds* PSS/Pb/H, $I = 0.02$ mol/dm^3; *closed triangles* PSS/Pb/Na, $I = 0.01$ mol/dm^3; *open triangles* PSS/Pb/H, $I = 0.01$ mol/dm^3 [46]

where $\tilde{c}_s$ is the mass density of solute s and $A_{s,s}$ and $A_{s,s,s}$ are the second and third osmotic virial coefficients, respectively, of solute s in the solvent. In nearly all light scattering experiments of aqueous solutions of a polyelectrolyte, the solvent was not only pure water but an aqueous solution of a salt, and the experimental results were interpreted using (10). Then, the evaluated data for the osmotic virial coefficients depend on the nature and the concentration of that salt. Table 6 gives a survey of literature sources for the second osmotic virial coefficients of a single polyelectrolyte in an aqueous solution from light scattering experiments. The second osmotic virial coefficient is determined by extrapolating experimental results to infinite dilution. Light scattering is particularly suited for the investigation of such highly diluted mixtures.

However, even when the polymer contains no ionic groups the extrapolation might be rather difficult (e.g., Hasse et al. [70], Kany et al. [71, 72]). Figure 7 shows a typical example of the influences of the molecular mass and the concentration of a salt on the second osmotic virial coefficient of a polyelectrolyte in water. The second osmotic virial coefficient increases considerably with decreasing salt concentration. The influence of the molecular mass is less distinct and often hidden by the scattering of the experimental data, particularly if that data is from different literature sources. In an aqueous solution of a strong electrolyte, the second osmotic virial coefficient of polyelectrolytes with different backbone monomers can vary by about one order of magnitude.

Table 6 Survey of literature data for the second osmotic virial coefficients of a single electrolyte in an aqueous solution from light scattering investigations

Polymer	Polymer concentration $\tilde{c}_p$ (g/dm^3)	Salt	Salt concentration c_s (mol/dm^3)	References
NaPSS	0.5–3	NaCl	0.005–4.2	[57]
Na/HPSS	< 0.8	Na/H/NO$_3$	0.005–3.7	[47]
NaPSS; Pb/ HPSS	0.4–3	NaNO$_3$, HNO$_3$, Pb(NO$_3$)$_2$	0.005–2	[58]
KPSS		KCl	0.1	[59]
NaPA	0.1–2	NaCl	0.01–1	[59]
HPAA	0.1–3	NaCl	0.01–1	[60, 61]
		NaBr	1.5	
		CaCl	0.1	
NaCMC	0.1–4	NaCl	0.001–0.5	[62]
NaCMC	0.2–0.8	NaCl	0.005–0.5	[63]
NaPAMS	–	NaCl	0.01–5	[64]
NaPAMA	–	NaCl	1	[65]
PDADMAC	–	NaCl	0.5	[66]
PDADMAC	–	NaCl	1	[67]
PAAm	1.87	NaCl	0.05–3	[68]
PTMAC	–	NaCl	1	[65]
PTMAC	–	NaCl	0.1–4	[69]

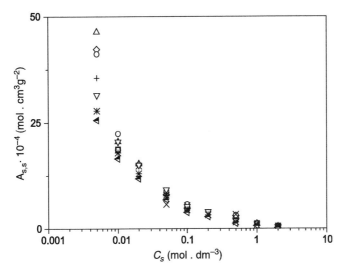

Fig. 7 Second osmotic virial coefficient of sodium poly(styrene sulfonate) of varying molecular mass (*M*) vs. the concentration of added NaCl: *squares* (23.4 × 10^5); *diamonds* (22.8 × 10^5); *triangles* (15.5 × 10^5); *circles* (10 × 10^5); *crosses* (3.9 × 10^5); *plus* (3.2 × 10^5) data from Takahashi et al., 25°C [57]; *half-closed triangles* (12.2 × 10^5); *stars* (7.3 × 10^5); *inverted triangles* (3.2 × 10^5) data from Nordmeier, 20°C [58]

4 Gibbs Energy of Aqueous Solutions of Polyelectrolytes

For several reasons, it is rather difficult to develop a reliable method for describing (i.e., correlating and predicting) the thermodynamic properties of aqueous solutions of polyelectrolytes. The thermodynamics of polymer solutions in nonaqueous systems as well as of aqueous electrolyte solutions are still major areas of research and, consequently, the situation is less satisfactory for aqueous solutions of polyelectrolytes, for which the dissociation reactions have to be taken into account. This section reviews the most important features of some methods of modeling the Gibbs energy of aqueous polyelectrolyte solutions. The Gibbs energy of an aqueous solution is the sum of contributions from all (solute plus solvent) species *i*:

$$G = \sum_i n_i \mu_i, \quad (11)$$

where n_i and μ_i are the number of moles and the chemical potential of component *i* (i.e., of the solvent and the solutes), respectively. It is common to split the Gibbs energy into two parts, a contribution from ideal mixing and an excess contribution:

$$G = G^{\text{id.mix.}} + G^E. \quad (12)$$

As the Gibbs energy is the sum of contributions from all components, $G^{\text{id.mix.}}$ and G^E are also sums of contributions by all components:

$$G^{\text{id.mix.}} = \sum_i n_i \mu_i^{\text{id.mix.}}, \tag{13}$$

$$G^E = \sum_i n_i \mu_i^E \tag{14}$$

or

$$G = \sum_i n_i (\mu_i^{\text{ref}} + RT \ln a_i). \tag{15}$$

Therefore, the following relation holds:

$$\mu_i = \mu_i^{\text{id.mix.}} + \mu_i^E = \mu_i^{\text{ref}} + RT \ln a_i. \tag{16}$$

By definition, component i experiences in an ideal mixture the same intermolecular forces as in the reference state and therefore all differences between $\mu_i^{\text{id.mix.}}$ and μ_i^{ref} are caused by differences in the concentration (i.e., dilution) only:

$$\mu_i^{\text{id.mix.}} = \mu_i^{\text{ref}} + RT \ln a_i^{\text{id.mix.}}. \tag{17}$$

Consequently, the activity of component i in an ideal mixture, $a_i^{\text{id.mix.}}$, is known from the composition of the real solution. However, the actual expression for $a_i^{\text{id.mix.}}$ depends on the choice of reference states and the concentration scale applied. The reference state for the solvent (in this case water) is usually the pure liquid at the temperature and pressure of the mixture:

$$\mu_s^{\text{ref}} = \mu_{s,\,\text{pure liquid}}(T, p). \tag{18}$$

However, various reference states are used for a dissolved component. One common reference state is a hypothetical solution of that component in water at a concentration of 1 mol/kg water (i.e., a one molal solution) where the solute experiences interactions only with water, i.e., as if infinitely diluted in water. With that reference state, it is also common practice to replace the activity of a solute species i by the product of molality m_i and activity coefficient $\gamma_i^{(m)}$:

$$a_i^{(m)} = m_i \gamma_i^{(m)}, \tag{19}$$

where superscript (m) indicates both the reference state and the concentration scale. The activity coefficient of a solute species i becomes unity in an ideal solution and,

Aqueous Solutions of Polyelectrolytes

consequently (as follows from the Gibbs–Duhem equation), the activity of the solvent (represented by subscript s) in an ideal mixture is:

$$\ln a_{s,\text{id.mix.}}^{(m)} = -\frac{1000}{M_s} \sum_{i \neq s} m_i, \qquad (20)$$

where i represents any (but only) solute species.

Another common reference state for a solute is a hypothetical solution of one mole of that solute in one liter of water (i.e., a one molar solution) where the solute experiences interactions only with water, i.e., as if infinitely diluted in water. With that reference state, it is also common practice to replace the activity of a solute species i by the product of molarity c_i and activity coefficient $\gamma_i^{(c)}$:

$$a_i^{(c)} = c_i \gamma_i^{(c)}, \qquad (21)$$

where superscript (c) indicates both the reference state and the concentration scale.

The activity coefficient of a solute species i becomes unity in an ideal solution and, consequently (following again from the Gibbs–Duhem equation), the activity of the solvent s is:

$$\ln a_{s,\text{id.mix.}}^{(c)} = -\frac{1}{\bar{\rho}_s^*} \sum_{i \neq s} c_i, \qquad (22)$$

where $\bar{\rho}_s^*$ is the molar density of water (in moles per liter).

As usual, the following relations also hold for the excess part of the chemical potential of a solute i and a solvent s:

$$\mu_i^E = RT \ln \left(\frac{a_i}{a_{i,\text{id.mix.}}} \right)_i, \qquad (23)$$

$$\mu_s^E = RT \ln \left(\frac{a_s}{a_{s,\text{id.mix.}}} \right). \qquad (24)$$

One has to keep in mind that the excess parts of the chemical potentials depend on the selection of the reference state for a solute component, as both the activity of a solute component and the activity of the solvent in an ideal mixture depend on the reference states of the solutes. The activity coefficients of a solute on molality scale, $\gamma_i^{(m)}$, and on molarity scale, $\gamma_i^{(c)}$, are related by:

$$\gamma_i^{(c)} = \gamma_i^{(m)} \frac{m_i}{c_i} \rho_s^*, \qquad (25)$$

where ρ_s^* is the specific density of the pure solvent in kg/dm^3.

Most methods assume that there are several contributions to the excess Gibbs energy:

$$G^E = \sum_k G_k^E,$$ (26)

where k represents such a contribution. More details on such contributions are given below.

For describing the thermodynamic properties of aqueous electrolyte solutions one often uses the osmotic pressure π:

$$\pi = -\frac{RT}{\bar{v}_{s,\text{pure}}} \ln a_s,$$ (27)

where $\bar{v}_{s,\text{pure}}$ is the molar volume of the pure solvent and the osmotic coefficient Φ_p:

$$\Phi_p = \frac{\pi}{\pi_{\text{id.mix.}}} = \frac{\ln a_s}{\ln a_{s,\text{id.mix.}}}$$ (28)

The numerical value of the osmotic coefficient depends on the selection of the reference state of the solutes, whereas the number for the osmotic pressure does not depend on that reference state.

5 Thermodynamic Models

The fundamentals of the thermodynamic modeling of aqueous solutions of polyelectrolytes were established by Lifson and Katchalsky [73, 74]. Their model was extended by various authors. For example, Dolar and Peterlin [75] extended it to polyelectrolytes with two different counterions. One of the most important extensions was presented by Manning in a series of papers. The new fundamental idea introduced by Manning is the so-called counterion condensation concept. That theory was further extended by Manning [76–78] and others. Manning's theory of counterion condensation was adopted in more recent work, where his results were applied in a more or less straightforward manner. Manning's concept has been supported by molecular dynamic simulations of polyelectrolyte solutions, showing the changes in the polymer backbone configuration and the counter ion condensation, e.g., by Stevens and Kremer [79].There are other examples for the solution of the Poisson–Boltzmann equation, which use other hypotheses about the boundary conditions for which the equation is solved. The examples are more or less related to the work cited before. The model by Feng et al. [80], who considered the presence of salts in the aqueous solution, is an interesting example. There are other interesting examples of extensions, such as those presented by Ospeck and

Fraden [81], who solved the Poisson–Boltzmann equation for a system of two cylinders confined between two plates, and by Dahnert and Huster [82, 83], who solved the Poisson–Boltzmann equation for a plate-like polyelectrolyte immersed in a salt solution. Rödenbeck et al. [84] solved the same equation using the approximation of elementary cells around a symmetrically charged central body. The use of cavity-correlation functions was investigated by Jiang et al. [85, 86]. However, using the Poisson–Boltzmann equation for such systems has also attracted some criticism. For example, Blaul et al. [87] compared results derived using the Poisson–Boltzmann equation with experimentally determined osmotic pressure data, and concluded that the difference between the predicted and the experimental behavior is due to some deficiencies of the model, for example, an insufficient treatment of ion–ion correlations. Deserno et al. [88] found that the cell model systematically overestimates the osmotic coefficient. Colby et al. [89] showed that, in the semidilute range of concentrations, the hypotheses used to solve the equations are no longer valid. Diehl et al. [90] mentioned that short-range interactions between the polymer backbones might not be negligible. Many other investigations, for example, by Monte Carlo simulations (Chang and Yethiraj [91]), by molecular dynamic simulations (Antypov and Holm [92]) and by field-theoretical methodologies (Baeurle et al. [93]) were conducted to achieve a better understanding of the behavior of polyelectrolyte solutions. Such investigations are important from a more theoretical point of view. However, it is very difficult either to apply them directly or to use their results in a more indirect way for engineering calculations. That statement particularly holds for aqueous solutions containing a polyelectrolyte and other compounds such as salts and/or neutral polymers. The difficulties are related to computational issues (which may still be an impediment), as well as to the absence of sufficient information. Therefore, despite the large amount of theoretical work, there is still a great need for simplified models that can be applied to the description of phase equilibrium in polyelectrolyte aqueous solutions at medium and high polyelectrolyte concentrations. A similar statement holds for the so-called scaling-law approach (cf. [94–100]).

This contribution is therefore restricted to the models introduced by Lifson and Katchalsky as well as by Manning on one side, and to the extensions and modifications of these models by Danner et al. [101, 102] and by members of our own research group as they seem to have the most potential for applications in chemical engineering.

5.1 Cell Model of Lifson and Katchalsky

The model of Lifson and Katchalsky [74] is an extension of the Debye–Hückel theory of highly diluted aqueous solutions of strong (low molecular weight) electrolytes to polyelectrolyte solutions. Lifson and Katchalsky start from the idea that an aqueous solution of a polyelectrolyte reveals a microscopic structure. That structure is caused by two competing effects: the electric charges on the backbone

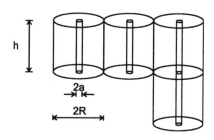

Fig. 8 Cell model of Lifson and Katchalsky [74] showing radii a and R

monomers (that tend to stretch the polyelectrolyte) and the tendency of the polyelectrolyte to increase its entropy by forming globular or entangled coils (at low or high polymer concentrations, respectively). As shown in Fig. 8, the polyelectrolyte backbone is modeled as a stretched cylinder of radius a and length h. That cylinder is surrounded by another cylindrical cell (radius R and length h). The electrical charge on the backbone is approximated by a uniform charge on the surface of the inner cylinder. The counterions are dissolved in the cylindrical space between radii a and R, where they form an ionic cloud. The radius R depends on the concentration of the polyelectrolyte. It is low in highly concentrated solutions and increases with decreasing concentration to reach infinity in an infinitely diluted solution. The electrostatics in that cloud are described by the Poisson–Boltzmann equation. In a manner analogous to the Debye–Hückel theory, the electrostatic potential caused by the interactions between the stretched backbone on one side and the surrounding counterions on the other side is calculated by solving the Poisson–Boltzmann differential equation. The electrostatic potential $\varphi(r)$ in the cylindrical space between the radii a and R ($a \leq r \leq R$) is:

$$\varphi(r) = \frac{kT}{e} \ln\left\{ \frac{2\lambda}{\beta^2} \frac{r^2}{(R^2 - a^2)} \sinh^2[\beta \ln(Ar)] \right\}, \tag{29}$$

where λ is a (dimensionless) charge density parameter that describes the charge density on the polyelectrolyte's backbone. When the repeating unit is a 1:1 electrolyte, that parameter becomes:

$$\lambda = \frac{l_B}{b}, \tag{30}$$

where l_B is the Bjerrum length:

$$l_B = \frac{e^2}{4\pi\varepsilon\varepsilon_0 kT}, \tag{31}$$

which characterizes the solvent through its relative dielectric constant ε. Parameters b, e, and ε_0 are the length of that repeating unit, the elementary charge, and the permittivity of vacuum, respectively, The two other parameters A (which is an

Aqueous Solutions of Polyelectrolytes

inverse length) and β (which is dimensionless) are determined from the condition of electroneutrality in the cylindrical cell of radius R and from the condition that at $r = R$, the electrostatic potential $\varphi(r)$ has to reach an extreme (for symmetry reasons). As long as the charge density parameter λ is "small" parameter β is a real number (between zero and one), whereas it is an imaginary number (between zero and 1.0i) for "large" charge densities. The distinction between "small" and "large" depends on the polyelectrolyte concentration. When β becomes imaginary, β has to be replaced by $|\beta|$ in (29). Consequently there are two different regions where the remaining parameters (A and β) have to be determined:

When β is real:

$$\lambda = \frac{1 - \beta^2}{1 + \beta \coth(\beta \gamma_{LK})}, \tag{32}$$

$$1 + \beta \coth[\beta \ln(AR)] = 0. \tag{33}$$

When β is imaginary:

$$\lambda = \frac{1 + |\beta|^2}{1 + |\beta| \cot(|\beta|\gamma_{LK})}, \tag{34}$$

$$\beta \ln A + |\beta| \ln R + \arctan|\beta| = 0, \tag{35}$$

where γ_{LK} is another dimensionless parameter:

$$\gamma_{LK} = \ln \frac{R}{a} \tag{36}$$

that is related to the volume fraction ϕ_p of the polyelectrolyte in the aqueous solution:

$$\phi_p = \ln \left(\frac{a}{R}\right)^2. \tag{37}$$

Unfortunately, there is no analytical solution to determine A and β. But at infinite dilution (i.e., when $\gamma_{LK} \to \infty$) one finds from (32) and (33):

$$\beta = 1.0 - \lambda \quad \text{for} \quad \lambda \leq 1 \quad \text{and} \quad \beta = 0.0 \quad \text{for} \quad \lambda > 1. \tag{38}$$

Figure 9 shows the results for $\beta(\lambda, \gamma_{LK})$ as calculated from (33) to (35).

Lifson and Katchalsky [74] determined the influence of the electrostatic potential $\varphi(r)$ on the thermodynamic properties of an aqueous solution of a single polyelectrolyte through an expression for the change of the Helmholtz energy ΔF_{LK} that is due to the presence of the electrostatic potential by:

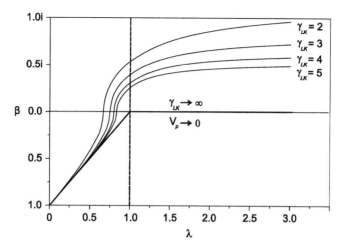

Fig. 9 Theory of Lifson and Katchalsky: Integration constant β as a function of the charge density parameter λ for several parameters γ_{LK}

$$\Delta F_{LK} = c_p V R T f^{el} \tag{39}$$

with:

$$f^{el} = \frac{v}{\lambda}\left[\lambda \ln \frac{\left((\exp \gamma_{LK})^2 - 1\right)\left((1-\lambda)^2 - \beta^2\right)}{2\lambda} - \lambda - \left(1 + \beta^2\right)\gamma_{LK} - \ln \frac{(1-\lambda)^2 - \beta^2}{1-\beta^2}\right], \tag{40}$$

where c_p is the molarity of the polyelectrolyte in the solution, and V is the volume of the solution (i.e., $c_p V$ is the number of moles of polyelectrolyte in the solution) and v is the number of repeating units of the polyelectrolyte.

The osmotic pressure π is split into two contributions:

$$\pi = \pi_{id.mix.} + \Delta \pi^{el}, \tag{41}$$

where the osmotic pressure of the ideal mixture $\pi_{id.mix.}$ is calculated assuming that the polyelectrolyte is completely dissociated. When the repeating unit is a 1:1 electrolyte that contribution is:

$$\pi_{id.mix.} = (1+v)c_p RT. \tag{42}$$

The second contribution $\Delta \pi^{el}$ is caused by the electrostatic potential. It is calculated from the contribution of the electrostatic forces to the Helmholtz energy:

$$\Delta \pi^{el} = -\left(\frac{\partial F^{el}}{\partial V}\right)_{T,composition}. \tag{43}$$

The final result for the osmotic coefficient $\Phi_p^{(c)}$ (on molarity scale) is:

$$\Phi_p^{(c)} = \frac{1-\beta^2}{2\lambda} \quad \text{for } \lambda \leq 1 \tag{44}$$

and:

$$\Phi_p^{(c)} = \frac{1+\beta^2}{2\lambda} \quad \text{for } \lambda > 1. \tag{45}$$

Figure 10 shows the osmotic coefficient of an aqueous solution of a single polyelectrolyte as a function of the molarity of the repeating units $c_m = v c_p$ and the charge density parameter λ. In a highly diluted aqueous solution (i.e., when $\gamma_{LK} \to \infty$), the final result is:

$$\Phi_p^{(c)} = 1 - \frac{\lambda}{2} \quad \text{for } \lambda \leq 1, \tag{46}$$

$$\Phi_p^{(c)} = \frac{1}{2\lambda} \quad \text{for } \lambda > 1. \tag{47}$$

For the calculation of the osmotic pressure of an aqueous solution of a single polyelectrolyte where the repeating unit is a 1:1 electrolyte one needs:

– For the polyelectrolyte: the radius of the hard polymer rod a, the length of the polymer rod h (or the length of a cylindrical monomer b and the number of such monomers in a polyelectrolyte molecule v)

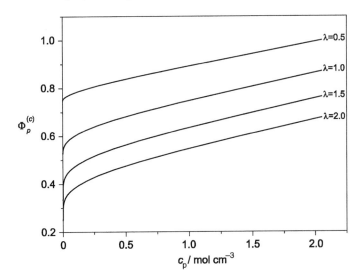

Fig. 10 Theory of Lifson and Katchalsky: Osmotic coefficient of an aqueous solution of a single polyelectrolyte as a function of the charge density parameter λ

– The concentration of the polyelectrolyte in the aqueous phase (or the radius R of a cylindrical cell surrounding each polyelectrolyte molecule)
– The relative dielectric constant of water ε

The equations only hold for an aqueous solution of a single polyelectrolyte that consists of monovalent repeating units (1:1 electrolytes). There are publications on extensions. For example, Dolar and coworkers treated polyelectrolytes with two counterions (Dolar and Peterlin, [75], Dolar and Kozak, [103]), and Katchalsky [104] extended the theory to aqueous solutions of a polyelectrolyte and a low molecular weight salt. Katchalsky just superimposed the contributions from the polyelectrolyte with those from the added salt. The osmotic pressure of an aqueous solution of a single polyelectrolyte then becomes:

$$\pi = \pi_p + \Delta\pi^{\text{el}} + \Delta\pi_s. \tag{48}$$

The first term on the right-hand side, π_p, is the contribution for an ideal aqueous solution (on molarity scale) of the undissociated polyelectrolyte:

$$\pi_p = c_p RT. \tag{49}$$

The second term, $\Delta\pi^{\text{el}}$, results from the dissociation of the polyelectrolyte. It is expressed by combining (42), (45) and (47). When β is real:

$$\Delta\pi^{\text{el}} = \frac{1 - \beta^2}{2\lambda} v c_p RT \tag{50}$$

and when β is imaginary:

$$\Delta\pi^{\text{el}} = \frac{1 + \beta^2}{2\lambda} v c_p RT. \tag{51}$$

The third term, $\Delta\pi_s$, is approximated by the osmotic pressure of an aqueous solution of the single, low molecular weight strong electrolyte S, that consists of v_M cations M and v_X anions X:

$$\Delta\pi_s = (v_M + v_X) c_s RT \Phi_S^{(c)}, \tag{52}$$

where c_s and $\Phi_S^{(c)}$ are the molarity of the strong electrolyte S and the osmotic coefficient (on molarity scale) of an aqueous solution of the single strong electrolyte S. Then, the osmotic coefficient of an aqueous solution of a polyelectrolyte P (of monovalent repeating units) and a low molecular weight strong electrolyte S becomes:

$$\Phi_{p+s}^{(c)} = \frac{\pi_p + \Delta\pi^{\text{el}} + \Delta\pi_s}{\pi_{\text{id.mixture}}} = \frac{c_p + \Phi_p^{(c)} v c_p + (v_M + v_X) c_s \Phi_S^{(c)}}{c_p + v c_p + (v_M + v_X) c_s}. \tag{53}$$

Aqueous Solutions of Polyelectrolytes

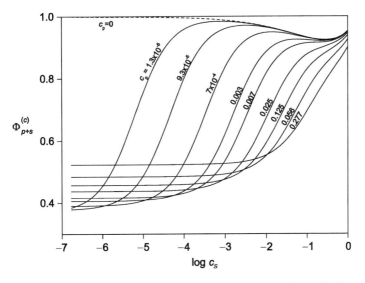

Fig. 11 Osmotic coefficient of an aqueous solution of a polyelectrolyte (charge density parameter $\lambda = 1.5$) and NaCl (salt molarity c_s) at 25°C for several values for the molarity c_p of the polyelectrolyte (the data for the polyelectrolyte-free solution are taken from [105])

As commonly $v \gg 1$ that equation can be simplified to:

$$\Phi_{p+s}^{(c)} = \frac{\Phi_p^{(c)} v c_p + (v_M + v_X) c_s \Phi_s^{(c)}}{v c_p + (v_M + v_X) c_s}. \quad (54)$$

Figure 11 shows a typical example for the osmotic coefficient of an aqueous solution of a polyelectrolyte and NaCl calculated with (54).

An extension to multisolute aqueous solutions with a polyelectrolyte, nonelectrolyte solutes, and low molecular weight salts might start from (48) using (49) together with (50) and (51) for the contributions of the polyelectrolyte, but replacing (52) by the osmotic pressure of an aqueous solution of the polyelectrolyte-free solutions, i.e., an aqueous solution of the low molecular weight salts and the other nonelectrolyte solutes. However, such an extension always suffers from neglecting the interactions between the other solutes and the polyelectrolyte.

5.2 Counterion Condensation Theory of Manning

Manning [76–78] modified and extended the Lifson–Katchalsky model to include the effects caused by the presence of strong, low molecular weight electrolytes in the aqueous polyelectrolyte solution. The polyelectrolyte is described as a linear chain of N monomers that contain an electrolyte group. Dissociation results in ionic groups on the polymer backbone, resulting in a charged polymeric backbone and counterions. The electric charge is considered to be evenly distributed over the whole backbone and the dissociated counterions are considered as point charges in the solvent of relative dielectric constant ε. Similarly to Lifson and Katchalsky, Manning also assumes that the number of repeating units in the polyelectrolyte chain is very large and, therefore, chain-end effects are neglected. The excess Gibbs energy results from interactions between the charged chain and all other ions (counterions as well as ions from some added low molecular weight strong electrolytes) in the surrounding solution. The properties of that surrounding electrolyte solution are approximated by the Debye–Hückel theory. Manning assumes that some counterions might form ion pairs with some monomers of the backbone. These ion pairs are not really fixed to the backbone but can move in a certain volume around the backbone, i.e., these "condensed counterions" have an additional mobility that increases the entropy of the system. Manning neglects all interactions between all backbone groups of the polyelectrolyte molecule.

In the following description, the molarity scale is used. The reference states are:

- For the solvent: the pure liquid
- For the polyion: the completely dissociated polyion that experiences no electrostatic interactions, but otherwise behaves like at infinite dilution in water
- For an added salt: the completely dissociated salt that also experiences interactions as if at infinite dilution in water.

As an example, we discuss here an aqueous solution of one polyelectrolyte P and one strong electrolyte S ($=M_{v_M}X_{v_X}$), where P and S share a common counterion X. Some of the counterions that originate from the polyelectrolyte are assumed to be located in a small volume V_p around the polyelectrolyte backbone (the phenomenon is called "counterion condensation"). The polyelectrolyte, "condensed" counterions, "free" counterions, free coions, and water contribute to the Gibbs energy of the solution:

$$G = G_p + G_{\text{cond.CI}} + G_{\text{freeCI}} + G_{\text{freeCOI}} + G_{\text{w}}. \tag{55}$$

Each contribution consists of a contribution from the reference state and a contribution from mixing:

$$G_i = n_i(\mu_i^{\text{ref}} + RT \ln a_i). \tag{56}$$

The reference state for a solute is always based on the molarity scale (at unit molarity in water but with interactions as if infinitely diluted in water) whereas for water the reference state is pure liquid water. The contributions are described below.

Aqueous Solutions of Polyelectrolytes

5.2.1 Contribution from the Polymer

Manning assumed that there are only contributions from electrostatic interactions. He approximated these contributions from the cylindrical version of the Debye–Hückel theory:

$$G_p = n_p \mu_p^{\text{ref}} + G_p^{E,\text{el}} \tag{57}$$

with

$$G_p^{E,\text{el}}/(RT) = -n_p v (1 - z_{\text{Cl}}\theta_z)^2 \lambda \ln[1 - \exp(-\kappa b)]. \tag{58}$$

n_p and v are the number of moles of the polyion and the number of dissociable repeating units in that polyion, respectively (i.e., $n_p v$ is the total mole number of dissociable electrolytic groups in the backbone). z_{Cl} is the (absolute) valency of the counterions of the polyelectrolyte. $z_{\text{Cl}}\theta_z$ is the ratio of charges carried by those counterions that are "condensed" to the backbone to the maximum number of charges on that backbone. That ratio is also called the "neutralization fraction of the polyion". Thus $(1 - z_{\text{Cl}}\theta_z)$ is the ratio of the actual number of charges q to the maximum number q_{max} of charges on the backbone of the polyelectrolyte.

$$\frac{q}{q_{\text{max}}} = (1 - z_{\text{Cl}}\theta_z). \tag{59}$$

Consequently, the number of moles of dissociated repeating units, $n_{p,\text{diss}}$ is:

$$n_{p,\text{diss}} = n_p v (1 - z_{\text{Cl}}\theta_z). \tag{60}$$

λ is the charge density parameter [cf. (30)]. When the charge density is small (i.e., the distance b between two dissociable groups is large so that $\lambda < 1$) the polyelectrolyte is completely dissociated. Thus, the first part on the right-hand side of (58) [i.e., $n_p v (1 - z_{\text{Cl}}\theta_z)^2 \lambda$] is the number of moles of dissociated polymer groups times the charge density parameter. Parameter κ is the inverse of the radius of the ionic cloud in the aqueous solution, as introduced in the Debye–Hückel theory:

$$\kappa^2 = 2N_A \frac{e^2}{\varepsilon\varepsilon_0 kT} I_S = 8\pi N_A I_S l_B, \tag{61}$$

where I_S is the ionic strength of the aqueous solution on the molarity scale.

When a single polyelectrolyte and a single low molecular weight salt $M_{v_M} X_{v_X}$ are dissolved in water, that ionic strength is:

$$I_S = \frac{1}{2}\left[\left(v_M z_M^2 + v_X z_X^2\right)c_s + z_{\text{Cl}} z_p (1 - z_{\text{Cl}}\theta_z)v c_p\right], \tag{62}$$

where z_M, z_X, and z_p are the charge numbers of cations M, anions X and the monomer electrolyte group, respectively and c_s and c_p are the molarity of the low molecular weight salt and of the polyelectrolyte, respectively. Thus, the last term on the right-hand side of (58) describes the influence of the ionic cloud of the solution on the Gibbs excess energy of the polyelectrolyte.

5.2.2 Contribution from Condensed Counterions

Those counterions that do not dissociate from the polyion are treated as a further solute:

$$G_{\text{cond.CI}} = n_{\text{cond.CI}}\mu_{\text{cond.CI}}^{\text{ref}} + \Delta G_{\text{cond.CI}}. \tag{63}$$

$\Delta G_{\text{cond.CI}}$ results from a transfer of the condensed counterions from the real solution (i.e., at molarity c_{CI}) to a volume V_p near the polyelectrolyte. In that volume, the concentration of the counterions differs from the concentration in the surrounding aqueous solution as that volume contains all condensed counterions, i.e., $n_p\theta_z z_p$ counterions. The molarity of the counterions in that volume is the "local" molarity $c_{\text{CI,local}}$:

$$c_{\text{CI,local}} = \frac{n_{\text{cond.CI}}}{V_p} = \frac{n_p v\theta_z z_p}{V_p}. \tag{64}$$

V_p is the (unknown) volume of the *condensate*. The change of the Gibbs energy encountered in that transfer is approximated by the corresponding change of the entropy:

$$\Delta G_{\text{cond.CI}}/(RT) = n_{\text{cond.CI}} \ln\left(\frac{c_{\text{CI,local}}}{c_{\text{CI}}}\right) \tag{65}$$

resulting in:

$$\Delta G_{\text{cond.CI}}/(RT) = n_p v z_p \theta_z \ln\left(\frac{\theta_z z_p}{v_p^* c_{\text{CI}}}\right), \tag{66}$$

where v_p^* is an unknown molar volume:

$$v_p^* = \frac{V_p}{n_p v}. \tag{67}$$

That molar volume v_p^* is estimated by Manning in the following way. As a decrease in the degree of dissociation (i.e., an increase of θ_z) results in an increase

Aqueous Solutions of Polyelectrolytes

of $G_p^{E,\text{el}}$ and a decrease of $\Delta G_{\text{cond.CI}}$, Manning assumed that the condensation process reaches an equilibrium when the sum $(G_p^{E,\text{el}} + \Delta G_{\text{cond.CI}})$ reaches a minimum:

$$\left(\frac{\partial \left(G_p^{E,\text{el}} + \Delta G_{\text{cond.CI}}\right)}{\partial \theta_z}\right) = 0. \tag{68}$$

This equation is evaluated for the aqueous solution of the single polyion, also assuming that the influence of θ_z on the ionic strength can be neglected. The molar volume v_p^* is calculated from:

$$1 + \ln\left(\frac{\theta_z z_p}{v_p^* c_{\text{CI}}}\right) + 2\frac{z_{\text{CI}}}{z_p}\lambda(1 - z_{\text{CI}}\theta_z)\ln[1 - \exp(-\kappa b)] = 0. \tag{69}$$

For low concentrations of the polyelectrolyte (i.e., when $c_{\text{CI}} \to 0$) that equation can only be fulfilled when:

$$\theta_z^{(0)} = \lim_{c_{\text{CI}} \to 0} \theta_z = \frac{1}{z_{\text{CI}}}\left(1 - \frac{z_p}{z_{\text{CI}}\lambda}\right). \tag{70}$$

As $\theta_z^{(0)} z_{\text{CI}}$ is the ratio of the number of actual charges on the polyelectrolyte backbone to the maximum number of charges on that backbone, i.e., $0 \leq (\theta_z^{(0)} z_{\text{CI}}) \leq 1$, then $\theta_z^{(0)} z_{\text{CI}}$ is positive as long as $z_p/(z_{\text{CI}}\lambda) < 1$. The polyelectrolyte is completely dissociated when $z_p/(z_{\text{CI}}\lambda) = 1$, i.e., $\theta_z^{(0)} z_{\text{CI}} = 0$. $\theta_z^{(0)}$ cannot be negative even if $z_p/(z_{\text{CI}}\lambda) > 1$. Therefore, two cases have to be distinguished:

Case A:

$$z_p/(z_{\text{CI}}\lambda) \leq 1, \quad \theta_z^{(0)} = \frac{1}{z_{\text{CI}}}\left(1 - \frac{z_p}{z_{\text{CI}}\lambda}\right). \tag{71}$$

Case B:

$$z_p/(z_{\text{CI}}\lambda) > 1, \quad \theta_z^{(0)} = 0. \tag{72}$$

v_p^* is calculated from (69) by replacing θ_z by $\theta_z^{(0)}$, which is taken from (71). As for the case when $\theta_z^{(0)} = 0$, no counterion condensation occurs and therefore there is no contribution from condensed counterions to the Gibbs energy.

The result for the molar volume v_p^* is:

$$v_p^* = 4\pi N_A z_p^2 b^3 \left(\frac{z_{\text{CI}}\lambda}{z_p} - 1\right)\exp(1). \tag{73}$$

5.2.3 Contribution from Free Counterions

The aqueous phase contains "free" (or "dissolved") counterions. These ions are either dissociated from the polyelectrolyte or result from the dissolution of the salt S. Their contribution to the Gibbs energy of the solution is:

$$G_{\text{freeCI}} = n_{\text{freeCI}} \mu_{\text{freeCI}}^{\text{ref}} + \Delta G_{\text{freeCI}}, \tag{74}$$

$$\Delta G_{\text{freeCI}}/(RT) = \left[n_{\text{freeCI}}^{(p)} + n_{\text{freeCI}}^{(s)} \right] \ln\left\{ c_{\text{freeCI}} \gamma_{\text{CI}}^{(c)} \right\}. \tag{75}$$

The mole number $n_{\text{freeCI}}^{(p)}$ of the counterions that originate from the polyelectrolyte is:

$$n_{\text{freeCI}}^{(p)} = \frac{n_p \nu (1 - z_{\text{CI}} \theta_z)}{z_{\text{CI}}} z_p \tag{76}$$

and the mole number of the same counterionic species from the added salt is:

$$n_{\text{freeCI}}^{(s)} = n_s \nu_{\text{CI}}, \tag{77}$$

where n_s and ν_{CI} (either ν_M or ν_X) are the mole number of the dissolved salt S and the stochiometric coefficient of the counterion of S, respectively. The molarity c_{freeCI} of the counterions in the aqueous solution surrounding the polyelectrolyte is:

$$c_{\text{freeCI}} = \left(n_{\text{freeCI}}^{(p)} + n_{\text{freeCI}}^{(s)} \right)/V \tag{78}$$

or:

$$c_{\text{freeCI}} = \frac{z_p \nu (1 - z_{\text{CI}} \theta_z)}{z_{\text{CI}}} c_p + \nu_{\text{CI}} c_s. \tag{79}$$

Therefore:

$$\Delta G_{\text{freeCI}}/(RT) = \left[\nu_{\text{CI}} n_s + \frac{z_p}{z_{\text{CI}}} (1 - z_{\text{CI}} \theta_z) \nu n_p \right]$$
$$\times \ln\left\{ \gamma_{\text{CI}}^{(c)} \left[\nu_{\text{CI}} c_s + \frac{z_p \nu}{z_{\text{CI}}} (1 - z_{\text{CI}} \theta_z) c_p \right] \right\}. \tag{80}$$

$\gamma_{\text{CI}}^{(c)}$ is the activity coefficient of the counterions in the aqueous solution of ionic strength I_s (on molarity scale) [cf. (62)]. That activity coefficient might be set to unity or be approximated by the Debye–Hückel theory.

Aqueous Solutions of Polyelectrolytes

5.2.4 Contribution from Coions

When a neutral salt S is dissolved in the aqueous polyelectrolyte solution there is also a contribution to the Gibbs energy of the aqueous solution by the other ions, here called coions. When ν_{COI} is the stoichiometric coefficient of that coion in S, following the same ideas as explained before for the free counterions, that contribution is:

$$G_{freeCOI} = n_{freeCOI}\mu^{ref}_{freeCOI} + \Delta G_{freeCOI}, \qquad (81)$$

where:

$$\Delta G_{freeCOI}/(RT) = \nu_{COI}n_s \ln\left(\nu_{COI}c_s\gamma^{(c)}_{COI}\right). \qquad (82)$$

The activity coefficient $\gamma^{(c)}_{COI}$ of the coions (on molarity scale) is treated in the same way as the activity coefficient $\gamma^{(c)}_{CI}$ of the counterions (i.e., it is either set to unity or expressed through the Debye–Hückel expression).

5.2.5 Contribution from Water

The final contribution to the Gibbs energy results from the presence of water (subscript w):

$$G_w = n_w\mu^{ref}_w + n_wRT \ln a_w. \qquad (83)$$

The activity of water is approximated by using the osmotic coefficient $\Phi^{(c)}_p$ on the molarity scale:

$$\Phi^{(c)}_p = \frac{\ln a_w}{\ln a^{(c)}_{w,id.mix.}} = -\frac{\ln a_w}{(c_{CI} + c_{COI})/\bar{\rho}^*_w}. \qquad (84)$$

where $\bar{\rho}^*_w$ is the molar density of water in the aqueous solution in moles per liter:

$$\ln a_w = -\Phi^{(c)}_p\left[(\nu_{CI} + \nu_{COI})n_s + \frac{z_p}{z_{CI}}(1 - z_{CI}\theta_z)\nu n_p\right]. \qquad (85)$$

The osmotic coefficient $\Phi^{(c)}_p$ is again either set to unity (that is the common approach) or taken from the Debye–Hückel theory for an aqueous solution containing n_s moles of salt S and $n^{(p)}_{freeCI}$ moles of counterions dissociated from the polyelectrolyte.

For highly diluted solutions, the results of Manning's theory agree with the results of Lifson and Katchalsky [cf. (46) and (47)]. For example, Manning [106] gives for the osmotic coefficient of an aqueous solution of a single polyion where the counterions have the (absolute) charge number z_{CI}:

$$\lim_{c_p \to 0} \Phi_p^{(c)} = 1 - \frac{\lambda}{2} z_p z_{CI} \quad \text{for } \lambda z_p z_{CI} \le 1 \tag{86}$$

and:

$$\lim_{c_p \to 0} \Phi_p^{(c)} = \frac{1}{2\lambda z_p z_{CI}} \quad \text{for } \lambda z_p z_{CI} > 1. \tag{87}$$

Manning [106] gives for the limiting activity coefficient of the counterions in such an aqueous solution:

$$\lim_{c_p \to 0} \ln \gamma_{CI}^{(c)} = -\frac{\lambda}{2} z_p z_{CI} \quad \text{for } \lambda z_p z_{CI} \le 1 \tag{88}$$

and:

$$\lim_{c_p \to 0} \ln \gamma_{CI}^{(c)} = -\frac{1}{2} - \ln(\lambda z_p z_{CI}) \quad \text{for } \lambda z_p z_{CI} > 1. \tag{89}$$

The equations (88) and (89) are only appropriate when a single polyelectrolyte is dissolved in an aqueous solution of a single salt and a single polyelectrolyte with a common counterion. Manning has also given extensions for cases in which several low molecular weight salts are dissolved and when those salts and the polyelectrolyte have no common ions [78].

5.3 Modifications of Manning's Theory

There have been some efforts (for example, by Nordmeier [107] and by Hao und Harvey [108]) to modify Manning's model. Here, only the modification by Hao and Harvey will be discussed. Hao and Harvey applied statistical thermodynamics for a linear lattice to derive an improved expression for the "neutralization fraction of the polyion" θ_z that can be used to avoid the approximation $\theta_z = \theta_z^{(0)}$. For an aqueous solution of a single salt and a single polyion (both having a common ion – the counterion), that result is:

$$\theta_z = \frac{1}{z_{CI}} \left(1 - \frac{z_p}{z_{CI}\lambda} \right) - \ln(f) \left[\frac{\ln(jc_{CI})^{-1}}{z_{CI}^2 \lambda} + \frac{\ln(jc_{CI})^{-2}}{(z_{CI}\lambda - 1)(z_{CI}^2\lambda - z_{CI}\lambda + 1)} \right], \tag{90}$$

Aqueous Solutions of Polyelectrolytes

where j stands for:

$$j = 4\pi N_A b^2 l_B \left(z_{CI}^2 v_{CI} + z_{COI}^2 v_{COI} \right) \tag{91}$$

and f is a "short-range" parameter. Hao and Harvey did not use the concept of a volume V_p where the condensed counterions are located, but introduced a binding constant to describe the counterion condensation phenomenon. They expressed that binding constant using an adjustable, dimensionless (positive) parameter f. For $f = 1$, (90) reduces to Manning's approximation ($\theta_z = \theta_z^{(0)}$), whereas for $f \neq 1$ the correction term on the right-hand side of (90) does not vanish.

5.4 NRTL Model of Nagvekar and Danner

Nagvekar and Danner [101] tried to overcome the limitations of the theoretical expression by combining Manning's result for highly diluted aqueous solutions of a polyion with the semiempirical electrolyte–NRTL (nonrandom two liquid) equation of Chen and Evans [109]. Their expression for the Gibbs energy of an aqueous solution of a polyion consists of three parts. The first part describes the ideal mixture, the two other parts describe the excess Gibbs energy G^E, which results from short-range (superscript SR) as well as from long-range (superscript LR) electrostatic interactions:

$$G = \sum_{\text{all components } j} n_j \mu_{j,\text{id.mix.}} + G^{E,\text{SR}} + G^{E,\text{LR}}. \tag{92}$$

The chemical potential of a component j in an ideal mixture $\mu_{j,\text{id.mix.}}$ is defined on the mole fraction scale using the unsymmetrical convention, i.e., the reference state for the solvent (water) is the pure liquid solvent. For any solute species, the reference state is a hypothetical pure liquid where the species experience interactions as if at infinite dilution in water.

As the activity coefficient $\gamma_j^{(x)}$ of component j is:

$$RT \ln \gamma_j^{(x)} = \left(\frac{\partial G^E}{\partial n_j} \right)_{n_{k \neq j}, p, T}, \tag{93}$$

$\gamma_j^{(x)}$ is a product of a short-range and a long-range contribution:

$$\gamma_j^{(x)} = \gamma_j^{\text{SR},(x)} \gamma_j^{\text{LR},(x)}. \tag{94}$$

Danner et al. express the short-range contribution using a modification of the electrolyte-NRTL equation of Chen and Evans [109] and take the long-range contribution from Manning's model (for the case of infinite dilution of a polyelectrolyte in water). For example, they treat an aqueous solution of a single polyelectrolyte as a three-component mixture consisting of the solvent, the counterion, and the polyion backbone that is approximated by its charged repeating units. As the electrolyte-NRTL model is a "local composition" model, such a solution is described by cells. There are as many types of cells as there are different species in the mixture. Each cell type consists of a single species surrounded by its nearest neighbors. There are three different cells in an aqueous solution of a single polyion, i.e., with a water molecule, a counterion, or a repeating unit, in the center The cell with water as the central species might be surrounded by other water molecules, counterions, and repeating units of the polyion. The nearest neighborhood of a cell with a central counterion also contains water and repeating units of the polyion, but it is assumed that there are no further counterions. The nearest neighborhood of a cell with a central repeating unit consists of two further repeating units (its neighbors in the polyion), counterions, and water molecules. In contrast to Chen and Evans, Danner and coworkers [101, 102] do not assume that the criterion of electroneutrality is fulfilled in each cell. Because the electrolyte-NRTL model is commonly given for a symmetrical convention, whereas polyelectrolyte systems are normalized according to the unsymmetrical convention, Danner et al. use the following expression for $G^{E,SR}$ of a multicomponent solution:

$$\frac{G^{E,SR}}{n_T RT} = \frac{G^{E,SR,sym}}{n_T RT} - \sum_{\text{all solutes } j} x_j \ln \gamma_j^{SR,(x),\infty}, \tag{95}$$

where $G^{E,SR,sym}$ is the excess Gibbs energy in the symmetrical convention, n_T is the total mole number of the solution:

$$n_T = \sum_{\text{all components } j} n_j \quad j = w, a, c \tag{96}$$

and $\gamma_j^{SR,(x),\infty}$ is the contribution of the short-range interactions to the activity coefficient of solute j (i.e., either a cation c or anion a, in the symmetrical convention, on the mole fraction scale at infinite dilution in water).

$$\gamma_j^{SR,(x),\infty} = \lim_{n_k \to 0} \gamma_j^{SR,(x)}, \tag{97}$$

where subscript k stands for all solutes and:

$$RT \ln \gamma_j^{SR,(x)} = \left(\frac{\partial G^{E,SR,sym}}{\partial n_j} \right)_{n_{k \neq j,p,T}}. \tag{98}$$

The mole fraction of species in a shell of nearest neighbors around a central species is expressed using a Boltzmann term as a weighting factor. Danner et al. give the following expression for the contributions of short-range forces to the

Aqueous Solutions of Polyelectrolytes

excess Gibbs energy of a multisolute and multisolvent mixture in the symmetrical convention:

$$\frac{G^{E,SR,sym}}{n_T RT} = \sum_s X_s \frac{\sum_j X_j G_{js} \tau_{js}}{\sum_k X_k G_{ks}} + \sum_c X_c \sum_a \frac{X_a}{\sum_{a'} X_{a'}} \frac{\sum_{j \neq c} X_j G_{jc,ac} \tau_{jc,ac}}{\sum_{k \neq c} X_k G_{kc,ac}}$$
$$+ \sum_a X_a \sum_c \frac{X_c}{\sum_{c'} X_{c'}} \frac{\sum_{j \neq a} X_j G_{ja,ca} \tau_{ja,ca}}{\sum_{k \neq a} X_k G_{ka,ca}}, \tag{99}$$

where subscript s refers to a solvent component, and subscripts c and a refer to anionic and cationic species, respectively, regardless of their source (either from a polyion or from an added salt). Subscripts j and k stand for any of the species in the mixture and primes are used to distinguish different species of the same type. The composition of the mixture is described by "modified mole fractions" X_j:

$$X_j = x_j z_j, \tag{100}$$

where x_j is the mole fraction of species j and z_j is (for a charged species) its (absolute) charge number, and for any uncharged species $z_i = 1$

There are two types of interaction parameters that are distinguished by the number of subscripts: G_{ji} and τ_{ji} on one side and $G_{ji,ki}$ and $\tau_{ji,ki}$ on the other side, which are expressed using binary parameters g_{ji} for interactions between species j and i and by binary and ternary nonrandomness parameters α_{ji} and $\alpha_{ji,ki}$:

$$G_{ji} = \exp(-\alpha_{ji} \tau_{ji}) \tag{101}$$

with:

$$\tau_{ji} = \frac{g_{ji} - g_{ii}}{RT} \tag{102}$$

and:

$$G_{ji,ki} = \exp(-\alpha_{ji,ki} \tau_{ji,ki}) \tag{103}$$

with:

$$\tau_{ji,ki} = \frac{g_{ji} - g_{ki}}{RT}. \tag{104}$$

Danner et al. used Manning's results for the long-range contribution to the activity coefficient of the counterions in an aqueous solution of a single polyion:

$$\ln \gamma_{CI}^{LR,(c)} = -\frac{\lambda}{2} z_p z_{CI} \quad \text{for } \lambda z_p z_{CI} \leq 1 \tag{105}$$

and

$$\ln \gamma_{\mathrm{CI}}^{\mathrm{LR},(c)} = -\frac{1}{2} - \ln(\lambda z_p z_{\mathrm{CI}}) \quad \text{for } \lambda z_p z_{\mathrm{CI}} > 1, \tag{106}$$

where λ is a charge density parameter [see (30) and (31)] and z_p and z_{CI} are the (absolute) numbers of elementary charges on a dissociated repeating unit and on the counterion, respectively. The osmotic coefficient on the molarity scale is:

$$\Phi_p^{(c)} = \frac{\ln a_w}{\ln a_{w,\mathrm{id.mix.}}^{(c)}} = \frac{\ln\left(x_w \gamma_w^{\mathrm{LR},(x)} \gamma_w^{\mathrm{SR},(x)}\right)}{\ln a_{w,\mathrm{id.mix.}}^{(c)}} = \frac{\ln\left(x_w \gamma_w^{\mathrm{LR},(x)}\right)}{\ln a_{w,\mathrm{id.mix.}}^{(c)}} + \frac{\ln \gamma_w^{\mathrm{SR},(x)}}{\ln a_{w,\mathrm{id.mix.}}^{(c)}}, \tag{107}$$

where, as in (22):

$$\ln a_{w,\mathrm{id.mix}}^{(c)} = -\frac{1}{\bar{\rho}_w^*} \sum_{i \neq w} c_i. \tag{108}$$

The long-range contribution is described using Manning's results and one obtains for the osmotic coefficient (on the molarity scale):

$$\Phi_p^{(c)} = \lim_{c_p \to 0} \Phi_{p,Ma}^{(c)} - \bar{\rho}_w^* \frac{\ln \gamma_w^{\mathrm{SR},(x)}}{\sum_{\text{all solutes } j} c_j}. \tag{109}$$

$$\Phi_p^{(c)} = \lim_{c_p \to 0} \Phi_{p,Ma}^{(c)} - \bar{\rho}_w^* \frac{\ln \gamma_w^{\mathrm{SR},(x)}}{c_p \left(1 + \frac{z_p}{z_{\mathrm{CI}}} v\right)}, \tag{110}$$

where

$$\lim_{c_p \to 0} \Phi_{p,Ma}^{(c)} = 1 - \frac{\lambda}{2} z_p z_{\mathrm{CI}} \quad \text{for } \lambda z_p z_{\mathrm{CI}} \leq 1 \tag{111}$$

and

$$\lim_{c_p \to 0} \Phi_{p,Ma}^{(c)} = \frac{1}{2\lambda z_p z_{\mathrm{CI}}} \quad \text{for } \lambda z_p z_{\mathrm{CI}} > 1. \tag{112}$$

As above, c_p and v are the molarity of the polyion and the number of repeating units of that polyion. The short-range part of the activity coefficient of water is calculated using (98).

The model needs numerical values for interaction parameters and nonrandomness parameters. Danner et al. mention that the nonrandomness parameters α_{ji} and

Aqueous Solutions of Polyelectrolytes

$\alpha_{ji,ki}$ were arbitrarily set to 0.20, except when the central species in the cell is a repeating unit and its nearest neighbors are also repeating units. Then, the nonrandomness parameter was set to 0.33. They finally adjusted four interaction parameters to experimental results for the osmotic coefficient of an aqueous solution of a single polyelectrolyte. However, no parameters have been published and all comparisons were given only in graphical form. But, the method is obviously suited for a good correlation of experimental data for the osmotic coefficient of aqueous solutions of a single polyelectrolyte.

Danner et al. did not report results from their method to describe the influence of an added salt on the osmotic coefficient of aqueous solutions that contain a single polyion.

5.5 Pessoa's Modification of the Pitzer Model

Pessoa and Maurer [110] assume that a polyion might not completely dissociate in an aqueous solution and that the degree of dissociation is independent of the composition of the aqueous solution. They propose the use of experimental data for the osmotic coefficient of an aqueous solution of the single polyelectrolyte at infinite dilution to determine that degree. On the molality scale the osmotic coefficient is:

$$\Phi_p^{(m)} = \frac{\pi}{\pi_{\text{id.mix.}}^{(m)}} = \frac{\ln a_w}{\ln a_{w,\text{id.mix.}}^{(m)}}, \tag{113}$$

where

$$\ln a_{w,\text{id.mix.}}^{(m)} = -M_w^* \sum_{\text{all solutes } j} \frac{m_j}{m^\circ}. \tag{114}$$

M_w^* is the relative molecular mass of water divided by 1,000 (i.e., $M_w^* = 0.01806$), m_j is the molality of species j and $m^\circ = 1 \text{ mol}/(\text{kg water})$. The ideal solution is defined so that all counterions are completely dissociated:

$$m_{\text{Cl}} = v m_p \frac{z_p}{z_{\text{Cl}}}, \tag{115}$$

where m_p is the molality of the polyion. The activity of water in an ideal aqueous solution of a single polyelectrolyte is:

$$\ln a_{w,\text{id.mix.}}^{(m)} = -M_w^* \left(1 + v^* \frac{z_p}{z_{\text{Cl}}}\right) \frac{m_p}{m^0}. \tag{116}$$

Taking into account that, in a real solution, the polyion is not completely dissociated, the equation gives at high dilution in water:

$$\ln a_w^{(m)} = -M_w^* \left(1 + v^* \frac{z_p}{z_{\text{Cl}}}\right) \frac{m_p}{m^0}.$$

(117)

The ratio v^*/v is the degree of dissociation of the repeating units of the polyion. Combining (116) and (117) results in:

$$\lim_{m_p \to 0} \Phi_p^{(m)} = \Phi_p^{0,(m)} = \frac{1 + v^* \frac{z_p}{z_{\text{Cl}}}}{1 + v \frac{z_p}{z_{\text{Cl}}}}.$$

(118)

When the repeating unit is a 1:1 electrolyte and the number of repeating units is large, the limiting value of the osmotic coefficients equals the degree of dissociation:

$$\Phi_p^{0,(m)} = \frac{v^*}{v}.$$

(119)

A real aqueous solution of a single polyion is considered to be a mixture of water, (partially dissociated) polymer chains, and the dissolved counterions. In an ideal mixture all solutes only experience interaction with water, whereas in a real solution there are also interactions between the solutes. The deviations that are caused by these interactions are taken into account through an expression for the excess Gibbs energy. Pessoa and Maurer [110] started from Pitzer's equation [105, 111] for the excess Gibbs energy of aqueous solutions of low molecular weight strong electrolytes. That method was extended previously to describe the Gibbs energy of aqueous solutions that contain both a strong electrolyte and a neutral polymer [112–114]. As in the work by Danner et al. [101, 102], the Gibbs energy of an aqueous solution is split into a contribution from ideal mixing and contributions from long-range and short-range interactions. The contributions are expressed using the unsymmetrical convention. However, Pitzer's equation applies the molality scale to express the composition of the aqueous solution:

$$G = \sum_{\text{all components } j} n_j \mu_{j,\text{id.mix.}} + G^{E,\text{SR}} + G^{E,\text{LR}}.$$

(120)

For a solute component j, the chemical potential in an ideal mixture $\mu_{j,\text{id.mix.}}$ is:

$$\mu_{j,\text{id.mix.}} = \mu_j^{\text{ref},(m)} + RT \ln \frac{m_i}{m^\circ},$$

(121)

where $\mu_j^{\text{ref},(m)}$ is the chemical potential of solute j in an one molal aqueous solution (i.e., $m_j = m^\circ = 1 \text{ mol}/(\text{kg water})$). In that reference state, the solute experiences similar interaction as in infinite dilution in water. For the solvent (i.e., water) the reference state is the pure liquid solvent and the difference between the chemical

Aqueous Solutions of Polyelectrolytes

potential in the real mixture and that of the pure liquid is expressed via the activity of water:

$$\mu_w = \mu_{w,\text{pure}} + RT \ln a_w^{(m)}. \tag{122}$$

Pitzer uses a modification of the limiting law by Debye and Hückel to account for long-range interactions that are caused by Coulomb forces:

$$\frac{G^{E,\text{LR}}}{n_w M_w^* RT} = -A_\varphi \frac{4I_m}{b} \ln(1 + b\sqrt{I_m}). \tag{123}$$

A_φ is the Debye-Hückel parameter (at 298.2 K $A_\varphi = 0.3914$), I_m is the ionic strength (on molality scale), and b is a numerical value ($b = 1.2$). This expression is very well suited to describe the activity coefficient of ions at high dilutions, but cannot directly be applied to polyelectrolyte solutions because the Debye-Hückel term was developed for punctual electric charges (such as small mobile ions). It is not valid for highly charged polymer backbones. Pessoa and Maurer [110] replaced the contribution of the polyion in the expression for the ionic strength:

$$I_m = \frac{1}{2} \sum_j \frac{m_j^*}{m^\circ} (z_j^*)^2, \tag{124}$$

where for all solute species (with the exception of the polyion) $m_j^* = m_j$ and $z_j^* = z_j$, whereas for a polyion ($j \equiv p$) $m_p^* = v^* m_p$ and $z_p^* = z_p$, i.e., for the calculation of the ionic strength the polyelectrolyte is replaced by its dissociated repeating units. The activity coefficient of a solute caused by the long-range interactions is:

$$\ln \gamma_i^{\text{LR},(m)} = -A_\varphi \sigma_i z_i^2 \left(\frac{2}{b} \ln\left(1 + b\sqrt{I_m}\right) + \frac{\sqrt{I_m}}{1 + b\sqrt{I_m}} \right), \tag{125}$$

where $\sigma_i = 1$ for all solutes, with the exception of the polymer where $\sigma_i = v^*$. The long-range contribution to the activity coefficient of the solvent is:

$$\ln \gamma_w^{\text{LR},(m)} = 2A_\varphi M_w^* \frac{I_m^{3/2}}{1 + b\sqrt{I_m}}. \tag{126}$$

The short-range contributions are described with a virial-type equation for the excess Gibbs energy that was adapted from Pitzer [105]. It is applied here neglecting ternary and higher interactions between solute species:

$$\frac{G^{E,\text{SR}}}{n_w M_w^* RT} = \sum_{i \neq w} \sum_{j \neq w} \lambda_{ij}(I_m) \frac{m_i^*}{m^\circ} \frac{m_j^*}{m^\circ}, \tag{127}$$

where n_w is the number of moles of water and $\lambda_{ij}(I_m)$ is an osmotic virial coefficient for interactions between solute species i and j that depends on the ionic strength I_m through:

$$\lambda_{ij}(I_m) = \lambda_{ij}^{(0)} + \lambda_{ij}^{(1)} \frac{2}{\alpha^2 I_m} \left(1 - (1 + \alpha\sqrt{I_m}) \exp(-\alpha\sqrt{I_m})\right). \tag{128}$$

Equation (127) applies the same definition for the molality m_i^* of a solute species i as (124) for the ionic strength. $\lambda_{ij}^{(0)}$ and $\lambda_{ji}^{(1)}$ are binary parameters for interactions between the solutes, for example, between a repeating unit of the polyion and a dissolved counterion. No distinction is made – as far as the interaction parameters are concerned – between neutral and dissociated repeating units. The binary parameters are symmetrical ($\lambda_{ij}^{(0)} = \lambda_{ji}^{(0)}$ and $\lambda_{ij}^{(1)} = \lambda_{ji}^{(1)}$) and α is a constant ($\alpha = 2$). For a solute species, the contribution of the short-range interactions to the activity coefficient is:

$$\ln \gamma_i^{SR,(m)} = 2\sigma_i \sum_{j \neq w} \lambda_{ij}(I_m) \frac{m_j^*}{m^\circ} + -\sigma_i z_i^2 M_w^*$$

$$\times \sum_{j \neq w} \sum_{k \neq w} \lambda_{ij}^{(1)} \frac{1}{\alpha^2 I_m^2} \left(1 - \left(1 + \alpha\sqrt{I_m} + \frac{\alpha^2 I_m}{2}\right) \exp(-\alpha\sqrt{I_m})\right) \frac{m_j^*}{m^\circ} \frac{m_k^*}{m^\circ}$$

$$\tag{129}$$

and for the solvent:

$$\ln \gamma_w^{SR,(m)} = -M_w^* \left(\sum_{i \neq w} \sum_{j \neq w} \left(\lambda_{ij}^{(0)} + \lambda_{ij}^{(1)} \exp(-\alpha\sqrt{I_m})\right) \frac{m_i^*}{m^\circ} \frac{m_k^*}{m^\circ}\right). \tag{130}$$

The final equation for the activity of a solute i and of the solvent (water) is obtained by coupling the above expressions through:

$$a_i = m_i \gamma_i^{LR} \gamma_i^{SR}. \tag{131}$$

$$a_w = \exp\left(-M_w^* \sum_{i \neq w} \frac{m_i}{m^\circ}\right) \gamma_w^{LR} \gamma_w^{SR}, \tag{132}$$

where the sum is over all solute species, i.e., in an aqueous solution of a single electrolyte, i stands for the polyion ($m_i \rightarrow m_p$) and for the counterion ($m_i \rightarrow v^* \frac{z_p}{z_{Cl}} m_p$).

Modeling the osmotic coefficient of an aqueous solution of a single homopolymer polyion (i.e., a polyion that consists of a single repeating unit and a single counterion) requires:

- The osmotic coefficient of an aqueous solution of the polyelectrolyte at infinite dilution $\Phi_p^{0,(m)}$
- The number of repeating units of the polyion v
- The binary interaction parameters $\lambda_{ij}^{(0)}$ and $\lambda_{ij}^{(1)}$

$\Phi_p^{0,(m)}$ is either determined from experimental results for the osmotic coefficient or estimated using the results of Manning's theory (in that case the length of a repeating unit has to be known). For a polyion that consists only of a single repeating unit, the number of repeating units v is calculated from the number-averaged molecular mass of the polyion and the molecular mass of the repeating unit. It is assumed that binary interaction parameters between species carrying electrical charges of the same sign can be neglected (i.e., they are set to zero). Therefore, there are only two, nonzero binary parameters for interactions between a repeating (subscript p) unit and the counterion $\lambda_{pCl}^{(0)}$ and $\lambda_{pCl}^{(1)}$. These interaction parameters are fitted to some experimental properties such as the osmotic coefficient. Figure 12 shows a typical example for a correlation. The model can be straightforwardly extended to aqueous solutions of a single polyion and a single low molecular weight strong electrolyte (cf. Fig. 13) but also to aqueous solutions of a polyion and a neutral polymer. Such mixtures often form aqueous two-phase systems. Figure 14 gives a typical example.

5.6 VERS-PE Model

Lammertz et al. [116] extended the Virial-Equation with Relative Surface Fractions (VERS) model of Großmann et al. [112–114] for the excess Gibbs energy of aqueous solutions of neutral polymers and low molecular weight electrolytes to the treatment of aqueous solutions that also contain polyions. That extension is

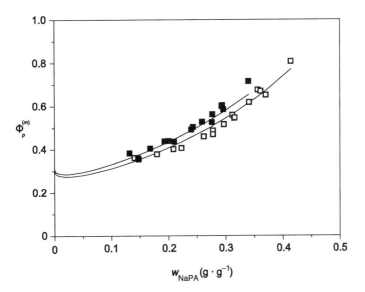

Fig. 12 Osmotic coefficient of aqueous solutions of NaPA at 298.15 K. Experimental data [41]: *closed squares* NaPA 5; *open squares* NaPA 15. *Lines* show the modeling

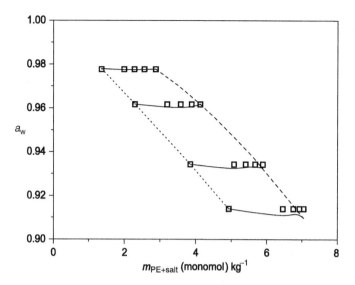

Fig. 13 Activity of water in aqueous solutions of NaPA 5 and NaCl. *Symbols* experimental data [28]; *dashed line* modeling of systems without salt; *dotted line* modeling of systems without polyelectrolyte; *solid lines* correlation results

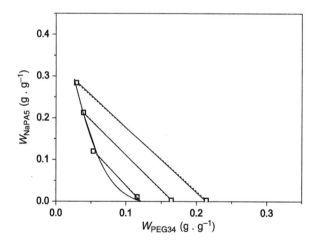

Fig. 14 Liquid–liquid equilibrium of aqueous solutions of NaPA 5 and PEG 34. *Symbols* experimental equilibrium compositions [115]; *dotted lines* experimental tie lines; *solid lines* correlation results

called VERS model for polyelectrolytes (VERS-PE model). Like the model of Pessoa and Maurer, the VERS model is based on Pitzer's equation [105] for the Gibbs excess energy of aqueous solutions of low molecular weight, strong electrolytes. Großmann et al. introduced two modifications to allow for the treatment of neutral polymers: the molality scale was replaced by a surface fraction scale, and

Aqueous Solutions of Polyelectrolytes

the interactions with a polymer are described via interactions with groups of that polymer, i.e., the polymer was split into groups. The groups commonly consist of the repeating units of the polymer. The extension of the model to polyions also considers the phenomenon of counterion condensation by a chemical reaction equilibrium approach. For convenience, the extension of the model for the excess Gibbs energy to aqueous solutions of polyions is described here first for an aqueous solution of a single polyelectrolyte (designated by subscript p) where only a single counterion might dissociate from a repeating unit and that repeating unit is a 1:1 electrolyte.

The reference state for the chemical potential of the solvent (water) is the pure liquid, whereas for the solute (polyelectrolyte) it is a hypothetical one molal solution of the undissociated polyelectrolyte in water $(m_p = m° = 1\ mol/(kg\ water))$, where it experiences interactions with water molecules only, i.e., in that reference state the undissociated polyelectrolyte is infinitely diluted in water ($m_p = 0$ in pure water). The difference between the chemical potential of the polyelectrolyte in the real solution $\mu_p(T, m_p)$ and in its reference state μ_p^{ref} is calculated in five steps:

$$\mu_p - \mu_p^{\text{ref}} = \Delta_{12}\mu_p + \Delta_{23}\mu_p + \Delta_{34}\mu_p + \Delta_{45}\mu_p + \Delta_{56}\mu_p. \tag{133}$$

In the first step (Δ_{12}), only the molality of the polyelectrolyte is changed to its molality m_p in the real solution:

$$\Delta_{12}\mu_p = RT \ln\left(\frac{m_p}{m°}\right). \tag{134}$$

This contribution accounts for the change from the reference state to an ideal dilution (assuming that at state 2 the interactions are the same as in the reference state) and there is still no dissociation.

The second contribution (Δ_{23}) describes the change in the chemical potential due to splitting the polyelectrolyte into its monomers. This change is the sum of two contributions: a free volume contribution (superscript fv) caused by the increase of the number of particles, and a combinatorial contribution (superscript comb) caused by the increase of the number of degrees of freedom:

$$\Delta_{23}\mu_p = \Delta_{23}\mu_p^{\text{fv+comb}}. \tag{135}$$

In state 3, the aqueous polyelectrolyte solution has been replaced by an aqueous solution of the nondissociated repeating units. The repeating units still experience only interactions with water. As one polyelectrolyte molecule consists of v monomer units (characterized by subscript A), the molality of species A, m_A, is:

$$m_A = v m_p \tag{136}$$

and the chemical potential of the polyelectrolyte in state 3 is:

$$\mu_{p,3} = v\left(\mu_A^{\text{ref}} + RT \ln\left(\frac{v m_p}{m°}\right)\right). \tag{137}$$

μ_A^{ref} is the chemical potential of nondissociated repeating units in their reference state. That reference state is defined in the same way as the reference state for a polyelectrolyte.

In the next step (Δ_{34}), all monomer units A are split into two groups. Subscript C designates all repeating units that will never dissociate while subscript D designates repeating units that are assumed to undergo a dissociation reaction. The condition of mass balance requires that for the chemical potential of the polyelectrolyte in state 4 is:

$$\mu_{p,4} = v_{C,4} \cdot \mu_{C,4} + v_{D,4} \cdot \mu_{D,4}. \tag{138}$$

For the sake of simplicity, $v_{C,4}$ and $v_{D,4}$ are expressed through a new property k, that is directly related to the degree of counterion condensation at infinite dilution of the polyelectrolyte in water.

$$k = \frac{v_{C,4}}{v}. \tag{139}$$

The chemical potential of the polyelectrolyte in state 4 is:

$$\begin{aligned} \mu_{p,4} = vk\left(\mu_C^{\text{ref}} + RT\ln\left(\frac{vkm_p\Gamma_{C,4}}{m^\circ} \right) \right) + v(1-k) \\ \times \left(\mu_D^{\text{ref}} + RT\ln\left(\frac{v(1-k)m_p\Gamma_{D,4}}{m^\circ} \right) \right). \end{aligned} \tag{140}$$

$\Gamma_{C,4}$ and $\Gamma_{D,4}$ are the activity coefficients (on molality scale) of species C and D, respectively, in state 4. But, as there is at this stage no difference between the natures of groups A, C, and D:

$$\Gamma_{C,4} = \Gamma_{D,4} = \Gamma_{A,4} \tag{141}$$

and

$$\mu_C^{\text{ref}} = \mu_D^{\text{ref}} = \mu_A^{\text{ref}}. \tag{142}$$

Consequently,

$$\mu_{p,4} = v\left(\mu_A^{\text{ref}} + RT\ln\left(\frac{vk^k(1-k)^{(1-k)}m_p\Gamma_{A,4}}{m^\circ} \right) \right). \tag{143}$$

The change of the chemical potential of the polyelectrolyte caused by the transition from step 3 to 4 is:

$$\Delta_{34}\mu_p = vRT\ln\left(k^k(1-k)^{(1-k)}\Gamma_{A,4} \right). \tag{144}$$

Aqueous Solutions of Polyelectrolytes

In the next step (Δ_{45}), the partial dissociation of species D is achieved (to account for the phenomenon of counterion condensation). The dissociation is expressed by a chemical reaction:

$$D \rightleftharpoons CI + F,$$

where CI and F stand for the counterion and for the dissociated monomer unit, respectively. The chemical potential of the polyelectrolyte in state 5 is (as only a single counterion dissociates from one monomer unit):

$$\mu_{p,5} = \left(kv + v_{D,5} \right) \mu_{A,5} + v_{CI,5}\, \mu_{CI,5} + v_{CI,5}\mu_{F,5}. \tag{145}$$

For convenience, the total degree of dissociation of the repeating units α is introduced:

$$\alpha = \frac{v_{CI,5}}{v} \quad \text{where } 0 < \alpha < (1-k). \tag{146}$$

As there is no difference between species C and D (all are designated by A): $\mu_{C,5} = \mu_{D,5} = \mu_{A,5}$:

$$\mu_{p,5} = v \cdot \left(\mu_{A,5} + \alpha\left(\mu_{CI,5} + \mu_{F,5} - \mu_{D,5} \right) \right). \tag{147}$$

Because in dissociation equilibrium:

$$\mu_{CI,5} + \mu_{F,5} - \mu_{D,5} = 0 \tag{148}$$

the chemical potential of the polyelectrolyte in state 5 is:

$$\mu_{p,5} = v \left(\mu_A^{\text{ref}} + RT \ln\left(vk^k (1-k-\alpha)^{(1-k)} \frac{m_p}{m^\circ} \Gamma_{A,5} \right) \right). \tag{149}$$

When furthermore (as another approximation), the difference between the activity coefficients of the undissociated repeating units in states 4 and 5 is neglected, the change of the chemical potential of the polyelectrolyte caused by the transition from 4 to 5 is:

$$\Delta_{45}\mu_p = v(1-k)RT \ln\left(1 - \frac{\alpha}{1-k} \right). \tag{150}$$

The fifth contribution to the chemical potential is to account for the repolymerization of the charged and noncharged monomers. This difference is approximated by reversing the change from state 2 to state 3, but applying a correction term $\Delta_{56}\mu_p^{\Delta conf}$ that accounts for the difference in the conformation of the polymer chain from a more globular structure in state 2 (where the polymer is neutral) to a more stretched structure in state 6 (where the polymer backbone is charged):

$$\Delta_{56}\mu_p = -\Delta_{23}\mu_p^{\text{fv+comb}} + \Delta_{56}\mu_p^{\Delta\text{conf}}. \tag{151}$$

It is assumed that the fraction of polyions in a stretched configuration equals the total degree of dissociation α of the repeating units. Furthermore, the difference between the chemical potentials of a polyion in its stretched and its globular structures is approximated using the combinatorial part of the UNIQUAC (universal quasichemical) model of Abrams and Prausnitz [117] for the excess Gibbs energy of nonelectrolyte solutions. In the UNIQUAC model, the shape of a molecule i is described by a volume parameter r_i and a surface parameter q_i. A change in the polyelectrolyte's conformation changes only its surface parameter resulting in:

$$\frac{\Delta_{56}\mu_p}{RT} = 5\alpha\left(\ln\left(\frac{(\Theta_{\text{st}})^{q_{\text{st}}}}{(\Theta_{\text{gl}})^{q_{\text{gl}}}}\Xi^{\left(q_{\text{gl}}-q_{\text{st}}\right)}\right) + (1-\Xi)\left(q_{\text{gl}}-q_{\text{st}}\right)\right). \tag{152}$$

q_{st} and q_{gl} are the polyelectrolyte's surface parameters in the stretched and the globular configuration, respectively. Similarly, Θ_{st} and Θ_{gl} are the polyelectrolyte's surface fractions in the stretched and the globular configuration, respectively, and Ξ is the volume fraction of the polyelectrolyte in the aqueous solution. The polyelectrolyte's surface fraction is:

$$\Theta_{ab} = \frac{\frac{m_p}{m^o}\cdot q_{ab}}{\frac{m_p}{m^o}\cdot q_{ab} + 55.5\cdot q_w} \quad \text{for "}ab\text{" either "}st\text{" or "}gl\text{",} \tag{153}$$

where q_w is the surface parameter of water. The surface parameter of the stretched polyion is calculated using the surface parameter q_{rp} of a repeating unit and the number v of repeating units which form that polyion:

$$q_{\text{st}} = vq_{\text{rp}}. \tag{154}$$

The surface parameter of the globular polyion is smaller than that of the stretched polyion. It is approximated by introducing a configurational parameter b^* (that is close to, but smaller than 0.5):

$$q_{\text{gl}} = v^{2b^*}q_{\text{rp}}. \tag{155}$$

The polyelectrolyte's volume fraction is:

$$\Xi = \frac{v\cdot\frac{m_p}{m^o}\cdot r_{\text{rp}}}{v\cdot\frac{m_p}{m^o}\cdot r_{\text{rp}} + 55.5r_w}, \tag{156}$$

where r_{rp} and r_w are the volume parameters of a repeating unit of the polyion and of water, respectively.

Aqueous Solutions of Polyelectrolytes

Summing up the contributions from the five steps gives the chemical potential of the polyion in an aqueous solution. The chemical potential of the polyion on the molality scale is also given by:

$$\mu_p = \mu_p^{\text{ref}} + RT \ln\left(\frac{m_p}{m^\circ}\gamma_p^{(m)}\right) \tag{157}$$

and the activity coefficient is:

$$\ln\gamma_p^{(m)} = v\ln\left(k^k(1-k-\alpha)^{(1-k)}\Gamma_{A,5}\right)$$
$$+ 5\alpha\left(\ln\left(\frac{(\Theta_{\text{st}})^{q_{\text{st}}}}{(\Theta_{\text{gl}})^{q_{\text{gl}}}}\Xi^{(q_{\text{gl}}-q_{\text{st}})}\right) + (1-\Xi)(q_{\text{gl}}-q_{\text{st}})\right). \tag{158}$$

The model requires pure-component surface (q_{rp} and q_w) and size (r_{rp} and r_w) parameters for the monomer unit and for water, the degree of counterion dissociation in infinite dilution (k), the total degree of dissociation of the repeating units (α), the configurational parameter (b^*), and interaction parameters (in the expressions for the activity coefficients in state 5 where the solution is a mixture of water, undissociated as well as dissociated repeating units and counterions).

Surface and size parameters are either available in the literature or are calculated following the proposals by Bondi [118]. The degree of counterion dissociation in infinite dilution is estimated from experimental data for the limiting osmotic coefficient of an aqueous solution of the polyion. Following the ideas outlined in the description of the Pessoa and Maurer model above, one finds when the repeating unit is a 1:1 electrolyte:

$$\Phi_p^{0,(m)} = \frac{1+kv}{1+v} \approx k. \tag{159}$$

The activity coefficients in state 5 ($\Gamma_{i,5}$, where i is any solute that is present in state 4, i.e., the neutral repeating unit A, the dissociated repeating unit F and the counterion CI) are calculated using the VERS model of Großmann et al. [112–114]. The activity coefficient Γ_i is assumed to consist of contributions from van der Waals-like interactions Γ_i^{vdW} and electrostatic interactions Γ_i^{el}:

$$\Gamma_i = \Gamma_i^{\text{vdW}}\Gamma_i^{\text{el}}. \tag{160}$$

The electrostatic contribution is expressed in a similar way as the long-range contribution in the model of Pessoa and Maurer [cf. (125)] from the Debye–Hückel parameter A_φ, the charge number z_i of groups/species i and the ionic strength I_m (on molality scale):

$$\ln\Gamma_i^{\text{el}} = -A_\varphi z_i^2 \frac{1}{2}\frac{1-k}{\alpha v}\left(\frac{2}{1.2}\ln(1+1.2\sqrt{I_m}) + \frac{\sqrt{I_m}}{1+1.2\sqrt{I_m}}\right). \tag{161}$$

The ionic strength is:

$$I_{\mathrm{m}} = \frac{1}{2} \sum_{k=\mathrm{Cl,F}} \frac{m_k}{m^\circ} z_k^2.$$

(162)

The contribution from short-range interactions to the activity coefficient of solute species (i.e., groups) i, Γ_i^{vdW}, is taken from the VERS model:

$$
\begin{aligned}
\ln \Gamma_i^{\mathrm{vdW}} = {} & \frac{2}{M_w^*} \frac{q_i}{q_w} \sum_{\text{all groups } L} \frac{\Theta_L}{\Theta_w} \left(a_{i,L}^{(0)} + a_{i,L}^{(1)} \cdot f_1(I_{\mathrm{m}}) \right) \\
& - \left(\frac{z_i}{M_w^*} \right)^2 f_2(I_{\mathrm{m}}) \sum_{\text{all groups } L} \sum_{\text{all groups } k} \frac{\Theta_L}{\Theta_w} \frac{\Theta_k}{\Theta_w} a_{L,K}^{(1)} \\
& + \frac{3}{(M_w^*)^2} \frac{q_i}{q_w} \sum_{\text{all groups } L} \sum_{\text{all groups } k} \frac{\Theta_L}{\Theta_w} \frac{\Theta_k}{\Theta_w} b_{i,L,K}
\end{aligned}
$$

(163)

with:

$$f_1(I_{\mathrm{m}}) = \frac{1}{2I_{\mathrm{m}}} \left[1 - \left(1 + 2\sqrt{I_{\mathrm{m}}} \right) \exp\{-2\sqrt{I_{\mathrm{m}}}\} \right] \quad \text{and}$$

(164)

$$f_2(I_{\mathrm{m}}) = \frac{1}{4I_{\mathrm{m}}^2} \left[1 - \left(1 + 2\sqrt{I_{\mathrm{m}}} + 2I_{\mathrm{m}} \right) \exp\{-2\sqrt{I_{\mathrm{m}}}\} \right]$$

(165)

The sum in (163) is over all solute species, i.e., nondissociated repeating units C, nondissociated repeating units D, dissociated repeating units F and counterions CI. M_w^* is the relative molar mass of water divided by 1,000 ($M_w^* = 0.018016$). Subscript w stands for water and q_i is the surface parameter of species i. The surface fraction of a group L is abbreviated by Θ_L. As the mixture consists of species C, D, F, CI and water, the following relative surface ratios are required:

$$\frac{\Theta_C}{\Theta_w} = M_w^* v k \frac{m_p q_{\mathrm{rp}}}{m^\circ q_w},$$

(166)

$$\frac{\Theta_D}{\Theta_w} = M_w^* v (1 - k - \alpha) \frac{m_p q_{\mathrm{rp}}}{m^\circ q_w},$$

(167)

$$\frac{\Theta_F}{\Theta_w} = M_w^* v \alpha \frac{m_p q_{\mathrm{rp}}}{m^\circ q_w},$$

(168)

$$\frac{\Theta_{\mathrm{CI}}}{\Theta_w} = M_w^* v \alpha \frac{m_p q_{\mathrm{CI}}}{m^\circ q_w}.$$

(169)

Aqueous Solutions of Polyelectrolytes

$a_{i,L}^{(0)}$ and $a_{i,L}^{(1)}$ denote binary interaction parameters between species (groups) i and L. These interaction parameters are symmetric, i.e., $a_{i,L}^{(0)} = a_{L,i}^{(0)}$ and $a_{i,L}^{(1)} = a_{L,i}^{(1)}$. They form a set of adjustable model parameters. The degree of dissociation α is calculated assuming chemical equilibrium between monomers D and its dissociation products F and CI in state 5:

$$K = \frac{m_{Cl} m_F}{m_D m^\circ} \frac{\Gamma_{Cl} \Gamma_F}{\Gamma_D} = \frac{v\alpha^2}{1 - k - \alpha} \frac{m_p}{m^\circ} \frac{\Gamma_{Cl}^{vdW} \Gamma_F^{vdW}}{\Gamma_D^{vdW}} \Gamma_{Cl}^{el} \Gamma_F^{el}, \qquad (170)$$

where all molalities are those in state 5. Chemical reaction equilibrium constant K is one of the adjustable parameters of the model.

When there is also an additional single 1:1 salt MX in the aqueous solution, (162) (for the ionic strength) and (163) (for the van der Waals contribution to the group activity coefficient) have to be extended; the sums must also include the ions M and X. The extension requires the relative surface ratios for M and X:

$$\frac{\Theta_M}{\Theta_w} = M_w^* \frac{m_{MX} q_M}{m^\circ q_w}, \qquad (171)$$

$$\frac{\Theta_X}{\Theta_w} = M_w^* \frac{m_{MX} q_X}{m^\circ q_w}. \qquad (172)$$

Furthermore, as well as the chemical potential of the polyelectrolyte, the chemical potential of MX is also required (for the calculation of the activity of water, see below). That chemical potential is given by the sum of the chemical potentials of cations M and anions X:

$$\mu_{MX} = \mu_M + \mu_X = \mu_M^{ref} + \mu_X^{ref} + RT \ln \left(\frac{m_M \Gamma_M}{m^\circ} \frac{m_X \Gamma_X}{m^\circ} \right). \qquad (173)$$

Finally, the activity of water a_w is calculated from the chemical potentials of the solutes (either a single polyelectrolyte or a binary solute mixture of a polyelectrolyte and a low molecular weight salt) by applying the Gibbs–Duhem equation:

$$d\mu_w = d(\mu_w - \mu_w^{pure\ liquid}) = -M_w^* \sum_{i \neq w} \frac{m_i}{m^\circ} \cdot d\mu_i. \qquad (174)$$

Integration at constant temperature for an aqueous solution containing a polyelectrolyte P and a salt MX results in:

$$\Delta \mu_w = RT \ln a_w = -M_w^* \int_{water}^{mix} \frac{m_p}{m^\circ} d\mu_p - M_w^* \int_{water}^{mix} \frac{m_{MX}}{m^\circ} \cdot d\mu_{MX}. \qquad (175)$$

The right-hand side is solved in two steps. In the first step, the integration is carried out starting from pure water to a polyelectrolyte-free but salt-containing solution:

$$\frac{\Delta\mu_{w,1}}{2RTM_w^*} = -\frac{m_{MX}}{m^o} + A_\varphi \frac{I_{m,MX}^{1.5}}{1 + 1.2\sqrt{I_{m,MX}}}$$
$$- \left(\frac{m_{MX}}{m^o}\right)^2 \left[a_{MX}^{(0)} + a_{MX}^{(1)} \exp\left(-2\sqrt{I_{m,MX}}\right)\right], \qquad (176)$$

where $a_{MX}^{(0)}$ and $a_{MX}^{(1)}$ are binary interaction parameters between ions M and X^- and $I_{m,MX}$ is the ionic strength (on molality scale) of the polyelectrolyte-free aqueous solution of MX.

In the second step, the molality of the salt is fixed at m_{MX} and the molality of the polyelectrolyte increases from zero to m_p:

$$\Delta\mu_{W,2} = -M_w^* \left[\int_{m_p=0}^{m_p} m_p \frac{\partial\mu_p}{\partial m_p} d\left(\frac{m_p}{m^o}\right)\right]_{m_{MX}}$$
$$- M_w^* \frac{m_{MX}}{m^o}\left[\mu_{MX}(m_{MX}, m_p) - \mu_{MX}(m_{MX}, m_p = 0)\right], \qquad (177)$$

where

$$\left[\int_{m_p=0}^{m_p} m_p \frac{\partial\mu_p}{\partial m_p} d\left(\frac{m_p}{m^o}\right)\right]_{m_{MX}} = RT\left[\frac{m_p}{m^o} + \int_{m_p=0}^{m_p} m_p \left(\frac{\partial\ln\gamma_p^{(m)}}{\partial m_p}\right)_{m_{MX}} d\left(\frac{m_p}{m^o}\right)\right]. \quad (178)$$

The integral is solved numerically using (158) for the activity coefficient of the polyelectrolyte.

The final equation for the activity of water in an aqueous solution of a strong electrolyte MX and a polyelectrolyte P (where both MX and the repeating unit of the polyion are 1:1 electrolytes) is:

$$\frac{1}{M_W^*} \ln a_w = -2\frac{m_{MX}}{m^o} - \frac{m_p}{m^o} + 2 \cdot A_\varphi \frac{I_{m,MX}^{1.5}}{1 + 1.2\sqrt{I_{m,MX}}}$$
$$- 2\left(\frac{m_{MX}}{m^o}\right)^2 \left[a_{MX}^{(0)} + a_{MX}^{(1)} \exp\left(-2\sqrt{I_{m,MX}}\right)\right]$$
$$- \frac{m_{MX}}{m^o RT}\left[\mu_{MX}(m_{MX}, m_p) - \mu_{MX}(m_{MX}, m_p = 0)\right]$$
$$- \left[\int_{m_p=0}^{m_p} m_p \left(\frac{\partial\ln\gamma_p^{(m)}}{\partial m_p}\right)_{m_{MX}} d\left(\frac{m_p}{m^o}\right)\right]. \qquad (179)$$

Aqueous Solutions of Polyelectrolytes

It is worth mentioning that the chemical potential $\mu_{MX}(m_{MX}, m_p)$ is calculated from (173) where the activity coefficients of both ions M and X are calculated for an aqueous solution in which the polyelectrolyte is cut into its repeating units and the species [nondissociated repeating units (D), counterions (CI), and dissociated repeating units (F)] are in chemical reaction equilibrium.

For the example treated here (an aqueous solution of a strong electrolyte MX and a polyelectrolyte P where the salt and the repeating unit of the polyion are 1:1 electrolytes) the activity of water in an ideal solution is:

$$\frac{1}{M_W^*} \ln a_{w,id.mix}^{(m)} = -2 \cdot \frac{m_{MX}}{m^o} - (1+v) \cdot \frac{m_p}{m^o} \tag{180}$$

The following parameters must be known when the activity of water (or the osmotic coefficient) of an aqueous solution of a single polyelectrolyte is to be calculated:

The number of monomer units v is estimated from the number-averaged molecular mass of the polymer and the molecular mass of a repeating unit.

- UNIQUAC surface (q_k) and volume (r_k) parameters of water and the nondissociated repeating units are calculated by the method of Bondi [118]. No distinction is made between those parameters for the dissociated and nondissociated repeating units. The surface parameter of water $(q_w = 1.4)$ is also assigned to all counterions.
- The degree of counterion condensation k at infinite dilution in water is determined from experimental data for the osmotic coefficient at infinite dilution (as for $v >> 1$) $\Phi_p^{0,(m)} = 1 - k$.
- The chemical reaction (dissociation) constant K is one of the adjustable parameters of the model. It is assumed that, at constant temperature, K is a constant for a certain repeating unit.
- Parameter b^* that is used to describe the configurational change from a globular to a stretched conformation of the polyelectrolyte is also an adjustable model parameter.
- Binary parameters $(a_{i,j}^{(0)}$ and $a_{i,j}^{(1)})$ are used for interactions between all solute species in water. As these parameters are symmetric and as there are three solute species, there are 12 such parameters. However, all parameters $a_{i,j}^{(1)}$ are neglected, $(a_{i,j}^{(1)} = 0)$ and all parameters $a_{i,j}^{(0)}$ for interactions with the counterion are also neglected $(a_{i,CI}^{(0)} = 0$ for all solutes $i)$. The parameter for interactions between dissociated repeating units is also neglected $(a_{F,F}^{(0)} = 0)$. With these assumption, there are only two parameters: one for interactions between nondissociated repeating units $(C$ or $A)$ and one for interactions between these nondissociated monomers and the dissociated repeating units. The distinction between these binary parameters is also neglected, resulting in a single, adjustable binary interaction parameter that characterizes the polyelectrolyte's repeating unit A: $a_{A,A}^{(0)} = a_{A,F}^{(0)} = a_{p,p}$

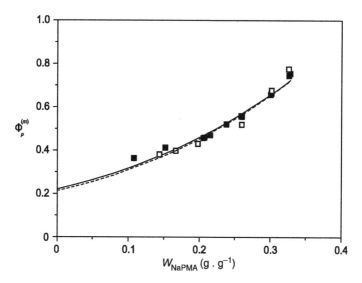

Fig. 15 Osmotic coefficient of aqueous solutions of poly(sodium methacrylate) at 298.2 K with two different molecular masses. Experimental results: *closed squares* NaPMA 6; *open squares* NaPMA 15. Correlation results: *solid line* NaPMA 6; *dashed line* NaPMA 15 [116]

Figure 15 shows a typical example for correlation of experimental results for the osmotic coefficient (on molality scale) of aqueous solutions of poly(sodium methacrylate).

For the calculation of the thermodynamic properties of an aqueous solution of a single polyion that additionally contains a low molecular weight strong electrolyte, some more model parameters are required. The volume and surface parameters of the ions of the strong electrolyte are also approximated by the parameters of water. Therefore, for an aqueous solution of the single salt the model does not differ from Pitzer's model, and for a large number of salts the binary interaction parameters $a_{MX}^{(0)}$ and $a_{MX}^{(1)}$ are available in the literature. All further interaction parameters (i.e., between cations and anions of the salt on one side and groups and counterions from the polyion on the other side) are also set to zero, with the exception of a single parameter. That parameter accounts for interactions between that ion of *MX* that carries an electrical charge of the opposite sign as the counterion of the polyion on one side, and the neutral group of the polyelectrolyte (i.e., *A* or *C*) on the other side. For example, if NaCl is added to an aqueous solution of poly(sodium methacrylate), the only additional interaction parameter is $a_{A,Cl}^{(0)}(=a_{p,Cl})$. Because the configuration of the polyion in the aqueous salt-containing solution might differ from that in the salt-free solution, it might be advantageous to consider the influence of the low molecular weight salt on the polyion's configuration parameter b^*. An empirical relation such as:

$$b^* = \varpi^{(0)} + \varpi^{(1)} \frac{m_{MX}}{m^o}, \qquad (181)$$

Aqueous Solutions of Polyelectrolytes

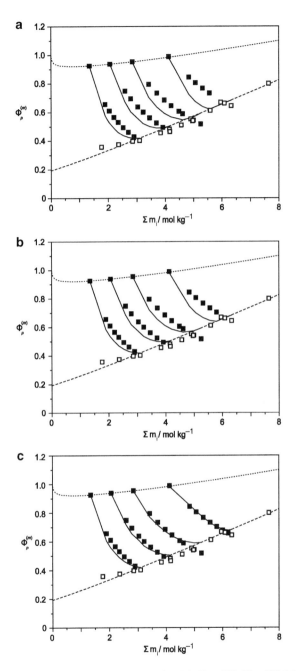

Fig. 16 Osmotic coefficient of aqueous solutions of NaPA 15 and NaCl at 298.2 K. Experimental results are shown with *symbols*. (**a**) Prediction results. (**b**) Correlation results setting $\varpi^{(1)} = 0$. (**c**) correlation results setting $\varpi^{(1)} \neq 0$. *Dashed lines* corresponds to systems without salt; *dotted lines* systems without polyelectrolyte; *solid lines* lines of constant water activity [116]

where $\varpi^{(0)}$ is the configurational parameter for the polyion when it is dissolved in pure water and $\varpi^{(1)}$ is an adjustable parameter, proved to be sufficient to describe that influence.

Predictions from the model for the osmotic coefficient can be made when the binary parameter between nondissociated repeating units and the counterion of the low molecular weight salt, as well as the influence of that salt on the configurational parameter $b*$ are neglected. Figure 16 shows comparisons between experimental data and calculation results for the osmotic coefficient for aqueous solutions of a sodium poly(acrylate) (NaPA 15) and NaCl. The osmotic coefficient (on molality scale) is plotted versus the "overall solute molality" $\sum \bar{m}_i$ that is defined as:

$$\sum \bar{m}_i = 2m_{MX} + (1+v)m_p. \tag{182}$$

The experimental results for the mixed solute systems are shown for a constant activity of water. The results extend from the polyelectrolyte (i.e., salt-free) system to the (NaCl + water) system. The top diagram of Fig. 16 shows the comparison with prediction results, i.e., the calculations were performed setting $a_{p,Cl} = 0$ and $\varpi^{(1)} = 0$. The middle diagram of Fig. 16 shows the comparison with correlation results when the influence of NaCl on the configurational parameter $b*$ is neglected (i.e., adjusting only $a_{p,Cl}$). The bottom diagram of Fig. 16 shows that the best agreement is achieved by adjusting both parameters. With those parameter an essential improvement is achieved, in particular at high concentrations (i.e., at low water activities). Figure 17 shows a comparison between the correlation results

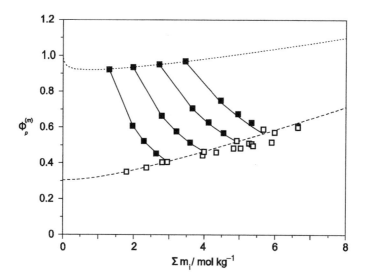

Fig. 17 Osmotic coefficient of aqueous solutions of NH$_4$PA 10 and NaCl at 298.2K. Experimental results are shown with *symbols*. *Dashed line* correlation results for system without salt; *dotted line* correlation results for system without polyelectrolyte; *solid lines* lines of constant water activity [116]

Aqueous Solutions of Polyelectrolytes

and the experimental data for the osmotic coefficient of aqueous solutions of ammonium poly(acrylate) (NH$_4$PA 10) and NaCl. For this particular system it was not necessary to consider an influence of NaCl on the configurational parameter. The comparisons reveal that the model is well suited for the correlation of the vapor–liquid equilibrium of aqueous solutions of polyelectrolytes with and without an added low molecular weight salt.

6 Summary

A literature review is given on the liquid–vapor phase equilibrium of aqueous solutions of polyelectrolytes. Experimental findings as well as selected thermodynamic models for the prediction and correlation of such phase equilibria are reviewed. The treatment of the thermodynamic models starts with theories and later focuses on combining the results from such theories with engineering models for the excess Gibbs energy. Such combinations allow for a good correlation of experimental data, for example, the osmotic coefficient (and related properties) of aqueous solutions of a single polyelectrolyte with and without an added salt.

References

1. Ohki A, Kawamoto K, Naka K, Maeda S (1992) Sensing of anionic polymers by an ion-selective electrode. Anal Sci 8:85–86
2. Ishii T, Bowen HK (1988) Dispersion and pressureless sintering of Al$_3$O$_3$ – SiC whiskers. Ceram Trans I:452–458
3. Pai SA, Kamat VS (1982) New dispersion aids in the service of surface coatings industry. J Colour Soc 21:23–29
4. Buscall R, Corner T, McGowan IJ (1982) Micro-electrophoresis of polyelectrolyte-coated particles. In: Tadros TF (ed) The effect of polymers on dispersion properties. Academic, London, pp 379–395
5. Hettche A, Trieselt W, Diessel P (1986) Co-Builder: Carboxylgruppenhaltige Verbindungen als Co-Builder in Waschmittelformulierungen. Tenside Surfact Det 23:12–19
6. Horn D (1989) Preparation and characterization of microdisperse bioavailable carotenoid hydrosols. Angew Makromol Chem 166/167:139–153
7. Philipp B, Gohlke U, Jaeger W, Kötz J (1989) Polymere lösen Umweltprobleme in der Wasserwirtschaft. Wissenschaft und Fortschritt 39:137–140
8. Kulicke WM, Lenk S, Detzner HD, Weiss T (1993) Anwendung von Polyelektrolyten bei der mechanischen Fest/Flüssig-Trennung. Chem-Ing-Tech 65:541–552
9. Westman L, Grundmark H, Petersson J (1989) Cationic polyelectrolytes as retention aids in newsprint production. Nord Pulp Pap Res J 4:113–116
10. Brouwer PH (1989) Kationische Kartoffelstärke und anionische Störsubstanzen. Wochenblatt für Papierfabrikation 117:881–889
11. Kennedy JP, Phillips GO, Williams A (1990) Cellulose sources & exploitation. Horwood, Chichester
12. Dautzenberg H, Jaeger W, Kötz J, Philipp B, Seidel C, Stscherbina D (1994) Polyelectrolytes: formation, characterization and application. Carl Hanser, München

13. Alleavitch J, Turner WA, Finch CA (1989) Gelatin. In: Elvers B (ed) Ullmann's encyclopedia of industrial chemistry, vol A12. VCH, Weinheim, pp 307–312
14. Petrie EM (2006) Handbook of adhesives and sealants. McGraw-Hill, New York
15. Putnam D, Kopeček J (1995) In Peppas NA, Langer RS (eds) Advances in polymer science, vol 122. Springer, Berlin, pp 55–123
16. Scranton AB, Rangarajan B, Klier J (1995) Biomedical applications of polyelectrolytes. In: Peppas NA, Langer RS (eds) Advances in polymer science, vol 122. Springer, Berlin, pp 1–54
17. Reddy M, Marinsky JA (1970) A further investigation of the osmotic properties of hydrogen and sodium polystyrenesulfonates. J Phys Chem 74:3884–3891
18. Chu P, Marinsky JA (1967) The osmotic properties of polystyrenesulfonates I. The osmotic coefficients. J Phys Chem 71:4352–4359
19. Oman S (1974) Osmotic coefficients of aqueous polyelectrolyte solutions at low concentrations, 1. Makromolekul Chem 175:2133–2140
20. Oman S (1974) Osmotic coefficients of aqueous polyelectrolyte solutions at low concentrations, 2. Makromolekul Chem 175:2141–2148
21. Oman S (1977) Osmotic coefficients of aqueous polyelectrolyte solutions at low concentrations, 3. Makromolekul Chem 178:475–485
22. Kozak D, Kristan J, Dolar D (1971) Osmotic coefficient of polyelectrolyte solutions 1. Polystyrenesulphonates with monovalent couterions. Z Phys Chem 76:85–92
23. Kozak D, Kristan J, Dolar D (1971) Osmotic coefficient of polyelectrolyte solutions 2. Polystyrenesulphonates with divalent counterions. Z Phys Chem 76:93–97
24. Ise N, Okubo T (1968) Mean activity coefficient of polyelectrolytes VIII. Osmotic and activity coefficients of polystyrenesulfonates of various gegenions. J Phys Chem 72: 1361–1366
25. Kakehashi R, Yamazoe H, Maeda H (1998) Osmotic coefficients of vinylic polyelectrolyte solutions without added salt. Colloid Polym Sci 276:28–33
26. Takahashi A, Kato N, Nagasawa M (1970) The osmotic pressure of polyelectrolyte in neutral salt solutions. J Phys Chem 74:944–946
27. Boyd GE (1974) Thermodynamic properties of strong electrolyte-strong polyelectrolyte mixtures at 25°C. In: Sélégny E (ed) Polyelectrolytes. D. Reidel, Boston, pp 135–155
28. Lammertz S, Pessôa Filho PA, Maurer G (2008) Thermodynamics of aqueous solutions of polyelectrolytes: experimental results for the activity of water in aqueous solutions of (a single synthetic polyelectrolyte and sodium chloride). J Chem Eng Data 53:1796–1802
29. Oman S, Dolar D (1967) Activity coefficients of counterions in solutions of polymethylstyrene-sulphonic acid and its salts 1. Monovalent counterions. Z Phys Chem 56:1–12
30. Oman S, Dolar D (1967) Activity coefficients of counterions in solutions of polymethylstyrene-sulphonic acid and its salts 2. Divalent counterions. Z Phys Chem 56:13–19
31. Ise N, Okubo T (1967) Mean activity coefficient of polyelectrolytes IV. Isopiestic measurements of sodium polyacrylates. J Phys Chem 71:1287–1290
32. Asai K, Takaya K, Ise N (1969) Mean activity coefficient of polyelectrolytes XI. Activity coefficients of various salts of polyacrylic acid and carboxymethylcellulose. J Phys Chem 73:4071–4076
33. Ise N, Okubo T (1965) Mean activity coefficients of polyelectrolytes I. Measurements of sodium polyacrylate. J Phys Chem 69:4102–4109
34. Okubo T, Nishizaki Y, Ise N (1965) Single-ion activity coefficient of gegenions in sodium polyacrylate. J Phys Chem 69:3690–3695
35. Ise N, Okubo T (1967) Mean activity coefficients of polyelectrolytes V. Measurements of polyvinyl sulfates of various gegenions. J Phys Chem 71:1886–1890
36. Ise N, Okubo T (1968) Mean activity coefficients of polyelectrolytes X. Measurements of polyphosphates of various gegenions. J Phys Chem 72:1370–1373
37. Ise N, Asai K (1968) Mean activity coefficients of polyelectrolytes IX. Measurements of polyethylene-sulfonates of various gegenions. J Phys Chem 72:1366–1369

Aqueous Solutions of Polyelectrolytes

38. Horváth J, Nagy M (2006) Role of linear charge density and counterion quality in thermodynamic properties of strong acid type polyelectrolytes: divalent transition metal cations. Langmuir 22:10963–10971
39. Nagy M (2004) Thermodynamic study of polyelectrolytes in aqueous solutions without added salts: a transition from neutral to charged macromolecules. J Phys Chem B 108: 8866–8875
40. Lipar I, Zalar P, Pohar C, Vlachy V (2007) Thermodynamic characterization of polyanetholesulfonic acid and its alkaline salts. J Phys Chem B 111:10130–10136
41. Lammertz S, Pessôa Filho PA, Maurer G (2008) Thermodynamics of aqueous solutions of polyelectrolytes: experimental results for the activity of water in aqueous solutions of some single synthetic polyelectrolytes. J Chem Eng Data 53:1564–1570
42. Koene RS, Nicolai T, Mandel M (1983) Scaling relations for aqueous polyelectrolyte-salt solutions 3. Osmotic pressure as a function of molar mass and ionic strength in the semidilute regime. Macromolecules 16:231–236
43. Dolar D, Bester M (1995) Activity coefficient of a polyelectrolyte from solubility measurements. J Phys Chem 99:4763–4767
44. Kakehashi R, Maeda H (1996) Donnan equilibria of simple electrolytes in polyelectrolyte solutions 2. Extension to polycation and the effect of charge densities. J Chem Soc Faraday Trans 92:4441–4444
45. Kakehashi R, Maeda H (1996) Donnan eqilibria of simple electrolytes in polyelectrolyte solutions. J Chem Soc Faraday Trans 92:3117–3121
46. Nordmeier E, Dauwe W (1991) Studies of polyelectrolyte solutions I. Counterion condensation by polystyrenesulfonate. Polym J 23:1297–1305
47. Nordmeier E, Dauwe W (1992) Studies of polyelectrolyte solutions II. The second virial coefficient and the persistence length. Polym J 24:229–238
48. Okubo T, Ise N, Matsui F (1967) Mean activity coefficient of polyelectrolytes in the ternary system water-sodium polyacrylate-sodium chloride. J Am Chem Soc 89:3697–3703
49. Bloys van Treslong CJ, Moonen P (1978) Distribution of counterions in solution of weak polyelectrolytes. A study to the effects of neighbour interaction between charged sites and the structure of the macromolecule. Recueil J Roy Netherlands Chem Soc 97:22–27
50. Alexandowicz Z (1960) Results of osmotic and of Donnan equilibria: measurements in polymethacrylic acid- sodium bromide solutions II. J Polym Sci 43:337–349
51. Bordi F, Cametti C, Biasio AD (1987) Osmotic pressure of aqueous rod-like polyelectrolyte solutions with mono- and divalent counterions. Ber Bunsen Gesell 91:737–740
52. Joshi YM, Kwak JCT (1979) Mean and single ion activity coefficients in aqueous mixtures of sodium chloride and sodium pectate, sodium pectinate, and sodium CMC. J Phys Chem 83:1978–1983
53. Kowblansky M, Zema P (1981) Effect of polyelectrolyte charge density on calcium ion activity coefficients and additivity in aqueous solutions of calcium acrylamide-acrylic acid copolymers. Macromolecules 14:1448–1451
54. Kowblansky M, Zema P (1981) Interactions of sodium ions with the sodium salts of poly(acrylic acid/acrylamide) copolymers of varying charge density. Macromolecules 14:166–170
55. Nordmeier E (1994) Studies of polyelectrolyte solutions V. Effects of counterion binding by polyions of varying charge density and constant degree of polymerization. Polym J 26:539–550
56. Baré W, Nordmeier E (1996) Studies of polyelectrolyte solutions VI. Effects of counterion binding by dextran sulfate and dextran phosphate in aqueous/organic solvents. Polym J 28:712–726
57. Takahashi A, Kato T, Nagasawa M (1967) The second virial coefficient of polyelectrolytes. J Phys Chem 71:2001–2010
58. Nordmeier E (1993) Studies of polyelectrolyte solutions III. Correlations between the second and the third virial coefficient. Polym J 25:1–17

59. Orofino TA, Flory PJ (1959) The second virial coefficient for polyelectrolyte. Theory and experiment. J Phys Chem 63:283–290
60. Hara M, Nakajima A (1980) Anomalies in light scattering from polyelectrolyte in semi-dilute solution region. Polym J 12:711–718
61. Hara M, Nakajima A (1980) Characteristic behaviors of light scattering from polyelectrolyte in dilute solution region. Polym J 12:701–709
62. Trap HJL, Hermans JJ (1954) Light-scattering by polymethacrylic acid and carboxymethyl-cellulose in various solvents. J Phys Chem 58:757–761
63. Schneider NS, Doty P (1954) Macro-ions IV. The ionic strength dependence of the molecular properties of sodium carboxymethylcellulose. J Phys Chem 58:762–769
64. Fisher LW, Sochor AR, Tan JS (1977) Chain characteristics of poly(2-acrylamido-2-methyl-propanesulfonate) polymers. 1. Light scattering and intrinsic-viscosity studies. Macromole-cules 10:949–954
65. Griebel T, Kulicke WM, Hashemzadeh A (1991) Characterization of water-soluble, cationic polyelectrolytes as exemplified by poly(acrylamido-co-trimethylammoniumethylmethacrylate) and the establishment of structure-property relationships. Colloid Polym Sci 269:113–120
66. Dautzenberg H, Rother G (1991) Light-scattering studies on supramolecular structures in polymer solutions. J Appl Polym Sci 48:351–369
67. Burkhardt CW, McCarthy KJ, Parazak DP (1987) Solution properties of poly(dimethyldial-lyl ammonium chloride). J Polym Sci Pol Lett 25:209–213
68. Ochiai H, Handa M, Matsumoto H, Moriga T, Murakami I (1985) Dilute solution properties of poly(allylammonium chloride) in aqueous sodium chloride solutions. Makromolekul Chem 186:2547–2556
69. Stickler M (1984) Molecular weight distribution and solution properties of poly(2-trimethyl-ammoniumethyl methacrylate chloride). In: Forschung d. Röhm GmbH. Hüthig & Wepf, Basel, pp. 85–117
70. Hasse H, Kany HP, Tintinger R, Maurer G (1995) Osmotic virial coefficients of aqueous poly (ethylene glycol) from laser-light scattering and isopiestic measurements. Macromolecules 28:3540–3552
71. Kany HP, Hasse H, Maurer G (1999) Thermodynamic properties of aqueous dextran solu-tions from laser-light scattering, membrane osmometry and isopiestic measurements. J Chem Eng Data 44:230–242
72. Kany HP, Hasse H, Maurer G (2003) Thermodynamic properties of aqueous poly(vinyl-pyrrolidone) solutions from laser-light scattering, membrane osmometry and isopiestic measurements. J Chem Eng Data 48:689–698
73. Katchalsky A, Lifson S, Mazur J (1953) The electrostatic free energy of polyelectrolyte solutions I. Randomly kinked macromolecules. J Polym Sci 11:409–423
74. Lifson S, Katchalsky A (1954) The electrostatic free energy of polyelectrolyte solutions II. Fully stretched macromolecules. J Polym Sci 13:43–55
75. Dolar D, Peterlin A (1969) Rodlike model for a polyelectrolyte solution with mono- and divalent counterions. J Chem Phys 50:3011–3015
76. Manning GS (1969) Limiting laws and counterion condensation in polyelectrolyte solutions I. Colligative properties. J Chem Phys 51:924–933
77. Manning GS (1978) The molecular theory of polyelectrolyte solutions with applications to the electrostatic properties of polynucleotides. Q Rev Biophys 11:179–246
78. Manning GS (1984) Limiting laws and counterion condensation in polyelectrolyte solutions 8. Mixtures of counterions, species selectivity and valence selectivity. J Phys Chem 88:6654–6661
79. Stevens MJ, Kremer K (1996) The nature of flexible linear polyelectrolytes in salt-free solution: a molecular dynamics study. J Chem Phys 103:1669–1690
80. Feng Z, Liu HL, Hu Y (1996) Study on thermodynamic properties of polyelectrolyte solutions. Acta Chim Sinica 54:1076–1083
81. Ospeck M, Fraden S (1998) Solving the Poisson–Boltzmann equation to obtain interaction energies between confined, like-charged cylinders. J Chem Phys 109:9166–9171

82. Dähnert K, Huster D (1999) Comparison of the Poisson–Boltzmann model and the Donnan equilibrium of a polyelectrolyte in salt solution. J Colloid Interf Sci 215:131–139
83. Dähnert K, Huster D (2000) Thermodynamics of the laminar Donnan system. J Colloid Interf Sci 228:226–237
84. Rödenbeck M, Müller D, Huster D, Arnold K (2001) Counterion condensation as saturation effect under the influence of ion hydration. Biophys Chem 90:255–268
85. Jiang JW, Liu HL, Hu Y, Prausnitz JM (1998) A molecular-thermodynamic model for polyelectrolyte solutions. J Chem Phys 108:780–784
86. Jiang JW, Liu HL, Hu Y (1999) Polyelectrolyte solutions with stickiness between polyions and counterions. J Chem Phys 110:4952–4962
87. Blaul J, Wittemann M, Ballauff M, Rehahn M (2000) Osmotic coefficient of a synthetic rodlike polyelectrolyte in salt-free solution as a test of the Poisson–Boltzmann cell model. J Phys Chem B 104:255–268
88. Deserno M, Holm C, Blaul J, Ballauff M, Rehahn M (2001) The osmotic coefficient of rod-like polyelectrolytes: computer simulation, analytical theory, and experiment. Eur Phys J E 5:97–103
89. Colby RH, Boris DC, Krause WE, Tan JS (1997) Polyelectrolyte conductivity. J Polym Sci B 35:2951–2960
90. Diehl A, Carmona HA, Levin Y (2001) Counterion correlations and attraction between like-charged macromolecules. Phys Rev E 64:011804-1–011804-6
91. Chang R, Yethiraj A (2005) Osmotic pressure of salt-free polyelectrolyte solutions: a Monte Carlo simulation study. Macromolecules 38:607–616
92. Antypov D, Holm C (2007) Osmotic coefficient calculations for dilute solutions of short stiff-chain polyelectrolytes. Macromolecules 40:731–738
93. Baeurle SA, Charlot M, Nogovitsin EA (2007) Grand canonical investigations of prototypical polyelectrolyte models beyond the mean field level of approximation. Phys Rev E 75:011804-1–011804-11
94. De Gennes PG, Pincus P, Velasco RM (1976) Remarks on polyelectrolyte conformation. J Phys (Paris) 37:1461–1473
95. Odijk T (1977) Polyelectrolytes near the rod limit. J Polym Sci Pol Phys 15:477–483
96. Odijk T (1979) Possible scaling relations for semidilute polyelectrolyte solutions. Macromolecules 12:688–692
97. Odijk T (1983) On the limiting law solution of the cylindrical Poisson-Boltzmann equation for polyelectrolytes. Chem Phys Lett 100:145–150
98. Koene RS, Mandel M (1983) Scaling relations for aqueous polyelectrolyte-salt solutions. 1. Quasi-elastic light scattering as a function of polyelectrolyte concentration and molar mass. Macromolecules 16:220–227
99. Koene RS, Mandel M (1983) Scaling relations for aqueous polyelectrolyte-salt solutions. 2. Quasi-elastic light scattering as a function of polyelectrolyte concentration and salt concentration. Macromolecules 16:227–231
100. Mandel M (1993) Some properties of polyelectrolyte solutions and the scaling approach. In: Hara M (ed) Polyelectrolytes. Marcel Dekker, New York, pp 1–76
101. Nagvekar M, Danner RP (1989) An excess free energy model for polyelectrolyte solutions. Fluid Phase Equilibr 53:219–227
102. Nagvekar M, Tihminlioglu F, Danner RP (1998) Colligative properties of polyelectrolyte solutions. Fluid Phase Equilibr 145:15–41
103. Dolar D, Kozak D (1970) Osmotic coefficients of polyelectrolyte solutions with mono- and divalent counterions. Proc Leiden Symp 11:363–366
104. Katchalsky A (1971) Polyelectrolytes. Pure Appl Chem 26:327–374
105. Pitzer KS (1973) Thermodynamics of electrolytes I. Theoretical basis and general equations. J Phys Chem 77:268–277
106. Manning GS (1974) Limiting laws for equilibrium and transport properties of polyelectrolyte solutions. In: Sélégny E (ed) Polyelectrolytes. D. Reidel, Boston, pp 9–37

107. Nordmeier E (1995) Advances in polyelectrolyte research: counterion binding phenomena, dynamic processes and the helix coil transition of DNA. Macromol Chem Phys 196:1321–1374
108. Hao MH, Harvey SC (1992) A lattice theory for counterion binding on polyelectrolytes. Macromolecules 25:2200–2208
109. Chen CC, Evans LB, Harvey C (1986) A local composition model for the excess Gibbs energy of aqueous electrolyte systems. AIChE J 32:444–454
110. Pessôa Filho PA, Maurer G (2008) An extension of the Pitzer equation to aqueous polyelectrolyte solutions. Fluid Phase Equilibr 269:25–35
111. Pitzer KS, Mayorga G (1973) Thermodynamics of electrolytes: IV. Activity and osmotic coefficients for strong electrolytes with one or both ions univalent. J Phys Chem 77:2300–2308
112. Großmann C, Tintinger R, Zhu J, Maurer G (1995) Aqueous two-phase systems of poly (ethylene glycol) and di-potassium hydrogen phosphate with and without partitioning biomolecules – experimental results and modeling of thermodynamic properties Ber. Bunsenges Phys Chem 99:700–712
113. Großmann C, Tintinger R, Zhu J, Maurer G (1995) Aqueous two-phase systems of poly (ethylene glycol) and dextran: experimental results and modeling of thermodynamic properties. Fluid Phase Equilibr 106:111–138
114. Großmann C, Tintinger R, Zhu J, Maurer G (1997) Partitioning of some amino acids and low molecular peptides in aqueous two-phase systems of poly(ethylene glycol) and di-potassium hydrogen phosphate. Fluid Phase Equilibr 137:209–228
115. Grünfelder T, Pessôa Filho PA, Maurer G (2009) Liquid-liquid equilibrium data of aqueous two-phase systems containing some synthetic polyelectrolytes and polyethylene glycol. J Chem Eng Data 54:198–207
116. Lammertz S, Grünfelder T, Ninni L, Maurer G (2009) Model for the Gibbs excess energy of aqueous polyelectrolyte solutions. Fluid Phase Equilibr 280:132–143
117. Abrams DS, Prausnitz JM (1975) Statistical thermodynamics of liquid mixtures: a new expression for the excess Gibbs energy of partly or completely miscible systems. AIChE J 21:116–128
118. Bondi A (1968) Physical properties of molecular crystals, liquids and glasses. Wiley, New York

Adv Polym Sci (2011) 238: 137–177
DOI: 10.1007/12_2010_83
© Springer-Verlag Berlin Heidelberg 2010
Published online: 14 July 2010

Gas–Polymer Interactions: Key Thermodynamic Data and Thermophysical Properties

Jean-Pierre E. Grolier and Séverine A.E. Boyer

Abstract Gas–polymer interactions play a pivotal role in the formation of different molecular organizations/reorganizations of polymeric structures. Such structural modifications can have a negative impact on the material properties and should be understood in order to prevent them or these modifications are of engineering interest and they should be purposely tailored and properly controlled. Two newly developed techniques, gas-sorption/solubility and scanning transitiometry, are shown to be well adapted to provide the necessary (key) data to better understand and monitor the polymeric modifications observed under the triple constraints of temperature, elevated pressure, and gas sorption. This article illustrates the major contribution of gas–polymer interactions in different interconnected applied and engineering fields of the petroleum industry, polymer science, and microelectronics.

Keywords Gas sorption · Glass transition · High pressure · Self-assembling · Solubility · Transitiometry · Vibrating-wire technique

Contents

1 Introduction 138
2 Experimental Techniques 140
    2.1 Gas Sorption and Solubility 140
    2.2 pVT–Calorimetry: Scanning Transitiometry 144
3 Gas-Polymer Interactions and Practical Applications 149

J.-P.E. Grolier (✉)

Laboratoire de Thermodynamique des Solutions et des Polymères, Université Blaise Pascal de Clermont-Ferrand, 63177 Aubière, France
e-mail: j-pierre.grolier@univ-bpclermont.fr

S.A.E. Boyer

Centre de Mise en Forme des Matériaux (CEMEF), Mines ParisTech, 06904 Sophia Antipolis, France
e-mail: severine.boyer@mines-paristech.fr; severine.boyer@univ-bpclermont.fr

3.1	Evaluation of Gas Solubility and Associated Swelling	150
3.2	Gas–Polymer Interaction Energy	153
3.3	Thermophysical Properties at High Pressures	153
3.4	Phase Transition at High Pressures	161
4	Conclusion	174
References		175

1 Introduction

Gas–polymer interactions play a pivotal role in polymer science for the development of new polymeric structures for specific applications. Typically, this is the case for polymer foaming [1] and for self-assembling of nanoscale structures [2, 3]. Not only the nature of the gas, but also the thermodynamic conditions, are essential factors in control of the processing operations. For this, the amount of gas solubilized has to be accurately determined together with the possible associated swelling of the polymer due to the gas sorption. Another important applied field in which gas sorption in polymers has to be documented through intensive investigations concerns the (non)controlled sorption of light gases in polymers that are used in industry for items such as seals, containers, flexible hosepipes, and pipelines. Nowadays, polymer-based materials are at the center of applications in which they are frequently subjected to temperature variations and also to gas pressures ranging from a few megapascal to 100 MPa or even more. An important example of the large-scale use of polymer materials is in the transport of petroleum fluids [4] using flexible hosepipes; these hosepipes are made of extruded thermoplastic or rubber sheaths and reinforcing metallic armor layers. The type of transported fluids (which might contain important amounts of dissolved gases) and the operating temperature and pressure dictate the composition of the hosepipe sheath. However, these thermoplastic polymers, like elastomers, are not entirely impermeable and undergo sorption/diffusion phenomena. A rupture of the thermodynamic equilibrium after a sharp pressure drop could eventually damage the polymer components. Gas concentration in the polymer, together with temperature gradients, can cause irreversible "explosive" deterioration of the polymeric structures. This blistering phenomenon, usually termed "explosive decompression failure" (XDF), is actually very dramatic for the material. The resistance to physical changes is related to the influence of the gas–polymer interactions on the thermophysical properties of the polymer. The estimation of the gas sorption and of the concomitant polymer swelling, as well as the measurement of the thermal effect associated with the gas–polymer interactions, provide valuable and basic information for a better understanding and control of polymer behavior in different applications in which temperature and pressure, in combination with gas sorption, might deeply affect polymer stability and properties. The striking effect of gas sorption is particularly observed when the gas is in a supercritical state, depending on the thermodynamic conditions.

Other numerous industrial activities deal with polymer modifications and transformations, through different processes like extrusion, injection, and molding. Polymer foaming, among others, is currently achieved in various ways, but typically involves elevated temperatures and pressures as well as the addition of chemicals, mostly gases that are used as blowing agents. Thermal, barometric, and/or chemical stress can shift, even permanently, the polymer glass transition temperature, T_g, which consequently modifies the physical properties of the material. Sorption of fluids such as gases in the supercritical state induces significant plasticization, resulting in a substantial decrease of T_g. If such an effect is rather weak when using helium or nitrogen, due to their low solubility in polymers, sufficiently high pressure should induce higher gas sorption by polymers. In this respect, gases such as carbon dioxide or hydrofluorocarbons (HFCs) are known to be good fluids for plasticization of a polymer like polystyrene (PS). As a result of international regulations, the blowing gases intensively used in the foaming industry have to be replaced by blowing agents that are less harmful to the ozone layer. Knowledge of the influence of gas sorption and concomitant swelling on the T_g of a {gas–polymer} system is of real importance in generating different types of foams. In the context of the above applications, the thermophysical properties of gas-saturated thermoplastic semicrystalline polymers are key elements for the development of several engineering applications.

Typically, thermophysical properties feature the most important information when dealing with materials submitted to thermal variations and/or mechanical constraints. The properties of interest are of two types: bulk properties and phase transition properties. The bulk properties are either caloric properties like heat capacities C_P, or mechanical properties like isobaric thermal expansivities α_P, isothermal compressibilities κ_T, and isochoric thermal pressure coefficients β_V. The need for accurate control of thermodynamic properties concerns the two main phase transitions: the first-order transitions of melting and crystallization, and the glass transition. All these properties are now accessible thanks to recent progress in various technologies that allow measurements in the three physical states over extended ranges of pressure (p) and temperature (T), including in the vicinity of the critical point. In this respect, knowledge (i.e., measurement) of the thermophysical properties of polymers over extended ranges of temperature and pressures and in different gaseous environments is absolutely necessary to improve the use and life-time of end-products made of polymeric materials.

Examples have been selected in three main domains: oil exploitation and transport, polymer foaming and modification, and self-assembling nanostructures. These examples are directly connected to industrial activities in the petroleum industry, the insulating material industry, and the microelectronic industry. In many cases, gases and polymers of different types intimately interact under external conditions of T and p. In the subsequent examples, the {gas–polymer} systems selected for a targeted industrial purpose (e.g., foaming materials and material processing) are polymeric materials in contact with {gas–liquid} systems (e.g., pipes or tanks in the gas and petroleum industry), or are used as intermediate materials to elaborate templates for making 3D electronic circuitry.

140 J.-P.E. Grolier and S.A.E. Boyer

The foaming materials industry is a rapidly growing area where constant innovation and added-value products are key factors for economic success in the face of high international competition. The mastering of polymer degradation (typically blistering) by high pressure dissolved gases is another key issue. Microelectronics is presently the most competitive industrial activity. The focus of the present article is thus on the behavior of {gas–polymer} systems from the point of view of gas solubility and associated thermal effects. Depending on the temperature and pressure ranges, polymers are either in the solid or molten state, i.e., at temperatures between T_g and the temperature of melting, T_m; in most cases, gases are supercritical fluids (SCFs). The present contribution, essentially based on current activities of the authors, is split into two main parts: experimental measurements (Sect. 2) and evaluation of gas-polymer interactions (Sect. 3) through experimental measurements of gas solubility (Sect. 3.1), thermal effects reflecting interaction energies (3.2), thermophysical properties of polymers (3.3) and phase transitions (3.4). In addition, the importance of such data for engineering applications is stressed.

This article illustrates the contribution of two techniques in providing accurate information to meet the demand for the data described above: the vibrating-wire (VW)–pressure-volume-temperature (pVT) technique for gas sorption and polymer swelling; and scanning transitiometry for simultaneous thermal and mechanical measurements. Two complementary thermodynamic approaches have been developed to characterize gas–polymer interactions in evaluating either gravimetric and volumetric changes or thermally energetic changes associated with gas sorption (up to saturation) in a polymer. The first approach is based on a "weighing technique" using a VW sensor coupled with a pVT method. The second approach is based on the coupling of a calorimetric detector with a p, V, or T scanning technique.

2 Experimental Techniques

2.1 Gas Sorption and Solubility

Gas solubility in polymers can be measured using different techniques, i.e., gravimetric techniques, including vibrating or oscillating techniques; pVT techniques with the pressure decay method; and gas-flow techniques. A brief review of existing techniques is given below, followed by the description of a technique we recently developed that couples a new gravimetric technique with a pVT–pressure decay technique.

2.1.1 Gravimetric Techniques

These techniques consist in precisely weighing a polymer sample during gas sorption. They are very sensitive at low-to-moderate gas pressures [5, 6], and use

Key Thermodynamic Data and Thermophysical Properties

of a magnetic coupling to transmit the weight to a balance [7, 8] has permitted the pressure range to be extended up to 35 MPa.

With vibrating or oscillating techniques, the change of mass of a polymer sample is calculated from the resonance characteristics of a vibrating support, either a piezoelectric crystal [9, 10] or a metal reed [11], to which the polymer sample is fixed (very often this support is a spherical quartz resonator on which a thin polymer film is wrapped). Depending on the type of oscillator, the maximum pressure can be between 15 and 30 MPa.

With the pVT techniques based on the pressure decay method [12, 13], a polymer sample is seated in a container of known volume acting as equilibrium cell; the quantity of gas initially introduced in this cell is evaluated by pVT measurements in a calibrated cell from which the gas is transferred into the equilibrium cell in a series of isothermal expansions. The pressure decay in the equilibrium cell during sorption permits evaluation of the amount of gas penetrating into the polymer. The pressure decay principle allows a sensitivity of few hundredths of milligram of absorbed gas per gram of polymer [14].

With the glass flow techniques, the solubility of gases in polymers is obtained from gas flow measurements by inverse gas chromatography [15]. In this procedure, the polymer sample (glassy or molten) acts as the chromatographic stationary phase to measure retention times.

2.1.2 Coupled VW–pVT Technique

In all the techniques where the polymer sample is immerged in the penetrating gas, the associated swelling of the polymer due to the gas sorption is an important phenomenon that needs to be accurately evaluated. Swelling can affect the buoyancy force exerted by the gas on the polymer sample in the case of gravimetric measurements, as well as the internal volume in the case of pVT measurements. Usually, swelling is determined separately by techniques using direct visual observation and estimation of the volume change and is in the order of 0.3% of the volume of the initially degassed polymer [16]. Alternatively, swelling has been estimated using a theoretical model like the Sanchez–Lacombe molecular theory [17].

Recently, Hilic et al. [18, 19] designed an original technique to evaluate the gas solubility in polymers that permits simultaneous determination in situ of the amount of gas penetrating the polymer and the concomitant change in volume of the polymer due to gas sorption. This technique associates a VW force sensor, acting as gravimetric device, and a pressure decay installation to evaluate the amount of gas penetrating into the polymer.

Vibrating-Wire Sensor

The VW sensor (Fig. 1) is employed as a force sensor to weigh the polymer sample during the sorption: the buoyancy force exerted by the pressurized fluid on the

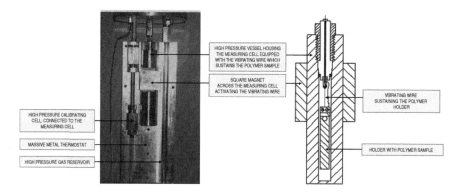

Fig. 1 The coupled VW–*pVT* technique. *Left*: Photograph of the inside of the experimental setup showing the three high-pressure cells. *Right*: Equilibrium cell that houses the VW sensor and the holder containing the polymer sample

polymer depends on the swollen volume of the polymer due to the gas sorption. This VW sensor is essentially a high-pressure cell in which the polymer sample is placed in a holder suspended by a thin tungsten wire (diameter 25 μm, length 30 mm) in such a way that the wire is positioned in the middle of a high magnetic field generated by a square magnet placed across the high-pressure cell. Through appropriate electric circuitry and electronic control, the tungsten wire is activated to vibrate. The period of vibration, which can be accurately measured, is directly related to the mass of the suspended sample. The working equation (1) for the VW sensor relates the mass m_{sol} of gas absorbed (solubilized) in the polymer to the change in volume ΔV_{pol} of the polymer. The natural angular frequency of the wire, through which the polymer sample holder is suspended, depends on the amount of gas absorbed. The physical characteristics of the wire are accounted for in (1) as:

$$m_{sol} = \rho_g \, \Delta V_{pol} + \left[\left(\omega_B^2 - \omega_0^2 \right) \frac{4L^2 R^2 \rho_S}{\pi \, g} + \rho \left(V_C + V_{pol} \right) \right]. \tag{1}$$

The volume of the degassed polymer is represented by V_{pol}, and ρ_g is the density of the fluid. The terms ω_0 with ω_B represent the natural (angular) frequencies of the wire in vacuum and under pressure, respectively, and V_C is the volume of the holder. The symbols L, R, and ρ_s are, respectively, the length, the radius, and the density of the wire.

pVT Method and Pressure Decay Measurements

For *pVT* measurements, the three-cell principle of Sato et al. [14] has been used (Fig. 2) to determine the amount of gas solubilized in the polymer. The experimental method consists of a series of successive transfers of the gas by connecting the

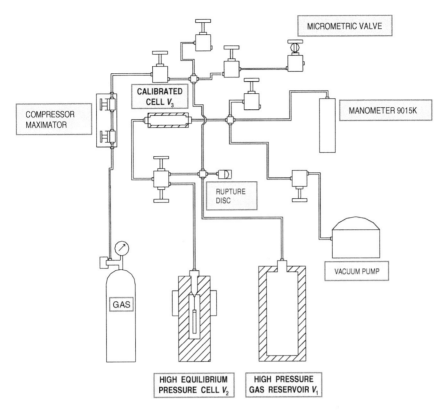

Fig. 2 Three-cell principle for *pVT* measurements, after Sato et al. [14]. The high-pressure line connects the three cells: the high-pressure reservoir cell (V_1), the high-pressure equilibrium cell housing the VW sensor (V_2), and the high-pressure calibrated transfer cell (V_3)

calibrated transfer cell V_3 to the equilibrium cell V_2, which contains the polymer. Initial p_i and final p_f pressures are recorded between each transfer. The initial methodology was based on the iterative calculation described by Hilic et al. [18, 19]. The (rigorous) working equation (2) for the *pVT* technique gives the amount of gas entering the polymer sample during the first transfer, once equilibration is attained:

$$m_{sol} = \frac{M_g}{R} \frac{p_f}{T_f\, Z_f} \Delta V_{pol} + \frac{M_g}{R} \left[\frac{p_i}{Z_i\, T_i} V_3 - \frac{p_f}{Z_f\, T_f} (V_2 + V_3 - V_{pol}) \right]. \qquad (2)$$

Equation 2 permits calculation of the mass m_{sol} of gas dissolved in the polymer. M_g is the molar mass of the dissolved gas. Z_i and Z_f are the compression factors of the gas entering the polymer at the initial and final equilibrium sorption conditions, respectively. Volume of the degassed polymer and the volume change due to sorption are represented by V_{pol} and ΔV_{pol}, respectively. The total amount of gas absorbed by the polymer after completion of the successive transfers is given by (3):

$$\Delta m_{sol}^{(k)} = \frac{M_g}{R} \frac{p_f^{(k)} \Delta V_{pol}^{(k)}}{Z_f^{(k)} T_f^{(k)}}$$

$$+ \frac{M_g}{R} \left[\frac{p_i^{(k)} V_3}{Z_i^{(k)} T_i^{(k)}} + \frac{p_f^{(k-1)} \left(V_2 - V_p - \Delta V_{pol}^{(k-1)} \right)}{Z_f^{(k-1)} T_f^{(k-1)}} - \frac{p_f^{(k)} \left(V_2 + V_3 - V_{pol} \right)}{Z_f^{(k)} T_f^{(k)}} \right], \tag{3}$$

where $\Delta m_{sol}^{(k)}$ is the increment in dissolved gas mass resulting from the transfer k, and $\Delta V_{pol}^{(k)}$ is the change in volume after transfer k.

2.2 pVT–Calorimetry: Scanning Transitiometry

Certainly, calorimetry is a major technique for measurement of the thermodynamic properties of substances and for following phase change phenomena. In most applications, calorimetry is carried out at constant pressure, while the tracked phenomenon is observed with increasing or decreasing temperature. The possibility of controlling the three most important thermodynamic variables (p, V, and T) during calorimetric measurements makes it possible to perform simultaneous measurements of both thermal and mechanical contributions to the thermodynamic potential changes caused by the perturbation. Calorimetric techniques provide valuable additional information on transitions in complex systems. Their contribution to the total change of thermodynamic potential not only leads to the complete thermodynamic description of the system under study, but also permits the investigation of systems with limited stability or systems with irreversible transitions. By a proper external change of the controlling variable, the course of a transition under investigation can be accelerated, impeded, or even stopped at any degree of its advancement and then taken back to the beginning, all with simultaneous recording of the heat and mechanical variable variations. The seminal presentation by Randzio [20] of thermodynamic fundamentals for the use of state variables p, V, and T in scanning calorimetric measurements opened the path [21–23] from pVT–calorimetry to the now well-established technique of scanning transitiometry [24, 25]. The main characteristics of scanning transitiometry are reviewed in this section.

Practically, the technique utilizes the principle of differential heat flux calorimetry, with which it is possible to operate under four thermodynamic situations where the perfectly controlled variation (or perturbation) of one of the three state variables (p, V, or T) is simultaneously recorded with the thermal effect resulting from the generated perturbation of the system under investigation. The principle of scanning transitiometry [23] offers the possibility to scan, in the measuring calorimetric cell, one of the three independent thermodynamic variables (p, V, or T)

while keeping another one constant. During this scan, the variation of the (third) dependable variable (i.e., the mechanical output) and the calorimetric energy generated (i.e., the thermal output) are recorded simultaneously in situ in the measuring cell. From these two quantities, associated to a given scan, two thermodynamic derivatives, mechanical and thermal, are thus determined. The derivatives perfectly characterize the evolution of the thermodynamic potential of the investigated system, particularly any undergone transition or state change induced by the variable scan. As illustrated in Fig. 3, making use of the rigorous Maxwell relations between thermodynamic derivatives, it is possible to directly obtain the ensemble of the thermophysical properties; undoubtedly this shows the potentiality of the technique. During measurements, it is essential that the different scans be performed with sufficiently slow rates in order to keep the investigated system at equilibrium over the entire scan and so that the (Maxwell) thermodynamic relations remain valid.

The four possible thermodynamic situations (Fig. 3) are obtained by simultaneous recording of both the heat flux (thermal output) and the change of the dependent variable (mechanical output). Then, making use of the respective related Maxwell relations, one readily obtains the main thermophysical properties as follows: (a) scanning pressure under isothermal conditions yields the isobaric thermal expansivity α_p and the isothermal compressibility κ_T as functions of pressure at a given temperature; (b) scanning volume under isothermal conditions yields the isochoric thermal pressure coefficient β_V and the isothermal compressibility κ_T as functions of volume at a given temperature; (c) scanning temperature under isobaric conditions yields the isobaric heat capacity C_p and the isobaric thermal expansivity α_p as functions of temperature at a given pressure; (d) scanning temperature under isochoric conditions yields the isochoric heat capacity C_V and the isochoric thermal pressure coefficient β_V as function of temperature at a given volume.

In the present work, two different operating modes were used: (1) the use of pressure as scanned variable along different isotherms while recording (versus time t)

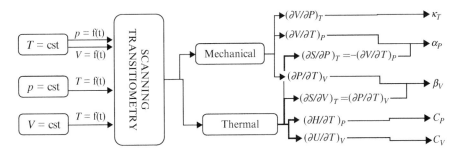

Fig. 3 Thermodynamic scheme of scanning transitiometry showing the four possible modes of scanning. Each of these modes delivers two output derivatives (mechanical and thermal), which in turn lead to four pairs of the different thermomechanical coefficients, namely α_p, κ_T, β_V, C_p, and C_V

simultaneously the associated thermal effect $(\delta Q/dt)_T$ and the mechanical effect $(\partial V/dt)_T$; and (2) the use of temperature as scanned variable along isobars while recording simultaneously the associated thermal effect $(\delta Q/dt)_p$ and the mechanical effect $(\partial V/dt)_p$.

In the case of the first mode, the straightforward thermodynamic relation [23]:

$$dH(T,p) = (\partial H/\partial T)_p dT + (\partial H/\partial p)_T dp \qquad (4)$$

with:

$$dH(T,p) = \delta Q + V dp, \qquad (5)$$

allows one to express finally that the thermal effect $q_T(p)$ along the scan is:

$$\begin{aligned} q_T(p) = (\delta Q/dt)_T &= a\left[(\partial H/\partial p)_T - V\right] = aT(\partial S/\partial p)_T \\ &= -aT(\partial V/\partial T)_p = -aTV\alpha_p, \end{aligned} \qquad (6)$$

where H, S, a, and α_p are the enthalpy, entropy, pressure scanning rate and isobaric thermal expansion, respectively. In addition, the associated mechanical effect $(\partial V/dt)_T$, (or equivalently $(\partial V/\partial p)_T$) allows one to obtain the isothermal compressibility κ_T. Similarly, in the second mode, from (4) and (5), at constant pressure (e.g., $dp = 0$) one obtains for the thermal effect $q_p(T)$ an equation equivalent to (4), C_p being the heat capacity:

$$q_p(T) = b(\partial H/\partial T)_p = bC_p. \qquad (7)$$

In the same way as above, the mechanical effect $(\partial V/\partial T)_p$ allows one to obtain the isobaric thermal expansion α_p.

The transitiometric technique can be used for fluids (gases and liquids) as well as for solid materials (polymers and metals). Remarkably, measurements can be performed in the vicinity of and above the critical point. Concretely, the investigated polymer samples are placed in ampoules, i.e., open mini test-tubes seated in the transitiometric measuring vessel in such a way that the sample is in direct contact with the pressurizing fluid. More details on the technique can be found elsewhere [24, 25]. The transitiometers (from BGR TECH, Warsaw) used in these studies of polymers, built according to the above principle, can be operated over the following ranges of temperature and pressure: 173 K $< T <$ 673 K and 0.1 MPa $< p <$ 200 MPa (or 400 MPa). A detailed description of a basic scanning transitiometer is given elsewhere [26].

A schematic representation of the instruments is shown in Fig. 4. The transitiometer itself is constructed as a twin calorimeter with a variable operating volume. It is equipped with high-pressure vessels, a pVT system, and Lab-VIEW-based virtual instrument software. Two cylindrical fluxmeters or calorimetric detectors (internal diameter 17 mm, length 80 mm), each made from 622

Key Thermodynamic Data and Thermophysical Properties

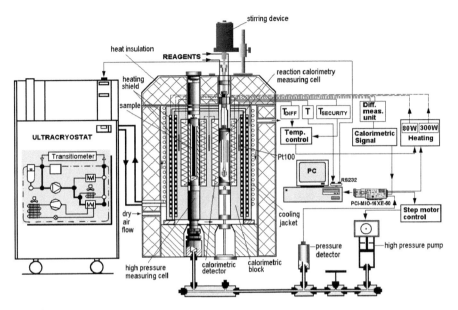

Fig. 4 Scanning transitiometry setup for in situ simultaneous determination of the thermal and mechanical derivatives. For convenience, two types of cells are shown: on the *left* is the standard high pressure cell and on the *right* is a reaction-type cell that can accommodate various accessories (stirrer, reagents feeding, capillaries, optical fibers or probes for UV/Vis/near-IR spectroscopic analysis)

thermocouples (chromel-alumel), are mounted differentially and connected to a nanovolt amplifier, which functions as a noninverting amplifier, whose gain is given by an external resistance (with 0.1% precision). The calorimetric detectors are placed in a metallic block, the temperature of which is directly controlled by a digital feedback loop of 22 bits resolution ($\sim 10^{-4}$ K), being part of the transitiometer software. The calorimetric block is surrounded by a cooling/heating jacket, which is connected to an ultracryostat (Unistat 385 from Huber, Germany). The temperature difference between the block and the heating/cooling jacket is set at a constant value. In addition, the jacketed calorimetric block is embedded in an additional electrically heated shield. The temperature difference between the block and the heated shield is set to a constant value (5, 10, 20, or 30 K) and is controlled by an analogue controller. The temperature measurements, both absolute and differential, are performed with calibrated 100 Ω Pt sensors; a Pt100 temperature sensor is placed between the sample and the reference calorimetric detectors. The heaters are homogeneously embedded on the outer surfaces of both the calorimetric block and the cooling/heating shield. The whole assembly is thermally insulated and enclosed in a stainless steel body.

The stainless steel body is fixed on a sliding support (Fig. 5 shows the main elements of a scanning transitiometer), which can be moved up and down along two guiding rails. This is part of a mechanical displacement system consisting of a winch and counterweight that allows the calorimetric body to easily move vertically

Fig. 5 Photograph of a standard scanning transitiometer (from BGR TECH, Warsaw). The calorimetric detector, which can be moved up and down over the measuring and reference calorimetric vessels (in twin differential arrangement), is shown in its upper position. In this position, the calorimetric vessels, which are firmly fixed on the stand table, are then accessible for loading

over the two calorimetric vessels (i.e., the measuring and reference vessels). These two vessels are firmly fixed on a stand, to which the displacement system is itself attached in such a way that the vessels always find the same positions inside the calorimetric detectors when the calorimetric body is moved down to its working position (see Fig. 5). When performing measurements near 0°C or below, dry air is pumped through the apparatus in order to prevent the condensation of water vapor from air.

The variable volume is realized with a stepping motor-driven piston pump. The resolution of the volume detection is ca. 5.24×10^{-6} cm^3 per step (as found by measurement of the piston displacement for given numbers of steps) and the total variable volume is 9 cm^3. The calorimetric block can then be lifted to load the sample into the cell, or for cleaning.

The pressure sensors are connected close to the piston pump. Pressure can be detected with a precision of ±4 kPa. The connection between the cryostat and the heating/cooling shield of the calorimetric block is made via two flexible thermo-isolated hoses. The Hüber cryostat/thermostat connected to the calorimeter can be operated over the temperature range from -90 to $200°C$, with $\pm 0.02°C$ temperature stability at $-10°C$, and has cooling power at 0, -20, -40, -60 and $-80°C$ of 5.2, 5, 4.2, 3.1, and 0.9 kW, respectively.

The maximum delivery of the circulating pump is 40 L min^{-1} and the maximal delivery pressure is 1.5 bars. The cryostat is microprocessor-controllable and equipped with an RS232 interface. The cryostat is PC-controlled thanks to Lab-worldsoft 3.01 graphical software. The software allows building the temperature program (up to 99 sequences), controls the temperature with high accuracy, and performs data acquisition into a file, with a selectable frequency.

A striking (patented [25]) feature of scanning transitiometry is, for the investigation of gas–polymer interactions, the possibility to use different pressurizing or pressure-transmitting hydraulic fluids. Depending on the type of measurement, the sample under investigation can either be confined in a closed supple ampoule, itself immersed in the hydraulic fluid, or positioned directly in contact with the hydraulic fluid. In the latter case, the energetic interaction upon the possible sorption of the fluid with the sample can be directly evaluated and documented.

Transitiometry is at the center of different types of utilization since, with such techniques, bulk properties, transitions, and reactions can all be advantageously studied. In the case of polymer synthesis, a scanning transitiometer was used as an isothermal reaction calorimeter, the advancement of a polymerization reaction being accurately monitored through rigorous control of the thermodynamic parameters [27, 28]. To gather additional information, the measuring cell can be coupled with other analytical devices (e.g., on-line FTIR, particle sizing probes, turbidity probes, pH or other ion selective probes) [29]. For studying chemical reactions, the scanning calorimeter has been also used as a temperature oscillation calorimeter, and the high-pressure cells replaced by specially designed reaction vessels. These vessels allow stirring, different dosing profiles for one or two reactants, and can accommodate a small optical probe coupled to a miniaturized spectrophotometer (for more details see [28–31]).

3 Gas-Polymer Interactions and Practical Applications

The performance and advantages of combining scanning transitiometry and the gas sorption–swelling technique are well demonstrated by typical applications in several important fields: (a) transitions of polymer systems under various constraints (temperature, pressure, gas sorption) including first-order phase and glass transitions; (b) polymer thermophysical properties and influence of gas sorption (blistering phenomena); (c) thermodynamic control of molecular organization in polymeric structures (foaming process, self-assembling nanostructures). Some illustrative examples have been chosen for their impact in polymer science, in the petroleum industry, and in microelectronics.

3.1 Evaluation of Gas Solubility and Associated Swelling

3.1.1 Coupled VW–pVT Method: Theory and Modeling

The VW–pVT procedure allows one, in principle, to obtain simultaneously two unknowns from the two rigorous equations (1) and (2): the gas solubility m_{sol} and the change in volume ΔV_{pol} (the swelling) of the polymer due to sorption at pressures up to 100 MPa and from room temperature to 473 K. However, despite its evident advantages, the coupled technique needs further improvement [32]. The change in volume associated with high pressure gas sorption is not a simple phenomenon. On the one hand, the chemical structures of both the polymer and the gas play a major role in terms of thermal energy of gas–polymer interactions during sorption; on the other hand, pressure also plays an important role, depending again on the polymer structure. For example, with the two polymers, medium density polyethylene (MDPE) and poly (vinylidene fluoride) (PVDF), it has been demonstrated (see Sect. 3.2) that supercritical carbon dioxide (scCO$_2$) substantially affects the cubic expansion coefficient of the polymers, especially at pressures ranging from 10 to 30 MPa, where the gas–polymer interactions are more marked. It appears that, at lower pressures, the main interactions correspond to the exothermic sorption of CO$_2$ by the surface and amorphous phase, and possibly by some interstitial sites of the crystalline part of the polymer. At higher pressures, CO$_2$ is forced to enter deeply inside the interstitial or other voids in the polymer and cause their mechanical distortion, which is associated with an endothermic effect. At high pressures (above 30 MPa), the polymers saturated by CO$_2$ behave as pseudohomogeneous phases and their cubic thermal expansion coefficients are only slightly higher, because of absorbed CO$_2$, than for pure polymers. Heats of interaction of CO$_2$ with PVDF are higher than with MDPE, demonstrating that CO$_2$ preferentially penetrates more into PVDF than into MDPE. Undoubtedly, gas solubility in polymers is a complex phenomenon and, most likely as a consequence, it has been observed that the two characteristic working equations (1) and (2) of the VW–pVT technique do not converge [32]; thus solubility and swelling cannot be obtained simultaneously by direct experimental determination. Effectively, a common term, the density ρ_g of the gas (8), appears in both working equations (1) and (2):

$$\rho_g = \frac{M_g}{R} \frac{p_f}{T_f \; Z_f} \tag{8}$$

and, despite the other terms being different, (1) and (2) can be both expressed by the same reduced (9), having the same slope given by (8) :

$$\Delta m_{sol}^{(k)} = \rho_g \, \Delta V_{pol} + d. \tag{9}$$

The term d represents the apparent concentration of gas in the polymer, i.e., when the change in volume ΔV_{pol} is zero. The main source of uncertainty in evaluating the gas concentration comes from the first term of (9), which contains

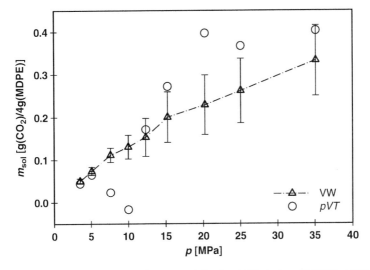

Fig. 6 Comparison of the total masses in grams of dissolved CO_2 in 4 g of MDPE at 333.15 K at different pressures, as obtained with the two techniques of *pVT* (*open circles*) and VW (*open triangles*). *Error bars* are shown for the VW results

the density of the gas and the change in volume of the polymer. It was thus necessary to elaborate a procedure to estimate the apparent solubility of the gas and the associated change in volume.

At this stage, it appears that the VW sensor technique is more precise than the *pVT* technique because there are no cumulative errors like in the case of the *pVT* method, when the successive transfers are performed during an isothermal sorption. Figure 6 compares the mass in grams of CO_2 dissolved in 4 g of MDPE at 333.15 K obtained with the two methods, VW and *pVT*. With the pressure decay method, after the critical zone (7.65 MPa), uncertainties in the mass dissolved become too large. Evidently, in the critical region, a small variation of pressure leads to a significant variation of the compressibility factor. The VW technique does not require extensive calibrations. Essentially, uncertainties come from the experimentally measured resonance frequencies. Errors are reduced in the data acquisition, which permits recording simultaneously the phase and the frequency: effectively, the phase angle is better suited than the amplitude (the half-width) to detect the natural resonant frequency [32]. Figure 7 shows as an example the data obtained by the VW technique with a MDPE polymer sample in the presence of $scCO_2$ at 338.15 K. Experimental amplitude and phase are correctly fitted by the fluid-mechanical theory [18] of the vibrating wire. Standard deviations for both amplitude and phase are also shown.

3.1.2 Selected Example: The {CO_2 + MDPE} System

To estimate the change in volume (swelling), ΔV_{pol}, the Sanchez–Lacombe equation of state SL-EOS [17, 33–35] (10, 11) using the equation of DeAngelis [36] (12)

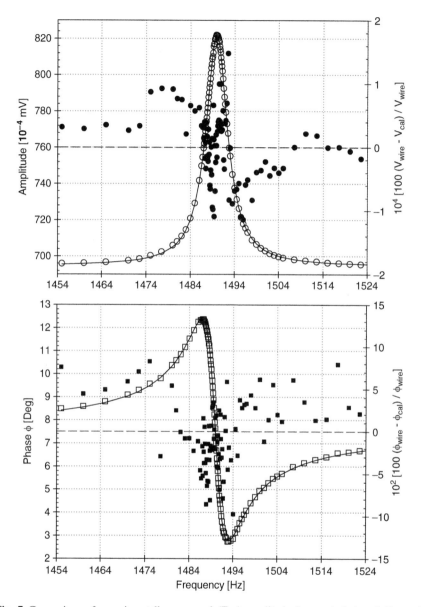

Fig. 7 Comparison of experimentally measured (*Top*) amplitude (*open circles*) and (*Bottom*) phase (*open squares*) of the VW sensor. *Solid lines* correspond to the curves calculated with the theoretical model (1) describing the characteristics of the VW sensor. *Dashed line* indicates the half-width amplitude. *Closed symbols* represent the standard deviation between experimental and calculated values (corresponding scales are shown on the *right*)

Key Thermodynamic Data and Thermophysical Properties

has been selected. Then, only one binary adjustable interaction parameter k_{12} of the SL-EOS has to be calculated by fitting the sorption data as follows:

$$\Delta p* = k_{12}\sqrt{p_1^* p_2^*}. \qquad (10)$$

$$w_1 = \frac{\varphi_1}{\varphi_1 + (1 - \varphi_1)\frac{\rho_2*}{\rho_1*}}, \qquad (11)$$

where $\Delta p*$ is the parameter characterizing the interactions in the mixture; w_1 is the mass fraction of permeant gas at equilibrium; ϕ_1 is the volume fraction of the gas in the polymer; and (ρ_1*, p_1*, T_1*) and (ρ_2*, p_2*, T_2*) are the characteristic parameters of pure compounds. The volume change is then calculated by following:

$$\frac{\Delta V_{pol}}{V_{pol}} = \frac{1}{\tilde{\rho}\rho*}\frac{1}{(1 - w_1)}\frac{1}{\hat{v}_2^0}, \qquad (12)$$

where $\rho*$ and $\tilde{\rho}$ are respectively the mixture characteristic and reduced densities, and $\hat{v}_2^0$ is the specific volume of the pure polymer at fixed temperature, pressure, and composition. According to the procedure, the solubility data are obtained through combined experimental measurements and theoretical estimation of the volume change of the polymer due to the sorption. Figure 8 shows the results obtained by the pVT technique using the SL-EOS for the sorption of CO_2 in MDPE at 333 K to estimate ΔV_{pol}. In this figure, comparison is made with literature values for a low density polyethylene LDPE at 308 K [5].

3.2 Gas–Polymer Interaction Energy

Scanning transitiometry has been used to determine the gas–polymer interaction energy, for instance upon CO_2 sorption in MDPE and in PVDF samples (Fig. 9).

Measurements have been made under either compression or decompression runs realized by pressure jumps Δp between 6 and 28 MPa. The most striking result is that CO_2–PVDF (exothermic) interactions are larger than CO_2–MDPE interactions for CO_2 pressures lower than 30 MPa, whereas above this pressure an inversion is observed, with CO_2–MDPE interactions being larger than CO_2–PVDF interactions.

3.3 Thermophysical Properties at High Pressures

As mentioned in the "Introduction" (Sect. 1), the thermophysical properties of thermoplastic semicrystalline polymers are essential for the development of numerous engineering applications. Such data have to be documented over extended

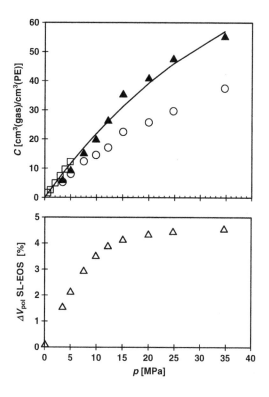

Fig. 8 (*Top*) Sorption of CO_2 in MDPE at 333 K as a function of pressure up to 35 MPa. Apparent concentrations C (cm^{-3} CO_2 per cm^{-3} PE) are indicated by *open circles*. Corrected concentrations calculated from the polymer swelling ΔV_{pol} are indicated by *closed triangles*. Literature values (*open squares*) from Kamiya et al. for LDPE at 308 K [5] are shown for comparison. (*Bottom*) Values of ΔV_{pol} (*open triangles*) calculated using the Sanchez–Lacombe equation of state

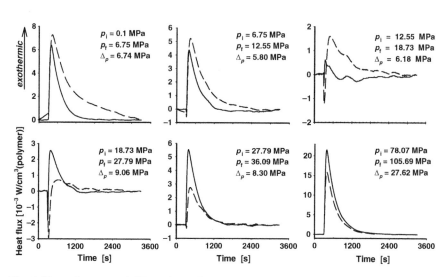

Fig. 9 Thermal energy of CO_2–polymer interactions at 353 K: comparison between the two polymers MDPE (*solid line*) and PVDF (*dashed line*) at increasing CO_2 pressures. Measurements have been made under pressure jumps Δp, between initial pressures p_i and final pressures p_f. The differential exothermal heat fluxes are of the order of a few milliwatts per cubic centimeter of polymer

Key Thermodynamic Data and Thermophysical Properties

ranges of p and T for polymers in the presence or absence of fluids (liquids, gases, or SCFs), which can enter the polymeric structures by natural (at atmospheric pressure) or forced sorption (under high pressures). As seen previously, the fluids will have, depending on their respective characteristics (inert, neutral, or chemically active), more or less significant impact on the polymer molecular organization. In this context, until recently most of the investigations of gas–polymer interactions have concerned sorption properties of glassy polymers. In such systems, the dual-mode sorption concept is generally accepted. According to this model, part of the sorbate is dissolved in a molecular environment described by the Henry law, whereas another part is absorbed (as described by a Langmuir-type sorption isotherm) in preexisting voids or free volume resulting from extremely long segmental relaxation times between chains in the polymer glassy state [37]. In semicrystalline polymers, it was widely accepted from the early studies of Michaels and Parkers [38] and Michaels and Bixler [39] that the gas sorption takes place only in the amorphous phase (following the Henry law), while the crystallites form impenetrable barriers that even prevent diffusion in the amorphous phase. More recent studies have established that low molecular diluents might also penetrate the crystalline regions, where interstitial free spaces could accommodate small molecules like CO_2 or methane (CH_4) [40–44]. However, all the above studies were realized at rather low pressures of a few megapascals. In the mid-1990s, thanks to scanning transitiometry, it was possible to initiate a systematic investigation of gas–polymer interactions.

A new experimental and theoretical approach has been proposed to study transitions in {gas–polymer} systems in terms of the heat involved [45]. Scanning transitiometry, which combines a calorimetric detector with a pVT scanning technique, offers advantageous features for such study. The differential mode of operation permits precise control of both temperature and pressure, keeping them exactly identical in the two calorimetric (reference and measuring) vessels. The pVT technique allows the scanning of pressure or volume during sorption (fluid-pressurization) and desorption (fluid-depressurization). The calorimetric detector measures the differential heat flux (between reference and measuring vessels) resulting from the physicochemical effects occurring during the sorption/desorption runs. From the determination of the heat involved in the measuring vessel (containing the polymer sample) and by virtue of the Maxwell relation, $(\partial S/\partial p)_T = -(\partial V/\partial T)_p$, the global cubic thermal expansion coefficient of the gas-saturated polymer $\alpha_{\text{pol-g-int}}$ is obtained at different isothermal conditions, according to (13):

$$\alpha_{\text{pol-g-int}} = \frac{\left(Q_{\text{diff, SS}} - Q_{\text{diff, pol}}\right) + V_{\text{SS, r}}\, \alpha_{\text{SS}}\, T\, \Delta p}{V_{\text{pol}}\, T\, \Delta p}. \tag{13}$$

Q_{pol} and Q_{ss} represent the heat fluxes corresponding to the polymer sample and to the inert sample (made of stainless steel and having the same size and geometry as the polymer sample), respectively placed in the measuring and reference vessels; α_{SS} is the cubic expansion coefficient of the stainless steel of which the vessels are made; and Δp is the variation of gas-pressure change under investigation at constant temperature T. Volumes V_{SS} and V_{pol} are those of the stainless steel inert reference and of the polymer sample, respectively. In (13), it was assumed for simplicity *faute de mieux* that the volume of the polymer did not change significantly upon gas sorption. This assumption may be justified in the sense that the pressure is much higher ($\sim$100 MPa) in calorimetric measurements than in the VW–pVT technique ($\sim$40 MPa); the hydrostatic pressure must probably compensate for a large part of the swelling effect due to gas sorption, as a result of the equilibrium between the plasticization effect and the hydrostatic effect.

Three differential modes were investigated, taking into account the differential principle of the instrument (Fig. 10): thermal I differential without reference sample, thermal II differential with reference sample, and thermal II differential comparative mode. With the thermal I differential mode, in an initial experiment the polymer sample is placed in the measuring cell, which is connected to the gas line. The reference cell, not connected to the gas line, acts as a thermal reference. An additional blank experiment (under identical conditions) is performed in which the polymer sample is replaced by an inert-metal (stainless steel) sample of similar volume. Then, the difference in the heat effects between polymer and blank experiments allow quantification of the thermal effect due to the gas–polymer interactions. In the thermal II differential mode, the polymer sample is placed in the measuring cell while an inert-metal sample of equal dimensions is seated in the reference cell, both cells being connected to the gas line which serves to pressurize. Then, under gas pressure, the calorimetric differential signal is proportional to the thermal effect due to the gas–polymer interactions. The third and last mode

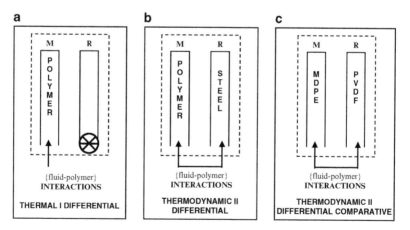

Fig. 10 Three differential modes of scanning transitiometry according to the differential principle of the calorimetric detector, taking into account the respective roles of the measuring (M) and reference (R) vessels and the content of the reference vessel. (**a**) Thermal I differential without reference sample mode. (**b**) Thermal II differential with reference sample mode. (**c**) Thermal II differential comparative mode: in this case a direct comparison between two polymers (MDPE and PVDF) samples is possible

Key Thermodynamic Data and Thermophysical Properties

corresponds to a validation of the two previous modes through the thermal II differential comparative mode. This allows direct comparison of the response and behavior of two polymer samples, MDPE and PVDF, in similar supercritical conditions. A MDPE polymer sample was placed in the measuring vessel while a PVDF polymer sample of equal size and volume was placed in the reference vessel. Both cells were connected to the gas line. The calorimetric signal, i.e., the differential heat flux, was thus directly proportional to the thermal effect due to the difference in the gas–polymer interactions between the two polymers interacting with the same gas. In that case, the differential heat flux between the measuring and the reference vessels is small, because calorimetric signals of {gas-MDPE} and {gas-PVDF} systems have relatively close amplitudes; the detection sensitivity of the apparatus was then optimal. For each thermal II differential with reference sample and thermal II differential comparative mode, the data were corrected through a blank standard calibration. Under identical conditions of T and p, and under the assumption that there were no interactions between the stainless steel rod and the gas, blank experiments were performed in which the polymer samples were replaced by a metal sample of identical dimensions.

Investigations of polymer behavior [4] consist typically of measuring the physicochemical properties in the solid state, i.e., at temperatures between T_g and T_m. MDPE and PVDF were submitted to gas pressure of either CO_2 or N_2 at different temperatures between 333 and 403 K, under pressure steps or scans in the range between 0.1 and 100 MPa. The polymer samples were extruded MDPE (reference Finathene 3802) and PVDF (reference Kynar 50HD, polymer without additives like plasticizers or elastomers). Their transitions temperatures T_g and T_m were, respectively, 163.0 K and 400.0 K for MDPE, and 235.0 and 440.9 K for PVDF. The two polymers had degrees of crystallinity X_c, of 49% and 48%, respectively. The masses of samples were about 2–5 mg, and thermograms were obtained under a continuous flow of N_2 at 15 mL min^{-1}. Measurements were performed on cylindrical rod samples (75.0 mm in height, 4.4 mm in diameter) having a relatively small mass, i.e., about 1.0 g for the MDPE sample and 1.9 g for the PVDF sample; measurements were taken from 352.38 to 401.50 K. For each investigation, a new sample was used. More details are given elsewhere [4]. Using the thermal II differential mode with reference sample, pressure changes of CO_2 and N_2 were performed on MDPE and PVDF samples at 352 and 372 K under pressure jumps of 6–28 MPa in the pressure range between 0.1 and 100 MPa. The CO_2-pressurizing pressure jumps manifest themselves by exothermic heat fluxes [29, 45], whereas CO_2-depressurization pressure jumps exhibit endothermic heat fluxes, both passing through a minimum around 20 MPa (see Fig. 11).

Interestingly, the heat flux minimum is reflected in the isotherms of $\alpha_{\text{pol-g-int}}$ coefficients of the fluid-saturated polymers plotted as functions of the feed pressure. The global cubic thermal expansion coefficients $\alpha_{\text{pol-g-int}}$ of saturated polymer were obtained through the procedure previously described [45]. Comparison of these coefficients for both polymers (MDPE and PVDF) under CO_2 and N_2, i.e., the corresponding curves for the {CO_2-MDPE}, {CO_2-PVDF} and {N_2-PVDF} systems, show a clear difference (Fig. 11). Additional investigations of {Hg-MDPE}

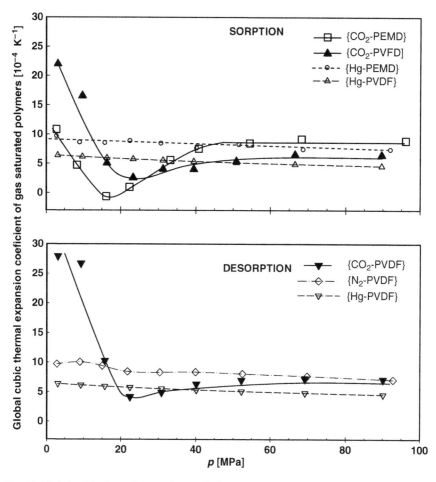

Fig. 11 Global cubic thermal expansion coefficients $\alpha_{\text{pol-g-int}}$ at 372.02 K of MDPE and PVDF samples, under either CO_2 or N_2 during (**a**) sorption and (**b**) desorption under jumps in pressure, obtained with the thermal II differential mode. The variations in $\alpha_{\text{pol-g-int}}$ in the presence of either N_2 or CO_2 are compared to those of samples in the presence of Hg (considered as an inert fluid)

and {Hg-PVDF} systems have been made using Hg as an "inert" pressure-transmitting fluid [4, 45]. High Hg-pressure runs permit the decoupling of hydrostatic pressure effects from solvent solubility effects, whereas high N_2-pressure runs permit separation of the preferential interaction effects between polymers with respect to CO_2. Under CO_2, the thermal expansivity $\alpha_{\text{pol-g-int}}$ shows minima around 14–18 and 21–25 MPa for MDPE and PVDF, respectively. This is in contrast to what is observed under N_2 or Hg, i.e., the isotherms of interaction vary "monotonously" (Fig. 11). Below 30 MPa, more energetic interactions are observed with PVDF compared to MDPE, which is demonstrated by higher global $\alpha_{\text{pol-g-int}}$ for the {CO_2-PVDF} system. Above 30 MPa, CO_2–MDPE interactions are larger than

Key Thermodynamic Data and Thermophysical Properties
159

CO_2–PVDF interactions, and the global $\alpha_{pol\text{-}g\text{-}int}$ for the {CO_2-MDPE} system is higher than the global $\alpha_{pol\text{-}g\text{-}int}$ for the {CO_2-PVDF} system. As shown by Fig. 11, in the case of PVDF, N_2 acts as a "relatively neutral" fluid like Hg, but with stronger interactions. The values of $\alpha_{pol\text{-}g\text{-}int}$ with N_2 are smaller than those with CO_2 [$\alpha_{pol\text{-}g\text{-}int}$ {N_2-PVDF} $<$ $\alpha_{pol\text{-}g\text{-}int}$ {CO_2-PVDF}], demonstrating that interactions of PVDF with N_2 are weaker than with CO_2. With N_2 (a relatively neutral fluid) the heat effects reflect the sorption under pressure and parallel the remarkable plasticization efficiency of N_2 in PS, particularly at elevated pressure [46, 47] (see Sect. 3.4.3). The PVDF values during decompression under N_2 and/or CO_2 are similar, which is satisfactory as regards the reversibility of the sorption/desorption phenomena. The minimum of $\alpha_{pol\text{-}g\text{-}int}$ observed with {CO_2-MDPE} and {CO_2-PVDF} systems at about 15 MPa corresponds to the supercritical domain of CO_2. The dependency of $\alpha_{pol\text{-}g\text{-}int}$ coefficients on the nature of the pure gas (i.e., a minimum corresponds in a mirror-image to the maximum in the temperature dependence of α_p for pure CO_2 gas) is a striking feature of previous studies [45]. This clearly shows the influence of supercritical sorption on the thermophysical properties of the polymers. With the semicrystalline polymers, low pressures most probably induce a first adsorption of CO_2 in the amorphous part and in some interstitial sites of the crystalline part, with the possible formation of a microorganized domain generated in the amorphous phase of the polymer [44] (see also Sect. 3.1.1). High pressures favor the absorption into the whole polymer matrix (i.e., deep inside the interstitial or other voids in the polymer) with a mechanical distension, the CO_2-saturated polymer behaving as a pseudohomogeneous state [45]. Furthermore, the minimum would mean that supercritical gas–polymer interactions are favored. The lowering of molecular polymer–polymer interactions is concomitantly associated with the ease of CO_2 dissolution into the polymer matrix, thus inducing an increase of free volume together with an increase in polymer chain mobility [48]. This plasticization effect is shown by the minimum of $\alpha_{pol\text{-}g\text{-}int}$ as a function of pressure. Quantitatively, this is confirmed by the net increase of gas sorption into the polymer and the swelling of the polymer due to the sorption around 15 MPa (as investigated by the gravimetric–volumetric VW–pVT method) [49, 50]. As a matter of fact, around this pressure there is compensation between plasticization and hydrostatic pressure effects upon high CO_2-pressure sorption into the polymer. The supercritical hydrostatic pressure corresponding to the minimum for MDPE is slightly smaller than that for PVDF.

The thermal II differential comparative mode is conveniently adapted to compare two different polymer samples submitted to the same gas under pressure. This mode was used to measure the differential heat flux obtained when a MDPE and a PVDF sample (of identical size and volume, each placed in one of the two calorimetric vessels) were simultaneously submitted to the same gas pressure at an identical temperature (372.59 K). The experimental signal, the differential heat flux $dQ_{\{MDPE\text{-}PVDF\}}$, compares directly the interactions of the two polymers in the same gas/supercritical environment at constant temperature. The calorimetric responses were collected during pressure jumps and during continuous volume

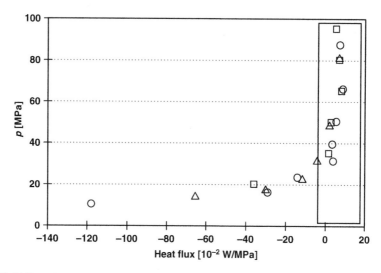

Fig. 12 Differential heat flux $dQ_{\{MDPE-PVDF\}}$ observed when two samples (MDPE and PVDF) are submitted to CO_2 at 372.59 K, with the thermal II differential comparative mode. Above about 30 MPa, positive values of $dQ_{(MDPE-PVDF)}$, shown in boxed region, indicate stronger interactions of CO_2 with MDPE than with PVDF. Measurements were taken during either sorption or desorption under jumps in pressure (*circles*), continuous changes in volume $dV = 1.364$ cm^3 (*triangles*), and continuous changes in pressure $dp = 15$ MPa (*squares*)

and pressure scans. Figure 12 shows the plots of heat flux (in 10^{-2} W MPa^{-1}) versus pressure for the two samples.

Below 30 MPa, calorimetric signals are endothermic, with $dQ_{\{MDPE-PVDF\}}/dp < 0$, i.e., PVDF exhibits higher interactions with CO_2 than does MDPE. Above 30 MPa, calorimetric signals become exothermic, with $dQ_{\{MDPE-PVDF\}}/dp > 0$, i.e., the differential heat flux of interactions for the {CO_2-MDPE} system becomes larger than for the {CO_2-PVDF} system. This direct comparative method, which permits differentiation of the interactions between both polymers (MDPE and PVDF) submitted to the same supercritical CO_2 pressure, reproduces exactly the results obtained with the two preceding methods. At low pressures, more energetic interactions are observed with PVDF than with MDPE.

The gas–polymer interactions being stronger than the interactions between the chains segments suggests that incorporation of CO_2 in PVDF is easier and stronger than in MDPE, which was confirmed with the experiments of sorption and of swelling using VW–pVT. In addition, this is confirmed by measurements at high pressure, which show that thermal expansion coefficients are smaller for highly condensed {CO_2-PVDF} systems than for less condensed {CO_2-MDPE} systems. As shown in Fig. 11, at high pressure, say above 30 MPa, the global cubic thermal expansion coefficient is smaller for {CO_2-PVDF} (for which the gas–polymer interactions are larger) than for {CO_2-MDPE}. All the above observations show that CO_2 sorption is higher in PVDF than in MDPE. Both polymers have the same volume fraction of amorphous state, $\phi_a = 0.53$ [48, 51], and thus solubility is

Key Thermodynamic Data and Thermophysical Properties

favored by the presence in the PVDF main chain of polar groups C–F, which can form strong dipolar interactions with polarizable CO_2 [51–54]. This explains why CO_2–PVDF interactions are stronger than CO_2–MDPE interactions. The extent of the gas–polymer interactions is fully documented through the thermophysical properties of gas-saturated polymers, directly measured, in conjunction with experimentally measured gas solubility in polymers.

3.4 Phase Transition at High Pressures

3.4.1 First-Order Transitions

Melting/Crystallization at High Pressures (Hydrostatic Effect)

The investigation of a classic first-order phase transition is illustrated by the melting/crystallization of a semicrystalline polymer like MDPE. Chemically inert Hg was used as pressure-transmitting fluid, the polymer sample being completely surrounded by the fluid inside the detecting calorimetric zone; in fact, the polymer sample was simply floating on the Hg.

Isobaric scans were performed at the temperature rate of 0.833 mK s^{-1}, both on heating and cooling, at different pressures from 50 to 200 MPa. Remarkably, the associated heat flux and volume variations were simultaneously recorded with a scanning transitiometer. Both the melting temperature T_m and crystallization temperature T_{cr} were ascribed from the conventional method, taking the onset transition temperature (namely the intercept of the largest slope of the measured signal with the baseline) for each recorded peak. Very good concordance of temperatures obtained by either heat flux signals or volume variation signals was observed; for example, at 200 MPa, heat flux and volume variation yield the same value of 456.2 K for T_m. Figure 13 shows the pressure effect on melting and crystallization temperatures, which are both shifted toward higher values by increasing pressure. The above measurements also allowed, at 200 MPa for example, evaluation of the variations of volume and of enthalpy for the melting transition, giving 0.0573 cm^3 g^{-1} and -88.54 J g^{-1}, respectively. The value of 0.297 K MPa^{-1} was found for the Clapeyron slope $\Delta T_m/\Delta p$, in good agreement with literature values [55, 56].

Gas-Assisted Melting/Crystallization at High Pressures (Plasticizing Effect)

Scanning transitiometry was also adapted to study the influence of a SCF on first-order phase transitions. In that case, the pressure-transmitting fluid was a gas in supercritical state. Inside the measuring vessel, the polymer sample is placed in an open ampoule (either glass or stainless steel) resting on top of a spring that maintains it in the central zone of the calorimetric detector and in direct contact with the SCF. The vessel is connected to a pressure detector and to the high pressure pump through a stainless high pressure capillary. The hydraulic fluid contained in

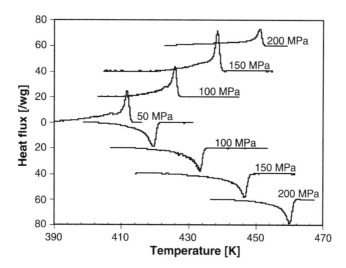

Fig. 13 Heat flux thermograms obtained under different pressures from 50 to 200 MPa during isobaric T scans at 0.833 mK s^{-1} on heating (*downward exothermic peaks*) and cooling (*upward endothermic peaks*) for a Hg-pressurized MDPE sample. The *base lines* are shifted for the sake of clarity to show the effect of pressure on melting/crystallization temperatures

the pump and in the pressure detector is separated from the SCF by a Hg column. This arrangement can conveniently be used to investigate pressure effects on transitions of the type (solid + fluid) to fluid.

The {CH$_4$-MDPE} system, the 1, 1-difluoroethylene + poly(vinylidene fluoride) {C$_2$H$_2$F$_2$-PVDF} system, and the {N$_2$-PVDF} system (all binary asymmetric systems) have been selected to illustrate the use of supercritical scanning transitiometry [26, 29, 56]. These systems are of interest because they exhibit a pronounced nonideal behavior at elevated pressures due to the large differences in the molecular sizes of components.

Interestingly, CH$_4$ modifies the MDPE structure but is easily removed from the modified polymer. Since the upper critical solution pressure of the {MDPE-CH$_4$} system is rather high (>250 MPa [57]), CH$_4$ can be a plasticizer of MDPE up to elevated pressures. Experimentally, an MDPE sample (density 938 kg m^{-3}; degree of crystallinity 0.55; number-average molar mass $M_n = 16.100 \times 10^3$ g mol^{-1} and weight-average molar mass $M_w = 83.720 \times 10^3$ g mol^{-1}, respectively) was placed in an open stainless steel ampoule, positioned in the high-pressure measuring vessel, and flushed with supercritical methane (scCH$_4$) for a few minutes. After closing the vessel, the scCH$_4$ was initially compressed to 25–30 MPa, and then the pressure modified up to 300 MPa. At a given pressure, isobaric scans (at 0.833 mK s^{-1}) were performed in heating and cooling. Remarkably, pressure remained constant within ±0.1%, even during the rapid volume changes occurring during phase transitions. It was then possible to perform several successive melting/crystallization experiments while recording the corresponding thermograms.

As illustrated in Fig. 14, typical thermograms show the influence of scCH$_4$ on the two (first-order) transitions by comparing "original" and "final" states of a

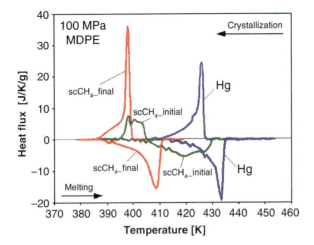

Fig. 14 Heat flux thermograms obtained under 100 MPa during isobaric T scans at 0.833 mK s^{-1} on heating (*downward exothermic peaks*) and cooling (*upward endothermic peaks*) for a MDPE sample pressurized under supercritical methane (*scCH$_4$ initial* and *scCH$_4$ final*) and under Hg (*original*)

MDPE sample being, respectively, native or saturated with scCH$_4$. Figure 14 shows recorded thermograms at 100 MPa: "Original" thermograms were obtained with a native (virgin) MDPE sample pressurized under inert Hg as pressure-transmitting fluid (see Sect. 3.4.1.1). "scCH$_4$-final" thermograms were obtained with the MDPE sample that was submitted to repeated melting/crystallization cycles. "scCH$_4$-initial" thermograms were obtained during the first heating and cooling of the native (virgin) MDPE sample under compressed scCH$_4$.

Comparison of the shapes and magnitudes between "original" and "scCH$_4$-initial" thermograms show the extent of the effect of scCH$_4$ on both melting and crystallization. The "scCH$_4$-final" thermograms were obtained as the very last thermograms after repeated melting/crystallization cycles under compressed scCH$_4$, when the thermograms no longer changed with subsequent melting/crystallization cycles. The striking result is the similarity between "scCH$_4$-final" and "original" thermograms, whereas the respective melting and crystallization temperatures of the scCH$_4$-saturated sample are significantly shifted toward lower values. Physical and textural analyses also show important differences between initial and modified samples [56], attesting to a permanent rearrangement of the organization of the long chain molecules. A simple qualitative explanation of such modification is the entropically better alignment of the polymeric structures favored by scCH$_4$, which acts as a "lubricant" between the chains.

For the polymer PVDF (in the solid state), the monomer C$_2$H$_2$F$_2$ is a good solvent even at high T and p (over 500 K and 200 MPa, respectively); in this respect the comparison with the solubilization thermodynamics of inert N$_2$ in PVDF is of practical interest since PVDF is a major polymeric material in numerous industrial applications. Furthermore, because the monomer C$_2$H$_2$F$_2$ is the major component of the PVDF polymerization, it is essential to document the thermodynamics of the {$C_2H_2F_2$-PVDF} system to better control the polymerization industrial process. Melting/crystallization under high pressure (i.e., supercritical) $C_2H_2F_2$ has been investigated through isobaric temperature scans on PVDF samples ($M_n = 113.100 \times 10^3$ g mol^{-1} and $M_w = 330.000 \times 10^3$ g mol^{-1}, respectively) at different pressures between 0.1 and 180 MPa [55]. Isobaric temperature scans on PVDF samples under high pressure N_2 have been performed between 0.1 and 30 MPa.

This study shows, like in the case of the {MDPE-CH_4} system, the significant influence of the "active" supercritical solvent on the melting/crystallization of the polymer. Figure 15 shows the influence of supercritical $C_2H_2F_2$ and of supercritical N_2 on the T_m and T_{cr} of PVDF. Obviously, both temperatures increase with increasing N_2 pressure. In the investigated pressure range (0.1–30 MPa), the (Clapeyron) slope of the two plots T_m/p and T_{cr}/p were 0.108 ± 0.002 K MPa^{-1} and 0.115 ± 0.002 K MPa^{-1}, respectively. By contrast, $C_2H_2F_2$ depresses first the melting/crystallization temperatures upon sorption of the gas by the polymer, up to 30 MPa. Then, the antiplasticization effect of the hydrostatic pressure of $C_2H_2F_2$ takes over above 30 MPa, which confirms the usual competition between plasticization and hydrostatic effects of a (chemically) "active" SCF on the melting/crystallization phenomena: the hydrostatic pressure increases the temperature of the first-order transitions, while the increase of solubility of the SCF in the polymer-rich phase depresses this temperature.

3.4.2 Isotropic Transitions (Self-Assembling of Polymeric Structures Under High-Pressure Gas Sorption)

Different fields of application require the knowledge of interfacial phase behavior between gaseous molecules and polymers. New application fields appear with the rapid growth of information technology, for which ongoing downscaling of microelectronics evolves into nanoelectronics. The development of highly ordered

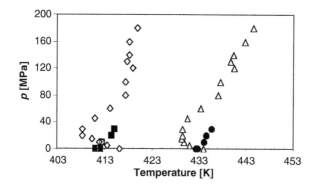

Fig. 15 Pressure–temperature phase diagram for the two systems {$C_2H_2F_2$-PVDF} (*open symbols*) and {N_2-PVDF} (*closed symbols*) showing the depression by about 20 K of melting temperature (*right*) and crystallization temperature (*left*) at pressures up to 30 MPa for PVDF under supercritical $C_2H_2F_2$

nanostructures in the macroscopic area has attracted increasing interest through nanoscience and nanotechnology breakthroughs in the new generation of microelectronic and optical devices. Since liquid crystals exhibit a rich variety of phases and phase transitions, block copolymer systems are promising candidates for building periodic nanostructures at low cost by simple self-assembly [58]. Modification of nanoordered structures formed by block copolymers is currently a key technology in nanoscience. An important feature of self-ordered structures is their possible reorganization by modification of the interface between the two components of block copolymers by a pressurizing fluid. In this context, we have investigated the interactions between diblock copolymers and different pressurizing fluids [2, 3, 59, 60]. For this purpose, our study was focused on phase diagrams of such systems as functions of the thermodynamic independent variables (p, T, V) and the respective volume fractions f_i of the two components of block copolymers of various types.

Liquid crystalline amphiphilic diblock copolymers poly(ethylene oxide)-*block*-11-[4-(4-butylphenyl-azo)phenoxy]-undecyl methacrylate, PEO_m-*b*-$PMA(Az)_n$, as shown in Fig. 16, prepared by atom transfer radical polymerization [61], were composed of hydrophilic PEO_m sequences and hydrophobic $PMA(Az)_n$, with azobenzene moieties such as mesogen connected by a flexible spacer. The synthesis of such amphiphilic liquid crystal block copolymers has been recently reported [62]. In diblock copolymers PEO_m-*b*-$PMA(Az)_n$, m and n indicate the degree of polymerization of PEO and PMA(Az) components, respectively. Differential scanning calorimetry (DSC) of PEO_m-*b*-$PMA(Az)_n$ gives a clear picture of the thermal properties of these liquid crystalline polymers, as shown in Fig. 17, for PEO_{114}-*b*-$PMA(Az)_{20}$ [58, 61].

Four phase transitions are ascribed to the melting of PEO, the glass transition of azobenzene moieties PMA(Az), the smectic (hardly visible), and the isotropic transitions.

High-pressure technology using gases plays an important role in nucleation of materials, and particularly interesting are current developments and applications in soft matter science with typical modifications and tailoring of liquid crystals, colloids, and polymers (including block copolymers) by means of supercritical gases [63–68]. In this respect, the thermodynamic investigation of diblock copolymers connecting incompatible polymers by covalent bonds is illustrative from both fundamental and applied aspects [69–73]. Typically, copolymers PEO_m-*b*-PMA

Fig. 16 Amphiphilic diblock copolymers of PEO_m-*b*-$PMA(Az)_n$, where m and n indicate the degree of polymerization of PEO and PMA(Az) components, respectively

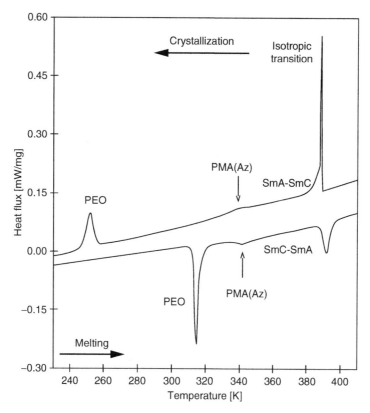

Fig. 17 DSC heating and cooling curves (heat flux in mW mg^{-1} vs temperature) for PEO$_{114}$-b-PMA(Az)$_{20}$ showing the high temperature isotropic transition

(Az)$_n$ generate at the interface between PEO and PMA(Az) moieties well-ordered structures of one in the other, depending on their respective volume fractions f_i. The ordered structures can be of three different types: spheres, cylinders, or lamellae, as illustrated in Fig. 18 for an AB-type diblock copolymer [A and B standing, respectively, for PEO and PMA(Az) components].

Obtaining a given molecular organization of these structures as regards their type, size, and arrangement is directly controlled by the thermodynamic conditions, i.e., p, T, and the nature of the hydraulic fluid used to pressurize. To this end, the isotropic transition of the diblock copolymer at which well-defined self-organized nanoscale structures form is the main thermodynamic property to document.

In the series of PEO$_m$-b-PMA(Az)$_n$ copolymers, PEO self-organized entities in the form of highly ordered periodic hexagonal-packed PEO cylinders are formed in the PMA domain by annealing at the isotropic state. This shows that controlling the phase changes at the interface allows tailoring of the nanoscale structures, as illustrated in Fig. 19.

Scanning transitiometry has been used to evaluate the pressure dependence of the isotropic transition temperature T_{tr}, as well as the transition enthalpy ΔH_{tr} and

Key Thermodynamic Data and Thermophysical Properties

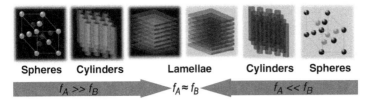

Fig. 18 Possible ordered structures of AB diblock copolymer, depending on the respective volume fractions f_A and f_B (by courtesy of Prof. H. Yoshida)

Fig. 19 Ordered periodic hexagonal-packed PEO cylinders formed in the PMA domain while CO_2 penetrates the interface

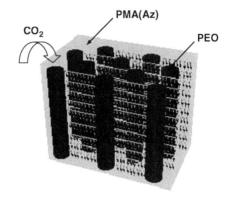

corresponding volume ΔV_{tr}. The role of the nature of the pressurizing fluid on the transition thermodynamics was also evaluated therefrom. For this, the rigorous Clapeyron equation was advantageously used to document the pressure effect because this equation relates the slope (dT/dp) of the phase boundary on the p–T surface to the changes in volume ΔV_{tr} and enthalpy ΔH_{tr} at the transition, as given by (14):

$$(dp/dT)_{tr} = \Delta H_{tr}/T_{tr}\Delta V_{tr} = \Delta S_{tr}/\Delta V_{tr}, \qquad (14)$$

where ΔS_{tr} is the change of entropy during the transition at temperature T_{tr}.

Remarkably, the transition entropy ΔS_{tr} decreases with increasing pressure when the pressurizing fluid is Hg; this is typically the manifestation of a pure hydrostatic effect, which restricts molecular motions under inert Hg. In complete contrast, ΔS_{tr} increases when the pressure is exerted by N_2 and CO_2. In this respect, as observed previously, N_2 is a "neutral" fluid as compared to "chemically active" CO_2 and, consequently, the large increase in ΔS_{tr} shows that the organization of nanostructures is easiest the more "active" is the fluid, in particular when the fluid is in supercritical state [2, 59, 60]. The influence of the pressure-transmitting fluid on the transition temperature T_{tr} is well illustrated (see Fig. 20) in the case of PEO_{114}-b-$PMA(Az)_{40}$ copolymer by the increase of the Clapeyron slope $(dp/dT)_{tr}$ in the

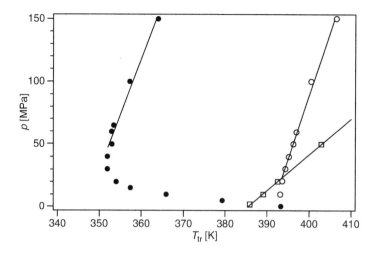

Fig. 20 Relationship between pressure and the temperature of the isotropic transition T_{tr} for PEO$_{114}$-b-PMA(Az)$_{40}$ under CO$_2$ (*closed circles*) and N$_2$ (*open circles*), and for PEO$_{114}$-b-PMA (Az)$_{20}$ under Hg (*open squares*) for comparison. The different *lines* represent the Clapeyron slopes, depending on the pressure-transmitting fluid. Note the significant shift by CO$_2$ of T_{tr} to a lower temperature

sequence Hg < CO$_2$ < N$_2$. In addition, the strong influence of supercritical CO$_2$ on the transition is spectacularly demonstrated by the significant shift of the isentropic transition temperature T_{iso} to lower temperatures. Figure 20 illustrates the relationship between the isotropic transition temperature and pressure for PEO$_{114}$-b-PMA (Az)$_{40}$ under N$_2$ [2] and CO$_2$ [3]. The isotropic transition temperature for PEO$_{114}$-b-PMA(Az)$_{20}$ under Hg pressure [60] is also shown for comparison. The hydrostatic effect under N$_2$ and CO$_2$ pressure is dominative above 20 and 40 MPa, respectively. The dP/dT values under N$_2$ and CO$_2$ pressure are 10.2 and 8.8 MPa K^{-1}, respectively. The dP/dT value of PEO$_{114}$-b-PMA(Az)$_{20}$ under Hg pressure is 2.85 MPa K^{-1}. The larger dP/dT (14) value under N$_2$ and CO$_2$ pressure than under Hg pressure suggests that the transition volume ΔV_{tr} under N$_2$ and CO$_2$ pressure is smaller than under Hg pressure. The N$_2$ and CO$_2$ adsorbed in PEO$_{114}$-b-PMA(Az)$_{40}$ reduces the free volume. Because the compressibility of gaseous molecules is much smaller than that of the free volume, the volume change under N$_2$ and CO$_2$ is smaller than the change under Hg pressure. The larger value of dP/dT under N$_2$ pressure than under CO$_2$ pressure shows that the volume change at the isotropic transition is larger under CO$_2$ pressure than under N$_2$ pressure. Furthermore, because the interaction between the PMA(Az) domain and CO$_2$ by dipole–quadrupole interactions is stronger than the physical interaction between the PMA (Az) domain and N$_2$, the space (the molecular distance) between the PMA(Az) domain and CO$_2$ is smaller than under N$_2$. Therefore, the free volume under CO$_2$ pressure is larger than under N$_2$ pressure with, consequently, a larger volume change and greater ease of molecular reorganization at the entropic transition.

Key Thermodynamic Data and Thermophysical Properties

3.4.3 Glass Transitions

Glass Transitions at High Pressures (Hydrostatic Effect)

Evidently, from the above observations, the T_g of semicrystalline polymers is similarly affected by pressure because an increase of pressure induces a decrease in the total volume and, as a consequence of the decrease of free volume, a shift to higher values of T_g is expected. This aspect is particularly important in engineering operations such as molding or extrusion, when operations at close to T_g can result in stiffening of the material. Investigation of the pressure effect on the T_g of polymers is thus of major importance in an industrial context. Particularly, the T_g of elastomers whose T_g are often well below the ambient temperature is of practical interest when performing experimental measurements by scanning transitiometry. In this case, Hg, which is conveniently utilized as pressure-transmitting fluid, must be replaced because its crystallization temperature is relatively high, i.e., 235.45 K. Selecting a substituting fluid is a challenging problem because the fluid should be chemically inert with respect to the investigated sample (with respect to all its constituents). Also, the values of its thermophysical properties, isothermal compressibility, κ_T, and isobaric thermal expansivity, α_p, should be smaller than those of investigated samples. Another difficulty in the investigation of second-order-type transformations is the relatively weak thermal effect measured. It is well known that the amplitude of the heat flux at T_g increases with the temperature scanning rate, whereas the time constant of differential heat flux calorimeters imposes relatively low temperature scans rates. However, using an ultracryostat coupled to the transitiometer, it was possible with the help of a temperature program to accurately determine T_g at relatively high scanning rates [29].

In a typical run (see Fig. 21a), the temperature of the thermostatic liquid is lower than that of the calorimetric block during the stabilization periods (isothermal segments), and higher during the dynamic segment. In such a way, the scanning rate can be increased up to 1.166 mK s^{-1}, always maintaining a minimal difference between the target and real temperatures of the calorimetric block. Because the temperature gradient between the thermostatic heating fluid and the calorimetric block is kept constant (20 K), the power uptake of the heating elements is quasi-constant, thus avoiding the interference of sudden changes of power uptake on the calorimetric signal. For the reported results, measurements were performed using silicon oil instead of Hg as the hydraulic pressurizing fluid, and the polymer sample was placed in a lead (soft metal) ampoule. Test measurements were made on polyvinyl acetate (PVA) for which the $\Delta T_g/\Delta p$ coefficient was found to be 0.212 $\pm$ 0.002 K MPa^{-1}, in good agreement with the literature value of 0.22 K MPa^{-1} [73]. The calorimetric traces obtained [31] with the same method for a poly (butadiene-*co*-styrene) vulcanized rubber during isobaric scans of temperatures ranging from 218.15 to 278.15 K at the rate of 0.666 mK s^{-1} are shown in Fig. 21b. This figure also shows the evolution of T_g at pressures of 0.25, 10, 30, 50, and 90 MPa: T_g increases linearly with pressure, with a $\Delta T_g/\Delta p$ coefficient of 0.193 $\pm$ 0.002 K MPa^{-1}.

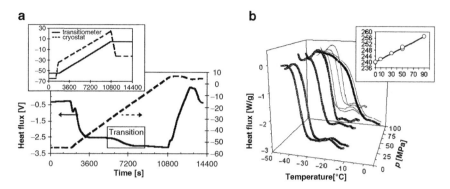

Fig. 21 Scanning transitiometry technique for the investigation of polymer T_g at low temperature and high pressure. (**a**) Experimental thermogram recorded during an isobaric temperature scan under 50 MPa (on a styrene–butadiene rubber sample of 1.56 g; scanning rate 0.666 mK s^{-1}). The *inset* shows the temperature programs for the transitiometer (*solid line*) and for the cryostat (*dashed line*). (**b**) Typical thermograms (heat flux vs temperature) for the transition domain of the vulcanized rubber are shown for different pressures. The *inset* shows the change of T_g with pressure, and the slope gives the pressure coefficient $\Delta T_g/\Delta p$

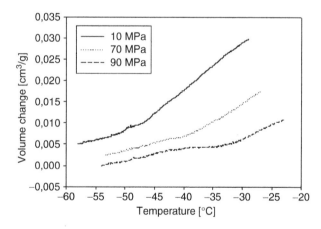

Fig. 22 Effect of pressure on the T_g of vulcanized rubber under isobaric conditions. Typical volume variations (ΔV vs T) are shown for the transition domain at 10, 70, and 90 MPa; the scanning conditions are the same as used for the measurements reported in Fig. 21

It should be noted that T_g is expressed as the temperature corresponding to the peak of the first derivative of the heat flux (i.e., at the inflexion point of the heat flux). The volume variations associated with the glass transition, which are also simultaneously measured by scanning transitiometric measurements, are depicted in Fig. 22. In accordance with the heat flux curve, T_g increases with increasing the pressure. Above T_g, there is an increase of the slope of the variation of the specific volume versus temperature. However, the change in the slope is gradual and T_g can be determined at the point where the two lines intersect.

Gas-Assisted Glass Transitions at High Pressures (Plasticizing Effect)

There is not much information available in the literature on calorimetric study of plasticization of polymers at high pressures, above say 50 MPa, induced by gases. Plasticization is well characterized by the shift to lower values of T_g. Actually, when pressure is induced by a gas, both plasticization and hydrostatic effects contribute to the shift of T_g. If plasticization tends to lower T_g because of the gain of mobility of the polymeric chains, the hydrostatic effect raises it in diminishing the free volume. CH_4 is assumed to be a nonplasticizing gas, but our results show that in the case of PS, at higher pressures, plasticization overtakes the hydrostatic effect, probably due to a higher solubility of the gas in PS at higher pressures; this kind of behavior has been suggested for high-enough pressures [74]. The plasticization of PS using CH_4 seems to be possible, but it is necessary to apply high pressure (i.e., 200 MPa) in order to obtain approximately the same shift of the T_g as with ethylene (C_2H_4) under 9.0 MPa! In this respect, CH_4 cannot be considered as a good plasticizing gas.

An important aspect of polymer foaming is certainly the "ease" with which the blowing agent can enter, dissolve, and diffuse into the polymer matrix. Control of two parameters, T and p, is essential to control these phenomena. The nature and properties of the polymer and of the fluid evidently play a major role. In this context, the physical state of the polymer must be appropriately modified to undergo plasticization; this optimal condition for having the "blowing" effect taking place depends upon the T_g. Plasticization depends on all the thermodynamic variables and parameters listed above. In particular, it is necessary to know to what extent T_g is advantageously decreased in order to optimize the foaming process. From a practical point of view, the ΔT_g shift should be accurately determined or predicted. Moreover, many properties can be correlated with the T_g depression ΔT_g due to plasticization. In order to predict the variation ΔT_g, the model of Chow [75] was selected. The calculations using the model of Chow were made using experimental data of solubilities directly measured with the new technique combining a VW weight sensor and a pVT setup [46], as described in Sect. 2.1.2.

Chow has proposed a relation based on the Gibbs and Di Marzio principle (the entropy of the glassy state is zero) [76, 77] to account for the change in T_g due to the sorbed component, as follows:

$$\ln\left(\frac{T_g}{T_{go}}\right) = \beta \left[(1-\theta)\ln(1-\theta) + \theta \ln\theta\right], \tag{15}$$

where:

$$\beta = \frac{zR}{M_p \Delta C_p}, \quad \theta = \frac{M_p}{z M_d} \frac{w}{1-w}.$$

T_g and T_{go} are the glass transition temperatures for the {gas–polymer} system and the pure polymer, respectively; M_p is the molar mass of the polymer repeat unit;

M_d is the molar mass of the (diluent) gas; R is the gas constant; w the mass fraction of the gas in the polymer; ΔC_p is the heat capacity change associated with the glass transition of the pure polymer; and z is the lattice coordination number. All parameters of the model have physical meanings, except the number z. The value of this parameter changes according to the state of the diluent: $z = 2$ when the diluent is in the liquid state and $z = 1$ when it is gas.

In order to compare the model calculations with experimental calorimetric data, PS samples were modified in a transitiometer used, in this case, as a small reactor to modify PS under equilibrium conditions in the presence of a chosen fluid. Modifications of PS have been done in the presence of N_2 and CO_2, along isotherms at a given pressure. For these two fluids, a final temperature of 398.15 K and a final pressure of 80 MPa have been attained. The T_g of modified and nonmodified PS samples were determined by temperature-modulated DSC (TMDSC). The solubilities of the different gases were measured using the VW–pVT sorption technique [48, 49] along different isotherms, and the mass fraction of the gas in the polymer was then determined with the following equation:

$$w = \frac{s}{s+1},\tag{16}$$

where s is the solubility of the fluid in the polymer, in milligrams of fluid per milligram of polymer.

Using the values of w determined for each gas–PS system, the Chow equation (15) allows estimation of the variation, ΔT_g, of the temperature of the glass transition with pressure, along the different isotherms of the sorption measurements.

The use of the Chow model is delicate because the choice of the value of z, i.e., the state of the diluent, significantly influences the results. The T_g shift under CO_2 pressures is spectacular, showing the high plasticizing effect of CO_2. The good agreement of the literature data for the $\{CO_2–PS\}$ system with the calculated values [78–80] (as seen in Fig. 23) can certainly be explained by the state of the diluent, which is most probably in the critical state in the ranges of T and p considered. Effectively, the critical temperature T_c and critical pressure p_c of CO_2 support the hypothesis of the gas being in the near-critical region. Depending on the experimental conditions in the vicinity of the critical point, the fluid can exist in one or the other state (gas or liquid), or even in both. In the present case, literature data for the $\{CO_2–PS\}$ system have been obtained under a pressure $p \leq p_c$ and at a temperature $T \geq T_c$ for CO_2; then two phases of the diluent can coexist in different proportions. Despite the difficulty in determining exactly the variation of T_g, particularly under supercritical conditions of a diluent fluid, the model of Chow is a useful guide for prediction of the variation of the glass transition of a polymer modified by a high pressure fluid. However, the exact determination of the glass transition depression, ΔT_g, becomes more difficult when the pressure increases, especially near and above the critical point of the diluent fluid. This means that when plotting ΔT_g as a

Key Thermodynamic Data and Thermophysical Properties

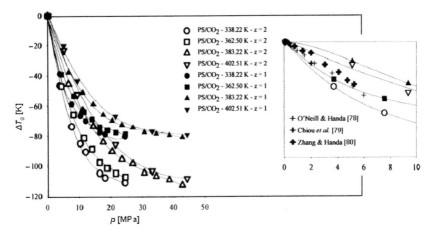

Fig. 23 Variation of T_g with pressure for the {CO$_2$–PS} system. Calculations have been made for 338.22, 362.50, 383.22, and 402.51 K. *Solid symbols* represent results for $z = 1$ and *open symbols* for $z = 2$. *Inset*: Literature values are represented by *crosses* in a magnified section of the graph (the same scale of temperature is kept). *Lines* are hand-drawn through the points

function of pressure, the temperature of measurement plays a major role. If we do not take into account this temperature, it is preferable to represent ΔT_g as a function of the mass fraction of the fluid in the polymer.

Compared to polar CO$_2$ and because of its non-polarizability, N$_2$ should be a weaker plasticizing agent although, as shown in Fig. 24, it induces significant shifts of T_g with increasing pressures [46]. However, N$_2$, which should also be a good foaming agent, is not currently used in the foaming industry because of the need of too high a pressure to attain the desired depression in T_g. Figure 25 shows the scanning electronic microscopy (SEM) images of PS microstructures modified by high pressure gases (CO$_2$ or N$_2$) in the VW–pVT technique instrument at a similar temperature (~315 K) close to, but below, T_g (380 K). The modified PS exhibits different patterns depending on the use of CO$_2$ or N$_2$ as blowing agents. For the {N$_2$–PS} system, there is no appearance of a foam structure; the surface is only damaged by the gas pressure. For the {CO$_2$–PS} system, a foam structure is apparent. Further increase of temperature has shown that the observed microcellular structure is highly temperature-dependent. Below T_g, the microcellular structure is obtained with perfect spherical bubbles, and the diameter of the bubbles tends to increase with increasing T. At temperatures higher than T_g, this organized structure disappears and the foam becomes more homogeneous.

The sorption of compressed gases in polymers can now be well documented. Our results with CH$_4$, CO$_2$, and N$_2$ confirm earlier studies of Condo et al. [81, 82] and more recent investigations of Handa et al. [83, 84] on retrograde vitrification of polymers observed when a decrease of T_g is observed at gas pressures high enough to overcome the purely hydrostatic effect.

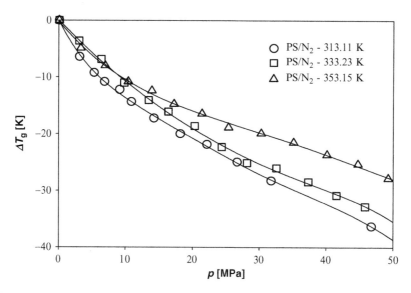

Fig. 24 Variation of T_g as a function of pressure for the {N_2–PS} system. Calculations have been made for 313.11, 333.23, and 353.15 K, using $z = 1$. *Lines* are hand-drawn through the points

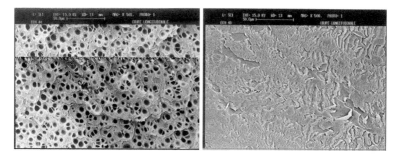

Fig. 25 SEM pictures (50 μm definition) of modified PS samples under CO_2 or N_2 gas pressure. The structure of the modified PS presents different aspects depending on the use of CO_2 or N_2 as blowing agent. *Left*: For the {CO_2–PS} system at 317.15 K, the structure of foam is apparent. *Right*: For the {N_2–PS} system at 313.12 K, there is no appearance of a foam structure; the surface of the PS is only damaged by the gas pressurization

4 Conclusion

An experimental setup coupling a VW detector and a *pVT* technique has been used to simultaneously evaluate the amount of gas entering a polymer under controlled temperature and pressure and the concomitant swelling of the polymer. Scanning transitiometry has been used to determine the interaction energy during gas sorption in different polymers. The technique was also advantageously used to determine the thermophysical properties (like isobaric thermal expansivity) of polymers in the presence and absence of gas sorption under pressure. Scanning transitiometry has also been used to evaluate the thermodynamic control of essential transitions in polymer science, first-order transitions, and glass transitions. The influence and the role of gas sorption on such transitions can be fully documented. Of particular interest is the tailoring of nanostructures at the isotropic transitions in amphiphilic diblock copolymers. The striking effect of gas sorption is particularly observed when the gas is in supercritical state, depending on the thermodynamic conditions. The main conclusion is that a rigorous thermodynamic approach is possible through appropriate experimental techniques in which the three main thermodynamic variables (p, V, and T) as well as the nature of pressurizing fluids are properly controlled. Evidently, applications of engineering interest are now at hand, as illustrated by examples taken from the petroleum and microelectronic industries.

References

1. Grolier JPE (2000) Polym Mater 1:7
2. Yamada T, Boyer SAE, Iyoda T, Yoshida H, Grolier JPE (2007) J Therm Anal Calorim 89:9
3. Yamada T, Boyer SAE, Iyoda T, Yoshida H, Grolier JPE (2007) J Therm Anal Calorim 89:717
4. Boyer SAE, Klopffer MH, Martin J, Grolier JPE (2007) J Appl Polym Sci 103:1706
5. Kamiya Y, Hirose T, Mizogushi K, Naito Y (1986) J Polym Sci B Polym Phys 24:1525
6. Wong B, Zhang Z, Handa YP (1998) J Polym Sci B Polym Phys 36:2025
7. Chaudary BI, Johns AI (1998) J Cell Plast 34:312
8. Schnitzler JV, Eggers R (1998) In: Proceedings of the 5th meeting on supercritical fluids, Nice, France, vol 1, p 93
9. Miura KI, Otake K, Kurosawa S, Sako T, Sugeta T, Nakane T, Sato M, Tsuji T, Hiaki T, Hongo M (1988) Fluid Phase Equilib 144:181
10. Wang NH, Takishima S, Matsuoka H (1994) Int Chem Eng 34:255
11. Briscoe BJ, Lo O, Wajs A, Dang P (1998) J Polym Sci B Polym Phys 36:243
12. Koros WJ, Paul DR (1976) J Polym Sci Polym Phys 14:1903
13. Stern SA, Meringo AHD (1978) J Polym Sci Polym Phys 16:735
14. Sato Y, Iketani T, Takishima S, Masuoka H (2000) Polym Eng Sci 40:1369
15. Sanchez IC, Rodgers PA (1990) Pure Appl Chem 62:2107
16. Wissinger RG, Paulaitis ME (1987) J Polym Sci B Polym Phys 25:2497
17. Sanchez IC, Lacombe RH (1976) J Phys Chem 80:2352
18. Hilic S, Padua AAH, Grolier JPE (2000) Rev Sci Instrum 71:4236
19. Hilic S, Boyer SAE, Padua AAH, Grolier JPE (2001) J Polym Sci B Polym Phys 39:2063
20. Randzio SL (1985) Thermochim Acta 89:215
21. Randzio SL, Eatough DJ, Lewis EA, Hansen LD (1988) J Chem Thermodyn 20:937
22. Randzio SL (1991) Pure Appl Chem 63:1409
23. Randzio SL, Grolier JPE, Quint JR (1994) Rev Sci Instrum 65:960
24. Randzio SL, Grolier JPE, Zaslona J, Quint JR (1991) French Patent 91–09227, Polish Patent P-295285
25. Randzio SL, Grolier JPE (1997) French Patent 97–15521. http://www.transitiometry.com. Accessed 4 May 2010
26. Randzio SL, Stachowiak Ch, Grolier JPE (2003) J Chem Thermodyn 35:639
27. Dan F, Grolier JPE (2002) Setaram News 7:13
28. Dan F, Grolier JPE (2004) Spectrocalorimetric screening for complex process optimization. In: Letcher T (ed) Chemical thermodynamics for industry. Royal Society of Chemistry, Cambridge, p 88

29. Grolier JPE, Dan F, Boyer SAE, Orlowska M, Randzio SL (2004) Int J Thermophys 25:297
30. Grolier JPE, Dan F (2006) Thermochim Acta 450:47
31. Grolier JPE, Dan F (2007) Advanced calorimetric techniques in polymer engineering, vol 259. In: Moritz HU, Pauer W (eds) Polymer reaction engineering, macromolecular symposia, p 371
32. Boyer SAE (2003) Ph.D. Thesis, Université Blaise Pascal, Clermont-Ferrand, France
33. Sanchez IC, Lacombe RH (1978) Macromolecules 11:1145
34. Boudouris D, Constantinou L, Panayiotou C (2000) Fluid Phase Equilib 167:1
35. Hariharan R, Freeman BD, Carbonell RG, Sarti GC (1993) J Appl Polym Sci 50:1781
36. DeAngelis MG, Merkel TC, Bondar VI, Freeman BD, Doghieri F, Sarti GC (1999) J Polym Sci B Polym Phys 37:3011
37. Koros WJ (1980) J Polym Sci Polym Phys 18:981
38. Michaels AS, Parkers RB (1959) J Polym Sci 41:53
39. Michaels AS, Bixler HJ (1961) J Polym Sci 50:393
40. Puleo AC, Paul DR, Wonk PK (1989) Polymer 30:1357
41. Manfredi C, Nobile MAD, Mensitieri G, Guerra G, Rapacciuolo M (1997) J Polym Sci Polym Phys 35:133
42. Guadagno L, Baldi P, Vittoria V, Guerra G (1998) Macromol Chem Phys 199:2671
43. Aletiev A, Drioli E, Gokzhaev M, Golemme G, Ilinich O, Lapkin A, Volkov V, Yampolskii Y (1998) J Memb Sci 138:99
44. Mogri Z, Paul DR (2001) Polymer 42:7781
45. Boyer SAE, Randzio SL, Grolier JPE (2006) J Polym Sci B Polym Phys 44:185
46. Boyer SAE, Grolier JPE (2005) Pure Appl Chem 77:593
47. Ismail AF, Lorna W (2002) Sep Purif Technol 27:173
48. Boyer SAE, Grolier JPE (2005) Polymer 46:3737
49. Sato Y, Fujiwara K, Takikawa T, Sumarno, Takishima S, Masuoka H (1999) Fluid Phase Equilib 162:261
50. Lorge O, Briscoe BJ, Dang P (1999) Polymer 40:2981
51. Flaconnèche B, Martin J, Klopffer MH (2001) Oil Gas Sci Techn Rev IFP· 56:261
52. Bos A, Pünt IGM, Wessling M, Strathmann H (1999) J Memb Sci 155:67
53. Hansen AR, Eckert CA (1991) J Chem Eng Data 36:252
54. Ghosal K, Chern RT, Freeman BD, Savariar RJ (1995) J Polym Sci B Polym Phys 33:657
55. Grolier JPE, Dan F (2004) Calorimetric measurements of thermophysical properties for industry. In: Letcher T (ed) Chemical thermodynamics for industry. Royal Society of Chemistry, Cambridge, p 144
56. Randzio SL, Grolier JPE (1998) Anal Chem 70:2327
57. Ehrlich P, Kurpen JJ (1963) J Polym Sci A 1:3217
58. Yoshida H, Wanatabe K, Wanatabe R, Iyoda T (2004) Trans Mater Res Soc Jpn 29:861
59. Boyer SAE, Grolier JPE, Pison L, Iwamoto C, Yoshida H, Iyoda T (2006) J Therm Anal Calorim 85:699
60. Boyer S.A.E, Grolier JPE, Yoshida H, Iyoda T (2007) J Polym Sci B Polym Phys 45:1354
61. Watanabe K, Tian Y, Yoshida H, Asaoka S, Iyoda T (2003) Trans Mater Res Sci Jpn 28:553
62. Tian Y, Watanabe K, Kong X, Abe J, Iyoda T (2002) Macromolecules 35:3739
63. Mukherjee PK, Rzoska SJ (2002) Phys Rev E 65:051705
64. Maeda Y, Niori T, Yamamoto J, Yokokawa H (2005) Thermochim Acta 57:428
65. Triolo R, Triolo A, Triolo F, Steytler DC, Lewis CA, Heenan RK, Wignall GD, DeSimone JM (2000) Phys Rev E 61:4640
66. Kawabata Y, Nagao M, Seto H, Komura S, Takeda T, Schwahn D, Yamada NL, Nobutou H (2004) Phys Rev Lett 92:056103
67. Seeger A, Freitag D, Freidel F, Luft G (2004) Thermochim Acta 424:175
68. Ryu DY, Lee JL, Kim JK, Lavery KA, Russell TP, Han YS, Seon BS, Lee CH, Thiyagarajan P (2003) Phys Rev Lett 90:235501
69. Pollard M, Russel TP, Ruzette AV, Mayes AM, Gallot Y (1998) Macromolecules 31:6493
70. Hamley IW (1998) The physics of block copolymers. Oxford University Press, Oxford

71. Masten MW, Bates FS (1996) Macromlecules 29:1091
72. Lazzari M, López Quintela MA (2003) Adv Mater 15:1583
73. Younglove B, Ely JF (1987) J Phys Chem Ref Data 16:577
74. Ribeiro M, Pison L, Grolier JPE (2001) Polymer 42:1653
75. Chow TS (1980) Macromolecules 13:362
76. Gibbs JH, Di Marzio EA (1958) J Chem Phys 28:373
77. Di Marzio EA, Gibbs JH (1963) J Polym Sci A 1:1417
78. O'Neill ML, Handa YP (1994) Plasticization of polystyrene by high pressure gases: a calorimetric study. In: Seyler RJ (ed) Assignment of the glass transition. ASTM, Philadelphia, p 165
79. Chiou JS, Barkow JW, Paul DR (1985) J Appl Polym Sci 30:263
80. Zhang Z, Handa YP (1998) J Polym Sci B Polym Phys 36:977
81. Condo PD, Sanchez IC, Panayiotou CG, Johnston KP (1992) Macromolecules 25:6119
82. Condo PD, Paul DR, Johnston KP (1994) Macromolecules 27:365
83. Handa YP, Zhang Z (2000) J Polym Sci B Polym Phys 38:716
84. Nawabi AV, Handa YP, Liao X, Yoshitaka Y, Tomohiro M (2007) Polym Int 56:67

Adv Polym Sci (2011) 238: 179–269
DOI: 10.1007/12_2010_81
© Springer-Verlag Berlin Heidelberg 2010
Published online: 13 July 2010

Interfacial Tension in Binary Polymer Blends and the Effects of Copolymers as Emulsifying Agents

Spiros H. Anastasiadis

Abstract The structure and the thermodynamic state of polymeric interfaces are important features in many materials of technological interest. This is especially true for multiconstituent systems such as blends of immiscible polymers, where the interface structure can affect greatly their morphology and, thus, their mechanical properties. In this article, we first present a review of the experimental and theoretical investigations of the interfacial tension in phase-separated homopolymer blends. We emphasize the effects of temperature and molecular weight on the behavior: interfacial tension γ decreases with increasing temperature (for polymer systems exhibiting upper critical solution temperature behavior) with a temperature coefficient of the order of 10^{-2} dyn/(cm °C), whereas it increases with increasing molecular weight. The increase follows a $\gamma = \gamma_\infty \left(1 - k_{int} M_n^{-z}\right)$ dependence (with $z \approx 1$ for high molecular weights), where γ_∞ is the limiting interfacial tension at infinite molecular weight and M_n the number average molecular weight. Suitably chosen block or graft copolymers are widely used in blends of immiscible polymers as compatibilizers for controlling the morphology (phase structure) and the interfacial adhesion between the phases. The compatabilitizing effect is due to their interfacial activity, i.e., to their affinity to selectively segregate to the interface between the phase-separated homopolymers, thus reducing the interfacial tension between the two macrophases. The experimental and theoretical works in this area are reviewed herein. The effects of concentration, molecular weight, composition, and macromolecular architecture of the copolymeric additives are discussed. An issue that can influence the efficient utilization of a copolymeric additive as an emulsifier is the possibility of micelle formation within the homopolymer matrices when the additive is mixed with one of the components. These micelles will compete with the interfacial region for copolymer chains. A second issue relates

---

S.H. Anastasiadis
Institute of Electronic Structure and Laser, Foundation for Research and Technology – Hellas, P. O. Box 1527, 711 10 Heraklion Crete, Greece
Department of Chemistry, University of Crete, P. O. Box 2208, 710 03 Heraklion Crete, Greece
e-mail: spiros@iesl.forth.gr

180   S.H. Anastasiadis

to the possible trapping of copolymer chains at the interface, which can lead to stationary states of partial equilibrium. The in-situ formation of copolymers by the interfacial reaction of functionalized homopolymers is also discussed.

Keywords Polymer interfaces · Interfacial tension · Compatibilizers · Interfacial partitioning · Emulsifying agents

Contents

1 Introduction 180
2 Methods of Measuring Interfacial Tension 183
3 Interfacial Tension in Binary Polymer Blends 189
    3.1 Experimental Studies of Polymer Interfacial Tension 189
    3.2 Theories of Polymer–Polymer Interfaces 196
4 Copolymers as Emulsifying Agents in Polymer Blends 225
    4.1 Copolymer Localization at the Polymer Blend Interface 225
    4.2 Experimental Studies on the Effect of Additives on Polymer–Polymer Interfacial
        Tension 228
    4.3 Theories of the Interfacial Behavior in Homopolymer/
        Homopolymer/Copolymer Blends 238
5 Concluding Remarks 254
References 258

1 Introduction

The increasing need of the modern world to create materials with new fascinating properties and better performance, that are more easily processable and, hopefully, more environmentally friendly has forced polymer scientists to face the challenge of developing new macromolecular systems with such characteristics. Realistically, however, industry would prefer to keep using the traditional commodity polymers because of the accumulated know-how and the significant investments made over the years. Between those two trends, scientists have found a way to satisfy both demands. Improving the performance of polymeric materials for many important scientific and industrial applications can be achieved by mixing different components with complementary properties. Polymer blending is a high-stakes game in the plastics industry, whereby basic resins are manipulated into becoming new polymer systems with properties beyond those available with the individual resin components [1, 2].

The development of compounds and blends of polymers dates back almost two centuries to the early rubber and plastics industry, when rubber was mixed with substances ranging from pitch [3] to gutta percha [4]. As each new plastic has been developed, its blends with previously existing materials have been explored. Thus, synthetic rubbers, in the early period of the plastics industry, were mixed into natural rubber and found to produce superior performance in tire components. Polystyrene (PS) was blended with natural and synthetic rubbers after its commercialization, and this led to high impact polystyrenes (HIPS), which now hold a

Interfacial Tension in Binary Polymer Blends and the Effects of Copolymers

stronger position in the market place than the bare plastic. Attention has been especially focused on blends since the commercialization of General Electric's Noryl, a blend of HIPS with poly(2,6-dimethyl-1,4-phenylene oxide). We are now in a period of investigating both the new blends and related compound systems and the scientific principles underlying blend characteristics [1, 2].

When the two (or more) blend components are compatible, the performance of the final product is straightforwardly controlled by the properties of the individual materials and their mixing ratio. In the most frequently encountered situation of immiscible polymer/polymer dispersions, however, one is faced with the problem of controlling the morphology (phase structure) and the interfacial adhesion between the phases in order to obtain an optimized product [1, 2]. The phase structure (e.g., the dispersed particle size) in such systems is controlled by the chemical character of the individual components and their rheological properties [5] as well as by the deformation and/or thermal history; these factors affect how the phase morphology evolves. A number of experimental investigations have clearly shown that the characteristic size of the dispersed phase in incompatible polymer blends is directly proportional to the interfacial tension [6], whereas the equilibrium adhesive bond strength between the two phases depends strongly on the interfacial tension. For example, the characteristic size of the dispersed phase obtained during melt extrusion of an incompatible polymer blend is related to the interfacial tension between the two phases (γ,), the viscosities of the dispersed phase and the matrix (η_d and η_m, respectively), and the process characteristics (shear rate, $\dot{g}$) by the empirical relationship [7]:

$$\frac{\dot{g}\eta_m d_n}{\gamma} = 4\left(\frac{\eta_d}{\eta_m}\right)^{\pm 0.84} \tag{1}$$

where d_n is the number-average particle diameter. The plus (+) sign applies for $p = \eta_d/\eta_m > 1$ and the minus (−) sign for $p < 1$. Moreover, the rate of phase growth during the later stages of phase separation increases with increasing interfacial tension [8]. It is noted that the size of the dispersed phase is an important factor that influences the mechanical properties of incompatible polymer blends.

Therefore, interfacial tension is an important, if not overriding, factor in the formation of a phase boundary and in the development of phase morphology in incompatible polymer blends. Interfacial tension, γ, is defined as the reversible work required to create a unit of interfacial area at constant temperature, T, pressure, P, and composition, n, i.e., [9–18]:

$$\gamma = \left(\frac{\partial G}{\partial A}\right)_{T,P,n} \tag{2}$$

where G is the Gibbs free energy of the system and A the interfacial area. Interfacial tension is, thus, a thermodynamic property of the system and can be calculated directly from statistical thermodynamic theories. Experimental investigation of interfacial tension is, therefore, a straightforward means for evaluating the validity of such theories.

For a certain polymer–polymer pair, interfacial tension generally decreases linearly with temperature with a temperature coefficient of the order of 10^{-2} dyn/(cm °C) [10, 19–24]. Increasing the molecular weight of either polymer leads to an increase in the interfacial tension; it is now recognized that, for high enough molecular weights, interfacial tension shows a M_n^{-z} dependence on the molecular weight [20, 21, 23–26] with $z \cong 1$, although there are reports for $z \cong 2/3$ or even 0.5 for lower molecular weights (M_n is the number average molecular weight). Moreover, interfacial tension was found to decrease with increasing polydispersity [22, 23, 26]. A number of thermodynamic theories have appeared from very early on [27–29] until more recently [25, 30–35], which predict the interfacial tension of blends of immiscible polymers and its temperature and molecular weight dependencies. Both the experimental and the theoretical investigations of polymer–polymer interfacial tension will be thoroughly reviewed in Sect. 3.

Suitably chosen block or graft copolymers are widely used by the polymer industry as emulsifiers in multiconstituent polymeric systems in order to improve the interfacial situation and, thus, obtain an optimized product [1, 2, 36]. This is due to their interfacial activity, i.e., to their affinity to preferentially segregate to the interface between the phase-separated homopolymers [37–44]. This partitioning of the block copolymers at the interface is responsible for the significant reduction of the interfacial tension between the two macrophases [45–59], aids droplet breakup, and inhibits coalescence of the dispersed phases [60, 61]. This leads to a finer and more homogeneous dispersion during mixing [52, 62–66], and improves interfacial adhesion [67, 68] and mechanical properties via the significant increase in the interfacial thickness between the macrophases [38, 69]. For a block or graft copolymer to be effective as an emulsifier, it is, thus, important that it is localized to the polymer–polymer interface [37, 38, 40–44], with each block preferentially extending into its respective homopolymer phase [39, 70–74]. Because block and graft copolymers are likely to be expensive, it is of great importance to maximize their efficiency so that only small amounts are required. The efficiency of interfacial partitioning is predicted to depend on the molecular weights of the copolymer blocks relative to those of the homopolymers [70, 75–79], on the macromolecular architecture/topology and composition of the copolymers [80–98], as well as on the interaction parameter balance between the homopolymers and the copolymer blocks [99, 100].

However, a crucial issue that could severely influence the efficient utilization of a copolymeric additive as an emulsifier is the possible formation of copolymeric micelles within the homopolymer phases when the additive is mixed with one of the components [101]. The micelles will compete with the interfacial region for copolymer chains, and the amount of copolymer at the interface or in micelles depends on the relative reduction of the free energy, with much of the premade copolymer often residing in micelles for high molecular weight additives. The effect of the existence of micelles on the interfacial partitioning of diblock copolymers at the polymer–polymer interface has received some attention in the literature [54, 56, 75, 77, 102–105]. As an alternative, in-situ formation of copolymers (usually grafts) is utilized [61, 106–117] in order to overcome "wasting" of the additive into micelles. A second issue relates to the possible trapping [71] of copolymer chains at the interface, which can lead to partial equilibrium situations. Finally, since in a typical preparation of homopolymer/copolymer blends the system can be diffusion-controlled, the optimal conditions for the molecular design of interfacially active copolymers obtained from equilibrium considerations might have to be modified.

The experimental and theoretical investigations of the effects of copolymeric additives on polymer–polymer interfacial tension will be reviewed in Sect. 4.

2 Methods of Measuring Interfacial Tension

Various techniques have been developed to measure surface and interfacial tensions of liquids and melts and an early extensive discussion was presented by Padday [118]. In principle, all the standard techniques can be used to measure the surface and/or interfacial tension of polymer liquids and melts; however, due to the high viscosity and viscoelastic character of the polymers, only a few methods are suitable. In general, equilibrium static techniques seem completely satisfactory. Due to the high equilibration times involved with polymeric materials, it has not been possible to demonstrate that pull, detachment, or bubble pressure measurements can always be made slowly enough to yield accurate results with highly viscous liquids. Extensive reviews on the suitability of the various methods applied to polymeric systems have been given by Frisch et al. [119], Wu [10, 120], Koberstein [121], Anastasiadis [122], Xing et al. [123], and Demarquette [124].

Only methods based on drop profiles are suitable for both surface and interfacial tension measurements. These include the pendant drop method [125–127], the sessile bubble or drop method [128, 129], and the rotating drop or bubble method [130, 131]. These methods are independent of the solid–liquid contact angle but require accurate knowledge of the density difference across the interface. The demand of accurate density data becomes even greater when the two phases have similar densities. The rotating drop or bubble method is particularly suited for the determination of very low surface and interfacial tensions.

Although the capillary rise [132, 133] is one of the static methods, the very slow attainment of equilibrium (because of the resistance to flow in the narrow capillary) makes it unsatisfactory for highly viscous materials. The Wilhelmy plate technique [134, 135] has the advantage that density data are not required; however, the requirement of zero contact angle makes it suitable only for surface tension measurements. Other standard techniques, such as the detachment methods (Du Noüy ring [136–138], drop weight methods [118]), and the maximum bubble pressure method [133] are severely limited by viscosity. Although these methods, except the drop weight methods, have been used to measure the surface tension of low-viscosity polymeric liquids, they are impractical for viscous fluids because of the extremely slow rates of attaining equilibrium. Most importantly, in this case, they are not suitable for measurements at the liquid–liquid interface.

Surface light scattering methods from thermally induced capillary waves at the interface [139–141] or from electric-field-induced surface waves [142, 143] have appeared. The technique is limited by the viscosities of the two phases; if the viscosities are too large, then the spatial damping of the surface capillary waves is too rapid to be detected by the technique. The applicability of this method for highly viscous polymeric interfaces has not been verified yet.

Two dynamic methods that have attracted the interest of the scientific community are the breaking thread method and the imbedded fiber retraction (IFR) method. Although they are dynamic methods and, thus, suffer from the high viscosities and viscoelastic character of polymers, they possess an important advantage in that they can be used to measure the interfacial tension between two phases of similar densities. The breaking thread method [144–147] involves the observation of the evolution of the shape of a long fluid thread imbedded in another fluid. Due to Brownian motion, small distortions of arbitrary wavelength are generated at the surface of the thread; this leads to a pressure difference between the inside and the outside of the thread, which induces important deformations caused by the effect of the interfacial tension that tends to reduce the interfacial area. It is possible to infer interfacial tension between the polymer forming the thread and the matrix from the study of the time evolution of the disturbances. However, the breaking thread method suffers from a major drawback related to residual stresses during the preparation of the threads; these fibers distort faster and lead to interfacial tension values much higher than the real value. Moreover, the fiber should be formed with the material that has the lowest viscosity and, at the same time, the higher softening temperature. Palmer and Demarquette [148] proposed a methodology for the improvement of the accuracy of the method by utilizing simultaneously the original theory of Tomotika [144], which evaluates the growth rate of the sinusoidal instabilities growing exponentially with time, with that of Tjahjadi et al. [146], which consists of fitting the dynamics of amplitude growth using curve-fitted polynomials, which are calculated from numerically generated results of the transient shape using boundary integral techniques.

The IFR method is a dynamic technique that has been widely used to measure the interfacial tension for blends comprising high molecular weight and/or high viscosity polymers, for which it is difficult or impossible to measure the interfacial tension using direct equilibrium techniques such as the sessile or pendant drop methods. The IFR method involves the analysis of the microscopic shape change of a fiber of one polymer embedded in a matrix of a second polymer [25, 149–151]. In general, the IFR studies are made on matrix polymers that are solid at room temperature and have high viscosities, which are obtained directly by compression molding or cut from large compression-molded samples. These systems require a melting and embedding step at a temperature below the retraction temperature. However, matrix polymers that are liquid at room temperature have been used as well [24].

The most versatile, convenient, and reliable technique for determining the surface and interfacial tension of polymer melts is the pendant drop method [19, 20, 45, 54, 56, 122, 126, 127, 152–155]. The results obtained by the pendant drop method constitute the bulk of the available data [10, 120]. The method is

Interfacial Tension in Binary Polymer Blends and the Effects of Copolymers

based on the principle that the shape of the profile of a drop of one fluid into a matrix of another is governed by a force balance between interfacial tension and gravity or buoyancy forces: a drop is pendant if the resultant force tends to pull it away from the supporting surface and sessile otherwise. The shape of the drop is described by the Bashforth–Adams equation [156], which in dimensionless form is given by:

$$\frac{d\phi}{dS} = \frac{2}{B} + Z - \frac{\sin \phi}{X}$$
$$\frac{dX}{dS} = \cos \phi \tag{3}$$
$$\frac{dZ}{dS} = \sin \phi$$
$$X(0) = Z(0) = \phi(0) = 0$$

where ϕ is the angle measured between the tangent to the drop profile at the point (x, z) and the horizontal axis and s the distance of point (x, z) from the drop apex along the drop contour. The dimensionless reduced parameters are defined as $X = x\sqrt{c}$, $Z = z\sqrt{c}$, and $S = s\sqrt{c}$. The shape of the drop is controlled by the shape parameter $B = a\sqrt{c}$, where a is the radius of curvature at the drop apex, g is the gravitational constant, γ is the interfacial tension, $\Delta\rho$ is the density difference across the interface, and $c = g\Delta\rho/\gamma$. Thus, the profile of a pendant drop at hydrodynamic and interfacial equilibrium provides the value of the interfacial tension.

Continuous monitoring of the drop profile can provide a criterion for hydrodynamic equilibrium of the drop by verifying conformity to the differential equation (3). The technique does not require any particular solid–liquid contact angle (except that the contact angle should be constant over the surface from which the drop is suspended, so that the drop shape will constitute a figure of revolution). Because of the minimal solid–liquid contact, the pendant drop offers the fastest equilibration among the various methods. One potential difficulty is that an initially stable drop might detach if sufficient reduction in interfacial tension occurs during the measurement.

Andreas et al. [157] first proposed that measurements of two diameters of the drop could be used to determine γ. Their procedure involved the determination of the maximum diameter d_e and a second diameter d_s located at distance d_e above the drop apex. The ratio $S = d_s/d_e$ was used to determine a correction factor H from tabulated values. The interfacial tension was then calculated by $\gamma = g\Delta\rho d_e^2/H$. More accurate tables of $1/H$ versus S were compiled later [158, 159] by numerical solution of the fundamental differential equations. Roe et al. [125] proposed the use of not one but several characteristic ratios for determining the drop shape by defining a series of diameters d_n ($n = 8, \ldots 12$) measured at heights $Z_n = (n/10)d_e$ ($n = 8, \ldots 12$) and the corresponding characteristic ratios $S_n = d_n/d_e$ ($n = 8, \ldots 12$). They suggested that, when a series of the $1/H$ values determined from the several S_n values are nearly identical, the drop can be considered to have attained its equilibrium shape and the interfacial tension can be calculated.

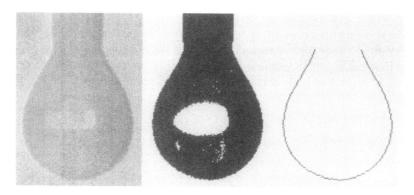

Fig. 1 Typical digitized drop images. Original gray level image (*left*); thresholded binary image (*center*); segmented drop profile (*right*) [155]

Advances in both data acquisition and analysis improved the precision and accuracy of axisymmetric fluid drop techniques. Digitizer palettes to efficiently record and store profile coordinates from enlarged photographic drop images were first used [19, 160, 161], but were eliminated entirely by the change to direct digitization of drop images with the aid of video frame grabbers or direct digital cameras [155, 162–164]. Digital processing of the drop images leads to rapid acquisition and analysis, thus, providing a simple means of detecting the attainment of equilibrium, a distinct advantage for viscous fluids such as polymer melts.

Figure 1 shows typical digital images of a pendant drop (left), the same drop following global thresholding (center) that reduced the 256 gray level image to a binary image, and the resultant segmented drop profile (right) [155].

Sophisticated algorithms for the analysis of drop profiles were developed concurrently [155, 160, 162, 165]. These methods either eliminate or minimize the necessity of specifying extremal drop dimensions, thereby reducing the inherent statistical error. Different optimization procedures have appeared. Girault et al. [162] and Huh and Reed [160] used a least squares optimization with exhaustive search through the shape parameter B, whereas Rotenberg et al. [165] utilized a sophisticated least squares optimization procedure using the Newton–Raphson method with incremental loading. Alternatively, Anastasiadis et al. [155] developed a robust shape analysis algorithm, which utilized the repeated median concept of Siegel and coworkers [166–168]. This algorithm has the advantages of robustness and resistance, namely that outlying points that are not consistent to the trend do not influence the fit. Such outlying points could result from inaccuracies in the computerized drop profile discrimination procedure.

The process of comparing the experimental drop profile to the theoretical profile, generated by numerical integration of (3), involves a five-parameter optimization. A total of three parameters are required for the alignment of the imaging system to the coordinate system of the dimensionless drop: an x- and y-translation, and a rotational angle. The two final parameters are the scale or magnification factor of the drop, $\sqrt{c}$, and the shape parameter B. As in all regression problems, the drop

Interfacial Tension in Binary Polymer Blends and the Effects of Copolymers

analysis procedure involves comparing N points $\{(x_i, y_i)\}$ from an experimental profile to N homologous points $\{(u_i, v_i)\}$ from a theoretical profile. The theoretical points must be rotated by an angle θ, translated by a vector (α, β), and scaled by a factor τ in order to effect this comparison. The transformed theoretical coordinates are given by:

$$\begin{pmatrix} u'_i \\ v'_i \end{pmatrix} = \begin{pmatrix} \alpha \\ \beta \end{pmatrix} + \tau \begin{pmatrix} \cos\theta & -\sin\theta \\ \sin\theta & \cos\theta \end{pmatrix} \begin{pmatrix} u_i \\ v_i \end{pmatrix} \tag{4}$$

and are compared to $\{(x_i, y_i)\}$ for each value of the shape parameter B. The value of the shape parameter, which yields the minimum overall error, provides the optimal fit. The interfacial tension is, then, obtained from the associated optimal scaling factor τ, recognizing that:

$$\tau = 1/\sqrt{c} = \left(\frac{\gamma}{g\Delta\rho}\right)^{1/2} \tag{5}$$

In least squares regression methods, the values of all the shape parameters (i.e., τ, α, β, θ) must be chosen simultaneously in order to minimize the sum of the squared residuals:

$$\text{sum} = \sum_{i=1}^{N} \left[(x_i - u'_i)^2 + (y_i - v'_i)^2 \right] \tag{6}$$

In contrast, with the robust shape comparison method, each of the optimal parameter values can be evaluated independently. In the case of rotation and magnification variables, this is accomplished using the concept of repeated medians as represented by the relationships:

$$\tau^* = \underset{i}{\text{med}} \left\{ \underset{i}{\text{med}} \{\tau_{ij}\} \right\} \tag{7a}$$

where

$$\tau_{ij} = \frac{\left[(x_j - x_i)^2 + (y_j - y_i)^2 \right]^{1/2}}{\left[(u_j - u_i)^2 + (v_j - v_i)^2 \right]^{1/2}} \tag{7b}$$

and

$$\theta^* = \underset{i}{\text{med}} \left\{ \underset{i}{\text{med}} \{\theta_{ij}\} \right\} \tag{8}$$

where θ_{ij} is the rotation required for the vector from point i to point j in the theoretical shape in order to have the same attitude as the homologous vector in the experimental profile. The translational parameters are calculated from the simple (nonrepeated) medians as:

$$\alpha^* = \underset{i}{\mathrm{med}} \{x_i - \tau^*[u_i \cos \theta^* - v_i \sin \theta^*]\} \tag{9a}$$

$$\beta^* = \underset{i}{\mathrm{med}} \{y_i - \tau^*[u_i \sin \theta^* + v_i \cos \theta^*]\} \tag{9b}$$

The advantages of double median robust techniques over traditional least squares regression methods have been discussed by Siegel et al. [166]. One particular advantage specific to the shape comparison problem can be understood by comparing (6) with (7)–(9). The least squares minimization is sensitive to local residuals between individual points, which are, however, only remotely related to the overall shapes of the two profiles being compared. The robust method affects a more global shape comparison, as can be seen from examining (7) and (8). Instead of comparing individual points of the two curves, the method compares vectors or line segments between all points i and j on the experimental profile with the corresponding vectors on the theoretical profile. In addition, the values of τ^*, θ^*, α^*, and β^* for each shape comparison (i.e., for each value of the shape parameter B) are specified directly by the robust relationships (7)–(9). Thus, the five-parameter optimization is reduced to a single variable optimization of the shape parameter B [155].

The application of the robust shape analysis algorithm is illustrated in Fig. 2 for a drop of polystyrene (PS, $M_n = 10{,}200$; $M_w/M_n = 1.07$) in a poly(ethyl ethylene)

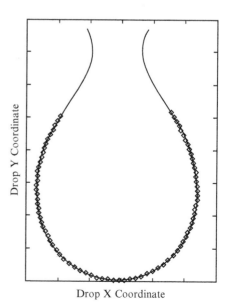

Fig. 2 Quality of the fit obtained by the application of the algorithm to an experimental profile for a PS 10,200 drop in a PBDH 4080 matrix at 147°C. *Solid line* is the theoretical profile, and the *data points* denote the original segmented experimental drop profile [20]. The interfacial tension is 2.6 dyn/cm

(PBDH $M_n = 4{,}080$; $M_w/M_n = 1.04$) matrix at 147°C. The analysis of the drop was performed using 23 data points for the shape comparison. The correspondence of the theoretical profile to these data points and to the original digitized drop profile is excellent [20].

The experimental setup, the digital image processing routines, and the robust shape analysis algorithm have been widely used to study the polymer–polymer interfacial tension [20], the effects of copolymeric additives on polymer–polymer interfacial tension [45, 48] and the influence of copolymer molecular weight [54] and architecture [56], the surface tension of homopolymers [169] and of miscible polymer blends [170], the effects of end-groups on the polymer surface tension and its molecular weight dependence [171], the effects of end groups on polymer–polymer interfacial tension [172], the work of adhesion of polymer–wall interfaces [173], etc. Moreover, the analysis algorithm was utilized by a different group in the development of another pendant drop instrumentation [164] and their measurements of polymer–polymer interfacial tension [21].

3 Interfacial Tension in Binary Polymer Blends

3.1 Experimental Studies of Polymer Interfacial Tension

Although knowledge of the interfacial tension in polymer/polymer systems can provide important information on the interfacial structure between polymers and, thus, can help the understanding of polymer compatibility and adhesion, reliable measurements of surface and interfacial tension were not reported until 1965 for surface tension [135, 138] and 1969 for interfacial tension [127, 154] because of the experimental difficulties involved due to the high polymer viscosities. Chappelar [145] obtained some preliminary values of the interfacial tension between molten polymer pairs using a thread breakup technique. The systems examined included nylon with polystyrene, nylon with polyethylene (PE), and poly(ethylene terephthalate) with PE; the values are probably only qualitatively significant [174].

Determinations by Roe [154] and Wu [127, 152, 153] using the pendant drop method and by Hata and coworkers [128, 175] using the sessile bubble technique have yielded values for a number of polymer pairs as a function of temperature. Gaines [174] and Wu [10, 120, 176] provided extensive reviews of the early work in the area of surface and interfacial tension of polymer liquids and melts.

In general, and for polymers that exhibit a miscibility gap at lower temperatures (blends that show upper critical solution temperature, UCST, behavior), interfacial tension is found to decrease linearly with increasing temperature, with temperature coefficients of the order of 10^{-2} dyn/(cm °C) [10]. This is about one half of the values observed for the temperature coefficients of polymer surface tension [10, 120, 176].

An increase in the molecular weight of either polymer leads, in general, to an increase in interfacial tension [10, 19, 20, 120, 176]; however, there are few systematic experimental studies of the effects of molecular weight on polymer–polymer interfacial tension. Bailey et al. [177] have examined the effect of molecular weight and functional end groups on the interfacial tension between poly(ethylene oxide), PEO, and poly(propylene oxide), PPO. The interfacial tension was found to increase with increasing PPO molecular weight but to decrease slightly with increasing the molecular weight of PEO. This dependence was attributed to the adsorption of the hydroxyl end groups of PPO to the interface. When these end groups were replaced by methoxy groups, the adsorption no longer took place and the interfacial tension increased with increasing molecular weight. Experimental interfacial tensions measured by Gaines and coworkers [178, 179] for the systems n-alkanes/perfluoroalkane $C_{12.5}F_{27}$, poly(dimethyl siloxanes)/$C_{12.5}F_{27}$ or C_8F_{18}, and alkanes/poly(ethylene glycols) all exhibited an increase with increasing molecular weight following an apparent $M_n^{-2/3}$ dependence, similar to that observed for homopolymer surface tension [10]. This similarity was predicted by several empirical theories that relate interfacial tension to the pure component surface tensions [153, 180], whereas no thermodynamic theory explicitly accounts for this dependence (see Sect. 3.2 below).

Anastasiadis et al. [20] utilized digital image processing of pendant fluid drops to investigate the effects of temperature and molecular weight on the interfacial tension for three blends of immiscible polymers. Interfacial tension was found to decrease almost linearly with increasing temperature for all systems (which exhibit a UCST behavior) and to increase with increasing molecular weight. The interfacial tension data for blends of polybutadiene (PBD 1000; $M_n = 980$; $M_w/M_n = 1.07$)

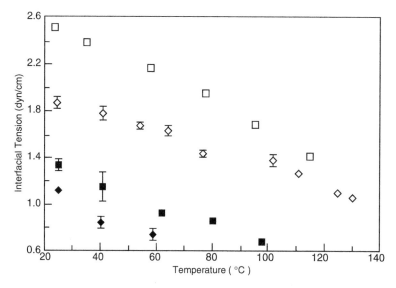

Fig. 3 Experimental interfacial tension as a function of temperature for PBD/PDMS pairs. *Open squares* PBD 1000/PDMS 3780; *open diamonds* PBD 1000/PDMS 2000; *filled squares* PBD 1000/PDMS 1250; *filled diamonds* PBD 1000/PDMS 770 [20]

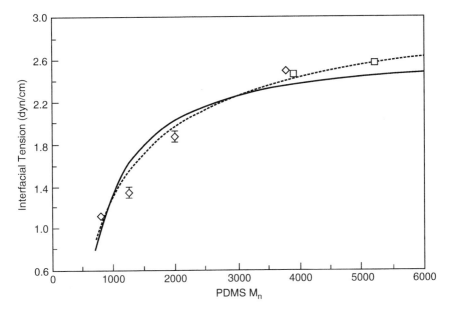

Fig. 4 Experimental interfacial tension at 25°C between PDMS and PBD 1000 as a function of the M_n of PDMS. *Solid line* represents the best fit to a M_n^{-1} dependence and the *dotted line* is the fit for a $M_n^{-0.5}$ dependence [19, 20]

with four poly(dimethyl siloxanes), PDMS, are shown in Fig. 3 as a function of temperature. Interfacial tension decreases almost linearly with temperature with temperature coefficients of 0.75×10^{-2} to 1.2×10^{-2} dyn/(cm °C).

The effect of PDMS molecular weight on the interfacial tension at constant temperature for a constant molecular weight of PBD ($M_n = 980$, $M_w/M_n = 1.07$) is illustrated in Fig. 4. The molecular weight dependence was obtained by performing nonlinear least-squares regression of the data to an expression of the form $\gamma = \gamma_\infty (1 - k_{int} M_n^{-z})$. This analysis yielded $z = 0.54$ for the present PDMS/PBD system of the specific range of low molecular weights.

The interfacial tension data for blends of PS of various molecular weights versus a poly(ethyl ethylene) (PBDH 4080; $M_n = 4800$, $M_w/M_n = 1.04$) exhibited a similar behavior with temperature, with temperature coefficients 0.9×10^{-2} to 1.5×10^{-2} dyn/(cm °C), and, qualitatively, with molecular weight. However, fitting the data to the expression $\gamma = \gamma_\infty (1 - k_{int} M_n^{-z})$ yielded $z = 0.68$ for PS molecular weights between 2200 and 10,200.

The measurements for the blends of PS and poly(methyl methacrylate) (PMMA; $M_n = 10,000$, $M_w/M_n = 1.05$) cover the broadest range of molecular weights (Fig. 5). For this system, nonlinear fit of the data to the expression $\gamma = \gamma_\infty (1 - k_{int} M_n^{-z})$ resulted in $z = 0.90$ for PS molecular weights between 2200 and 43,700.

These values for the exponent z should be taken with caution because of experimental errors. However, it was pointed out [20] that the smallest value for z

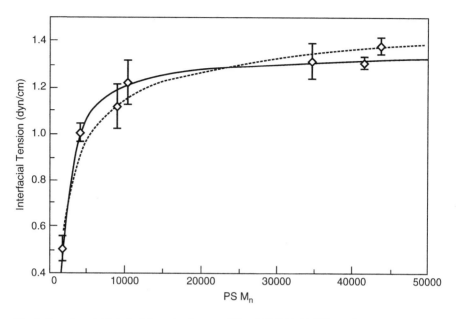

Fig. 5 Experimental interfacial tension between PS and PMMA 10,000 as a function of PS M_n at 199°C. The *solid line* represents the best fit to a M_n^{-1} dependence and the *dotted line* is the fit for a $M_n^{-0.5}$ dependence [20]

($z = 0.54$) was obtained for the system with the lowest molecular weights and highest polydispersities ($M_w/M_n = 2$ for the PDMS samples), whereas the largest value for z ($z = 0.9$) was observed for the system with the highest molecular weights. A smaller exponent for PDMS/PBD could be explained by the occurrence of surface segregation of the polydisperse PDMS according to molecular weight. Surface tension data for mixtures of PDMS oligomers suggest that the lower molecular weight species are concentrated at the surface [176]. Alternatively, the PDMS/PBD system is closest to its critical point, where a $M_n^{-0.5}$ dependence of interfacial tension has been predicted [181] (discussed in Sect. 3.2.4). The intermediate molecular weight system of PS/PBDH shows good correspondence with the $M_n^{-2/3}$ dependence. A similar dependence for the surface tension was explained by using a simple lattice analysis [182] that incorporated the contribution of the end groups at the interface. For these moderate molecular weights, the end-group effects are important and a $M_n^{-2/3}$ dependence might be expected.

The PS/PMMA blends, on the other hand, contain the highest molecular weight constituents and should, thus, conform best to the limit of high molecular weights. In this limit, the exponent z apparently approached unity. The fact that the estimated exponent is 0.90 probably suggests that the asymptotic regime (the M_n^{-1} behavior) was not yet reached even for those molecular weights. The nonlinear regression results, therefore, suggest that the exponent z of the molecular weight dependence of polymer–polymer interfacial tension increases as the molecular weights of the constituents increase.

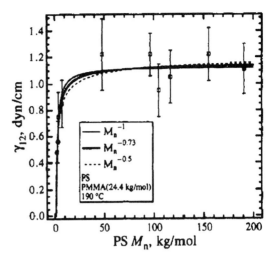

Fig. 6 Experimental interfacial tension between PS and PMMA 24,400 as a function of PS M_n at 190°C. The *unbroken curve* represents the three-parameter nonlinear fit to an expression $\gamma = \gamma_\infty(1 - k_{int}M_n^{-z})$. The *dashed curves* represent two-parameter fits assuming M_n^{-1} or $M_n^{-0.5}$ dependencies [25]

Ellingson et al. [25] investigated the molecular weight dependence of the interfacial tension between a PMMA ($M_n = 24{,}400$, $M_w/M_n = 1.10$) and a series of polystyrenes ($M_n = 2140$ to $191{,}000$, $M_w/M_n = 1.06$ to 3.26) utilizing an IFR method, which allowed them to study even larger molecular weights. Figure 6 shows the experimental data at a temperature of 190°C (slightly lower than for the measurements of Anastasiadis [20]). The data were analyzed with the expression $\gamma = \gamma_\infty(1 - k_{int}M_n^{-z})$, yielding a best fit value of $z = 0.73 \pm 0.24$; however, equal quality fits were obtained for $z = 0.50$ or $z = 1.0$.

Kamal et al. [21, 22] used a similar pendant drop apparatus to determine the interfacial tension between polypropylene (PP, $M_n = 54{,}000$ and $M_w/M_n = 5.54$) and a series of polystyrenes (M_n from 1600 to 380,000 and $M_w/M_n = 1.04$–1.06). Interfacial tension decreased almost linearly with temperature (Fig. 7) for this UCST-type system, with temperature coefficients of 3.7×10^{-2} to 4.4×10^{-2} dyn/(cm °C).

Figure 8 shows the effect of the PS molecular weight on the interfacial tension with PP [21, 22]. The interfacial tension increases as the PS molecular weight increases, in agreement with earlier works. The precision of the data, however, does not allow the unequivocal determination of the functional form of the molecular weight dependence. The data can be equally well fitted with the expression $\gamma = \gamma_\infty(1 - k_{int}M_n^{-z})$, with the exponent z being 0.5, or 0.68 or 1.

Arashiro and Demarquette [23] investigated the effects of temperature, molecular weight, and molecular weight polydispersity on the interfacial tension between low density PE and PS. Figure 9 shows the temperature dependence for three PE/PS pairs; interfacial tension decreases linearly with temperature for all three UCST-type systems. The temperature coefficient [3.0×10^{-2} to 4.4×10^{-2} dyn/(cm °C)] was found to decrease with increasing molecular weight, whereas it was higher for the polydisperse than for the monodisperse system, in agreement with earlier studies [21, 22].

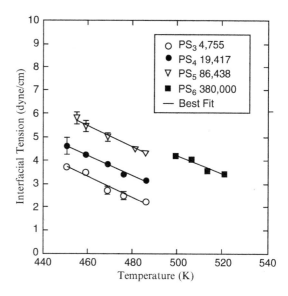

Fig. 7 Experimental interfacial tension as a function of temperature for PP/PS pairs. *Filled squares* PP/PS 380,000; *open inverse triangles* PP/PS 86,438; *filled circles* PP/PS 19,417; *open circles* PP/PS 4755 [21, 22]

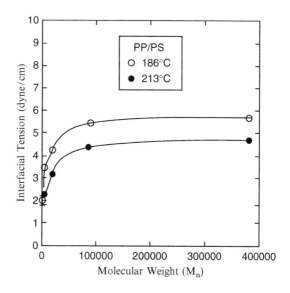

Fig. 8 Experimental interfacial tension for PP/PS pairs as a function of PS molecular weight. *Open circles* 186°C; *filled circles* 213°C [21, 22]

The effect of molecular weight polydispersity is shown in Fig. 10 for blends of one PE with two different series of polystyrenes with constant M_n (18,100 and 107,200) and different polydispersities. The interfacial tension decreased with increasing polydispersity in both cases, and the influence of polydispersity was higher for lower PS molecular weights. The decrease in interfacial tension could be

Fig. 9 Experimental interfacial tension as a function of temperature for PE/PS pairs. *Filled triangles* PE 82,300 ($M_w/M_n = 4.00$)/PS 200,600 ($M_w/M_n = 1.11$); *open diamonds* PE 82,300 ($M_w/M_n = 4.00$)/PS 18,100 ($M_w/M_n = 1.07$); *open triangles* PE 82,300 ($M_w/M_n = 4.00$)/PS 18,100 ($M_w/M_n = 2.68$) [23]

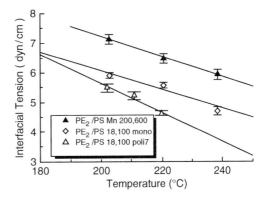

Fig. 10 Experimental interfacial tension as a function of molecular weight polydispersity, M_w/M_n. *Open triangles* PE 82,300 ($M_w/M_n = 4.00$)/PS 18,100; *open circles* PE 82,300 ($M_w/M_n = 4.00$)/PS 107,200 [23]

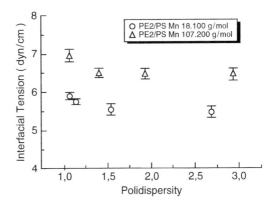

due to the migration of the short chains of the polydisperse systems to the interface (see Sect. 3.2.3). Thus, the short chains act similarly to a surfactant in that they lower the interfacial tension and broaden the thickness of the interface. Similar results have been shown by Nam and Jo [26] for PBD ($M_n = 4100$, $M_w/M_n = 1.4$) and PS (average $M_n \approx 5500$). Nam and Jo [26] also showed that the temperature coefficients increased linearly with increasing polydispersity in the range 1.1–1.5.

The interfacial tension between PE and PS increased with increasing PS molecular weight, whereas the influence of molecular weight decreased significantly when the PS molecular weight exceeded a certain value of the order of 45,000 [23]. The experimental data of interfacial tension as a function of molecular weight could be fitted to a type of power law if two molecular weight ranges were considered: one below and the other above this characteristic molecular weight. Moreover, the influence of PS molecular weight on the interfacial tension between PE and PS was shown to be smaller for lower molecular weights than for higher molecular weights of PE [23]. These are clearly shown in Fig. 11.

3.2 Theories of Polymer–Polymer Interfaces

It is not adequate to describe the junction between two homogeneous bulk phases as a simple two-dimensional plane without thickness. Because of the finite range of intermolecular forces, the interface can more properly be regarded as a region of finite thickness across which the density, the energy, or any other thermodynamic property changes gradually. Because this region has both area and thickness, it may be considered as an interphase that exists in either the solid or the liquid states. These interphases are usually referred to as two-dimensional phases, since the thickness parameter cannot be varied at will by the experimenter; indeed, it is controlled by the thermodynamics of the system [9].

Consider two homogeneous bulk phases, α and β, and separating them is an interfacial layer or interphase S (Fig. 12) [9, 183]. The boundary between the interphase and the bulk phase α is the plane AA', and that between the interphase and the bulk phase β is the plane BB'. The properties of the interphase are assumed to be uniform in any plane parallel to AA' or BB', but not in any other plane in the

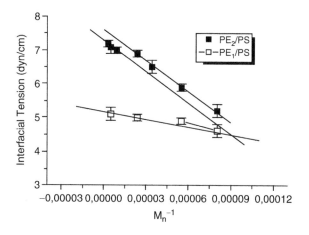

Fig. 11 Experimental interfacial tension as a function of temperature for PE/PS pairs. *Filled squares* PE 82,300 (M_w/M_n = 4.00)/PS (M_w/M_n = 1.03–1.12); *open squares* PE 3500 (M_w/M_n = 2.00)/PS (M_w/M_n = 1.03–1.12). The *lines* indicate the fits for the two different molecular weight regimes [23]

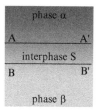

Fig. 12 Definition of an interphase [9, 183]

interphase. At and near the plane AA', the properties of S are identical with those of the bulk phase α. However, moving from AA' to BB' within the region S represents a gradual change in the properties of the interphase, from those corresponding to phase α to those corresponding to phase β.

In the bulk phases, the force across any unit area is equal in all directions, as is the hydrostatic pressure P. In the interphase, the force is not the same in all directions. However, if a plane of unit area is chosen parallel to AA' or BB', the force across the plane is the same for any position of the plane whether it lies in α, β, or S, because hydrostatic changes are assumed negligible. In contrast, the force balance for planes that cross the interphase, i.e., perpendicular to AA', is altered by the inclusion of an additional term due to the interfacial tension, γ. This force is associated with the anisotropic nature of intermolecular forces that result from the concentration gradient within the interphase.

The influence of the interfacial tension term on the thermodynamics can be illustrated by considering the work, W, performed on the interphase when additional interphase is formed. If the interphase volume increases by dV^S, i.e., a thickness increase of dx and an area increase of dA^S, the force balance leads to:

$$W = PA^S\,dx - (Px - \gamma)\,dA^S \tag{10a}$$

or:

$$W = -P\,dV^S + \gamma\,dA^S \tag{10b}$$

This last expression is the analogous work term for an interphase, which corresponds to the three-dimensional $-PdV$ term for a bulk phase. Incorporation of this term into the first and second laws of thermodynamics for multiconstituent open systems results in:

$$dU = T\,dS - P\,dV^S + \gamma\,dA^S + \sum_i \mu_i\,dn_i \tag{11}$$

where T is the thermodynamic temperature, S is the entropy, U is the internal energy, and μ_i and n_i are the chemical potential and number of moles of type i. Integration of the above equation, at constant intensive variables, produces the corresponding Euler relationship:

$$\gamma = \left(U + PV^S - TS - \sum_i \mu_i\,dn_i\right)\Big/A^S \tag{12}$$

Therefore, γ is the excess free energy per unit area arising from the formation of the interphase; it is equal to the difference between the Gibbs free energy of the system with the interphase, $(U + PV^S - TS)$, and that of an identical system without the interphase, $\sum_i \mu_i \, d\, n_i$, divided by the interfacial area A^S. Substitution of (11) into the total derivative of (12) results in:

$$-d\gamma = \left(S\,dT - V^S\,dP + \sum_i n_i\,d\,\mu_i \right)/A^S \tag{13}$$

which is a modified Gibbs–Duhem equation for the interphase.

The quantity of the components adsorbed at the interphase is a significant parameter, whereas the relationship between the extent of adsorption and the interfacial tension is particularly of interest; this is studied in terms of the Gibbs adsorption isotherm. At constant temperature and pressure, the Gibbs–Duhem relationship for an interphase is:

$$-d\gamma = \sum_i n_i\,d\,\mu_i/A^S = \sum_i \Gamma_i\,d\,\mu_i \tag{14}$$

where $\Gamma_i = n_i/A^S$ is the quantity of the i-th constituent contained per unit area of the interphase. Equation (14) indicates that spatial partitioning of constituents occurs at an interface (i.e., one constituent adsorbs preferentially at the interface) and that the extent of this adsorption is a function of the interfacial tension. The definition of Γ_i, however, is not exact because it depends on the concentration gradients present within the interphase, and its magnitude depends on the choice of the dividing boundary, often referred to as the Gibbs dividing surface.

For a two component system, the Gibbs adsorption isotherm is written as:

$$-d\gamma = \Gamma_1\,d\,\mu_1 + \Gamma_2\,d\,\mu_2 \tag{15}$$

Although recognizing that the interfacial region is best considered as an interphase, the alternative mathematical model is to consider the interface as a plane of infinitesimal thickness situated between AA′ and BB′ of Fig. 12. This dividing surface can be considered to be positioned so as to give rise to a simplification of (15). Gibbs [183] defined the position of the dividing surface such that the surface excess of constituent 1 is zero, and hence:

$$-d\gamma = \Gamma_2'\,d\,\mu_2 \tag{16}$$

where Γ_2' is the surface excess of constituent 2 with the dividing surface so defined. The equation relates the reduction in interfacial tension directly to the enrichment of one component within the interphase.

Although the thermodynamic description of an interphase is an invaluable tool, it is rarely used. The traditional approach of Gibbs requires the use of a dividing surface to which interfacial properties are referenced. This method is burdened with notational and conceptual difficulties [184]. As alternative but equivalent method of treating interphase thermodynamics was developed by Cahn [185], which avoided

Interfacial Tension in Binary Polymer Blends and the Effects of Copolymers

the pitfalls of the traditional approach by eliminating the arbitrary selection of the dividing surface. The development was based upon writing the Gibbs-Duhem relationship in a manner that made it independent of the definition of the dividing surface. However, Cahn's approach has not received much attention, although Sanchez [184] suggested that it would be useful because of its conceptual simplicity.

3.2.1 Semiempirical Theories of Polymer Interfaces

A number of semiempirical treatments have appeared over the years to develop "theories" relating the interfacial tension between a pair of incompatible substances to the surface tensions of the pure components. The first attempt to present a theory for interfacial tension is attributed to Antonoff [186–188]. He proposed an empirical rule that states that the interfacial tension, γ, is equal to the difference between the pure component surface tensions, σ_1 and σ_2:

$$\gamma = \sigma_1 - \sigma_2 \tag{17}$$

when $\sigma_1 \geq \sigma_2$. This can be correct only when phase 2 spreads on phase 1, and phase 2 is a small-molecule liquid. This empirical relationship is not applicable to polymer systems [120].

It is more appropriate to write the interfacial tension as:

$$\gamma = \sigma_1 + \sigma_2 - W_a \tag{18}$$

W_a is the work of adhesion, which is equivalent to the Gibbs free energy decrease (per unit area) when an interface is formed from two pure component surfaces. The work of adhesion increases as the interfacial attraction increases, leading to a decrease in interfacial tension. It is apparent from (18) that, if the two components are identical, an expression can be obtained that relates the surface tension σ_i to the work of cohesion (W_{ci}) for component i:

$$W_{ci} = 2\sigma_i \tag{19}$$

The interfacial tensions can, then, be related to the pure-component surface tensions by expressing W_a in terms of the Good–Girifalco [180, 189–192] interaction parameter ϕ_{GG}:

$$\phi_{GG} = W_a / (W_{c_1} W_{c_2})^{1/2} \tag{20}$$

The resulting equation of Good–Girifalco is:

$$\gamma = \sigma_1 + \sigma_2 - 2\phi_{GG}(\sigma_1 \sigma_2)^{1/2} \tag{21}$$

The interaction parameter, ϕ_{GG}, can be given [189–192] in terms of the molecular constants of the individual phases, including polarizabilities, ionization potentials, dipole moments, and molar volumes. The utility of the approach is limited by the lack of information about those molecular parameters for most polymer systems. Another difficulty arises from the fact that a $\sim 10\%$ error in ϕ_{GG} will result in a $\sim 50\%$ error in calculating γ, because for polymers the surface tensions are very similar. Thus, the ϕ_{GG} values must be accurately known. Values of ϕ_{GG} between some polymer pairs have been calculated from the measured interfacial and surface tensions [193, 194], and are found to be in the range 0.8–1.0. Empirically, it has been shown [194] that:

$$\frac{\partial \phi_{GG}}{\partial T} = 0 \tag{22}$$

An alternative treatment [153, 195] is based upon (18), where the work of adhesion is calculated using the theory of fractional polarity. Intermolecular energies are assumed to consist of additive nonpolar (i.e., dispersive) and polar components. Thus, the work of adhesion and the pure-component surface tensions can be separated into their dispersive (superscript d) and the polar (superscript p) components, such that:

$$\sigma_i = \sigma_i^d + \sigma_i^p \tag{23}$$

and:

$$W_a = W_a^d + W_a^p \tag{24}$$

The various polar interactions (including dipole energy, induction energy, and hydrogen bonding) are combined into one polar term.

Relationships between (23) and (24) have been obtained for two limiting cases. For low energy surfaces, characteristic of most polymer systems, the harmonic-mean approximation is valid for both the dispersive and the polar terms. This, combined with (18) gives:

$$\gamma = \sigma_1 + \sigma_2 - \frac{4\sigma_1^d \sigma_2^d}{\sigma_1^d + \sigma_2^d} - \frac{4\sigma_1^p \sigma_2^p}{\sigma_1^p + \sigma_2^p} \tag{25}$$

which has been found to give good results for polymers. Equation (25) can be rewritten in terms of (21); the interaction parameter ϕ_{GG} is then given by [195]:

$$\phi_{GG} = \frac{2x_1^d x_2^d}{g_1 x_1^d + g_2 x_2^d} + \frac{2x_1^p x_2^p}{g_1 x_1^p + g_2 x_2^p} \tag{26}$$

The fractional polarity is defined as $x_i^P = \sigma_i^P/\sigma = 1 - x_i^d$ with $g_1 = (\sigma_1/\sigma_2)^{1/2} = 1/g_2$.

For interfaces between a low and a high energy material, the geometric-mean approximation was used to give:

$$\gamma = \sigma_1 + \sigma_2 - 2\sqrt{\sigma_1^d \sigma_2^d} - 2\sqrt{\sigma_1^P \sigma_2^P} \tag{27}$$

When polar contributions are neglected, (27) reduces to the Fowkes equation [196]. In terms of the Good–Girifalco equation (21), the interaction parameter is given by:

$$\phi_{GG} = \sqrt{x_1^d x_2^d} + 2\sqrt{x_1^P x_2^P} \tag{28}$$

The generalized Good–Girifalco equation provides a framework for calculating the temperature and molecular weight dependence of interfacial tension. Differentiation of (21) with respect to temperature, taking into account (22), results in [120]:

$$\frac{d\gamma}{dT} = \frac{d\sigma_1}{dT} + \frac{d\sigma_2}{dT} - \phi_{GG}\left[g_1\frac{d\sigma_1}{dT} + g_2\frac{d\sigma_2}{dT}\right] \tag{29}$$

Although good agreement has been found for most of the cases originally reported by Wu [193–195], (29) should only be used for guiding the plots of interfacial tension versus temperature [120]. The molecular weight dependence derives directly from the incorporation of the empirically found relationship for the pure-component surface tensions [197, 198], $\sigma = \sigma_\infty - k/M^z$ (where M is the number-average molecular weight) to the Good–Girifalco relationship. One then obtains:

$$\gamma = \gamma_\infty - \frac{k_1}{M_1^z} - \frac{k_2}{M_2^z} \tag{30}$$

where M_1, M_2 are the number-average molecular weights of the two polymers, and the term γ_∞ is given by:

$$\gamma_\infty = \sigma_1 + \sigma_2 - 2\phi_{GG}\left[\sigma_1 - \frac{k_1}{M_1^z}\right]^{1/2}\left[\sigma_2 - \frac{k_2}{M_2^z}\right]^{1/2} \tag{31}$$

and is practically independent of molecular weight [178].

Although these semiempirical treatments can be useful in predicting interfacial tensions, they are not successful from a fundamental standpoint and cannot be used to predict the interfacial composition profile. Furthermore, these theories neglect the entropy effects associated with the configurational constraints on polymer chains in the interfacial region. These effects are unique in polymers and arise because the typical thickness of the interfacial region between polymer phases is less than the unperturbed molecular coil dimensions of a high polymer. Major perturbations of the spatial arrangement of polymer molecules must, then, occur in order for the interfacial thickness to become less than the unperturbed chain dimensions. Chain perturbations will also occur at a polymer–air interface for the same reason, i.e., the thickness of a polymer surface region (the region between unperturbed bulk polymer molecules and air) will also typically be less than the chain dimensions of the polymer molecules. Such chain perturbations contribute to the excess energy of surfaces or interfaces, and are reflected in the values of surface and/or interfacial tension. Since there is no direct relationship between the chain perturbations that occur at the polymer–air surfaces of the two individual polymers and the perturbations that would occur at the interface in a demixed polymer blend, there can be no direct fundamental relationship between the properties of polymer surfaces (surface tension) and polymeric interfaces (interfacial tension). Therefore, "theories" that attempt to present relationships for polymeric systems must be looked upon only as empirical.

3.2.2 Microscopic Theories of Polymer Interfaces

A number of thermodynamic theories have appeared that take a more fundamental approach, and, specifically, address the question of interfacial structure and its relation to interfacial tension.

Helfand and Tagami [27] formulated a statistical mechanical theory of the interface between two immiscible polymers, A and B. The approach is based on a self-consistent field, which determines the configurational statistics of the polymer molecules in the interfacial region. At the interface, energetic forces (determined essentially by the polymer A/polymer B segmental interaction parameter, χ) tend to drive the A and B molecules apart. This separation, however, must be achieved in such a way as to prevent a gap from opening between the polymer phases. The energetic force on, say, an A molecule must be balanced by an entropic force describing the tendency of A molecules to penetrate into the B phase, because of the numerous configurations of the A molecule which do so.

The theory was originally developed for symmetric systems, i.e., for similar polymers A and B that possess identical degrees of polymerization (Z), effective lengths of the monomer units (b), monomer number densities (ρ_0), and isothermal compressibilties (κ). The authors recommended the use of the geometric mean when these properties are not actually the same.

In the Helfand–Tagami mean field formulation, the effective mean field $W_A(\mathbf{r})$ on a segment of polymer A, which is the reversible work of adding the segment at position $\mathbf{r}$, where the densities are $\rho_A(\mathbf{r})$ and $\rho_B(\mathbf{r})$, less the work of adding the segment to bulk A, is given by:

$$\frac{W_A(\mathbf{r})}{kT} = \chi \frac{\rho_B(\mathbf{r})}{\rho_0} + \zeta \left[\frac{\rho_A(\mathbf{r})}{\rho_0} + \frac{\rho_B(\mathbf{r})}{\rho_0} - 1 \right] \tag{32}$$

with $\zeta = (\kappa \rho_0 k_B T)^{-1} - Z^{-1}$, where k_B is the Boltzmann constant. The first term arises from the relatively unfavorable interaction of the A polymer segments with

Interfacial Tension in Binary Polymer Blends and the Effects of Copolymers 203

the B polymer segments it encounters. The second term comes from the tendency of the system to pull polymer into regions where the total density $(\rho_A + \rho_B)$ is less than ρ_0, and push it out of regions of density greater than ρ_0. The inverse compressibility has been proven to be the proper account for this tendency [27, 28].

To obtain the self-consistent configurations of this system, one should focus on the quantity $q_A(\mathbf{r},t)$, which is essentially the ratio of the density at $\mathbf{r}$ of the ends of polymer molecules of type A and length Ztb with $0 < t < 1$, to the end density in the bulk A. Since the segment at Zt is the origin of two independent random walks, one of length Ztb and one of length $Z(1 - t)b$, the relative density is $q_A(\mathbf{r}, 1 - t) q_A(\mathbf{r},t)$. By summing over all segments, or integrating over t, the overall segment density of A at $\mathbf{r}$ is:

$$\rho_A(\mathbf{r}) = \int_0^1 q_A(\mathbf{r}, 1 - t)q_A(\mathbf{r}, t)dt \tag{33}$$

The quantity $q_A(\mathbf{r},t)$ can also be regarded as the ratio of the partition function of a polymer molecule that starts at $\mathbf{r}$ and has $Z t$ steps in the effective mean field $W_A(\mathbf{r})$, to that of a polymer in a zero field region, i.e., in the bulk phase. This ratio satisfies a modified diffusion equation, which, for the dividing surface at $x = 0$ and the A-rich phase at $x > 0$, can be written as:

$$\frac{1}{Z}\frac{\partial q_A(\mathbf{r}, t)}{\partial t} = \frac{b^2}{6}\nabla^2 q_A(\mathbf{r}, t) - \left[\chi\frac{\rho_B(\mathbf{r})}{\rho_0} + \zeta\left(\frac{\rho_A(\mathbf{r})}{\rho_0} + \frac{\rho_B(\mathbf{r})}{\rho_0} - 1\right)\right]q_A(\mathbf{r}, t) \tag{34}$$

By symmetry, the equation for $q_B(\mathbf{r},t)$ is:

$$\frac{1}{Z}\frac{\partial q_B(\mathbf{r}, t)}{\partial t} = \frac{b^2}{6}\nabla^2 q_B(\mathbf{r}, t) - \left[\chi\frac{\rho_A(\mathbf{r})}{\rho_0} + \zeta\left(\frac{\rho_A(\mathbf{r})}{\rho_0} + \frac{\rho_B(\mathbf{r})}{\rho_0} - 1\right)\right]q_B(\mathbf{r}, t) \tag{35}$$

with initial conditions:

$$\begin{aligned} q_A(\mathbf{r}, 0) &= q_B(\mathbf{r}, 0) = 1 \\ q_A(\infty, t) &= q_B(\infty, t) = 1 \\ q_A(-\infty, t) &= q_B(-\infty, t) = 0 \end{aligned} \tag{36}$$

where it is assumed that the asymptotic regions are pure A or B. Thus, (33)–(36) are the self-consistent set of equations for the density profile calculation.

These equations have been solved asymptotically for effective infinite degree of polymerization of the chains, and low isothermal compressibility $(\chi/\zeta \to 0)$, to yield the density profiles:

$$\begin{aligned} \frac{\rho_A(x; \chi)}{\rho_0} &= \frac{1}{2}[1 + \tanh(2x/a_1)] \\ \frac{\rho_B(x; \chi)}{\rho_0} &= \frac{1}{2}[1 - \tanh(2x/a_1)] \end{aligned} \tag{37}$$

where a_I is a measure of the effective interfacial thickness and:

$$a_I = \frac{\rho_0}{d\rho_A/dx|_{x=0}} = \frac{2b}{(6\chi)^{1/2}} \tag{38}$$

The interfacial tension was calculated as:

$$\gamma = \frac{k_B T}{\rho_0} \int_0^1 d\lambda \int_{-\infty}^{+\infty} dx \chi \rho_A(x; \lambda\chi)\rho_B(x; \lambda\chi) \tag{39}$$

where $\rho_i(x;\lambda\chi)$ is determined from the self-consistent equations by replacing χ with $\lambda\chi$. Using (37), one gets:

$$\gamma = \left(\frac{\chi}{6}\right)^{1/2} \rho_0 b k_B T \tag{40}$$

The theory was originally compared to three polymer pairs, namely PS/PMMA; PMMA/poly(n-butyl methacrylate), PnBMA; and PnBMA/poly(vinyl acetate), PVA. The calculated interfacial tension agreed exactly with the experimental value for PnBMA/PVA; it compared well for PMMA/PnBMA and differed by 50% for PS/PMMA. Helfand and Tagami suggested that, if χ is too large, then the characteristic interfacial thickness is too small for the mean-field theory to be appropriate. The theory has been widely used to estimate the interfacial tension in many different polymer–polymer systems with acceptable success.

However, the theory cannot be used if the asymmetry between A and B is too severe. Helfand and Sapse [29] refined the theory of Helfand and Tagami so as to remove the restrictive approximation of property symmetry of the two polymers. For a Gaussian random walk in a mean field, they obtained:

$$\gamma = k_B T \alpha^{1/2} \left[\frac{\beta_A + \beta_B}{2} + \frac{1}{6} \frac{(\beta_A - \beta_B)^2}{\beta_A + \beta_B} \right] \tag{41}$$

α is the mixing parameter, $\alpha = \chi(\rho_{0A}\rho_{0B})^{1/2}$ and $\beta_i^2 = \rho_{0i} b_i^2/6$. It was assumed that there was no volume change upon mixing and that the isothermal compressibility was small and independent of composition. The theory makes reasonable predictions, which are slightly improved when nonlocal interactions are considered. Inclusion of these nonlocal interactions gave:

$$\gamma = k_B T \alpha^{1/2} \left[\left(\frac{\beta_A + \beta_B}{2} + \frac{1}{6} \frac{(\beta_A - \beta_B)^2}{\beta_A + \beta_B} \right) + \frac{1}{18} \psi^2 \alpha \left(\frac{2}{\beta_A + \beta_B} - \frac{2}{5} \frac{(\beta_A - \beta_B)^2}{(\beta_A + \beta_B)^3} \right) + \cdots \right] \tag{42}$$

where ψ is a measure of the range of nonlocality, with ψ^2 being the second moment of the direct correlation function [199]. The interfacial thickness is predicted to be:

$$a_1 = 2\left(\frac{\beta_A^2 + \beta_B^2}{2\alpha}\right)^{1/2} \tag{43}$$

Tagami [200, 201] extended the theories of Helfand and coworkers to the case of compressible nonsymmetric polymer mixtures. A slight decrease in the predicted interfacial tension was found, due to the presence of finite compressibility of the polymers. This tendency was particularly apparent in the case of nearly symmetric polymer pairs, when the intersegmental interactions are of nonlocal nature. The results reduce to the results of Helfand and Sapse in the appropriate limits. However, the resulting equations are much too complicated, although the results do not differ significantly from those predicted by (41).

The difficulty in applying the above-mentioned theories is the paucity of accurate data for the physical parameters required by the theories. In particular, data for χ or α are not generally available, and the Hildebrand regular solution theory expression:

$$\alpha = \frac{(\delta_1 - \delta_2)^2}{kT} \tag{44}$$

has frequently been used, where δ_i is the solubility parameter of the i-th constituent. The fact that solubility parameters are normally available at only one temperature necessitates the additional assumption that they are temperature independent. Use of this expression for α yields interfacial tensions of reasonable magnitude, but gives the wrong sign for the temperature coefficient. Indeed, substitution of (44) in (40) or (41) results in an effective $T^{1/2}$ dependence, whereas a linear decrease with temperature is experimentally observed. However, a proper temperature dependence can be obtained if a small entropic term is added to the expression for α [19]; an apparent interaction density parameter of the form $\alpha = \alpha_H/T + \alpha_S$ gives a good agreement between theory and experiment.

The Gaussian random coil model is appropriate when the scale of inhomogeneity (e.g., the interfacial thickness) is large compared with the length of a bond, b, and the range of interactions, ψ. To handle the case where this is not true, a lattice model has been proposed by Helfand [202–205], in the spirit of the Flory-Huggins approach [206]. For infinite molecular weights, he obtained:

$$\gamma = \frac{1}{2}\frac{k_B T}{a}(\chi m)^{1/2}\left[1 + (1 + \chi)\chi^{-1/2} \arctan\left(\chi^{1/2}\right)\right] \tag{45}$$

where a is the cross-sectional area of a lattice cell and m is a lattice constant, defined such that the number of nearest neighbors of a cell in the same layer parallel to the interface is $z(1 - 2m)$ and in each of the adjacent layers is zm, where z is the number of nearest neighbors (coordination number) of a cell. Neglecting the nonlocality of interactions, Helfand obtained:

$$\gamma = \frac{1}{2}\frac{k_B T}{a}(\chi m)^{1/2} \tag{46}$$

which is consistent with the self-consistent mean field (SCMF) theory in that they both predict $\gamma \propto \chi^{1/2}$.

Roe [207] used a quasi-crystalline lattice model to determine the properties of the interface between two coexisting liquid phases, where one or both of the components are of polymeric nature. For χ much larger than the critical value χ_c, the composition transition at the interface is expected to be sharp. In this limit, Roe predicted that:

$$\gamma = \frac{k_B T}{a}\left[\chi m - 2(1 - 1/r)\ln(l + m) - 2\phi_2^0\left(1 + \frac{(1 - 1/r)m^2}{l + m}\right)\right] \tag{47}$$

where $l = 1 - 2m$, r is the degree of polymerization, and ϕ_2^0 is given by:

$$- \ln \phi_2^0 = l\chi + (1 - 1/r)\left[\frac{1}{l + m} - \ln\{m/(l + m)\}\right] \tag{48}$$

When $\chi - \chi_c \ll 1$, the interface is diffuse and the composition varies smoothly across the interface. In the limit, $\chi - \chi_c \ll \chi_c^2$, Roe found:

$$\gamma = \frac{k_B T}{a}\left(\frac{m}{2}\right)(\chi - \chi_c)^{3/2}\chi^{1/2}r \tag{49}$$

For an interface between polymers, and assuming infinite molecular weight, Roe obtained:

$$\gamma = \frac{4}{3}2^{-1/4}\frac{k_B T}{a}m^{1/2}\chi^{3/4} \tag{50}$$

that predicts $\gamma \propto \chi^{3/4}$, which is different from the results of Helfand and coworkers. For the thickness of the interfacial thickness, Roe predicted that for infinite molecular weight:

$$a_1 = 4 \times 2^{-1/4}dm^{1/2}\chi^{-1/4} \tag{51}$$

where d is the separation between adjacent lattice layers.

Helfand [202, 208] suggested that Roe's work contained a number of assumptions, which made it difficult to appraise the applicability of the theory. Helfand suggested that Roe's lattice theory did not treat the conformational entropy properly by assuming that the chances of going from a cell site to any empty neighboring cell

Interfacial Tension in Binary Polymer Blends and the Effects of Copolymers

were equally likely, thus neglecting the fact that bond orientations are inherently anisotropic. As a measure of the effect of Roe's assumptions on the qualitative nature of his results, Helfand pointed out that, when a gradient expansion of Roe's equation was made, the Gaussian random-walk equations [27, 28] were not recovered. Experimental verification of the lattice theories, however, has not been possible, because the lattice parameters a, m, and d are unknown a priori.

Kammer [209] examined the interfacial phenomena of polymer melts from a thermodynamic point of view. A system of thermodynamic equations has been derived to describe the temperature, pressure, and composition dependence of interfacial structure. Starting from the fundamental equations of Guggenheim [210], Kammer employed the Gibbs–Duhem equation of intensive parameters (13) to find that the interfacial composition is given by:

$$x_2^S = \frac{(d\gamma/dT)_\mathrm{P}+(d\sigma_1/dT)_\mathrm{P}}{(d\sigma_1/dT)_\mathrm{P}+0.5(d\sigma_2/dT)_\mathrm{P}} \tag{52}$$

where x_2^S is the molar fraction of component 2 at the interfacial region, and σ_1 and σ_2 are the surface tensions of the two components against air. Assuming that the interfacial layer is predominantly occupied by component 2 (i.e., $x_1^S \rightarrow 0$), he obtained:

$$\gamma = \frac{\mu_2^S}{A} \tag{53}$$

where μ_2^S is the chemical potential of component 2 and A is the molar area of the interface. Use of the Flory-Huggins formula of the chemical potential leads to:

$$\gamma = \gamma^0 + \frac{RT}{A}\left[\ln\phi_2^S + (1 - r_2/r_1)\phi_1^S + r_2\chi\left(\phi_1^S\right)^2\right] \tag{54}$$

where γ^0 is a constant, and r_i, χ, and ϕ_i^S are the degrees of polymerization, the Flory-Huggins interaction parameter, and the volume fraction of component i at the interphase. The interfacial thickness was shown to be:

$$a_1 = \frac{\upsilon}{RT} \frac{\gamma - \sigma_2}{r\chi\left(\phi_1^S\right)^2 + \ln\phi_2^S} \tag{55}$$

with υ the mean molar volume of the polymers.

Hong and Noolandi [211] have developed a theory for an inhomogeneous system, starting from the functional integral representation of the partition function as developed by Edwards [212], Freed [213], and Helfand [199]. The theory has been used to determine the interfacial properties and microdomain structures of a combination of homopolymers, block copolymers, monomers, and solvents. In that approach, the general free energy functional was optimized by the saddle-function method, subject to constraints of no volume change upon mixing and constant number of the individual particles, to obtain equations for the mean field acting on the polymers. They used this general theory to calculate the interfacial tension for polymer/polymer/solvent systems [214], where good agreement was obtained with experiment [215] for the PS/PBD/styrene ternary system. Their final results, however, are in integral form, which requires numerical integration. The application of the theory to the ternary systems polymer A/polymer B/diblock copolymer AB will be presented in Sect. 4.3.1.

Helfand and coworkers [30] responded to the experimental interest on the molecular weight effects on interfacial tension (see Sect. 3.1) by solving the equations they had derived earlier [27, 28, 199, 208] for the case of finite molecular weights; these equations were solved only in the infinite molecular weight limit earlier [27, 28]. The leading correction to the interfacial tension, which is of order r^{-1} (where r is the degree of polymerization of the two polymers in a symmetric system), is solely due to the placement entropy, i.e., it originates from the gain in translational entropy for finite chains, which can penetrate slightly more into the other phase. The interfacial tension for a symmetric system (polymers A and B with the same properties when pure) of large but finite molecular weights is, thus, calculated as:

$$\gamma = \gamma_\infty \left[1 - \ln 2 \frac{2}{\chi r} \right] \tag{56}$$

The leading correction to the concentration profile is also of the order of r^{-1} and is due to the entropic attraction of the chain ends to the interfacial region and the necessary readjustment of the remainder of the molecule. The authors gave a nonanalytic expression for the interfacial width. The concentration correction does not contribute to the interfacial tension at leading order because the free energy is calculated within a mean field approximation, where any change in the concentration can affect it in the second order, producing in this case a correction to the interfacial tension of the order of r^{-2}.

Tang and Freed [32] used density functional theory to investigate the effects of molecular weight on polymer–polymer interfacial tension. They considered possible reasons for the discrepancy between the theories available at that time and the experimental investigations on interfacial tension and concentration profiles across the interface. They postulated that certain approximations in the density functional previously used might be appropriate only in certain limited domains and, consequently, that higher order contributions to free energy functionals could contribute significantly to interfacial properties. Moreover, they considered the possible composition dependence of the Flory-Huggins interaction parameter. Tang and Freed calculated the interfacial tension for a symmetric blend for the entire two-phase region (from the weak to the strong segregation regime); it is given as:

$$\gamma = \gamma_\infty \left[1 - 0.90 \frac{2}{\chi r} - 0.10 \left(\frac{2}{\chi r} \right)^2 \right]^{3/2} \tag{57a}$$

Interfacial Tension in Binary Polymer Blends and the Effects of Copolymers

whereas the interfacial width is given by:

$$a_1 = a_{1\infty}\left[1 - \frac{5}{4}\frac{2}{\chi r} + \frac{1}{4}\left(\frac{2}{\chi r}\right)^2\right]^{-1/2} \tag{58a}$$

It is interesting that (57) shows an almost linear dependence on $(\chi r)^{-2/3}$ over a wide range of $2.5 < \chi r < 20$, consistent with experimental observations for intermediate molecular weights.

The asymptotic expressions for very high molecular weights are:

$$\gamma \cong \gamma_\infty\left[1 - 1.35\frac{2}{\chi r}\right] \tag{57b}$$

$$a_I \cong a_{I\infty}\left[1 + \frac{5}{8}\frac{2}{\chi r}\right] \tag{58b}$$

The coefficient $5/8 \approx 0.625$ in (58b) is very close to that ($\ln 2 \approx 0.693$) of Broseta in (87) (to be discussed later), whereas the coefficient of 1.35 in (57b) is about 50% larger than that ($\pi^2/12 \approx 0.82$) in (86) and about twice as large as the value of $\ln 2 \approx 0.693$ in (56).

The respective equations in the weak segregation limit (WSL) are:

$$\gamma \cong \gamma_\infty\frac{2}{\sqrt{3}}\left[\frac{\chi r}{2} - 1\right]^{3/2}; (\text{WSL}) \tag{57c}$$

$$a_I \cong a_{I\infty}\frac{2}{\sqrt{3}}\left[1 - \frac{2}{\chi r}\right]^{-1/2}; (\text{WSL}) \tag{58c}$$

3.2.3 Square-Gradient Approach

A conceptually different approach to the calculation of interfacial tensions is the use of the generalized square-gradient approach as embodied in the work of Cahn and Hilliard [216]. The Cahn–Hilliard theory provides a means for relating a particular equation of state, based on a specific statistical mechanical model, to surface and interfacial properties. The local free energy, g, in a region of nonuniform composition will depend on the local composition as well as the composition of the immediate environment. Thus, g can be expressed in terms of an expansion in the local composition and the local composition derivatives. Use of an appropriate free energy expression derived from statistical mechanics permits calculation of the surface or interfacial tension.

Generalized Gradient Theory of Fluids

Ideas that go back to van der Waals [217, 218] and Lord Rayleigh [219] on inhomogeneous systems were applied by Cahn and Hilliard [216] to the interface problem. In inhomogeneous fluids, the Helmholtz free energy is a functional of the component density distributions. Although exact formal expressions for this functional have been derived [220, 221] from statistical mechanics, they are impractical without approximation [222]. In the gradient approximation, this functional has been expressed as the sum of two contributions: one is a function of the local composition and the other is a function of the local composition derivatives [216, 223, 224]. The free energy for a binary system is postulated to have the form:

$$G = \int g(\phi, \nabla\phi, \nabla^2\phi, \ldots) \, dV \tag{59}$$

where the free energy density, g, is assumed to be a function of the local composition, ϕ, and all its derivatives, $\nabla\phi$, $\nabla^2\phi$, etc. Assuming that the composition gradient is small compared to the reciprocal of the intermolecular distance, g can be expanded in a Taylor series about $\nabla^k\phi = 0$, $k = 1, 2,\ldots$ and, truncating the expansion after terms of order $\nabla^3\phi$, one obtains for a fluid:

$$
\begin{aligned}
g &= g(\phi, \nabla\phi, \nabla^2\phi, \ldots) \\
&= g_0 + \sum_i L_i \frac{\partial\phi}{\partial x_i} + \sum_{i,j} \kappa_{ij}^{(1)} \frac{\partial^2\phi}{\partial x_i \partial x_j} + \frac{1}{2} \sum_{i,j} \kappa_{ij}^{(2)} \frac{\partial\phi}{\partial x_i} \frac{\partial\phi}{\partial x_j}
\end{aligned} \tag{60}
$$

where g_0 is the free energy density of a uniform system of composition ϕ, and:

$$
\begin{aligned}
L_i &= \left. \frac{\partial g}{\partial(\partial\phi/\partial x_i)} \right|_0 \\
\kappa_{ij}^{(1)} &= \left. \frac{\partial g}{\partial(\partial^2\phi/\partial x_i \partial x_j)} \right|_0 \\
\kappa_{ij}^{(2)} &= \left. \frac{\partial^2 g}{\partial(\partial\phi/\partial x_i)\,(\partial\phi/\partial x_j)} \right|_0
\end{aligned} \tag{61}
$$

For an isotropic medium, g is invariant to the symmetry operations of reflections $(x_i \rightarrow -x_i)$ and of rotation about a fourfold axis $(x_i \rightarrow -x_j)$. Therefore:

$$L_i = 0$$

$$
\kappa_{ij}^{(1)} = \begin{cases} \kappa_1 = \left. \partial g/\partial\nabla^2\phi \right|_0 & \text{for } i = j \\ 0 & \text{for } i \neq j \end{cases}
$$

$$
\kappa_{ij}^{(2)} = \begin{cases} \kappa_2 = \left. \partial^2 g/\partial|\nabla\phi|^2 \right|_0 & \text{for } i = j \\ 0 & \text{for } i \neq j \end{cases} \tag{62}
$$

Hence, (60) reduces to:

$$g = g_0(\phi) + \kappa_1 \nabla^2 \phi + \kappa_2 (\nabla \phi)^2 + \ldots \tag{63}$$

Integrating over a volume V of the system, the total free energy of this volume is:

$$G = \int dV \left(g_0(\phi) + \kappa_1 \nabla^2 \phi + \kappa_2 (\nabla \phi)^2 + \ldots \right) \tag{64}$$

Application of the divergence theorem, results in:

$$\int dV \left(\kappa_1 \nabla^2 \phi \right) = - \int (d\kappa_1 / d\phi)(\nabla \phi)^2 \, dV + \int_S (\kappa_1 \nabla \phi \cdot \mathbf{n}) \, dS \tag{65}$$

where S is an external surface with a normal vector n. Since one is not concerned with effects at the external surface, by choosing a boundary of integration in (65) such that $\nabla \phi \cdot \mathbf{n} = 0$ at the boundary, the surface integral vanishes. Using (65) to eliminate the term $\nabla^2 \phi$ from (64) one obtains:

$$G = \int dV \left(g_0(\phi) + \kappa (\nabla \phi)^2 + \ldots \right) \tag{66}$$

where:

$$\kappa = -d\kappa_1 / d\phi + \kappa_2 = -\frac{\partial^2 g}{\partial \phi \partial \nabla^2 \phi}\bigg|_0 + \frac{\partial^2 g}{\partial |\nabla \phi|^2}\bigg|_0 \tag{67}$$

$\kappa (\nabla \phi)^2$ is the additional positive contribution to the free energy, which arises from the local composition gradient. The coefficient of the square gradient term is related to the inhomogeneous fluid structure [220, 221]. It is essentially the second moment of the Ornstein–Zernike direct correlation function, $C(s,\phi)$, of a uniform fluid of composition ϕ. The relationship is:

$$\kappa(\phi) = \frac{4\pi kT}{6} \int s^4 C(s, \phi) \, ds \tag{68}$$

$C(s,\phi)$ depends on the range of correlation and is a function of the composition ϕ of the system.

Following the derivation of Cahn–Hilliard, the total free energy for the case of a one-dimensional composition gradient and a flat interface of area A becomes:

$$G = A \int_{-\infty}^{+\infty} \left[g_0(\phi) + \kappa \left(\frac{d\phi}{dx} \right)^2 \right] dx \tag{69}$$

Interfacial tension is, by definition, the difference per unit interfacial area between the actual free energy of the system and that which it would have if the properties of the phases were homogeneous throughout. Thus, the interfacial tension is given by:

$$\gamma = \int_{-\infty}^{+\infty} \left[\Delta g(\phi) + \kappa \left(\frac{d\phi}{dx} \right)^2 \right] dx \tag{70}$$

where $\Delta g(\phi)$ is the free energy density of the uniform system of composition ϕ with respect to a standard state of an equilibrium mixture of the two phases, α and β, without the interface, and is given by:

$$\Delta g(\phi) = \Delta g_0(\phi) - [n_A \Delta \mu_A(\phi_e) + n_B \Delta \mu_B(\phi_e)] \tag{71}$$

where n_A and n_B are the number densities of molecules of type A and B, $\Delta \mu_A$ and $\Delta \mu_B$ are the changes in the chemical potentials of A and B, and ϕ_e is equal to either of the compositions ϕ_α and ϕ_β of the two phases α and β at equilibrium.

According to (70), the more diffuse the interface is, the smaller will be the contribution of the gradient energy term to γ. But this decrease in energy can only be achieved by introducing more material at the interface of nonuniform composition and, thus, at the expense of increasing the integrated value of $\Delta g(\phi)$. At equilibrium, the composition variation will be such that the integral in (70) is a minimum. Substitution of the integrand of (70) into the Euler equation will produce the differential equation whose solution is the composition profile corresponding to the stationary values (i.e., minima, maxima, or saddle points) of the integral. Since the integrand does not explicitly depend on x, the appropriate form [225] of the Euler equation is:

$$I - \frac{d\phi}{dx} \left(\frac{\partial I}{\partial(d\phi/dx)} \right) = 0 \tag{72}$$

where I represents the integrand. Thus:

$$\Delta g(\phi) - \kappa \left(\frac{d\phi}{dx} \right)^2 = \text{const.} \tag{73}$$

The constant in this equation must be zero, since both $\Delta g(\phi)$ and $d\phi/dx$ tend to zero as $x \to \pm\infty$. Hence:

$$\Delta g(\phi) = \kappa \left(\frac{d\phi}{dx} \right)^2 \tag{74}$$

Interfacial Tension in Binary Polymer Blends and the Effects of Copolymers 213

Equation (74) can be used to calculate the composition profile across the interface. Using (74) with (70), and changing the integration variable from x to ϕ, interfacial tension is given as:

$$\gamma = 2 \int_{\phi_\alpha}^{\phi_\beta} [\kappa \Delta g(\phi)] \, d\phi \tag{75}$$

where ϕ_α and ϕ_β are the compositions of the two coexisting phases at equilibrium. The square-gradient approach has been widely used to model the surface tension of liquids [220, 223, 224] and polymer melts [226, 227], diffusion at interfaces and thin films [222], polymers at the liquid–liquid interface of binary regular solutions [228], interfacial tensions between low and high molecular weight liquid mixtures [229] and demixed polymer solutions [230], and spinodal decomposition in polymer blends [231–235]. Sanchez [184] has shown that the gradient theory is "in harmony with the microscopic theory of Helfand and coworkers [27–29, 200, 201] although the latter treats polymer interfaces from a completely different point of view."

The Square-Gradient Theory Applied to Polymer Interfaces

The gradient approach was first applied to calculate the interfacial tension between demixed polymer solutions by Vrij [230]. The polymer solution model, used by Debye [236] in his calculation of the light scattering from a polymer solution near the critical point, was used with the assumption of an interfacial thickness of the order of a polymer coil, thus misrepresenting the change in configurational entropy for the chains in the interface. Assuming that the Gaussian statistics is not distorted by the overlapping of the different polymer coils, even in the interfacial region, and for $T \ll T_c$ (where T_c is the critical temperature of demixing), he also predicted the interfacial tension between two homopolymers to be given by:

$$\gamma = r^{1/2} \frac{\Omega b}{6} \left(\frac{\pi}{32^{1/2}} - 0.426 \frac{T}{T_c} \right) \tag{76}$$

where Ω is a form of the interaction parameter, which is related to the Flory-Huggins interaction parameter, χ, by $\Omega = 2\chi k_B T / v$, with v being the segment volume. Because of the many inappropriate assumptions, the theory has not been utilized to predict polymer–polymer interfacial tensions. The theory predicts that $\gamma \propto r^{1/2}$, which is not followed by the experimental data.

Kammer [209] used the Cahn–Hilliard approach with the Flory-Huggins free energy of mixing and the assumption of a symmetric system to obtain:

$$\gamma = 2 \frac{RT}{v} (\kappa \chi) \left(\frac{\pi}{8} - \frac{0.602}{\chi r} - \frac{0.459}{(\chi r)^2} \right) \tag{77}$$

where v is the mean molar volume, r is the degree of polymerization, and κ is the square gradient coefficient, which is considered a constant given by $\kappa = \chi r b^2/6$. To estimate r in the interfacial region, he chose the expression $\chi r = 2.093 T_0/T$, with T_0 being an adjustable parameter. Thus, (74) leads to:

$$\gamma = 0.464 \frac{RT}{v} b\chi^{1/2} \left(\frac{T_0}{T}\right)^{1/2} \left[1 - 0.733\frac{T}{T_0} - 0.267\left(\frac{T}{T_0}\right)^2\right] \tag{78}$$

However, Kammer incorrectly assumed that the interfacial tension is the free energy of mixing per unit area, instead of the correct expression that defines interfacial tension as the *excess* free energy per unit interfacial area. Although interfacial tension is predicted to decrease with temperature, the results are not accurate fundamentally and the derivations should be recalculated.

Poser and Sanchez [229] used the generalized density gradient theory of interfaces [216] in conjunction with the compressible lattice fluid model of Sanchez and Lacombe [237–240] to approximate the interfacial tension and thickness between two immiscible high molecular weight polymer liquids. The theory is not expected to apply near the critical point, where the lattice fluid theory incorrectly describes the coexistence curve, or for highly polar polymers. Furthermore, the theory neglects intramolecular correlation effects present in long polymer chains, as well as changes in the configurational entropy at the interface. Due to the fact that the calculated phase diagrams, using the lattice fluid model, are extremely sensitive to the values of the two interaction parameters inherent in the model, and the assumption that the entropy in the interfacial region is independent of concentration gradients, Poser and Sanchez suggested that "*in its present form, the theory is being pushed to its limits when applied to a polymer–polymer interface.*"

The resultant equations yield predictions comparable to those of Helfand and Sapse [29]. Formally the two theories look quite similar. Conceptually, however, they are quite different. Gradient effects arise only from energetic considerations in the Poser–Sanchez theory, whereas they arise from the intrinsic connectivity of the polymer chain in the theory of Helfand–Sapse. In the simplest version of the Helfand–Sapse theory, compressibility effects are ignored whereas they play an important role in the Poser–Sanchez formulation. Poser and Sanchez suggested that a proper theory for polymeric interfaces should not ignore the compressible nature of polymer liquids (even though it is very small), nor can it ignore the intrinsic connectivity of a polymer chain.

Anastasiadis et al. have also developed a theory for polymer–polymer interfacial tension [20, 122], based upon the generalized square-gradient theory of Cahn and Hilliard [216] in conjunction with the Flory-Huggins theory of the free energy of mixing [206]. The free energy is calculated as:

$$\frac{\Delta g_0(\phi)}{k_B T} = \frac{\phi}{r_A v_A} \ln\phi + \frac{1 - \phi}{r_B v_B} \ln(1 - \phi) + \frac{\chi}{v_A}\phi(1 - \phi) \tag{79}$$

Interfacial Tension in Binary Polymer Blends and the Effects of Copolymers 215

where ϕ is the volume fraction of polymer A, χ is the Flory-Huggins interaction parameter, r_A and r_B are the degrees of polymerization of polymers A and B, respectively, υ_A and υ_B their specific monomeric volumes, T is the thermodynamic temperature, and k_B the Boltzmann constant. Moreover, the changes in chemical potentials, calculated from (79), are:

$$\frac{\Delta\mu_A(\phi)}{k_B T} = \ln\phi + (1-\phi)\left(1 - \frac{r_A \upsilon_A}{r_B \upsilon_B}\right) + \chi r_A (1-\phi)^2 \tag{80a}$$

$$\frac{\Delta\mu_B(\phi)}{k_B T} = \ln(1-\phi) + \phi\left(1 - \frac{r_B \upsilon_B}{r_A \upsilon_A}\right) + \chi r_B \phi^2 \frac{\upsilon_B}{\upsilon_A} \tag{80b}$$

The compositions ϕ^α and ϕ^β of the coexisting phases α and β at equilibrium were calculated by equating the chemical potentials, such that:

$$\begin{aligned} \Delta\mu_A(\phi^\alpha; \chi) &= \Delta\mu_A(\phi^\beta; \chi) \\ \Delta\mu_B(\phi^\alpha; \chi) &= \Delta\mu_B(\phi^\beta; \chi) \end{aligned} \tag{81}$$

The coefficients of the square gradient terms were derived using linear response theory within the framework of the random phase approximation [231, 241, 242]. de Gennes [242] suggested that the coils remain nearly ideal on the scale of one coil, even in the case of a dense mixture of interacting chains. Therefore, ideal single chain approximations can be employed to the calculation of the scattering function, $S(q)$, where q is the scattering vector. The scattering function is related to the volume fractions and the chain lengths by [231, 242, 243]:

$$\frac{1}{S(q)} = \frac{1}{\phi\upsilon_A r_A f_D(r_A, q)} + \frac{1}{(1-\phi)\upsilon_B r_B f_D(r_B, q)} - \frac{2\chi}{\upsilon_A} \tag{82}$$

where $f_D(r,q)$ is the Debye function [244], $f_D(r,q) = 2u^{-2}[u + \exp(-u) - 1]$, with $u = q^2 r b^2/6 = q^2\langle r_0^2\rangle/6 = q^2 R_G^2$; b is the Kuhn statistical segment length; $\langle r_0^2\rangle$ the mean-squared end-to-end distance; and R_G the radius of gyration of the coil.

Two limiting expressions for $S(q)$ can be calculated for $q R_G \gg 1$, and for $q R_G \ll 1$. The first corresponds to a sharp interface, and the second to a relatively diffuse interface. For $q R_G \gg 1, f_D(r,q) \approx 2/u^2 = 12/(q^2\langle r_0^2\rangle)$ and:

$$\frac{1}{S(q)} = \frac{q^2}{12}\left(\frac{\langle r_0^2\rangle_A}{\phi\upsilon_A r_A} + \frac{\langle r_0^2\rangle_B}{(1-\phi)\upsilon_B r_B}\right) - \frac{2\chi}{\upsilon_A} \tag{83a}$$

whereas for $q R_G \ll 1, f_D(r,q) \approx r(1 - q^2\langle r_0^2\rangle/18)$, and using the equation for the spinodal curve $2\chi_s(\phi)/\upsilon_A = 1/[\phi r_A \upsilon_A] + 1/[(1-\phi)r_B \upsilon_B]$, the scattering function is given by:

$$\frac{1}{S(q)} = \frac{2(\chi_s(\phi) - \chi)}{\upsilon_A} + \frac{q^2}{18}\left(\frac{\langle r_0^2\rangle_A}{\phi\upsilon_A r_A} + \frac{\langle r_0^2\rangle_B}{(1-\phi)\upsilon_B r_B}\right) \tag{83b}$$

Using the Ornstein–Zernike relationship between the direct correlation function and the static structure factor, the gradient terms for the two different limits are given by:

$$\frac{\kappa^{(i)}}{kT} = \frac{\langle r_0^2 \rangle_A}{24\phi v_A r_A} + \frac{\langle r_0^2 \rangle_B}{24(1-\phi)v_B r_B}; \quad qR_G \gg 1 \tag{84a}$$

$$\frac{\kappa^{(ii)}}{kT} = \frac{\langle r_0^2 \rangle_A}{36\phi v_A r_A} + \frac{\langle r_0^2 \rangle_B}{36(1-\phi)v_B r_B}; \quad qR_G \ll 1 \tag{84b}$$

for sharp and broad interfaces, respectively. Equation (84b) and its equivalent for a symmetric system were widely used by de Gennes [231], Ronca and Russell [232], Pincus [233], and Binder and coworkers [234] to model the dynamics of concentration fluctuations near the critical point, whereas (84a) had been used by Roe [243] to study the micelle formation in homopolymer/copolymer mixtures.

There are different objections to the application of the square-gradient approach that arise from the assumptions inherent to the theory. Halperin and Pincus [228] pointed out that, because the Cahn–Hilliard theory is a mean field theory, its validity near the critical point can only be qualitative. On the other hand, the theory assumes weak composition gradients that may be realized only close to the critical region. Binder [234, 235] suggested that, for $qR_G \gg 1$, an additional correction term should be included in the gradient terms, which arises from the finite range of interactions and is proportional to $\chi\psi^2(\nabla\phi)^2$, where ψ is the range of interactions. For $\chi < 1$, however, this correction is negligible, as suggested by de Gennes. [231]. Moreover, de Gennes argued [231] that (84b) describes well the additional positive contribution to the free energy from the local concentration gradients, even in the case $\chi \gg \chi_c$ (χ_c is the value of the interaction parameter at the critical point), i.e., when the attention is focused on the strong segregation regime.

The expressions for interfacial tension thus obtained were, in principle, similar to those of Helfand and Sapse [29]; however, the correct temperature coefficient was obtained, and the molecular weight effects were included via the use of the Flory-Huggins expression for the free energy [206] and the random phase approximation [231, 241, 242] for the gradient terms.

Numerical evaluation of the theoretical expression for the interfacial tension allowed the comparison of the theory to the experimental data of Anastasiadis et al. [20]. In general, a good agreement was obtained between theory and experiment for the interfacial tension and its temperature dependence, especially for higher molecular weights. Figure 13 shows the comparison for a blend of a polystyrene with $M_n = 10,200$ (PS 10200; $M_w/M_n = 1.07$) and a poly(ethyl ethylene) with $M_n = 4080$ (PBDH 4080; $M_w/M_n = 1.04$), which was prepared by hydrogenation of poly(vinyl ethylene), PVE. The interaction parameter values used, $\chi = 0.0057 + 21/T$, were evaluated by analyzing small-angle X-ray scattering data from homogeneous PS-b-poly(ethyl ethylene) diblock copolymers [245]; the blend exhibits a UCST behavior.

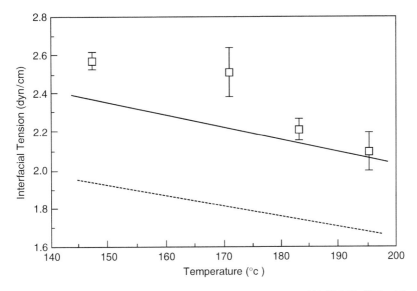

Fig. 13 Comparison of experimental interfacial tension for PS 10,200/PBDH 4080 with the square-gradient theory where the square-gradient coefficient is given by (84a) (*solid line*) and (84b) (*dotted line*). Adjustable parameters were not allowed in this comparison [20]

The predictions of the theory with respect to the molecular weight dependence of interfacial tension are compared to the experimental data for PS/PBDH 4080 data in Fig. 14. The representation in terms of the $M_n^{2/3}$ dependence was adopted [20] because it conformed closely to the result from nonlinear regression for this particular range of molecular weights. At high molecular weights, the theoretical curve corresponds well with the extrapolated empirical relationship for the experimental data when (84b) is used for the square-gradient coefficient, while use of (84a) leads to an overestimation of interfacial tension by ca. 20%. The theory does predict an apparent dependence of interfacial tension on $M_n^{2/3}$; however, it deviates considerably from the experimental data for low molecular weights. The theory erroneously indicates complete miscibility (i.e., $\gamma = 0$) for a PS molecular weight of ca. 2400, whereas two phases were always present under these conditions and appreciable mixing was not observed. It was discussed that the discrepancy was probably due to the inappropriate use of the interaction parameter determined from diblock copolymers to describe the interactions in polymer blends within the framework of Flory-Huggins theory.

Broseta et al. [31] extended the work of Anastasiadis et al. [20] and provided analytical expressions for the finite molecular weight corrections to the interfacial tension and interfacial thickness, and also studied the effects of polydispersity. Broseta first considered two strongly segregated monodisperse homopolymers A and B with comparable (high) incompatibility degrees $w_A = \chi r_A$ and $w_B = \chi r_B$, with each of the two phases at equilibrium being nearly pure in one of the two

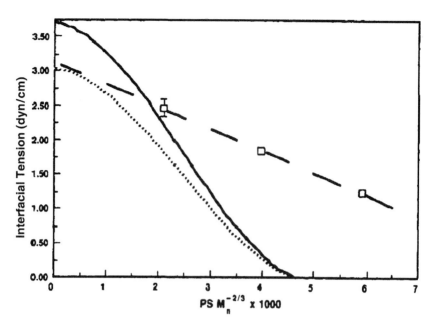

Fig. 14 Comparison of experimental interfacial tension for PS/PBDH 4080 at 171°C with the square-gradient theory where the square-gradient coefficient is given by (84a) (*solid line*) and (84b) (*dotted line*). The *dashed line* is a linear fit to the data. Adjustable parameters were not allowed in this comparison [20]

polymer species. Actually, Broseta estimated that the compositions of the two coexisting phases are given by:

$$\phi_A^\alpha = 1 - \phi_B^\alpha \approx 1 - \exp(-\chi r_B) \tag{85a}$$

$$\phi_A^\beta = 1 - \phi_B^\beta \approx \exp(-\chi r_A) \tag{85b}$$

The interfacial tension for high but finite molecular weights was, then, calculated to be given as:

$$\gamma = \gamma_\infty \left[1 - \frac{\pi^2}{12} \left(\frac{1}{\chi r_A} + \frac{1}{\chi r_B} \right) + \ldots \right] \tag{86}$$

where γ_∞ is the interfacial tension for infinite molecular weights as calculated by Helfand–Tagami, (40). As discussed by Broseta, this equation should be the asymptotic behavior of the theory of Anastasiadis and coworkers [20]; however, that regime was apparently not explored in the numerical calculations of Anastasiadis [20]. Moreover, the interfacial width was estimated as:

$$a_I = a_{I\infty} \left[1 + \ln 2 \left(\frac{1}{\chi r_A} + \frac{1}{\chi r_B} \right) + \ldots \right] \tag{87}$$

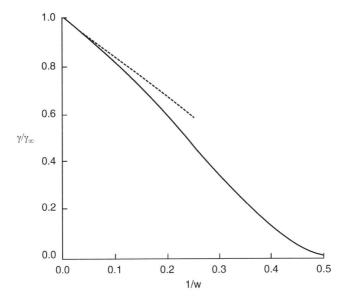

Fig. 15 Reduced interfacial tension γ/γ_∞ as a function of the inverse incompatibility $1/w$ (*solid line*). The *dashed line* is the asymptotic linear behavior (86) valid for large incompatibilities. Near the critical point ($1/w = 0.5$), the *dotted line* represents the more exact solution of Joanny and Leibler [246] (see Sect. 3.2.4)

where $a_{I\infty}$ is the interfacial thickness of Helfand–Tagami (38). The results are expected to be valid for strongly incompatible systems where the interface is smaller than the chain radii of gyration, whereas the analysis should not hold for weakly incompatible systems where the interface becomes of the order of R_G or larger.

Figure 15 shows the numerically calculated interfacial tension plotted as a function of the inverse incompatibility $1/w$, assumed to be the same for both polymers. The asymptotic behavior of (86) is a good approximation for a wide range of incompatibilities $w > 5$ (or $1/w < 0.2$). However, the increase in interfacial tension with molecular weight is predicted to be weaker for smaller molecular weights, in agreement with the experimental data of Anastasiadis [20].

Broseta also calculated the effect of molecular weight polydispersity on the interfacial tension [31]. He considered a specific case of polydispersity where the two polymer melts are binary mixtures with the same bimodal distribution of molecular weights, with r_1 being the length of the small chains, r_2 the length of the long chains ($r_1 < r_2$), and x_0 the volume fraction of monomers belonging to small chains. Broseta analyzed the strong segregation regime, i.e., large values of $w_i = \chi r_i$. The theory predicted a selective partitioning of the small chains to the polymer–polymer interface, which leads to a reduction of the interfacial tension. The enrichment of the small chains to the interface decreases when the chain length ratio w_2/w_1 decreases to 1 and when both chain lengths simultaneously increase.

When both chains become very long, the difference in chain lengths does not play any role. When the incompatibilities w_i are large but the difference Δw is small, interfacial tension was predicted to be given by:

$$\gamma = \gamma_\infty \left[1 - \frac{\pi^2}{6w_n} + \cdots \right] \tag{88}$$

where $w_n = [x_0/w_1 + (1 - x_0)/w_2]^{-1}$ is the number-averaged incompatibility degree. Since the number-average molecular weight is most heavily weighted by the smallest molecular weights, (88) shows that the interfacial tension is lowered by the presence of small chains, with the small chains in fact acting as surfactants.

Ermonskin and Semenov [33] utilized the square gradient approach in combination with the Flory-Huggins model for calculation of the structure of the interface between two immiscible polymers. They derived the conformational free energy including a correction of the order of $1/r$ to the dominant gradient term following the lines first proposed by Lifshitz [247]. The interfacial tension was obtained by minimization of the interfacial free energy. For strong segregation ($\chi r_i \gg 1$) and sharp interfaces, interfacial tension is given by:

$$\gamma = \gamma_\infty \left[1 - 2\ln 2 \left(\frac{1}{\chi r_A} + \frac{1}{\chi r_B} \right) \right] \tag{89}$$

where γ_∞ is the interfacial tension for infinite molecular weights as calculated by Helfand–Tagami, (40). Moreover, they derived an approximate analytical expression for the free energy of an inhomogeneous blend of two homopolymers valid for both high and moderate values of χr_i and they calculated numerically the dependence of interfacial tension on homopolymer molecular weight. Semenov pointed out that the prefactor $2\ln 2 \approx 1.39$ is very similar to the one predicted by Tang and Freed, $3[1 - (1/6)^{1/3}] \approx 1.35$ [32] in (57b), whereas it can be compared to the value of 0.82 of Broseta et al. [31] in (86) and $\ln 2 \approx 0.69$ of Helfand et al. [30] in (56), with the difference being due to the various approximations used.

Figures 16–18 show the comparison of the Semenov theory to the interfacial tension data of Anastasiadis et al. [20] for three different polymer systems, PDMS/ PBD 1000, PS/PBDH 3080 and PS/PMMA 10,000. The agreement is very good for the PDMS/PBD and PS/PBDH systems, whereas it is poor for the PS/PMMA blend. Actually, Semenov argued that the disagreement for PS/PMMA is far beyond possible errors due to approximations of the theory and that it might indicate that the model based on the Flory-Huggins interaction term may be inadequate for the PS/PMMA system, with higher order terms being important in the excess free energy of interaction.

Kamal et al. [22] compared the predictions of these thermodynamic theories to experimental data on the effect of temperature, molecular weight, and molecular weight polydispersity on the interfacial tension for polypropylene/polystyrene blends. Once more, the importance of an accurate estimation of the Flory-Huggins

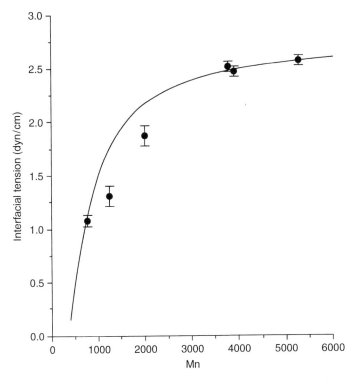

Fig. 16 Comparison of experimental interfacial tension for PDMS/PBD 1000 at 25°C [20] with the theory of Ermonskin–Semenov [33]. The interaction parameter $\alpha = \chi/v$ was adjusted to 3.35×10^{-3} mol/cm^3 (v is the effective monomer volume)

interaction parameter χ emerged. It was shown again that the relationship correlating χ to the Hildebrand solubility parameter (44) was not suitable for evaluating the theoretical predictions. The theoretical interfacial tensions of Broseta et al. [31] or Helfand et al. [30] were found to increase with increasing temperature, which is opposite to the behavior of the experimental interfacial tension data; this discrepancy was also observed earlier [19]. Alternatively, the interaction parameter was expressed as a sum of an enthalphic and an entropic contribution, $\chi = \chi_H/T + \chi_S$, as suggested earlier by Anastasiadis [19]. The two coefficients were evaluated by fitting the interfacial tension data at two different temperatures to the expression of Broseta (83); these coefficients were then used to predict the interfacial tension for other temperatures and different molecular weights with moderate success. Finally, the theoretical predictions on the effects of molecular weight polydispersity on interfacial tension [31] are in qualitative agreement with the data.

Lee and Jo [34] proposed a square-gradient theory combined with the Flory–Orwoll–Vrij equation of state theory [248]. The theory was used to calculate the interfacial tension between PS and PBD, and between PS and PMMA. For the PS/PBD system, they utilized an experimental cloud point curve to determine

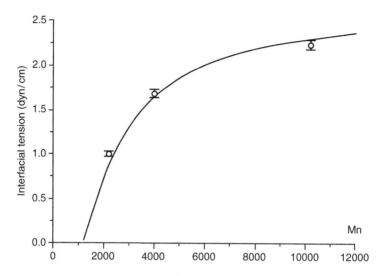

Fig. 17 Comparison of experimental interfacial tension for PS/PBDH 4080 at 184°C [20] with the theory of Semenov [33]. The interaction parameter $\alpha = \chi/\upsilon$ was adjusted to 0.93×10^{-3} mol/cm^3 (υ is the effective monomer volume)

the equation of state interaction parameter. The authors calculated the temperature and molecular weight dependence of interfacial tension for different molecular weights of PS (5000–30,000) and a fixed molecular weight of PBD (PBD 1000). The dependence of interfacial tension on temperature shows a linear decrease, except near the upper critical solution temperature. The interfacial tension increases with increasing PS molecular weight and approaches an asymptotic limit. The predicted interfacial tension follows a $M_n^{-2/3}$ dependence for moderate molecular weights, whereas it follows the M_n^{-1} dependence for high molecular weights. The theory was compared to the experimental data of Anastasiadis [20] for the PS/PMMA system: although the apparent trend with molecular weight is correctly predicted, the theory overestimates the values of interfacial tension when the interaction parameter was determined by fitting the equation of state theory for the binodal curve to the maximum temperature of an experimental cloud point curve.

3.2.4 Theories Near the Critical Point

The theories discussed up to now do not hold rigorously near the critical point of demixing, and an alternative approach is, thus, required. Nose [249] studied the interfacial behavior for both polymer mixtures and polymer solutions near the critical point. The theory was based on the Cahn–Hilliard theory [216] and takes into account the dimensions of the polymer coils at the interfacial region. For a

Interfacial Tension in Binary Polymer Blends and the Effects of Copolymers 223

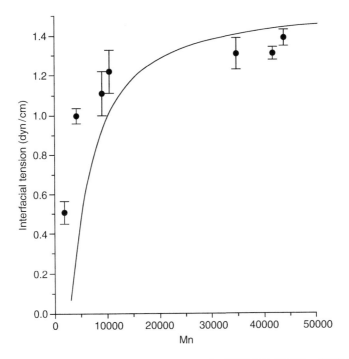

Fig. 18 Comparison of experimental interfacial tension for PS/PMMA 10,000 at 199°C [20] with the theory of Semenov [33]. The interaction parameter $\alpha = \chi/\upsilon$ was adjusted to 0.45×10^{-3} mol/cm^3 (υ is the effective monomer volume)

symmetric polymer/polymer system, as the temperature, T, approaches the critical temperature, T_c, the interfacial tension and interfacial width behave as:

$$\gamma \propto T_c r^{-1/2} \varepsilon^{3/2} \tag{90a}$$

$$a_I \propto r^{1/2} \varepsilon^{-1/2} \tag{91}$$

where r is the number of polymer segments and $\varepsilon = (T_c - T)/T_c$, with T_c the critical point of demixing. Because for a symmetric system, T_c varies with molecular weight as $T_c \propto r$, (90a) reduces to:

$$\gamma \propto r^{1/2} \varepsilon^{3/2} \tag{90b}$$

Thus, both interfacial tension γ and interfacial width a_I were predicted to vary with molecular weight to the 1/2 power, i.e., proportionally to the unperturbed dimension of the polymer coil. Furthermore, the theory predicts the classical mean field exponents of 3/2 and $-1/2$ for the dependencies of γ and a_I, respectively, on reduced temperature. Besides, Nose predicted a first order transition from a diffuse to a relatively sharp interface that results in a change in the slope of the γ versus

T curve. The transition temperature reduced by T_c increases with increasing molecular weight.

Joanny and Leibler [246] predicted the same critical exponents for the temperature dependence; however, they found that the interfacial tension decreases with increasing chain length r as $r^{-1/2}$, while the dependence of the interfacial width on chain length is the same as that of Nose [249]. For a symmetric system, their final expressions were:

$$\gamma = \frac{2}{3} k_B T b^{-2} r^{-1/2} \varepsilon^{3/2} \tag{92}$$

$$a_I = \frac{1}{3} b r^{1/2} \varepsilon^{-1/2} \tag{93}$$

where b is the Kuhn statistical segment length of the polymers.

Sanchez [181] used a Taylor expansion of the Flory-Huggins equation for the free energy density, and the Cahn–Hilliard theory with a constant coefficient for the gradient terms. He found the same classical mean field exponents for the temperature dependence of interfacial tension and thickness, but he predicted that, for the symmetric case, both the interfacial tension and the thickness are independent of chain length. Sanchez explained this result to be due to the fact that, in his approach, chain connectivity was only implicitly taken into consideration through the entropy of mixing. The theories of Nose [249] and Joanny and Leibler [246] take explicitly into account chain connectivity in various approximations.

Ronca and Russell [232] calculated the interfacial tension near the critical point. They used the Cahn–Hilliard expansion of the free energy with the Flory-Huggins approximation in modeling the spinodal decomposition in polymer mixtures. For a symmetric system, the interfacial tension was found to follow the classical dependence:

$$\gamma \propto T r^{-1/2} \varepsilon^{3/2} h(r) \tag{94}$$

where the function $h(r)$ depends on the chain length [232].

de Gennes [250] has argued that a polymer blend should behave nearly classically; thus, the predicted classical behavior of $\gamma \propto \varepsilon^{3/2}$ and $a_I \propto \varepsilon^{-1/2}$ may be very close to being correct. With respect to the molecular weight dependence, the situation is not clear. The results of Joanny and Leibler [246] and Ronca and Russell [232] would be similar to those of Nose if the temperature, T, appearing in (92) and (94), respectively, were equated to the critical temperature, T_c, as suggested by Sanchez [181]. Our opinion is that Sanchez's suggestion is correct. In that case, the theories would predict that, near the critical point, the interfacial tension increases with molecular weight to the 1/2 power, as:

$$\gamma \propto r^{1/2} \varepsilon^{3/2} \tag{95}$$

except for a correction introduced in the Ronca and Russell derivation [232].

4 Copolymers as Emulsifying Agents in Polymer Blends

4.1 Copolymer Localization at the Polymer Blend Interface

It is widely understood that the use of block or graft copolymers as emulsifying agents or compatibilizers in polymer blends is due to their affinity to selectively partition to the polymer–polymer interface. The segments of the compatibilizer can be chemically identical with those in the respective homopolymer phases [37, 38, 40, 45, 48, 54, 56] or can be miscible with or adhering to one of the homopolymer phases [251–254]. Figure 19 depicts ideal configurations of copolymer chains at the interface, with each block preferentially extending into the respective homopolymer phase [39, 70, 71, 73, 74]. Other conformational models are possible, such as segments adsorbed onto the surface of one polymer rather than penetrating it. Conformational restraints are important [255, 256], and, on this basis, a block copolymer is expected to be superior to a graft [257, 258]. A graft with one branch is shown in Fig. 19 for the case of graft copolymers; however, multiple branches restrict the opportunities of the backbone to penetrate its homopolymer phase. This, of course, would not preclude adhesion of the backbone to this phase. For the same reasons, diblock copolymers are more effective than triblocks [87]. The block or graft copolymer can localize itself at the blend interface only if it has the propensity to segregate into two phases. It is the repulsion of the unlike segments of the copolymer and the two homopolymers that leads to the localization of the copolymer at the interface. Therefore, the tendency in block and graft copolymers to migrate at the interface depends on the balance of the interaction parameters as well as on their molecular weights.

Fayt et al. [259, 260] used transmission electron microscopy (TEM) to study the localization of the copolymer at the polymer–polymer interface. Staining a short mid-block (isoprene) with OsO_4 permitted the direct observation of the location of the added PS-b-PI-b-PBDH triblock copolymer to the interface between PS and low density PE; TEM images showed the localization of the copolymer to the blend

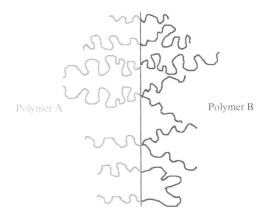

Fig. 19 Ideal location of block and graft copolymers at the interface between the homopolymer phases formed by the immiscible polymers A and B

interface. In contrast, when a PS-*b*-PMMA copolymer was added to the polyvinyl chloride (PVC)/PS blend (the PMMA sequences are miscible with PVC), dispersion of the copolymer within the PVC phase was observed instead of a preferential adsorption to the interface. It was pointed out that a properly tailored decrease in the interaction of the PMMA block with PVC (i.e., by controlling microstructure, molecular weight, and composition) would restore a more favorable situation but at the cost of a long optimization process. Thus, an important requirement for the copolymer is that it should not be miscible as a whole molecule within one of the homopolymer phases, because this would increase the amount of the copolymer required to reach interesting sets of properties.

Shull et al. [38] used forward recoil spectrometry to quantify the interfacial segregation of diblock copolymers consisting of deuterated polystyrene (dPS) and poly(2-vinylpyridine) (P2VP) at interfaces between PS and P2VP homopolymers. Figure 20 shows the equilibrium distribution of the diblock copolymer to the PS–P2VP interface after the appropriate annealing. The interfacial excess, estimated as the hatched area in Fig. 20, increased with increasing copolymer concentrations within the PS layer and was compared to mean-field theory predictions, which were quantitatively accurate for copolymer concentrations below the limiting value associated with the formation of block copolymer micelles. The segregation behavior in the regime where micelles were present was complicated by a strong tendency for micelles to segregate to the free PS surface and by a weaker tendency for micelles to segregate to the interfacial region. The effects of micelle formation within the bulk homopolymer phases on the interfacial behavior will be discussed further in the following sections.

Elastic recoil detection (ERD) was used by Green et al. [40] to study the segregation of low molecular weight symmetric copolymers of PS, and PMMA to the interface between PS and PMMA homopolymers. Bilayer films of PS and

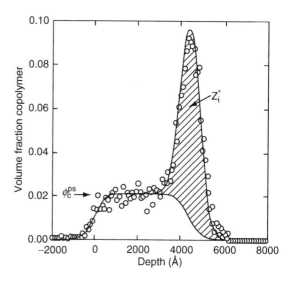

Fig. 20 Copolymer distribution for a dPS-*b*-P2VP 391–68 diblock copolymer added at constant concentration to the top PS layer (overlaid onto a P2VP layer) following by an 8-h annealing at 178°C. For this sample, the final equilibrium copolymer concentration in the PS phase was 2.1%. The interface copolymer excess, zi^*, corresponding to the *shaded area*, was equal to 100 Å [38]

PMMA mixed with a few percent of diblock copolymer were spin coated separately and, using the floating method, the assembly was built. The films were annealed to let the copolymer migrate to the interface and reach thermodynamic equilibrium. Figure 21 shows the volume fraction profile of a dPS-*b*-PMMA copolymer (262 segments) with a deuterated PS sequence at the interface between PS (18,000 segments) and PMMA (13,000 segments). At the concentrations studied, the excess number of copolymer chains per unit area at the polymer–polymer interface varied linearly with ϕ_c, the volume fraction of copolymer chains in the bulk. The results were compared with predictions based on a modification of the mean field arguments of Leibler [75] (discussed in Sect. 4.3.3). For low density of copolymer chains at the interface, the predictions are in a good agreement with the experimental behavior.

Neutron reflectivity was used to investigate the segment density distribution of symmetric diblock copolymers of PS and PMMA [39] (molecular weights of about 100,000) at the interface between PS and PMMA homopolymers (molecular weights of about 100,000). Selective deuterium labeling of either a block of the PS-*b*-PMMA or of the PS or PMMA homopolymers provided the contrast necessary to isolate the distribution of the segments of the individual components at the interface. Results from a series of experiments were used simultaneously to yield the density profiles of the PS and PMMA segments of the homopolymers, and of the copolymer blocks at the interface (Fig. 22).

It was found that the effective width of the interface between the PS and PMMA segments was 75 Å, i.e., it was 50% broader than that found between the PS and PMMA homopolymers in the absence of the diblock copolymer (50 ± 5 Å [261]) and between the PS and PMMA lamellar microdomains of the pure PS-*b*-PMMA in the bulk (50 ± 4 Å) [261, 262]. The area occupied by the copolymer at the interface between the homopolymers is 30% larger than that of the copolymers in the bulk lamellar microstructure [39]. In that study, the amount of diblock copolymer at the interface was (approximately) equivalent (~200 Å) to half of the long period of the neat ordered copolymer. The same PS/PS-*b*-PMMA/PMMA system was subsequently investigated by a lattice-based self-consistent field model that was extended

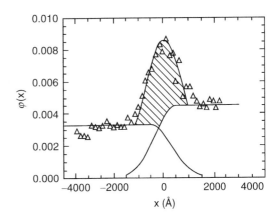

Fig. 21 Volume fraction profile of dPS-*b*-PMMA copolymer chains segregated to the PS–PMMA interface. The *shaded region* indicates the interface excess. The volume fraction of copolymer chains in the PS phase ($x < 0$) is $\phi_c^{PS} = 0.0033$. The volume fraction of copolymer chains in the PMMA phase ($x > 0$) is $\phi_c^{PMMA} = 0.0044$. The sample was annealed at 162°C for 100 h [40]

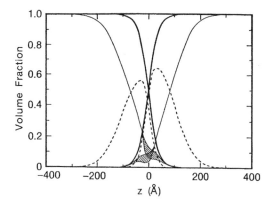

Fig. 22 Volume fraction profiles of the PS and PMMA homopolymers (*thin solid lines*), the PS and PMMA blocks of the PS-*b*-PMMA copolymer (*dashed lines*), and the total PS and PMMA segments summed over the homopolymer and the respective copolymer blocks (*thick solid lines*). The results were obtained by simultaneous analysis of neutron reflectivity experiments with different deuterium labeling of copolymer and homopolymer segments [39]

to incorporate chain conformational stiffness [73]. Excellent qualitative and quantitative agreement with the experimental data (Fig. 22) was obtained for the volume fraction profiles of both homopolymers and of both blocks of the copolymer at the interface [73].

In a subsequent study [263], it was shown that the width of the interface between the PS and PMMA segments broadened as the number of PS-*b*-PMMA chains added to the interface between PS and PMMA homopolymers increased. The width varied from the 50 Å thick interface between the PS and PMMA homopolymers up to ~85 Å at interfacial saturation (effective copolymer thickness of ~256 Å).

The organization of PMMA-*b*-PS-*b*-PMMA triblock copolymers at the interface of immiscible homopolymers [87] was studied by dynamic secondary ion mass spectrometry. Selective labeling of either the two end blocks or the central block provided the contrast necessary to determine the spatial arrangements of the blocks at the interfaces. It was found that the triblock copolymer chains were organized such that the central block preferentially segregated to one homopolymer, whereas the end blocks segregated to the other, thus adopting a "hairpin" type of conformation as indicated in Fig. 19.

4.2 Experimental Studies on the Effect of Additives on Polymer–Polymer Interfacial Tension

The effective interfacial tension between the two homopolymer phases in blends of immiscible homopolymers can be altered appreciably by adding different types of materials that can behave as interfacially active agents.

Interfacial tension between two incompatible homopolymers can be reduced by adding homopolymers containing functional side or end groups. In 1971, Patterson et al. [264] investigated the effect of functionalized poly(dimethyl siloxane), PDMS, additives on the interfacial tension between a commercial methyl-terminated PDMS and a commercial polyoxyethylene/polyoxypropylene copolymer, P(OE-OP). Starting with a high interfacial tension (8.3 dyn/cm), the presence of 10% carboxyl groups on alkyl side chains attached to the PDMS molecules reduced the interfacial tension by 63%. Doubling the number of carboxyl groups made this additive slightly less, rather than more, effective (57% reduction). Incorporation of carboxyl end groups on the PDMS chain provided a material that was capable of reducing interfacial tension in the same system by 49%. In contrast, hydroxyl end groups had no significant effect on the interfacial tension. Amino groups on the silicone additives had only a small effect on the interfacial behavior: 1% amino groups on alkyl side chains reduce interfacial tension by 28%, whereas increasing the amount of polar substituents to 6% produced a higher rather than lower interfacial tension value (18% reduction). In general, the interfacial activity of these additives is probably due to specific interactions between the additive and the homopolymers; these interactions increase compatibility and, consequently, reduce interfacial tension.

Patterson et al. [264] reported the effect of addition of PDMS-b-POE copolymers on the interfacial tension between PDMS and P(OE-OP) as well. A 72% reduction in interfacial tension was obtained with the addition of 2% of a 60/40 PDMS-b-POE block copolymer, as shown in Fig. 23. Increasing the level of polar polyether substitution from 40 to 75% did not result in any further reduction; it rather showed less interfacial activity (64% reduction). This agreed with the proposed maximum efficiency of symmetric copolymers [257, 258, 265]. Substitution of a POP for the POE in the 25/75 copolymer additive reduced its capability for reducing interfacial tension (51% reduction).

The effect of the concentration of the copolymer emulsifier was studied for the 60/40 PDMS-b-POE (Fig. 23). A major reduction in interfacial tension (55%) took

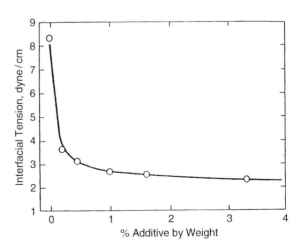

Fig. 23 Effect of the addition of a 60/40 PDMS-b-POE copolymer on the interfacial tension between PDMS and a P(OE-OP) copolyether fluid [264]

place with addition of 0.17% of the copolymer, whereas 68% reduction was observed with 1% additive. Increasing the copolymer amount to 2% led only to a 4% further reduction. That is, only a few percent of block copolymer additive is required to essentially saturate the interface and reach the limiting interfacial tension. A linear correlation was obtained when interfacial tension was plotted versus the logarithm of the concentration of the additive, expressed as grams of additive per liter of mixed liquids.

Gailard and coworkers [215, 266] demonstrated the surface activity of block copolymers by studying the interfacial tension reduction in demixed polymer solutions. Addition of a PS-b-PBD diblock copolymer to the PS/PBD/styrene ternary system showed first a characteristic decrease in interfacial tension followed by a leveling off, which is similar to the evolution of interfacial tensions for oil–water systems in the presence of surfactants. The early investigations were more of case studies that demonstrated the phenomenon without giving the fundamental detail required to help the understanding of the emulsification process and the factors that govern it.

Anastasiadis et al. [45] investigated the compatibilizing effect of an anionically synthesized model PS-b-PVE diblock copolymer on the interfacial tension between PS and PVE model polymers as a function of the concentration of the copolymer additive. They utilized the pendant drop method [155] to measure the interfacial tension between the immiscible polymer fluids. A sharp decrease in interfacial tension was observed with the addition of small amounts of copolymer (Fig. 24),

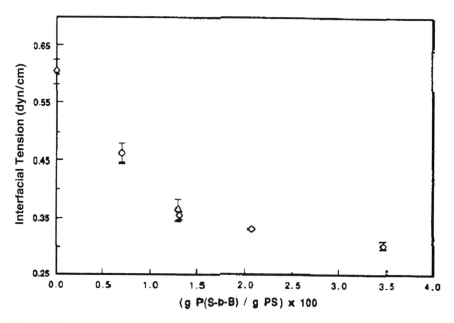

Fig. 24 Effect of the addition of a PS-b-PVE copolymer on the interfacial tension between PS and PVE at 145°C [45]

followed by a leveling off as the copolymer concentration increased above a certain concentration. This was attributed to an apparent critical micelle concentration (CMC). For concentrations lower than this critical concentration, the interfacial tension reduction was essentially linear with the copolymer content, a behavior that compared well with that predicted by Noolandi and Hong [70, 267].

Hu et al. [48] studied the addition of PS-*b*-PDMS diblock copolymer to the PS/PDMS blend. A maximum interfacial tension reduction of 82% was achieved at a critical concentration of 0.002% diblock added to the PDMS phase. At a fixed PS homopolymer molecular weight, the reduction in interfacial tension increases with increasing the molecular weight of PDMS homopolymer. Moreover, the degree of interfacial tension reduction was found to depend on the homopolymer the diblock is mixed with: when the copolymer was mixed into the PS phase, the interfacial tension reduction was much less than that when the copolymer was blended into the PDMS phase. This behavior suggested that the polymer blend interface may act as a kinetic trap that limits the attainment of global equilibrium in these systems.

Retsos et al. [54] investigated the effects of the molecular weight and concentration (ϕ_{add}) of compositionally symmetric PS-*b*-PI diblock copolymer additives on the interfacial tension between PS and PI immiscible homopolymers. The dependence of the interfacial tension on the additive concentration agreed with previous investigations: a sharp decrease with addition of a small amount of copolymer followed by a leveling off at higher copolymer concentrations (illustrated in Fig. 25). However, the reduction of the interfacial tension was a non-monotonic function of the copolymer additive molecular weight at constant copolymer concentration in the plateau region. The emulsifying effect, $\Delta\gamma = \gamma_0 - \gamma$, increased by increasing the additive molecular weight for low molecular weights,

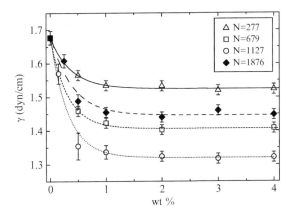

Fig. 25 Interfacial tension for the PS/PS-*b*-PI/PI systems as a function of copolymer concentration (wt%) added to PS at constant temperature (140 ± 1°C) for different diblock molecular weights with different numbers of segments (*N*) as shown. *Filled square* denotes the PS–PI interfacial tension in the absence of the diblock, γ_0. The *lines* are fits to an expression $\gamma = (\gamma_0 - \gamma_{sat}) \exp(-w_{add}/w_{char}) + \gamma_{sat}$ [65], where γ_{sat} is the interfacial tension at the plateau and w_{char} is the concentration needed to achieve the 1/*e* of the maximum reduction $\gamma_0 - \gamma_{sat}$ [54]

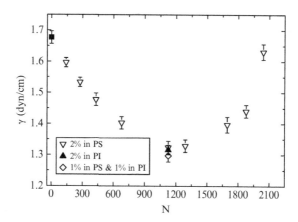

Fig. 26 Interfacial tension for the PS/PS-*b*-PI/PI systems as a function of the number of segments (*N*) of the copolymers at a constant temperature of 140 ± 1°C and constant 2 wt% copolymer added to the PS phase (*open inverse triangles*). For the $N = 1127$ diblock, data are also shown when 2 wt% copolymer is added to PI (*filled triangle*), and when 1 wt% is added to PS and 1 wt% is added to PI (*open diamond*). The PS–PI interfacial tension in the absence of the diblock is denoted by a *filled square* [54]

whereas it decreased by further increasing the copolymer molecular weight, thus going through a maximum (Fig. 26).

The results were understood by considering the possibility of micelle formation as the additive molecular weight increased, leading to a three-state equilibrium between copolymer chains adsorbed at the interface, chains homogeneously mixed with the bulk homopolymers, and copolymer chains at micelles within the bulk phases. A simple model was presented that qualitatively showed a similar behavior (see Sect. 4.3.3). The presence of micelles for high molecular weight additives and their absence for low molecular weights was supported by small-angle X-ray scattering data [55, 268].

Wagner and Wolf [46] investigated the effects of the addition of PDMS-*b*-PEO-*b*-PDMS triblock copolymers on the interfacial tension between PDMS and PEO homopolymers. In agreement with earlier investigations, interfacial tension was found to fall rapidly to ~10% of its initial value and level off as the effective CMC was surpassed. Moreover, the effect of the molecular weight of the PDMS block of the triblock copolymer was studied; this effectively studied the effect of copolymer composition without, however, keeping the copolymer molecular weight constant. The data (Fig. 27) showed that the interfacial tension decreased as the molecular weight of the PDMS block approached that of the PEO block.

Subsequently, Wolf and coworkers [49] investigated the effect of copolymer architecture on the interfacial tension reduction for the PDMS/PEO blend utilizing PDMS-*b*-PEO diblocks, PDMS-*b*-PEO-*b*-PDMS triblocks, and "bottle-brush" copolymers consisting of PDMS backbone and PEO brushes. The study showed that for the range of molecular weights investigated, the total number of PDMS

Interfacial Tension in Binary Polymer Blends and the Effects of Copolymers 233

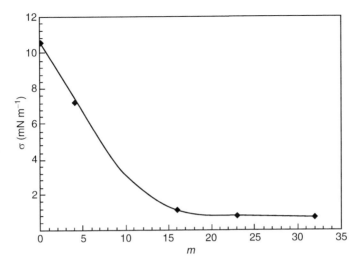

Fig. 27 Interfacial tension of a phase-separated mixture of PEO 35 containing 2 wt% PDMS$_m$-b-PEO$_n$-b-PDMS$_m$ and PDMS 100 as a function of m for $n = 37$ at 100°C; n and m are the numbers of monomeric units of the copolymer blocks [46]

segments was the most important parameter in determining the efficiency of the copolymer, irrespectively of their architecture or of the size of the PEO block.

Retsos et al. [56] investigated the effect of the macromolecular architecture and composition (f) of block copolymer additives on the interfacial tension between immiscible homopolymers. The systems investigated were PS/PI blends in the presence of PI$_2$PS (I$_2$S) and PS$_2$PI (S$_2$I) graft copolymers. The series of grafts possessed constant molecular weight and varying composition. A decrease in interfacial tension was observed with the addition of small amounts of copolymer followed by a leveling off (plateau) as the copolymer concentration (ϕ_{add}) increased, illustrating the surfactant-like behavior of the graft copolymers. The interfacial tension at interfacial saturation (plateau regime) was found to be a nonmonotonic function of the copolymer composition f exhibiting a minimum versus f (Fig. 28). The dependence on f was understood as a competition between the decreased affinity of the copolymer within the homopolymer phase when the size of the "other" constituent increased, which increased the driving force of the copolymer towards the interface, and the possibility of micellar formation. These ideas were supported by small-angle X-ray scattering measurements, which indicated the formation or absence of micelles.

Another observation in Fig. 28 concerns the fact that the interfacial tension for the I$_2$S graft with $f_{PI} = 0.36$ is lower than that for the symmetric linear diblock copolymer of the same total molecular weight (all in the plateau region of the interfacial tension reduction). It appeared that the old rule of thumb "diblocks better than triblocks better than grafts" should be reconsidered in the general case. The graft with $f_{PI} = 0.36$ had very similar composition with diblock SI ($f_{PI} = 0.41$) and

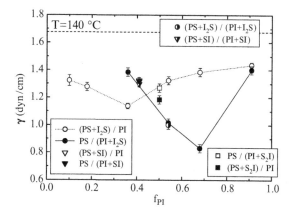

Fig. 28 Interfacial tension for the PS/I$_2$S/PI systems as a function of the composition of the graft copolymers at a constant temperature of 140°C and constant 2 wt% copolymer added to the PS (*open circles*), or to the PI phase (*filled circles*), or when 1 wt% was added to the PS and 1 wt% to the PI phases (*black and white circle*). Also shown are the interfacial tension data for PS/SI/PI at 140°C, i.e., with the addition of 2 wt% of the SI diblock copolymer to the PS (*open inverse triangle*), or to the PI phase (*filled inverse triangle*), or when 1 wt% was added to PS and 1 wt% to the PI phases (*black and white inverse triangle*). The *squares* are the interfacial tension data for PS/S$_2$I/PI at 140°C when 2 wt% of S$_2$I was added to the PS phase (*filled square*) or to the PI phase (*open square*). The *dashed line* indicates the PS/PI interfacial tension in the absence of additives [56]

very similar molecular weight, but it was more interfacially active, which most probably was an architecture effect. The better efficiency of the graft copolymer versus that of the diblock was not anticipated theoretically [85] (when micelles were not considered) but it was in agreement with an early study [269] on PS(PEO)$_2$ grafts versus PS-*b*-PEO diblocks of similar molecular weights added to water/organic solvent systems. It is believed that this is due to the higher tendency of the diblock to form micelles.

Furthermore, an important finding was that the final interfacial tension at saturation depended on the side of the interface to which the I$_2$S graft copolymer was added. When the I$_2$S was added to the PI homopolymer, the interfacial tension reduction was more significant, i.e., the apparent interfacial activity of the additive was higher. This pointed to a local equilibrium that can only be attained in such systems: the copolymer reaching the interface from one homopolymer phase does not diffuse to the other phase. For the symmetric SI diblock, the interfacial tension at saturation does not depend on whether the additive is premixed with the PS or to the PI phase, i.e., in that case adding the copolymer to the drop or the matrix phase did not make any difference. Thus, the SI data allowed the authors to rule out one of the possible explanations discussed by Hu et al. [48], who had suggested that such an effect could be due to the presence of a larger reservoir of diblock when added to the matrix phase, versus a depletion when it is added to the drop phase. When using the respective S$_2$I graft copolymers, a mirror image behavior was obtained, i.e., addition of the S$_2$I graft to the PS side followed the behavior of the I$_2$S added to PI and vice versa. The result signified that this behavior was not a kinetic effect but was rather due to a trapping of the system to a stationary state of local equilibrium, with the additive not crossing to the other side of the interface. Such an explanation was also discussed by Hu et al. for the PS-*b*-PDMS case [48]. It was suggested that, in the case of graft copolymers, it is the asymmetric architecture of the graft copolymers that leads to the great disparity between the two cases, whereas it should probably be the asymmetry in the statistical segment lengths of the two blocks in the PS-*b*-PDMS case that leads to an asymmetry in the CMC and, thus, in the interfacial tension reduction.

Wedge and Wolf [54] discussed similar "stationary states" to be due to larger thermodynamic driving forces and more pronounced back-damming when the PEO-*b*-PPO-*b*-PEO triblock was added to the PPO phase. This was attributed to a lower affinity of the additive to the PPO. Actually, the authors generalized their finding by suggesting that, in order to achieve the highest possible reduction of the interfacial tension by means of a given amount of compatibilizer, it should be added to the phase with the lower affinity to this component. The study of Retsos et al. [56] agreed with the statement that the effectiveness of the interfacial modifiers is controlled by the unfavorable interactions, which drive more of the additive towards the interface and thus reduce the interfacial tension further. However, the study pointed to the important effect of the formation of micelles within the bulk phase to which the compatibilizer is added, which is specifically important for nonsymmetric copolymer architectures. One should aim at adding the compatibilizer to the phase where it would form micelles with greater difficulty [56].

As was pointed out in the article of Retsos et al. [55], it should be noted that the concentration dependence of the surface tension in solvent/additive systems has been traditionally used for the estimation of the CMC in either small-molecule [270] or polymeric [265, 269, 271, 272] surfactant solutions. In those measurements [265, 269, 271, 272], the surface tension decreases with increasing concentration for concentrations up to a certain value, and then attains an almost constant value. The break in the γ_{surf} versus log c (c is the additive concentration) curve is used to denote the CMC. In the studies discussed above, however, it was found that even for concentrations in the plateau region (higher than the break) of the interfacial tension (or surface tension [55]) versus concentration curve, micelles are not present for low additive molecular weights, whereas they are present only for higher molecular weights (or equivalently for the graft copolymer case [56]). Therefore, it is apparent that the break in the interfacial tension versus concentration curve should denote interfacial saturation and not necessarily micellization. This statement is supported by an early study of solutions of PS-*b*-poly(hexyl methyl siloxane)-*b*-PS triblock copolymer in benzene [273], where, although the surface tension data exhibited the break discussed above, no micellization was established by static light scattering. No aggregation was expected since benzene is a good solvent for both blocks. The situation when both surface segregation (adsorption at a solid surface) and micellization might occur was investigated theoretically [274]. It was found that, depending on the incompatibility of the surface active block with the (monomeric or polymeric) solvent and its (attractive) interactions with the surface, one may have only adsorption onto the surface, only micellization, or an equilibrium of chains adsorbed onto the surface and chains in a micelle. This competition has not been investigated in detail in copolymer/solvent and copolymer/homopolymer systems.

Chang et al. [58] investigated recently the effect of copolymer composition on the interfacial tension reduction in PI/PDMS blends utilizing PI-*b*-PDMS additives. The authors utilized a series of diblock copolymers possessing constant molecular weight of the PI block and different molecular weights of the PDMS block (thus, different compositions) added to the PDMS phase. Ultralow values of interfacial tension of the order of 10^{-3} dyn/cm were obtained for almost symmetric diblock copolymers for additive concentrations in the plateau region. Such low interfacial tensions had never been measured previously in polymeric systems, whereas they had been obtained in systems of balanced small molecule surfactants, for which the thermodynamically preferred form of aggregation is a surfactant monolayer with no spontaneous curvature. The interfacial tension increased with increasing PDMS block, going from a symmetric to asymmetric diblocks. At certain copolymer composition, a discontinuity was observed with the interfacial tension exhibiting a jump. For highly asymmetric additives, the behavior was accounted for by a theory [105] that considered equilibrium between a PDMS phase containing swollen spherical micelles and a phase of nearly pure PI. The self-consistent field theory (SCFT) discussed the behavior of systems of nearly balanced copolymers, which tend to form highly swollen micelles, within the context of the Helfrich theory of interfacial bending elasticity [275], using elastic constants obtained from SCFT simulations of weakly curved monolayers.

Besides the considerations regarding the thermodynamic factors that determine the efficiency of a compatibilizer, the question of how and whether a state of equilibrium is reached in such systems is still open. In principle, in all experimental measurements, interfacial tension data are taken for long periods of time; "equilibrium" is considered to have been accomplished when the extracted values of the interfacial tension do not change with time. These times can be very or extremely long in the case of polymer–polymer interfaces due to the normally very high viscosities of the components of the mixtures. Actually, in these systems, one can study the kinetics with which time-independent interfacial tensions are established. Note that in the ternary systems it is the combined influence of hydrodynamic relaxation and interfacial segregation of the additive that determines the kinetics of equilibration measured. The time-dependent interfacial tension data of Stammer and Wolf [276] for random copolymers added to the polymer–polymer interfaces were fitted with a double exponential function, with the two characteristic times attributed to the viscoelastic relaxation and the compatibilizer transport to the interface. Cho et al. [53] studied the segregation dynamics of PS-*b*-PDMS diblock copolymer to the PS/PDMS polymer blend interface. The data were analyzed within diffusion-limited segregation models proposed by Budkowski et al. [277] and Semenov [278], as modified to treat interfacial tension data. The estimated apparent block copolymer diffusion coefficients obtained were close to the estimated self-diffusion coefficient of the PDMS homopolymer matrix. Shi et al. [57]

studied the time evolution of the interfacial tension when polyisobutylene (PIB)-*b*-PDMS was introduced to PIB/PDMS blend, with the copolymer added to the PIB phase; in that study both homopolymers were polydisperse. The time dependence of the interfacial tension was fitted with an expression that allowed the evaluation of the characteristic times of the three components. The characteristic time of the copolymer was the longest, whereas the presence of the additive was found to delay the characteristic times of the blend components from their values in the binary system. The possible complications of slow diffusivities on the attainment of a stationary state of "local equilibrium" at the interface were thoroughly discussed by Chang et al. [58] within a theoretical model proposed by Morse [279]. Actually, Morse [279] suggested that the optimal system for measuring the equilibrium interfacial tension in the presence of a nearly symmetric diblock copolymer would be one in which the copolymer tracer diffusivity is much higher in the phase to which the copolymer is initially added than in the other phase because of the possibility of a quasi-steady nonequilibrium state in which the interfacial coverage is depleted below its equilibrium value by a continued diffusion into the other phase.

In order to avoid the complications of micelle formation or the diffusion of the copolymer to the opposite side of the interface, the in-situ formation of copolymers has been utilized [61, 106, 107, 109, 112, 117, 172]. In a review article, Jérôme and coworkers [106] wrote that they found no evidence of commercial blends compatibilized with premade block copolymers, and indicated that the in-situ method is superior in compatibilization. Macosco and coworkers [107] have compared directly the effects of premade versus reactively formed compatibilizers; it was concluded that the premade copolymers are less capable of compatibizing polymer blends than the in-situ formed ones because of the possibility of micelle formation by the former.

Fleischer et al. [172] measured the interfacial tension reduction credited to the complexation between carboxy-terminated PBD and amine-terminated PDMS, which were added to an immiscible blend of PBD and PDMS. The changes in interfacial tension resembled the behavior observed for block copolymer addition to homopolymer blends: there is initially a linear decrease in interfacial tension with the concentration of functional homopolymer up to a critical concentration, at which the interfacial tension becomes invariant to further increases in the concentration of functional material. However, the formation of interpolymer complexes depends on the equilibrium between associated and dissociated functional groups and, thus, the ultimate plateau value for interfacial tension reduction is dependent on the functional group stoichiometry. A reaction model for end-complexation was developed in order to reproduce the interfacial tension reduction data with Fourier transform infrared spectroscopy applied to determine the appropriate rate constants. The model provided a reasonable qualitative description of the interfacial tension results, but was not able to quantitatively predict the critical compositions observed experimentally.

Recently, the kinetics of interfacial reaction between two end-functionalized homopolymers was investigated by Chi et al. [117] utilizing interfacial tension measurements. The authors measured the changes in the interfacial tension between PDMS and PBD during the reaction between amino-terminated PDMS and carboxyl-terminated PBD, which can react at the interface and form diblock copolymers that compatibilize the blend. The concentration of the reaction product was inferred from an application of Gibbs adsorption equation, justified for an insignificant degree of conversion of reactants in either phase. The obtained time-dependent copolymer concentration was found to follow a single-exponential growth function at low copolymer coverage, indicating first order kinetics.

Favis and coworkers [51, 52] critically examined the relationship between the interfacial tension reduction in the presence of diblock copolymer additives and the dispersed phase morphology evolution as a function of the concentration of the interfacial modifier. Blends of PS/PE in the presence of PS-*b*-hydrogenated polybutadiene-*b*-PS (Kraton, SEBS) [51] and of PE/PVC in the presence of PI-*b*-poly (4-vinyl pyridine) or PS-*b*-poly(acrylic acid) [52] were investigated. The authors unambiguously confirmed directly the relationship between interfacial tension and phase size, as predicted by the Taylor theory [280].

4.3 Theories of the Interfacial Behavior in Homopolymer/ Homopolymer/Copolymer Blends

Statistical thermodynamic theories have been formulated to understand and predict the emulsifying behavior of block copolymers at the polymer–polymer interface [70–75, 77–80, 95, 98, 99, 105, 267, 279, 281, 282]. Noolandi and Hong [70, 71, 281] utilized their theory of inhomogeneous systems in order to investigate the segment density profiles at the interface for the system homopolymer A/homopolymer B/diblock copolymer AB/common solvent. They investigated the effect of the molecular weight and the concentration of the diblock on the interfacial tension, under the assumption that the copolymer is either localized at the interface or is randomly distributed in the bulk homopolymer phases, i.e., for concentrations below the CMC. Shull and Kramer [77] developed and applied the Noolandi–Hong theory for the case without solvent and also discussed the possibility of micelle formation in view of their earlier experimental observations [38, 102]. Semenov [103] developed an analytical mean-field theory for the equilibrium of block copolymers in a homopolymer layer between an interface with another homopolymer and the free surface, and the results were compared to the data of Shull et al. [38]. Semenov also analyzed the situation for concentrations above CMC and found that micelles are attracted to both the free surface and (more weakly) to the polymer–polymer interface, but he did not investigate the interfacial tension reduction due to copolymer segregation to the polymer–polymer interface.

The effects of copolymer architecture on the interfacial efficiency of the compatibilizers have been investigated in a series of papers by Balazs and coworkers [80, 85] using a combination of SCMF calculations, analytical theory, and Monte Carlo simulations as well as by Dadmun [95, 98] using computer simulations.

Lyatskaya et al. [85] investigated the interfacial tension reduction due to the localization of AB block copolymers at the interface between two immiscible homopolymers A and B as a function of the copolymer architecture. For the same total copolymer molecular weight, for symmetric copolymers ($f = 0.5$) and for very high molecular weight homopolymers, both analytical arguments and SCMF theory agreed in that diblock copolymers are the most efficient at reducing the interfacial tension, followed by the simple grafts, the four-armed stars, and the n-teeth combs. The trade-off between total molecular weight and number of teeth was discussed when combs and diblocks of different molecular weights were compared, i.e., long combs are more efficient than short diblocks.

Retsos et al. [55, 56] made an attempt to provide a semiquantitative analysis of the interfacial activity of block copolymers at the polymer–polymer interface; the emphasis was on understanding the nonmonotonic dependence of the interfacial tension reduction on diblock molecular weight as well as the effects of macromolecular architecture and composition when graft copolymers were utilized as additives. The attempt was based on a modification of the analysis of Leibler [75], where the possibility of micellar formation was also taken into account. The thermodynamic equilibrium under consideration was, thus, that between copolymer chains adsorbed at the interface, chains homogeneously distributed in the bulk homopolymers, and chains at micelles formed within the homopolymer phases.

4.3.1 The Noolandi and Hong Theory

Hong and Noolandi constructed a general theory [211, 283] of inhomogeneous systems, beginning with the functional integral representation of the partition function as introduced by Edwards [212]. The free energy functional is minimized by the saddle-function method (including the constraints of no volume change upon mixing and a constant number of molecules of each component) to obtain the mean-field equations for the fundamental probability distribution functions that characterize a system of two immiscible homopolymers A and B diluted with solvent in the presence of a diblock copolymer AB. These equations were, then, solved numerically to obtain the polymer density profiles through the interfacial region. The difference between the total free energy and that of the bulk polymers was used to evaluate the interfacial tension.

For homopolymer A/homopolymer B/diblock copolymer AB/solvent system, six distribution functions were needed [70, 267] to describe the mixture: two for the two homopolymers A and B, and four for the copolymer. However, the expressions for the mean-field simplified to two functions [267] if the volume fractions of the homopolymer and the respective block of the copolymer were added together. The mean-field expressions then reduce to those for a ternary system: homopolymer A/homopolymer B/solvent [211, 283]. The assumption was made that the part of the copolymer that does not localize itself at the interface will be randomly distributed in the bulk of the homopolymers.

There are a number of factors that determine the state of the block copolymer in a phase-separated system. The entropy of mixing of the block copolymers with the homopolymers favors a random distribution of the copolymers. On the other hand, localization of the block copolymers to the interface displaces the homopolymers away from each other, thus, lowering the enthalpy of mixing. In addition, each block of the copolymer will prefer to extend into its compatible homopolymer to lower the block copolymer–homopolymer enthalpy of mixing. Besides suffering an entropy loss as a whole because of the confinement to the interphase, there is a further entropy loss for the blocks of the copolymer arising from the restriction of the blocks into their respective homopolymer regions. Finally, extension of the copolymer chains, as well as the effect of the excluded volume at the interphase for the homopolymers, lead to further loss of entropy. In their theoretical development, Noolandi and Hong [70] included the contributions to the free energy from all these effects, and obtained the concentration of the block copolymer at the interface as well as the associated reduction in the interfacial tension.

It is clear that similar considerations for the enthalpy and entropy of mixing of block copolymers could favor micellar aggregation rather than random distribution in the bulk of the homopolymers. In this case, the micelles could compete with the interfacial region for copolymer chains and the amount in each state would depend on the relative reduction in the free energy as well as the surface area. Since no complete treatment of this complicated case was given in the Noolandi and Hong paper [70], their results should be reliable only for low copolymer concentrations below the CMC. Their mean-field calculation cannot adequately describe the critical crossover regime from a random copolymer distribution to aggregation (micelle formation) and, thus, they only gave a rough estimate of the CMC.

The reduction in interfacial tension with increasing block copolymer concentration was calculated for a range of copolymer and homopolymer weights as well as for different initial concentrations of solvent in their systems. The calculated interfacial density profiles showed greater exclusion of the homopolymers from the interfacial region as the molecular weight of the copolymer increased. This greater localization of the copolymer resulted in a greater reduction in the interfacial tension as the block molecular weight increased for both infinite and finite molecular weights of the corresponding homopolymers.

The theory, however, generally overestimates the interfacial tension reduction upon addition of the copolymer. An attempt to model the exact polymer system studied by Gaillard et al. [215, 266] (PS/PBD/styrene/PS-b-PBD, discussed in Sect. 4.2) showed a disagreement between theory and experiment. The calculated interfacial tension fell to zero for a copolymer concentration (weight fraction with respect to one of the two homopolymers of equal weight) of ca. 10^{-4}, while the measurements indicated that interfacial tension decreased much more slowly with increasing block copolymer concentration, and reached a constant value for ca. 5%. Possible reasons for this discrepancy were discussed in the original paper [70]. The use of the spinning drop method to measure the interfacial tension for the demixed polymer solutions and the effect of the rotational speed on possible shift in the position of the block copolymer at the interphase were emphasized, together with

Interfacial Tension in Binary Polymer Blends and the Effects of Copolymers

the assumption of the random distribution of the copolymer chains in the bulk copolymers. The fact that the theory was unable to describe the crossover regime from a random distribution to the micelle formation was also discussed [70].

In another study, Noolandi and Hong [71] attempted to identify the relative importance of the various contributions that affect the interfacial tension reduction (as discussed earlier). The equations of their model were solved numerically in a "computer experiment" and the various contributions to the free energy and the interfacial tension were evaluated to determine their relative importance. The results were also discussed in another publication [281]. For a symmetric diblock copolymer, homopolymers of infinite molecular weight, and a symmetric solvent, they found that the interfacial tension reduction, $\Delta\gamma$, with increasing copolymer molecular weight and concentration arose mainly from the energetically preferred orientation of the blocks at the interface into their respective compatible homopolymers. The main counterbalancing term in the expression for $\Delta\gamma$ was the entropy loss of the copolymer that localizes at the interface. The loss of conformational or "turning back" entropy of both copolymer and homopolymer chains at the interface was shown to contribute little to $\Delta\gamma$.

Neglecting the loss of conformational entropy, Noolandi and Hong were able to obtain an analytical expression for the interfacial tension reduction for infinite homopolymer molecular weights, given by:

$$\Delta\gamma = \gamma - \gamma_0 = \frac{d}{b}\phi_c\left\{\frac{\chi\phi_p}{2} + \frac{1}{N} - \frac{1}{N}\exp\left(N\chi\phi_p/2\right)\right\} \tag{96}$$

whereas the amount of copolymer at the interface is:

$$\phi_c(0) = \phi_c\exp\left(N\chi\phi_p/2\right) \tag{97}$$

where d is the full width at half height of the copolymer profile and b is the Kuhn statistical segment length. Numerical calculations showed that d was almost constant for varying copolymer molecular weight. $\phi_c = \phi(\infty) = \phi(-\infty)$ is the copolymer volume fraction in the bulk homopolymer phases, which is very close to the nominal amount of the block copolymer present because the material segregated to the interface is negligible [71] compared to the total amount for a large system; $\phi_c(0)$ is the copolymer volume fraction at the interface. ϕ_p is the bulk volume fraction of polymer A or B (assumed equal), N is the degree of polymerization of the symmetric copolymer, and, χ is the Flory-Huggins interaction parameter between A and B segments. It was assumed that the interaction parameters between segments A and B and the solvent are $\chi_{AS} = \chi_{BS} = 0$, respectively. d is a parameter that was not determined by the simplified theory. For $N\chi\phi_p \ll 1$, (96) reduces to:

$$\Delta\gamma = \gamma - \gamma_0 = -\frac{d}{b}\phi_c N\chi^2\phi_p^2/8; \ N\chi\phi_p \ll 1 \tag{98}$$

The important features of the approximate relationships (96) and (98) were verified by the exact numerical calculations. An exponential dependence of the interfacial tension reduction on the block copolymer molecular weight as well as on the total homopolymer volume fraction was predicted that can explain the remarkable effectiveness of using large molecular weight diblocks as surfactants for concentrated mixtures of immiscible homopolymers. For small N, a linear dependence of $\Delta\gamma$ on N (98) was also predicted by the exact numerical calculations. Moreover, a linear dependence of $\Delta\gamma$ on the block copolymer volume fraction was predicted by the exact numerical solution, as shown by (96) and (98).

The homopolymer profile thickness was calculated numerically to increase exponentially with copolymer molecular weight and linearly with copolymer concentration. The increasing width (or decreasing slope) of the homopolymer profiles, as compared to the total polymer profiles (homopolymer plus copolymer segments), reflected the necessity to accommodate the increased amount of the copolymer at the interface.

Noolandi (personal communication) suggested that the theory can be applied to the experimental system PS/PS-b-PVE/PVE of [45], i.e., to a concentrated system without solvent, by letting the total polymer volume fraction, ϕ_p, go to 1 in (96) and (98). For the temperature of 145°C in the experiments, $\chi = 0.0388$ [245], and for the degree of polymerization of the diblock ($N = 261$), (98) becomes:

$$\Delta\gamma \cong -0.583 \frac{d}{b} \phi_c \tag{99}$$

with d being the width at the half height of the copolymer profile, which is a parameter related to the thickness of the interface but it was not determined by the simplified theory.

In order to compare the data with the theory, Noolandi and Hong, Anastasiadis et al. [45] assumed that the same volume fraction of copolymer exists in both bulk phases and, by using the bulk densities of PS and PB, they plotted the interfacial tension increment, $\Delta\gamma = \gamma - \gamma_0$, as a function of the copolymer bulk volume fraction, as shown in Fig. 29.

The interfacial tension increment, $\Delta\gamma = \gamma - \gamma_0$, was linear with the copolymer volume fraction, calculated for low concentration of the copolymer additive as suggested by theory for concentrations below the CMC. The slope of the fitted line was -37.0, and thus d was estimated to be 38 nm, or $63.5b$ when the geometric mean of the Kuhn statistical segment lengths of the two segments was used as 0.6 nm. This value of d (~ 63.5 monomer units) was about 24% of the contour length of the copolymer chains and, thus, indicated an extended configuration of the copolymer chains.

Noolandi and Hong [71, 281] pointed out that both copolymer concentration and molecular weight are equally important in reducing the interfacial tension. They noted, however, that the interfacial tension surface (γ plotted against N and ϕ_c) is bounded by a CMC curve because blocks of large molecular weights tend to form micelles in the bulk of the homopolymers rather than segregating to the

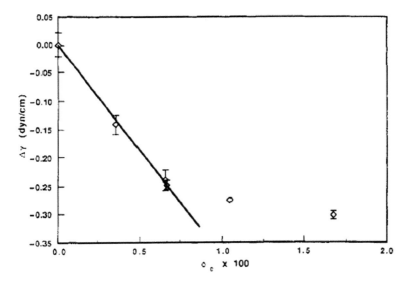

Fig. 29 Interfacial tension increment ($\Delta\gamma$) versus copolymer volume fraction (ϕ_c) for the PS/PS-b-PVE/PVE system at 145°C. *Solid line* is the linear fit of the data for concentrations below the CMC, according to the theory of Noolandi and Hong. From [45]

interface. Their theoretical treatment is valid for concentrations well inside the CMC boundary.

Whitmore and Noolandi [101] derived the structural parameters of monodispersed AB diblock copolymer micelles within an A homopolymer by minimizing a simple free energy functional. The CMC was calculated and shown to be dominated by an exponential dependence on χN_B (χ is the Flory-Huggins interaction parameter and N_B the degree of polymerization of the B block of the copolymer). The importance of diblock copolymer composition was emphasized as well. The CMC was calculated as:

$$\phi_c^{crit} = 0.30 \frac{\chi N}{(\chi N_B)^{2/3}} \exp(\Xi) \tag{100}$$

where:

$$\Xi = -\chi N_B + 1.65(\chi N_B)^{1/3} + \frac{1}{2}\left[1.65(\chi N_B)^{1/3} + \frac{1.56}{(\chi N_B)^{1/6}} - 3\right]$$

$$+ \frac{1}{2}\left[\alpha_A^2 + \frac{2}{\alpha_A} - 3\right] \tag{101}$$

N is the total degree of polymerization of the copolymer and α_A the stretching parameter for the block A of the copolymer, which is related to the molecular weights and the interactions and is calculated by the model. Good agreement was observed between the predicted micelle core radii and experimental data [284] for PS-*b*-PBD within PBD, obtained using small angle neutron scattering.

Shull and Kramer [77] developed and applied the Noolandi–Hong theory for the case of polymer A/polymer B/diblock copolymer AB, but without solvent. They found that, at a given value of the chemical potential of the copolymer in the bulk phases μ_c, the ability of a copolymer to reduce γ is highest for small N and small χ. However, at a given value of ϕ_c, higher values of N result in much higher values of $\Delta\gamma$ due to the exponential dependence of μ_c on χN and because an increase in μ_c results in an increase in the density of copolymer chains at the interface. Theoretical determination of the limiting value of μ_c associated with the formation of micelles was made separately [38], since the possibility of micelle formation was not explicitly introduced in the theory. A good agreement was found with the experimental data [38] for the total amount of copolymer segregating to a polymer–polymer interface for concentrations below CMC using only χ as an adjustable parameter. Using the best-fit value of χ, they estimated $\Delta\gamma$ for concentrations when micelles are not present. For concentrations higher than the CMC, more micelles will be formed without, however, significantly increasing the copolymer chemical potential; thus, the interfacial tension will not decrease further. For the copolymer molecular weights used, a significant increase in the total copolymer amount adsorbed at the interface was observed at higher copolymer concentrations, which was attributed [38, 102] to segregation of micelles to the polymer–polymer interface (as well to the polymer–air surface [38, 102, 285]). The location of the upturn was used to estimate the copolymer chemical potential at the CMC, which was in good agreement with a full self-consistent-field theoretical estimate [286].

4.3.2 Leibler Theory for Nearly Compatible Systems

Leibler [282] developed a simple mean-field formalism to study the interfacial properties of nearly compatible mixtures of two homopolymers, A and B, and a copolymer AB. The free energy was expressed in terms of monomer concentration correlation functions, which were calculated in a self-consistent way within the random phase approximation introduced by de Gennes [242]. For the very broad interface of nearly miscible systems, a gradient expansion was carried out giving a generalization of the Cahn–Hilliard theory [216]. As mentioned by Noolandi and Hong [71], with the gradient expansion in the theory of Leibler, the diblock copolymer was effectively treated as a small-molecule solvent compared to the large width of the interfacial region, and the structure of the copolymer became irrelevant. The system, thus, behaved as a mixture of two homopolymers driven to the consolute point by the addition of an excess of solvent. As pointed out by Leibler, for nearly compatible species ($2 < \chi N < 4\sqrt{2}$), two mechanisms of the interfacial activity of the copolymer chains had to be distinguished: (1) the species A and B are more closely mixed as copolymer chains and are present in both phases,

Interfacial Tension in Binary Polymer Blends and the Effects of Copolymers

and (2) the copolymer chains have a certain tendency to localize to the interface. In the case of nearly miscible species (near the consolute point), the first mechanism dominates whereas, for the highly incompatible case, the second dominates. Thus, for nearly miscible systems, the mechanism involved is quite different from that invoked for highly immiscible species. The dominant effect is the presence of copolymer chains in both the A-monomer-rich and B-monomer-rich phases: in consequence, when the copolymer amount increases, the difference between the total volume fractions of monomer A in the B-rich phase and B-rich phase decreases. The interfacial tension was found to consist of two parts:

$$\gamma = \gamma_0 - \gamma_1 \tag{102}$$

The first term, γ_0, represents the interfacial energy due to the inhomogeneity of the overall concentration of B monomer:

$$\gamma_0 = \gamma_0(0)R(\phi) \tag{103}$$

where:

$$\gamma_0(0) = \frac{2^{1/2}k_B T N^{-1/2}}{6b^2}(\chi N - 2)^{3/2} \tag{104}$$

and:

$$R(\phi) = (1 - \phi)\left[\chi N(1 - \phi) - 2\right]^{3/2}(\chi N - 2)^{-3/2}$$
$$\times \left[1 - \chi^2 N^2 \phi(1 - \phi)/8\right]^{1/2} \tag{105}$$

Here χN is the degree of incompatibility of the species, k_B is the Boltzmann constant, T is the absolute temperature, ϕ is the average copolymer volume fraction, and b is the Kuhn statistical segment length. Formally, the same expression for γ_0 would be obtained if there were no copolymer chains in the system.

The second contribution, γ_1, expresses a decrease in the interfacial tension due to the effect of the preferential localization of the copolymers at the interface. Near the critical region, γ_1 may be approximated by:

$$\gamma_1 \cong \frac{3}{2}\phi\gamma_0 \tag{106}$$

Calculations showed that, near the critical point, the contribution of γ_1 to the interfacial tension was almost negligible. However, for higher incompatibility degrees, i.e., higher values of χN, the term γ_1 could be comparable with γ_0. Therefore, it is the localization of the copolymer at the interface that is important.

To summarize, the mechanisms involved in the two different cases of highly immiscible systems and nearly compatible blends are quite different. In the first case, it is the surfactant activity of the block copolymer chains that cause the interfacial tension reduction whereas, in the second case, it is the presence of copolymer molecules in the bulk homopolymer phases that causes the compatibilizing behavior.

4.3.3 Leibler Theory for Strongly Incompatible Systems and Its Modification

Diblock Copolymer Additives

Leibler [75] considered a flat interface with surface area A between phase-separated A and B homopolymers. The thickness of the interfacial region $\alpha_I = b(6\chi)^{-1/2}$ and the interfacial tension $\gamma_0 = k_B T b^{-2} (\chi/6)^{1/2}$ are independent of the number of segments P_A and P_B of the two homopolymers [27] for a highly incompatible situation of $\chi P_i \gg 1$ (where $k_B T$ is the thermal energy). It was assumed that both types of links have the same segmental volume $\upsilon = b^3$. Suppose that Q copolymer chains with number of segments $N = N_A + N_B$ and composition $f_i = N_i/N$ are adsorbed at the A–B interface (for most practical situations, $\chi N_i \gg 1$). It was expected that the copolymer joints will be localized in a thin interfacial layer of thickness [103] $d' = (\pi/2)a_I$ (independent of N_i and P_i); d' is equal to the semiempirical parameter d of Noolandi et al. [71, 281] in (98), as discussed by Semenov [103]. The blocks A and B extend towards the respective bulk layers and form two "adsorbed layers" of thicknesses L_A and L_B, respectively. Since $d' \ll L_i$, each side of the interfacial film resembles a layer of polymers anchored by one end onto a wall. The free energy of the interfacial film can, thus, be approximated as [75]:

$$F_{\text{interf.film}} = \gamma_0 A + Q(g_A + g_B) \tag{107}$$

where γ_0 is the A–B interfacial tension in the absence of the additive, A is the interfacial area, and g_A, g_B represent the free energies per A–B chain of the A and B layers, respectively. The number of copolymer chains per unit interfacial area is given by $\sigma = Q/A$.

In most of the experimental studies, the copolymer chains are not so long relative to the homopolymers. Thus, mixing of the copolymer and homopolymer chains should be taken into account due to the penetration of homopolymers into the layer of chains anchored at the interface, whereas the copolymer chains can be either stretched (wet brush regime) or not (wet mushroom). Neglecting the composition gradients in the brush (Flory approximation), g_i is given by [40, 75, 287]:

$$\frac{g_i}{k_B T} = \ln(N_i b^2 \sigma) + L_i \frac{1}{\sigma b^3} \frac{1}{P_i} (1 - \eta_i) \ln(1 - \eta_i) + \frac{3}{2} \frac{L_i^2}{N_i b^2} \tag{108a}$$

where $\eta_i = \sigma N_i b^3 / L_i$ is the average volume fraction of monomers of the A block in the layer and $(1 - \eta_A)$ is that of the P_A monomers. The first two terms in (108a) approximate the entropy of mixing between copolymer and homopolymer chains, which tend to swell the copolymer blocks; the first term is associated with the translational freedom of the copolymers in the two-dimensional film, whereas the second term originates from the translational entropy of the homopolymer chains and has a standard excluded volume form [287]. The last term represents the elastic entropy term, which limits the swelling. For low values of $\eta_A \ll 1$, (108a) can be written as [40, 75]:

$$\frac{g_i}{k_B T} = \ln(N_i b^2 \sigma) + \frac{1}{2} \frac{N_i \eta_i}{P_i} + \frac{3}{2} \frac{L_i^2}{N_i b^2} \qquad (109a)$$

For stretched chains, the brush thickness L_i and the block monomer concentration η_i are obtained from (109a) by minimization with respect to L_i. In that case, $L_i = 6^{-1/3} N_i b (\sigma b^2)^{1/3} P_i^{-1/3}$, $\eta_i = 6^{1/3} (\sigma b^2)^{2/3} P_i^{1/3}$ and:

$$\frac{g_i}{k_B T} = \ln(N_i b^2 \sigma) + \frac{3^{4/3}}{2^{5/3}} (\sigma b^2)^{2/3} N_i P_i^{-2/3} \quad \text{(wet brush)} \qquad (110a)$$

which is valid for $P_i N_i^{-3/2} < \sigma b^2 < P_i^{-1/2}$. For nonstretched chains, $L_i \approx N_i^{1/2} b$ and the last term of (109a) can be neglected; this applies for $\sigma b^2 < P_i N_i^{-3/2}$ [287]. Then, $\eta_i = \sigma b^2 N_i^{1/2}$ and:

$$\frac{g_i}{k_B T} = \ln(N_i b^2 \sigma) + \frac{1}{2} \frac{N_i^{3/2} \sigma b^2}{P_i} \quad \text{(wet mushroom)} \qquad (111a)$$

The interfacial tension in the presence of the copolymer is calculated as[1]:

$$\gamma = \left. \frac{\partial F_{\text{interf.film}}}{\partial A} \right|_Q = \gamma_0 - \sigma^2 \left(\frac{\partial g_A}{\partial \sigma} + \frac{\partial g_B}{\partial \sigma} \right) \qquad (112)$$

Therefore, the interfacial tension reduction, $\Delta \gamma = \gamma_0 - \gamma$, is given by:

$$\frac{\Delta \gamma}{k_B T} = \frac{\gamma_0 - \gamma}{k_B T} = \begin{cases} \sigma \left[2 + \frac{3^{1/3}}{2^{2/3}} (\sigma b^2)^{2/3} \left(N_A P_A^{-2/3} + N_B P_B^{-2/3} \right) \right] & \text{(wet brush)} \\ \sigma \left[2 + \frac{1}{2} \sigma b^2 \left(N_A^{3/2} P_A^{-1} + N_B^{3/2} P_B^{-1} \right) \right] & \text{(wet mushroom)}. \end{cases}$$

$$(113a)$$

At equilibrium, σ is determined by equating the chemical potential of the copolymer chains at the interface with that of the copolymer chains either homogeneously mixed with the homopolymers or at micelles formed within the

---

[1]It is noted that Noolandi [288] objects to the use of (107) and (112) because he claims that the main contribution to the interfacial tension reduction is of enthalpic and not entropic origin (as (112) suggests), i.e., that it is due to the favorable energetics of the orientation of the copolymer blocks into their respective homopolymers and that entropic effects are second order. He suggests that (107) should be corrected by adding the contributions of the orientational entropy of the blocks and their entropy of localization. The latter was introduced by Shull and Kramer [77] by replacing γ_0 by $\gamma_0' = \gamma_0 + \sigma k_B T \ln[(L_A + L_B)/d']$. In the present analysis, the expression of Leibler [75, 76, 40] is utilized.

homopolymer phases. The chemical potential of a copolymer chain at the interface is calculated using (107) as:

$$\mu_{\text{int}} = \frac{\partial F_{\text{interf.film}}}{\partial Q}\bigg|_A = g_A + g_B + \sigma\left(\frac{\partial g_A}{\partial \sigma} + \frac{\partial g_B}{\partial \sigma}\right) \tag{114}$$

Therefore, with (110a) and (111a):

$$\frac{\mu_{\text{int}}}{k_B T} = 2 + \ln\left(N_A \sigma b^2\right) + \ln\left(N_B \sigma b^2\right)$$

$$+ \begin{cases} 2.271(\sigma b^2)^{2/3}\left(N_A P_A^{-2/3} + N_B P_B^{-2/3}\right) & \text{(wet brush)} \\ \sigma b^2\left(N_A^{3/2} P_A^{-1} + N_B^{3/2} P_B^{-1}\right) & \text{(wet mushroom)} \end{cases} \tag{115a}$$

The free energy density of a homogeneous mixture of an AB copolymer with a B homopolymer is [278]:

$$\frac{F_{\text{bulk}}}{k_B T} = \frac{\phi}{N}\ln\left(\frac{\phi}{e}\right) + \frac{1-\phi}{P_B}\ln\left(\frac{1-\phi}{e}\right) + \chi\phi f_A(1 - f_A\phi) \tag{116}$$

irrespective of the copolymer architecture. Thus, the chemical potential of a copolymer chain homogeneously distributed within the bulk B homopolymer, $\mu_{\text{bulk}} = N[(1-\phi)\partial F_{\text{bulk}}/\partial\phi + F_{\text{bulk}}]$, is:

$$\frac{\mu_{\text{bulk}}}{k_B T} = \ln\phi - \phi - (1-\phi)\frac{N}{P_B} + \chi N f_A\left(1 - 2f_A\phi + f_A\phi^2\right) \tag{117a}$$

where $\phi = \phi(\infty)$ is the copolymer volume fraction in the B-rich homopolymer phase.

The chemical potential of a copolymer chain in a micelle was evaluated by Semenov [278] for long homopolymer chains ($P > N$), which do not penetrate the micelles. Depending on the diblock copolymer composition, the micelle morphology could be spherical, cylindrical, or lamellar [278, 289]. The chemical potential of a diblock copolymer chain in a micelle formed within the B phase is then given by [55, 56, 278]:

$$\frac{\mu_{\text{mic}}^{\text{spherical}}}{k_B T} = (3/2)^{4/3} f_A^{4/9}\left[1.74 f_A^{-1/3} - 1\right]^{1/3}(\chi N)^{1/3}$$

$$\frac{\mu_{\text{mic}}^{\text{cylindrical}}}{k_B T} = 1.19(\chi f_A N)^{1/3}[1.64 - \ln f_A]^{1/3}$$

$$\frac{\mu_{\text{mic}}^{\text{lamellar}}}{k_B T} = 0.669(\chi N)^{1/3}(5.64 - f_A)^{1/3} \tag{118a}$$

Interfacial Tension in Binary Polymer Blends and the Effects of Copolymers 249

The expression for spherical micelles is the same as equation A-8 of Shull et al. [38] and is consistent with equation 21 of Lyatskaya et al. [85] and equation 36 of Semenov [103]. That for cylindrical micelles is the same as equation A-12 of Shull et al. [38] and as equation 36 of Semenov [103]. Finally, the equation for the chemical potential of lamellar micelles is the same as equation A-12 of Shull et al. [38] and somehow different from equation 36 of Semenov [103], as is also acknowledged by him [103].

When micelles are not present, the equilibrium is established between copolymer chains homogeneously distributed within the homopolymer phase and copolymer at the interface. The surface density σ is, then, determined by:

$$\mu_{\text{int}}(\sigma; N) = \mu_{\text{bulk}}(\phi_{\pm}; N) \tag{119a}$$

where, in this case, it is assumed that $\phi = \phi_{\pm} \approx \phi_{\text{add}}$. When micelles are present, then at thermodynamic equilibrium σ is determined by the equation:

$$\mu_{\text{int}}(\sigma; N) = \mu_{\text{mic}}(N) = \mu_{\text{bulk}}(\phi_{\pm}; N) \tag{119b}$$

which also determines the volume fraction $\phi_{\pm}$ of copolymers remaining homogeneously distributed in the bulk A or B phases.

For calculation of the interfacial tension reduction, one evaluates first the chemical potentials μ_{mic} and μ_{bulk} for $\phi = \phi_{\pm} = \phi_{\text{add}}$. If $\mu_{\text{bulk}}(\phi_{\text{add}}) < \mu_{\text{mic}}$, then the equilibrium is established between copolymers at the interface and copolymers homogeneously mixed within the B-rich phase. The interfacial excess σ is, then, determined by (119a) together with (115a) and (117a), and the interfacial tension reduction $\Delta\gamma$ by (113a). If $\mu_{\text{bulk}}(\phi_{\text{add}}) > \mu_{\text{mic}}$, equilibrium is established among the three different states of the copolymer and σ and $\phi_{\pm}$ are determined by (119b) together with (115a), (117a), and (118a); $\Delta\gamma$ is evaluated by (113a).

The semiquantitative model was compared with the data on the effects of the molecular weight of symmetric diblock copolymers on the polymer–polymer interfacial tension; the data showed a nonmonotonous dependence of the interfacial tension increment on the additive molecular weight in the plateau region. Although the assumptions involved in the model do not allow a quantitative comparison, the behavior of $\Delta\gamma$ when the copolymer molecular weight increases at constant additive concentration resembles the response seen experimentally. Figure 30 shows the estimated surface density of copolymers at the A–B interface, σ, together with the interfacial tension reduction, $\Delta\gamma = \gamma_0 - \gamma$, as a function of the number of segments of the copolymeric additive for $\phi_{\text{add}} = 0.02$. The parameters used were $P_A = P_{\text{PI}} = 81$, $P_B = P_{\text{PS}} = 112$, and $\chi = 0.04$. Moreover, for the present range of values of P_i and N_i, the wet-mushroom configuration for the adsorbed copolymer chains was assumed, which was then verified by the extracted σ values.

It was found that the magnitude of $\Delta\gamma$ increases with copolymer molecular weight, as long as the copolymer chains at the interface are at equilibrium with only homogeneously mixed chains and micelles do not exist (regime I). At higher molecular weights, when micelles are also present, $\Delta\gamma$ decreases with further

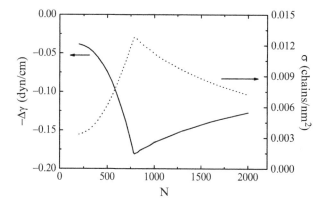

Fig. 30 Theoretically estimated interfacial tension reduction, $\Delta\gamma = \gamma_0 - \gamma$, (*solid line*), and estimated surface density (σ) of copolymer chains adsorbed at the interface (*dotted line*), for the PS/PS-*b*-PI/PI systems as a function of the number of segments (*N*) of the copolymer at constant 2 wt% copolymer concentration and for constant $\chi = 0.04$ [55]

increasing molecular weight (regime II). The values for $\Delta\gamma$ are in the range of the experimental values, although the functional form of the curve is different from the experimental one. For example, the copolymer molecular weight at the minimum is underestimated, indicating that micelles are calculated to form earlier than in the experimental system, whereas the minimum is much sharper than in the experiment; both are related to the functional form used for the free energy of the micelles (assumed lamellar) and the inherent assumptions made therein. The value of the interaction parameter used affects both the location of the minimum (with respect to *N*) and the values of $\Delta\gamma$; no fitting was attempted because the aim of the theoretical analysis was to obtain only the trends in order to understand the behavior of the experimental data. Indeed, the calculation indicated a behavior very similar to that seen experimentally. The origin of this trend is evidently related to the behavior of the estimated interfacial density of adsorbed chains, σ (shown in Fig. 30). Increasing the copolymer molecular weight when micelles are not present (for the low molecular weight side, regime I) rapidly drives more copolymer chains to the interface (σ increases), thus leading to an increase in $\Delta\gamma$. On the other hand, further increase inf the copolymer molecular weight when micelles are present (regime II) leads to a decrease in the surface density of copolymers, σ, thus reducing $\Delta\gamma$.

Graft Copolymer Additives

Lyatskaya et al. [85] extended the arguments of Leibler [75] for the case of comb and star copolymers. The homopolymers were considered to be highly incompatible, whereas the copolymer chains were assumed to form dry brushes at the interface and to be at equilibrium with chains homogeneously distributed in the bulk. For the case of simple graft copolymers, which were denoted as T-grafts and were considered as combs with $n = 1$ teeth, the interfacial tension reduction $\Delta\gamma$ for a AB_2 graft copolymer (one A block and two B blocks) was predicted as:

$$\frac{\Delta\gamma b^2}{k_B T} = \frac{(\gamma_0 - \gamma)b^2}{k_B T} = \left(\frac{2}{\pi}\right)\left(\frac{2}{3}\right)^{3/2} N^{-1/2}\mu_{bulk}^{3/2}(4 - 3f_A)^{-1/2} \quad (120)$$

where γ_0 is the A–B interfacial tension in the absence of the additive, χ is the Flory-Huggins interaction parameter, N is the number of segments of the graft copolymer, $f = f_{tooth} = f_A$ is the volume fraction of the tooth block A, b is the statistical segment length (it is assumed that both types of links have the same segmental volume $\upsilon = b^3$), and $k_B T$ is the thermal energy. μ_{bulk} is the chemical potential in the bulk and $\mu_{bulk} \approx \ln \phi_+ + \chi N f_A$ if the copolymer is added to the B-homopolymer phase, where ϕ_+ is the copolymer volume fraction in the bulk B-homopolymer phase (which is very close to the nominal amount of copolymer present, ϕ_{add}). Note that within the same assumptions, the respective interfacial tension reduction for a diblock copolymer is:

$$\frac{\Delta\gamma_{diblock}b^2}{k_B T} = \frac{(\gamma_0 - \gamma_{diblock})b^2}{k_B T} = \left(\frac{2}{\pi}\right)\left(\frac{2}{3}\right)^{3/2} N^{-1/2}\mu_{bulk}^{3/2} \quad (121)$$

Retsos et al. [56] made an attempt to extend these arguments for finite homopolymer molecular weights of simple graft copolymers by allowing for mixing of the graft copolymer and homopolymer chains (wet brush or mushroom regimes) and by explicitly including in the considerations the possibility of micelle formation, similarly to the earlier attempt for diblock copolymers [55].

In accordance with the case of diblock copolymer additives, the free energy of the interfacial film is calculated from (107), where now g_A and g_B represent the free energies per AB_2 chain of the A and B layers. For the case of AB_2 simple graft copolymers, the expressions (108a)–(111a) hold for layer A, which is formed by the single A block (with i = A). However, the analysis for the B layer should reflect the fact that the B layer is formed by two B blocks per AB_2 chain. Thus, g_B should be given by:

$$\frac{g_B}{k_B T} = \ln(N_B b^2 \sigma) + L_B \frac{1}{\sigma b^3} \frac{1}{P_B}(1 - \eta_B)\ln(1 - \eta_B) + 2\frac{3}{2}\frac{L_B^2}{(N_B/2)b^2} \quad (108b)$$

with $\eta_B = \sigma N_B b^3 / L_B$ being the average volume fraction of monomers of B chains. For low values of $\eta_B \ll 1$, (108b) can be written as [40, 75]:

$$\frac{g_B}{k_B T} = \ln(N_B b^2 \sigma) + \frac{1}{2}\frac{(N_B/2)\eta_B}{P_B} + 2\frac{3}{2}\frac{L_B^2}{(N_B/2)b^2} \quad (109b)$$

As before, for stretched chains L_B and η_B are obtained from (109b) by minimization with respect to L_B; thus, $L_B = (1/2)6^{-1/3}N_B b(\sigma b^2)^{1/3}P_B^{-1/3}$, $\eta_B = (2)$ $6^{1/3}(\sigma b^2)^{2/3}P_B^{1/3}$, and:

$$\frac{g_B}{k_B T} = \ln\left(N_B b^2 \sigma\right) + \frac{3^{4/3}}{2^{5/3}}\left(\sigma b^2\right)^{2/3} N_B P_B^{-2/3} \quad \text{(wet brush)} \tag{110b}$$

which is valid for $P_B(N_B/2)^{-3/2} < \sigma b^2 < P_B^{-1/2}$. For non-stretched chains, $L_B \approx (N_B/2)^{1/2}b$ and the last term of (109b) can be neglected; this applies for $\sigma b^2 < P_B(N_B/2)^{-3/2}$ [287]. Then, $\eta_B = \sigma b^2(2N_B)^{1/2}$ and:

$$\frac{g_B}{k_B T} = \ln\left(N_B b^2 \sigma\right) + \frac{1}{2^{3/2}}\frac{N_B^{3/2}\sigma b^2}{P_B} \quad \text{(wet mushroom)} \tag{111b}$$

The interfacial tension will then be calculated from (112), which in the AB_2 case becomes:

$$\frac{\Delta\gamma}{k_B T} = \frac{\gamma_0 - \gamma}{k_B T}$$

$$= \begin{cases} \sigma\left[2 + \frac{3^{1/3}}{2^{2/3}}\left(\sigma b^2\right)^{2/3}\left(N_A P_A^{-2/3} + N_B P_B^{-2/3}\right)\right] & \text{(wet brush)} \\ \sigma\left[2 + \frac{1}{2}\sigma b^2\left(N_A^{3/2}P_A^{-1} + N_B^{3/2}P_B^{-1}\right)\right] & \text{(wet mushroom)} \end{cases} \tag{113b}$$

The chemical potential of a copolymer chain at the interface is calculated using (114). Therefore, with (110b) and (111b):

$$\frac{\mu_{\text{int}}}{k_B T} = 2 + \ln\left(N_A \sigma b^2\right) + \ln\left(N_B \sigma b^2\right)$$

$$+ \begin{cases} 2.271(\sigma b^2)^{2/3}\left(N_A P_A^{-2/3} + N_B P_B^{-2/3}\right) & \text{(wet brush)} \\ \sigma b^2\left(N_A^{3/2}P_A^{-1} + N_B^{3/2}P_B^{-1}\right) & \text{(wet mushroom)} \end{cases} \tag{115b}$$

The free energy density of a homogeneous mixture of an AB copolymer (irrespectively of architecture) with a B homopolymer is given by (116) and the chemical potential of an AB_2 copolymer chain homogeneously distributed within the bulk B homopolymer by (117a). Note that, if the copolymer chain is homogeneously distributed within the bulk A homopolymer, its chemical potential is:

$$\frac{\mu_{\text{bulk}}}{k_B T} = \ln\phi - \phi - (1 - \phi)\frac{N}{P_A} + \chi N f_B\left(1 - 2f_B\phi + f_B\phi^2\right) \tag{117b}$$

where $\phi = \phi(-\infty)$ is the copolymer volume fraction in the A-rich homopolymer phase.

Interfacial Tension in Binary Polymer Blends and the Effects of Copolymers

For the calculation of the chemical potential of the AB_2 graft copolymer chains in a micelle, one has to distinguish two different cases: (1) when the micelle is formed within the B homopolymer, i.e., when the "tooth" A block forms the core of the micelle and the two B blocks form the corona; and (2) when the micelle is formed within the A homopolymer phase, i.e., when the two B blocks form the core and the A block forms the corona of the micelle. The two cases were considered by Retsos et al. [56] along the lines of Leibler [75] and Semenov [103] for the three different cases of formation of spherical, cylindrical, or lamellar micelles. The chemical potential of an AB_2 chain in a micelle formed within the B phase is, thus, [56]:

$$\frac{\mu_{mic}^{spherical}}{k_B T} = (3/2)^{4/3} f_A^{4/9} \left(4.74 f_A^{-1/3} - 4 \right)^{1/3} (\chi N)^{1/3}$$

$$\frac{\mu_{mic}^{cylindrical}}{k_B T} = 1.89 (\chi f_A N)^{1/3} (0.41 - \ln f_A)^{1/3}$$

$$\frac{\mu_{mic}^{lamellar}}{k_B T} = 1.75 (\chi N)^{1/3} (1.26 - f_A)^{1/3} \tag{118b}$$

whereas the chemical potential of an AB_2 chain in a micelle formed within the A phase is given by [56]:

$$\frac{\mu_{mic}^{spherical}}{k_B T} = (3/2)^{4/3} (1 - f_A)^{4/9} \left[3.96 (1 - f_A)^{-1/3} - 1 \right]^{1/3} (\chi N)^{1/3}$$

$$\frac{\mu_{mic}^{cylindrical}}{k_B T} = 1.19 (\chi f_A N)^{1/3} [6.57 - \ln(1 - f_A)]^{1/3}$$

$$\frac{\mu_{mic}^{lamellar}}{k_B T} = 1.57 (\chi N)^{1/3} (1.44 - f_A)^{1/3} \tag{118c}$$

When micelles are not present, the equilibrium between copolymer chains homogeneously mixed with the respective homopolymer and chains at the interface is established (119a) whereas, when micelles are present, (119b) determines the thermodynamic equilibrium. The equations result in the surface density of copolymer chains at the interface, σ. Again, it is assumed that $\phi_{\pm} \approx \phi_{add}$, with $\phi_{\pm}$ being the volume fraction of copolymers remaining homogeneously mixed in the bulk A or B phases.

Therefore, for the calculation of the interfacial tension reduction, one again evaluates the chemical potentials μ_{mic}, μ_{bulk} for $\phi = \phi_{\pm} = \phi_{add}$. If $\mu_{bulk}(\phi_{add}) < \mu_{mic}$, then the equilibrium is established between copolymers at the interface and copolymers homogeneously mixed within the B-rich (or A-rich) phase. The interfacial excess σ is, then, determined by (119a) together with (115b) and (117a) (or 117b, respectively), and the interfacial tension reduction $\Delta \gamma$ by (113b). If $\mu_{bulk}(\phi_{add}) > \mu_{mic}$, equilibrium is established among the three different states of the copolymer and σ and $\phi_{\pm}$ are determined by (119b) together with (115b), (117a) (or 117b), and (118b) (or (118c), respectively); $\Delta\gamma$ is evaluated by (113b). Due to the asymmetric architecture of the graft copolymers and in view of the respective experimental data [56], two situations were considered: (1) when the AB_2 diblock is added to the B homopolymer (e.g., an I_2S graft added to the PI phase [56]); and (2) when it is added to the A homopolymer (I_2S added to PS [56]).

Although the assumptions involved did not allow a quantitative comparison with the data [56], the behavior of the estimated $\Delta\gamma$ when graft copolymers of varying compositions were introduced into the PI or PS homopolymer phase (at constant additive concentration) resembled the experimental data for the molecular parameters of the experimental systems [56]. When the I_2S graft copolymers are added to the PI homopolymer, there are no micelles formed for high values of f_{PI}, and the copolymer chains at the interface are at equilibrium with chains homogeneously mixed within the PI phase. The surface density of chains increases with decreasing f_{PI} (from its high value) and the interfacial tension decreases. At lower values of f_{PI}, micelles are also present and σ does not increase (and even decreases) as f_{PI} decreases further; as a result the interfacial tension does not decrease further (and even increases). Similarly, when the I_2S copolymer is added to the PS homopolymer, there are no micelles formed for low values of f_{PI}, and the copolymer chains at the interface are at equilibrium with chains homogeneously mixed with PS. The surface density of chains increases with increasing f_{PI} and the interfacial tension decreases. At higher values of f_{PI}, micelles are also present and σ ceases to increase as f_{PI} increases further; as a result the interfacial tension does not decrease further. The $\Delta\gamma$ values were more or less in the range of the experimental values, although the apparent functional forms of the curves were different from the experimental ones [56]. For example, the dependencies in the region where micelles are present are apparently different to the experimental values. This is most probably due to the assumptions involved in the estimation of the chemical potentials for the copolymer chains in micelles (dry brush behavior was assumed). Even more, the value of the interaction parameter used affects both the location of the minimum (with respect to f_{PI}) and the values of $\Delta\gamma$. No fitting was attempted because the aim of the theoretical analysis was to obtain only the trends in order to understand the behavior of the experimental data. Indeed, it is evident that the calculation indicates a behavior very similar to that shown by the experimental data, with the origin of this trend evidently related to the behavior of the estimated surface density of adsorbed chains.

5 Concluding Remarks

Mixing two or more components that have complementary properties is largely utilized to improve the performance of polymeric materials for many important industrial applications. In spite of the great interest in homogeneous blends, a more desirable situation is that of a non-miscible system, i.e., a heterophase mixture wherein each of the constituents retains its own properties. In addition, the final

Interfacial Tension in Binary Polymer Blends and the Effects of Copolymers

product might also display some new features triggered by the particular phase morphology.

In such systems, a satisfactory overall physico-mechanical behavior will crucially depend on two demanding structural parameters: (1) a proper interfacial tension leading to a phase size that is small enough to allow the material to be considered as macroscopically "homogeneous"; and (2) an interfacial adhesion strong enough to assimilate stresses and strains without disruption of the established morphology. Both these structural parameters critically depend on the interfacial tension between the two macroscopic phases. Block or graft copolymers are widely used as emulsifying agents or compatibilizers in blends of immiscible polymers due to their affinity to selectively partition to the polymer–polymer interface, thus reducing the interfacial tension. In this article, an attempt has been made to present a review of the experimental and theoretical investigations of polymer–polymer interfacial tension in the absence and in the presence of block copolymer emulsifying agents.

The variety of experimental methods that have been utilized to efficiently measure the polymer–polymer interfacial tension have been briefly reviewed, with emphasis on the static methods (pendant drop, with the approach being very similar to the case of sessile drop) that have been widely used for polymeric liquids. The breaking thread method and the IFR method have been frequently used as well, especially for high molecular weight polymers.

Polymer–polymer interfacial tension measurements showed that interfacial tension decreases with increasing temperature (for polymer systems that exhibit USCT behavior[2]), with a temperature coefficient of the order of 10^{-2} dyn/(cm °C). Interfacial tension increases with increasing molecular weight and exhibits a $\gamma = \gamma_\infty \left(1 - k_{int} M_n^{-z}\right)$ dependence, with γ_∞ being the interfacial tension in the limit of infinite molecular weight. It is generally found that the exponent $z \to 1$ in the limit of high molecular weights.

We have reviewed the theories of polymer–polymer interfaces. We began by presenting the early semiempirical attempts. Then, we discussed in some detail the microscopic theories of polymer interfaces, with emphasis on the theories of Helfand and coworkers as well as on subsequent theories. One should emphasize here the significant influence of the original Helfang–Tagami theory on the field of polymer interfaces. The expression for the interfacial tension in the limit of infinite molecular weights, $\gamma = (\chi/6)^{1/2} \rho_0 b k_B T$ (40), has been utilized extensively for evaluation of the polymer–polymer interfacial tension; the same holds for the expression for the width of the interface (38), again in the limit of infinite molecular weights. The rest of the theoretical section on polymer–polymer interfaces focused on the square-gradient approach and its utilization to predict the temperature and

---

[2]For polymer blends exhibiting lower critical solution temperature (LCST) behavior, e.g., the system polystyrene/poly(vinyl methyl ether), one may anticipate the opposite behavior for purely phenomenological reasons. Interfacial tension should increase with increasing temperature in the two-phase region since the tie lines become longer with increasing temperature in that case

molecular weight effects on polymer interfacial tension. All the theories that address the effect of molecular weight predict that interfacial tension exhibits a $\gamma = \gamma_\infty \left(1 - k_{int}M_n^{-1}\right)$ dependence for high molecular weights, i.e., it increases with increasing molecular weight, with the exponent $z = 1$, in agreement with the current view from experiment. We then presented briefly the theoretical works that addressed the behavior of interfacial tension near the critical point of demixing, where interfacial tension is predicted as $\gamma \propto r^{1/2}\varepsilon^{3/2}$, with r being the chain length and ε the reduced distance from the critical temperature.

The emulsifying effect of diblock copolymers additives on the interfacial tension between two immiscible homopolymers was then reviewed. Early studies as well as studies on model systems demonstrated the surfactant-like behavior of the block copolymers added to the polymer–polymer systems: a sharp decrease with addition of a small amount of copolymer followed by a leveling off at higher copolymer concentrations. The dependence of the interfacial tension reduction on the copolymer molecular weight for symmetric diblocks apparently exhibits two different regimes: (I) for low molecular weights, the interfacial tension increment, $\Delta\gamma = \gamma_0 - \gamma$, at saturation (in the plateau region) increases by increasing the additive molecular weight, and (II) it decreases by further increasing the copolymer molecular weight, thus going through a maximum. This was understood by considering the possibility of micelle formation for high molecular weights, leading to a three-state equilibrium between copolymer chains adsorbed at the interface, chains homogeneously mixed in the bulk phases, and copolymers at micelles within the bulk phases. The effects of copolymer architecture and composition were also investigated utilizing triblock, graft, and comb copolymers. For a systematic series of I_2S simple graft copolymers, with constant molecular weight and varying composition, the interfacial tension at interfacial saturation was found to be a nonmonotonic function of the copolymer composition f_{PI}. This was understood to be due to the competition between the decreased affinity of the copolymer within the homopolymer phase when the size of the "incompatible" block increases, which increases the driving force of the copolymer towards the interface, and the possibility of micelle formation. Moreover, in certain systems possessing architectural, molecular, or interactional asymmetry, the interfacial tension at saturation was found to depend on the side of the interface the copolymer is added; such was the behavior of the I_2S graft copolymers (asymmetric architecture), the PS-b-PDMS copolymer (asymmetry in segment lengths), and the PEO-b-PPO-b-PEO (asymmetric affinity). These examples point to a local equilibrium that can only be achieved in such systems: the copolymer reaching the interface from one homopolymer phase most probably does not diffuse to the other phase. It was emphasized that this behavior was not a kinetic effect but it was rather due to the attainment of a stationary state of local equilibrium and a lack of global equilibrium in these interfacial systems.

It is, therefore, evident that the effectiveness of the interfacial modifiers is controlled by the unfavorable interactions, which drive the additive towards the interface, and by the formation of micelles, which reduces the emulsifying activity.

For nonsymmetric copolymer systems, the latter is affected by the side of the interface to which the additive is introduced and, thus, in most practical situations this should be taken seriously into consideration. Moreover, we briefly discussed the recent observations of ultralow interfacial tension in polymer systems obtained utilizing balanced copolymeric surfactants.

Finally, we have reviewed the theoretical approaches developed to address the compatibizing effect of block copolymers at polymer–polymer interfaces. The elegant theory of Noolandi and Hong was first presented in detail due to its significance in the area of copolymers at homopolymer interfaces (originally developed in the presence of solvent). Subsequently, we discussed the theory of Leibler for nearly compatible systems. Finally, we discussed in detail the theory of Leibler for highly incompatible systems with copolymer layers forming dry brushes at the interface, and presented its extensions that allow mixing of the homopolymer chains with the copolymer layer as well as take into account the possibility of micelle formation within the homopolymer phases.

The later theoretical model showed the same qualitative behavior as the experimental data for both the nonmonotonic dependence of interfacial tension increment on copolymer molecular weight as well as for the influence of the composition of graft copolymers. In both cases, however, the model fails to quantitatively account for the effect because of the assumptions involved in the estimation of the free energies of the various chain conformations and especially the free energies of those conformations within the micelles of various morphologies. The importance of the presence of micelles, either for high diblock copolymer molecular weights or for high composition of the incompatible block, is greatly emphasized.

Finally, we would like to emphasize that, in most applications, in-situ formed copolymers are utilized, which are formed by the reaction of appropriately functionalized homopolymer additives at the polymer–polymer interface. A review article [106] cites not a single case where a premade copolymer had been used in a real application. Therefore, the interfacial behavior in such systems should be investigated fundamentally in greater detail in order to probe the effects of the characteristics of the reactive species on the kinetics of interfacial partitioning and the subsequent reaction, as well as on the effect of the resultant (diblock or graft or comb) copolymer on the interfacial tension and, thus, on the morphology of the macrophase-separated polymer blend.

Acknowledgments The author wishes to acknowledge H. Retsos for his help in the preparation of the present article. He would also like to thank I. Gancarz, N. Hadjichristidis, H. Watanabe, K. Adachi, J. W. Mays, M. Pitsikalis, S. Pispas, H. Iatrou, and K. Hong for synthesizing and kindly supplying the diblock copolymers and some of the homopolymers used in the previous works by the author on polymer–polymer interfacial tension. J. T. Koberstein is acknowledged for introducing the author to the area of polymer interfaces, as are T. P. Russell, S. K. Satija, C. F. Majkrzak, and G. Felcher with whom the author studied the structure of polymer interfaces. This research was sponsored by NATO's Scientific Affairs Division in the framework of the Science for Peace Programmes (projects SfP-974173 and SfP-981438), by the Greek General Secretariat of Research and Technology in the framework of the ΠΕΝΕΔ Programme (projects 01ΕΔ587 and 03ΕΔ581), and by the European Union (projects G5RD-CT-2002-00834, NMP3-CT-2005-506621).

References

1. Paul DR, Newman S (eds) (1978) Polymer blends. Academic, New York
2. Paul DR, Bucknall CB (eds) (2000) Polymer blends set: formulation and performance. Wiley, New York
3. Hancock T (1823) English Patent 6:768
4. Hancock T (1823) English Patent 11:147
5. White JL, Min K (1985) Processing and phase morphology of incompatible polymer blends. In: Walsh DJ, Higgins JS, Maconnachie A (eds) Polymer blends and mixtures. Mortinus Nijhoff, Dordrecht, The Netherlands, pp 413–428
6. Utracki LA (ed) (1986) Polyblends-'86, NRCC/IMRI polymers symposium series. NRCC/IMRI, Montreal
7. Wu S (1986) Formation of dispersed phase in incompatible polymer blends: interfacial and rheological effects. In: Utracki LA (ed) Polyblends-'86, NRCC/IMRI polymers symposium series. NRCC/IMRI, Montreal, pp 4/1–42
8. Fortený I, Živný A, Juza J (1999) Coarsening of the phase structure in immiscible polymer blends. Coalescence or Ostwald ripening? J Polym Sci B Polym Phys 37:181–187
9. Jaycock MJ, Parfitt GD (1985) Chemistry of interfaces. Ellis Horwood, Chichester, UK
10. Wu S (1982) Polymer interface and adhesion. Marcel Dekker, New York
11. Miller CA, Neogi P (1985) Interfacial phenomena: equilibrium and dynamic effects. Marcel Dekker, New York
12. Feast WJ, Munro HS (eds) (1987) Polymer surfaces and interfaces. Wiley, New York
13. Feast WJ, Munro HS, Richards RW (eds) (1993) Polymer surfaces and interfaces II. Wiley, New York
14. Richards RW, Peace SK (eds) (1999) Polymer surfaces and interfaces III. Wiley, New York
15. Fleer GJ, Cohen Stuart MA, Scheutjens JMHM, Cosgrove T, Vincent B (1993) Polymers at interfaces. Chapman and Hall, London
16. Jones RAL, Richards RW (1999) Polymers at surfaces and interfaces. Cambridge University Press, Cambridge, UK
17. Karim A, Kumar SK (eds) (2000) Polymer surfaces, interfaces and thin films. World Scientific, Singapore
18. Stamm M (ed) (2008) Polymer surfaces and interfaces: characterization, modification and applications. Springer, Berlin
19. Anastasiadis SH, Chen JK, Koberstein JT, Sohn JE, Emerson JA (1986) The determination of polymer interfacial tension by drop image processing: comparison of theory and experiment for the pair PDMS/PBD. Polym Eng Sci 26:1410–1418
20. Anastasiadis SH, Gancarz I, Koberstein JT (1988) Interfacial tension of immiscible polymer blends: temperature and molecular weight dependence. Macromolecules 21:2980–2987
21. Kamal MR, Lai-Fook R, Demarquette NR (1994) Interfacial tension in polymer melts. Part II: Effects of temperature and molecular weight on interfacial tension. Polym Eng Sci 34:1834–1839
22. Kamal MR, Demarquette NR, Lai-Fook R, Price TA (1997) Evaluation of thermodynamic theories to predict interfacial tension between polystyrene and polypropylene melts. Polym Eng Sci 37:813–825
23. Arashiro EY, Demarquette NR (1999) Influence of temperature, molecular weight, and polydispersity of polystyrene on interfacial tension between low-density polyethylene and polystyrene. J Appl Polym Sci 74:2423–2431
24. Biresaw G, Carriere CJ, Sammler RL (2003) Effect of temperature and molecular weight on the interfacial tension of PS/PDMS blends. Rheol Acta 42:142–147
25. Ellingson PC, Strand DA, Cohen A, Sammler RL, Carriere CJ (1994) Molecular weight dependence of polystyrene/poly(methyl methacrylate) interfacial tension probed by imbedded-fiber retraction. Macromolecules 27:1643–1647

26. Nam KH, Jo WH (1995) The effect of molecular weight and polydispersity of polystyrene on the interfacial tension between polystyrene and polybutadiene. Polymer 36:3727–3731
27. Helfand E, Tagami Y (1972) Theory of the interface between immiscible polymers II. J Chem Phys 56:3592–3601
28. Helfand E, Tagami Y (1972) Theory of the interface between immiscible polymers. J Chem Phys 57:1812–1813
29. Helfand E, Sapse AM (1975) Theory of unsymmetric polymer–polymer interfaces. J Chem Phys 62:1327–1331
30. Helfand E, Bhattacharjee SM, Fredrickson GH (1989) Molecular weight dependence of polymer interfacial tension and concentration profile. J Chem Phys 91:7200–7208
31. Broseta D, Fredrickson GH, Helfand E, Leibler L (1990) Molecular weight and polydispersity effects at polymer–polymer interfaces. Macromolecules 23:132–139
32. Tang H, Freed KF (1991) Interfacial studies of incompressible binary blends. J Chem Phys 94:6307–6322
33. Ermonskin AV, Semenov AN (1996) Interfacial tension in binary polymer mixtures. Macromolecules 29:6294–6300
34. Lee HS, Jo WH (1998) Prediction of interfacial tension of immiscible polymer pairs using a square gradient theory combined with the FOV equation-of-state free energy expression. Polymer 39:2489–2493
35. Jo WH, Lee HS, Lee SC (1998) Temperature and molecular weight dependence of interfacial tension between immiscible polymer pairs by the square gradient theory combined with the Flory-Orwoll-Vrij equation-of-state theory. J Polym Sci B Polym Phys 36:2683–2689
36. Koning C, Van Duin M, Pagnoulle C, Jérôme R (1998) Strategies for compatibilization of polymer blends. Prog Polym Sci 23:707–757
37. Fayt R, Jérôme R, Teyssié Ph (1986) Molecular design of multicomponent polymer systems. XII. Direct observation of the location of a block copolymer in low-density polyethylene-polystyrene blends. J Polym Sci Polym Lett 24:25–28
38. Shull KR, Kramer EJ, Hadziioannou G, Tang W (1990) Segregation of block copolymers to interfaces between immiscible homopolymers. Macromolecules 23:4780–4787
39. Russell TP, Anastasiadis SH, Menelle A, Felcher G, Satija SK (1991) Segment density distribution of symmetric diblock copolymers at the interface between two homopolymers as revealed by neutron reflectivity. Macromolecules 24:1575–1582
40. Green PF, Russell TP (1991) Segregation of low molecular weight symmetric diblock copolymers at the interface of high molecular weight homopolymers. Macromolecules 24:2931–2935
41. Dai KH, Kramer EJ (1994) Molecular weight dependence of diblock copolymer segregation at a polymer/polymer interface. J Polym Sci B Polym Phys 32:1943–1950
42. Dai KH, Norton LJ, Kramer EJ (1994) Equilibrium segment density distribution of a diblock copolymer segregated to the polymer/polymer interface. Macromolecules 27:1949–1956
43. Reynolds BJ, Ruegg ML, Mates TE, Radke CJ, Balsara NP (2005) Experimental and theoretical study of the adsorption of a diblock copolymer to interfaces between two homopolymers. Macromolecules 38:3872–3882
44. Eastwood E, Viswanathan S, O'Brien CP, Kumar D, Dadmun MD (2005) Methods to improve the properties of polymer mixtures: optimizing intermolecular interactions and compatibilization. Polymer 46:3957–3970
45. Anastasiadis SH, Gancarz I, Koberstein JT (1989) Compatibilizing effect of block copolymers added to the polymer/polymer interface. Macromolecules 22:1449–1453
46. Elemans PHM, Janssen JMH, Meijer HEH (1990) The measurement of interfacial tension in polymer/polymer systems: the breaking thread method. J Rheol 34:1311–1325
47. Wagner M, Wolf BA (1993) Effect of block copolymers on the interfacial tension between two 'immiscible' homopolymers. Polymer 34:1460–1464

48. Hu W, Koberstein JT, Lingelser JP, Gallot Y (1995) Interfacial tension reduction in polystyrene/ poly(dimethyl siloxane) blends by the addition of poly(styrene-*b*-dimethylsiloxane). Macromolecules 28:5209–5214

49. Jorzik U, Wagner M, Wolf BA (1996) Effect of block copolymer architecture on the interfacial tension between immiscible polymers. Prog Colloid Polym Sci 101:170–171

50. Jorzik U, Wolf BA (1997) Reduction of the interfacial tension between poly(dimethylsiloxane) and poly(ethylene oxide) by block copolymers: effects of molecular architecture and chemical composition. Macromolecules 30:4713–4718

51. Mekhilef N, Favis BD, Carreau PJ (1997) Morphological stability, interfacial tension, and dual-phase continuity in polystyrene–polyethylene blends. J Polym Sci B Polym Phys 35:293–308

52. Liang H, Favis BD, Yu YS, Eisenberg A (1999) Correlation between the interfacial tension and dispersed phase morphology in interfacially modified blends of LLDPE and PVC. Macromolecules 32:1637–1642

53. Cho D, Hu W, Koberstein JT, Lingelser JP, Gallot Y (2000) Segregation dynamics of block copolymers to immiscible polymer blend interfaces. Macromolecules 33:5245–5251

54. Welge I, Wolf BA (2001) Reduction of the interfacial tension between 'immiscible' polymers: to which phase one should add a compatibilizer. Polymer 42:3467–3473

55. Retsos H, Margiolaki I, Messaritaki A, Anastasiadis SH (2001) Interfacial tension in binary polymer blends in the presence of block copolymers: effects of additive MW. Macromolecules 34:5295–5305

56. Retsos H, Anastasiadis SH, Pispas S, Mays JW, Hadjichristidis N (2004) Interfacial tension in binary polymer blends in the presence of block copolymers: II. Effects of additive architecture and composition. Macromolecules 37:524–537

57. Shi T, Ziegler VE, Welge IC, An L, Wolf BA (2004) Evolution of the interfacial tension between polydisperse 'immiscible' polymers in the absence and in the presence of a compatibilizer. Macromolecules 37:1591–1599

58. Chang K, Macosko CW, Morse DC (2007) Ultralow interfacial tensions of polymer/polymer interfaces with diblock copolymer surfactants. Macromolecules 40:3819–3830

59. Seo Y, Kang T, Choi HJ, Cho J (2007) Electrocapillary wave diffraction measurement of interfacial tension reduction between two oligomers (poly(dimethyl siloxane) and poly (ethylene glycol)) by a block copolymer and the mean-field theory prediction. J Phys Chem C 111:5474–5480

60. van Puyvelde P, Velankar S, Moldenaers P (2001) Rheology and morphology of compatibilized polymer blends. Curr Opin Colloid Interface Sci 6:457–463

61. Sundararaj U, Macosco CW (1995) Drop breakup and coalescence in polymer blends: the effects of concentration and compatibilization. Macromolecules 28:2647–2657

62. Heikens D, Barentsen WM (1977) Particle dimensions in polystyrene/polyethylene blends as a function of their melt viscosity and of the concentration of added graft copolymer. Polymer 18:69–72

63. Fayt R, Jérôme R, Teyssié Ph (1981) Molecular design of multicomponent polymer systems. I. Emulsifying effect of poly(hydrogenated butadiene-*b*-styrene) copolymers in LDPE/PS blends. J Polym Sci Polym Lett 19:79–84

64. Thomas S, Prud'homme RE (1992) Compatibilizing effect of block copolymers in heterogeneous polystyrene/poly(methyl methacrylate) blends. Polymer 33:4260–4268

65. Tang T, Huang B (1994) Interfacial behaviour of compatibilizers in polymer blends. Polymer 35:281–285

66. Macosco CW, Guegan P, Khandpur A, Nakayama A, Marechal P, Inoue T (1996) Compatibilizers for melt blending: premade block copolymers. Macromolecules 29:5590–5598

67. Brown H, Char K, Deline VR, Green PF (1993) Effects of a diblock copolymer on adhesion between immiscible polymers. 1. Polystyrene (PS)-PMMA copolymer between PS and PMMA. Macromolecules 26:4155–4163

68. Dai C-A, Jandt KD, Iyengar DR, Slack NL, Dai KH, Davidson WB, Kramer EJ, Hui C-Y (1997) Strengthening polymer interfaces with triblock copolymers. Macromolecules 30:549–560

69. Russell TP, Menelle A, Hamilton WA, Smith GS, Satija SK, Majkrzak CF (1991) Width of homopolymer interfaces in the presence of symmetric diblock copolymers. Macromolecules 24:5721–5726

70. Noolandi J, Hong KM (1982) Interfacial properties of immiscible homopolymer blends in the presence of block copolymers. Macromolecules 15:482–492

71. Noolandi J, Hong KM (1984) Effect of block copolymers at a demixed homopolymer interface. Macromolecules 17:1531–1537

72. Duke TAJ (1989) Ph.D. dissertation, University of Cambridge, Cambridge, UK

73. Fischel LB, Theodorou DN (1995) Self-consistent field model of the polymer/diblock copolymer/polymer interface. J Chem Soc Faraday Trans 91:2381–2402

74. Werner A, Schmid F, Binder K, Muller M (1996) Diblock copolymers at a homopolymer–homopolymer interface: a Monte Carlo simulation. Macromolecules 29:8241–8248

75. Leibler L (1988) Emulsifying effects of block copolymers in incompatible polymer blends. Makromol Chem Macromol Symp 16:1–17

76. Leibler L (1991) Block copolymers at interfaces. Physica A 172:258–268

77. Shull KR, Kramer EJ (1990) Mean-field theory of polymer interfaces in the presence of block copolymers. Macromolecules 23:4769–4779

78. Israels R, Jasnow D, Balazs AC, Guo L, Krausch G, Sokolov J, Rafailovich M (1995) Compatibilizing A/B blends with AB diblock copolymers: effect of copolymer molecular weight. J Chem Phys 102:8149–8157

79. Kim SH, Jo WHA (1999) Monte Carlo simulation of polymer/polymer interface in the presence of block copolymer. I. Effects of the chain length of block copolymer and interaction energy. J Chem Phys 110:12193–12201

80. Gersappe D, Harm PK, Irvine D, Balazs AC (1994) Contrasting the compatibilizing activity of comb and linear copolymers. Macromolecules 27:720–724

81. Gersappe D, Balazs AC (1995) Random copolymers as effective compatibilizing agents. Phys Rev E 52:5061–5064

82. Israels R, Foster DP, Balazs AC (1995) Designing optimal comb compatibilizers: AC and BC combs at an A/B interface. Macromolecules 28:218–224

83. Lyatskaya J, Jacobson SH, Balazs AC (1996) Effect of composition on the compatibilizing activity of comb copolymers. Macromolecules 29:1059–1061

84. Lyatskaya J, Balazs AC (1996) Using copolymer mixtures to compatibilize immiscible homopolymer blends. Macromolecules 29:7581–7587

85. Lyatskaya J, Balazs AC, Gersappe D (1995) Effect of copolymer architecture on the efficiency of compatibilizers. Macromolecules 28:6278–6283

86. Lyatskaya J, Gersappe D, Gross NA, Balazs AC (1996) Designing compatibilizers to reduce interfacial tension in polymer blends. J Phys Chem 100:1449–1458

87. Russell TP, Mayes AM, Deline VR, Chung TC (1992) Hairpin configurations of triblock copolymers at homopolymer interfaces. Macromolecules 25:5783–5789

88. Brown HR, Krappe U, Stadler R (1996) Effect of ABC triblock copolymers with an elastomeric midblock on the adhesion between immiscible polymers. Macromolecules 29:6582–6588

89. Guo HF, Packirisamy S, Mani RS, Aronson CL, Cvozdic NV, Meier DJ (1998) Compatibilizing effects of block copolymers in low-density polyethylene/polystyrene blends. Polymer 39:2495–2505

90. Cigana P, Favis BD (1998) The relative efficacy of diblock and triblock copolymers for a polystyrene/ethylene–propylene rubber interface. Polymer 39:3373–3378

91. Dai C-H, Dair BJ, Dai KH, Ober CK, Kramer EJ (1994) Reinforcement of polymer interfaces with random copolymers. Phys Rev Lett 73:2472–2475

92. Kulasekere R, Kaiser H, Ankner JF, Russell TP, Brown HR, Hawker CJ, Mayes AM (1996) Homopolymer interfaces reinforced with random copolymers. Macromolecules 29:5493–5496
93. Smith GD, Russell TP, Kulasekere R, Ankner JF, Kaiser H (1996) A Monte Carlo simulation of asymmetric random copolymers at an immiscible interface. Macromolecules 29:4120–4124
94. Benkoski JJ, Fredrickson GH, Kramer EJ (2001) Effects of composition drift on the effectiveness of random copolymer reinforcement at polymer–polymer interfaces. J Polym Sci Polym Phys 39:2363–2377
95. Dadmun M (1996) Effect of copolymer architecture on the interfacial structure and miscibility of a ternary polymer blend containing a copolymer and two homopolymers. Macromolecules 29:3868–3874
96. Eastwood EA, Dadmun MD (2002) Multiblock copolymers in the compatibilization of polystyrene and poly(methyl methacrylate) blends: role of polymer architecture. Macromolecules 35:5069–5077
97. Tsitsilianis C, Voulgas D, Kosmas D (1998) Heteroarm star copolymers as emulsifying agents in polymer blends. Polymer 39:3571–3575
98. Dadmun M (2000) Importance of a broad composition distribution in polymeric interfacial modifiers. Macromolecules 33:9122–9125
99. Vilgis TA, Noolandi J (1988) On the compatibilization of polymer blends. Makromol Chem Macromol Symp 16:225–234
100. Vilgis TA, Noolandi J (1990) Theory of homopolymer–block copolymer blends. The search for a universal compatibilizer. Macromolecules 23:2941–2497
101. Whitmore MD, Noolandi J (1985) Theory of micelle formation in block copolymer–homopolymer blends. Macromolecules 18:657–665
102. Shull KR, Winey KI, Thomas EL, Kramer EJ (1991) Segregation of block copolymer micelles to surfaces and interfaces. Macromolecules 24:2748–2751
103. Semenov AN (1992) Theory of diblock–copolymer segregation to the interface and free surface of a homopolymer layer. Macromolecules 25:4967–4977
104. Adedeji A, Lyu S, Macosco CW (2001) Block copolymers in homopolymer blends: interface vs micelles. Macromolecules 34:8663–8668
105. Chang K, Morse DC (2006) Diblock copolymer surfactants in immiscible homopolymer blends: swollen micelles and interfacial tension. Macromolecules 39:7746–7756
106. Konig C, van Duin M, Pagnoulle C, Jérôme R (1998) Strategies for compatibilization of polymer blends. Prog Polym Sci 23:707–757
107. Nakayama A, Inoue T, Guegan P, Macosko, CW (1993) Compatibilizers for melt blending: premade vs. reactively formed block copolymers. Polym Prepr 34:840–841
108. Jeon HK, Zhang J, Macosco CW (2005) Premade vs. reactively formed compatibilizers for PMMA/PS melt blends. Polymer 46:12422–12429
109. Orr CA, Adedeji A, Hirao A, Bates FS, Macosco CW (1997) Flow-induced reactive self-assembly. Macromolecules 30:1243–1246
110. Schulze JS, Cernohous JJ, Hirao A, Lodge TP, Macosco CW (2000) Reaction kinetics of end-functionalized chains at a polystyrene/poly(methyl methacrylate) interface. Macromolecules 33:1191–1198
111. Schulze JS, Moon B, Lodge TP, Macosco CW (2001) Measuring copolymer formation from end-functionalized chains at a PS/PMMA interface using FRES and SEC. Macromolecules 34:200–205
112. Lee Y, Char K (2001) Enhancement of interfacial adhesion between amorphous polyamide and polystyrene by in-situ copolymer formation at the interface. Macromolecules 27:2603–2606
113. Pagnoulle C, Konig C, Leemans L, Jérôme R (2000) Reactive compatibilization of SAN/EPR blends. 1. Dependence of the phase morphology development on the reaction kinetics. Macromolecules 33:6275–6283
114. Pagnoulle C, Jérôme R (2001) Reactive compatibilization of SAN/EPR blends. 2. Effect of type and content of reactive groups randomly attached to SAN. Macromolecules 34:965–975

115. Yin Z, Koolic C, Pagnoulle C, Jérôme R (2001) Reactive blending of functional PS and PMMA: interfacial behavior of in situ formed graft copolymers. Macromolecules 34:5132–5139
116. Kim HY, Ryu DY, Jeong U, Kim DH, Kim JK (2005) The effect of chain architecture of in-situ formed copolymers on interfacial morphology of reactive polymer blends. Macromol Rapid Commun 26:1428–1433
117. Chi C, Hu YT, Lips A (2007) Kinetics of interfacial reaction between two polymers studied by interfacial tension measurements. Macromolecules 40:6665–6668
118. Padday JF (1969) In: Matijevic E (ed) Surface and colloid science, vol 1. Wiley, New York, p 111
119. Frisch HL, Gaines GL Jr, Schonhorn H (1976) In: Hannay NB (ed) Treatise on solid state chemistry, vol 6B. Plenum, New York, p 343
120. Wu S (1974) Interfacial and surface tensions of polymers. J Macromol Sci Rev Macromol Chem C10:1–73
121. Koberstein JT (1987) In: Mark HF, Bikales NM, Overberger CG, Menges G (eds) Encyclopedia of polymer science and engineering, vol 8, 2nd edn. Wiley, New York, p 237
122. Anastasiadis SH (1988) Interfacial tension of immiscible polymer blends. Ph.D. dissertation, Princeton University
123. Xing P, Bousmina M, Rodrigue D, Kamal MR (2000) Critical experimental comparison between five techniques for the determination of interfacial tension in polymer blends: model system of polystyrene/polyamide-6. Macromolecules 33:8020–8034
124. Demarquette NR (2003) Evaluation of experimental techniques for determining interfacial tension between molten polymers. Int Mater Rev 48:247–269
125. Roe R-J, Bacchetta VL, Wong PMG (1967) Refinement of pendent drop method for the measurement of surface tension of viscous liquid. J Phys Chem 71:4190–4193
126. Roe R-J (1968) Surface tension of polymer liquids. J Phys Chem 72:2013–2017
127. Wu S (1969) Surface and interfacial tensions of polymer melts: I. Polyethylene, polyisobutylene, and polyvinyl acetate. J Colloid Interface Sci 31:153–161
128. Sakai T (1965) Surface tension of polyethylene melt. Polymer 6:659–661
129. Hata T (1968) Hyomen (Surface, Japan) 6:281
130. Vonnegut B (1942) Rotating bubble method for the determination of surface and interfacial tensions. Rev Sci Instrum 13:6–9
131. Princen HM, Zia IYZ, Mason SG (1967) Measurement of interfacial tension from the shape of a rotating drop. J Colloid Interface Sci 23:99–107
132. Schonhorn H, Ryan FW, Sharpe LH (1966) Surface tension of a molten polychlorotrifluoroethylene. J Polym Sci A-2 4:538–542
133. Edwards H (1968) Surface tensions of liquid polyisobutylenes. J Appl Polym Sci 12:2213
134. Wilhelmy L Ueber die Abhängigkeit der Capillaritäts-Constanten des Alkohols von Substanz und Gestalt des benetzten festen Körpers. Ann Phys 195:177–217
135. Dettre RH, Johnson RE Jr (1966) Surface properties of polymers: I. The surface tensions of some molten polyethylenes. J Colloid Interface Sci 21:367–377
136. Du Noüy PL (1919) A new apparatus for measuring surface tension. J Gen Physiol 1:521–524
137. Newman SB, Lee WL (1958) Surface tension measurements with a strain-gauge-type testing machine. Rev Sci Instrum 29:785–787
138. Schonhorn H, Sharpe LH (1965) Surface energetics, adhesion, and adhesive joints. III. Surface tension of molten polyethylene. J Polym Sci A-2 3:569–573
139. Löfgren H, Neuman RD, Scriven LE, Davis HT (1984) Laser light-scattering measurements of interfacial tension using optical heterodyne mixing spectroscopy. J Colloid Interface Sci 98:175–183
140. Jon DI, Rosano HL, Cummins HZ (1986) Toluene/water/1-propanol interfacial tension measurements by means of pendant drop, spinning drop, and laser light-scattering methods. J Colloid Interface Sci 114:330–341

141. Sauer BB, Yu H, Tien CF, Hager DF (1987) A surface light scattering study of a poly (ethylene oxide)–polystyrene block copolymer at the air–water and heptane–water interfaces. Macromolecules 20:393–400
142. Sauer BB, Skarlupka RJ, Sano M, Yu H (1987) Dynamic interfacial-tensions of polymer-solutions by an electrocapillary wave technique. Polym Prepr 28, 20–21
143. Ito K, Sauer BB, Skarlupka RJ, Sano M, Yu H (1990) Dynamic interfacial properties of poly (ethylene oxide) and polystyrene at toluene/water interface. Langmuir 6:1379–1384
144. Tomotika S (1935) On the instability of a cylindrical thread of a viscous liquid surrounded by another viscous fluid. Proc R Soc Lond A *150*:322–337
145. Chappelar DC (1964) Interfacial tension between molten polymers. Polym Prepr 5:363–364
146. Tjahjadi M, Ottino JM, Stone HA (1994) Estimating interfacial tension via relaxation of drop shapes and filament breakup. AIChE J 40:385–394
147. Machiels AGC, Van Dam J, Posthuma de Boer A, Norder B (1997) Stability of blends of thermotropic liquid crystalline polymers with thermoplastic polymers. Polym Eng Sci 37:1512–1513
148. Palmer G, Demarquette NR (2003) New procedure to increase the accuracy of interfacial tension measurements obtained by breaking thread method. Polymer 44:3045–3052
149. Carriere CJ, Cohen A, Arends CB (1989) Estimation of interfacial tension using shape evolution of short fibers. J Rheol 33:681–689
150. Cohen A, Carriere CJ (1989) Analysis of a retraction mechanism for imbedded polymeric fibers. Rheol Acta 28:223–232
151. Carriere CJ, Cohen A (1991) Evaluation of the interfacial tension between high molecular weight polycarbonate and PMMA resins with the imbedded fiber retraction technique. J Rheol 35:205–212
152. Wu S (1970) Surface and interfacial tensions of polymer melts. II. Poly(methyl methacrylate), poly(n-butyl methacrylate), and polystyrene. J Phys Chem 74:632–638
153. Wu S (1971) Calculation of interfacial tension in polymer systems. J Polym Sci C 34:19–30
154. Roe R-J (1969) Interfacial tension between polymer liquids. J Colloid Interface Sci 31:228–235
155. Anastasiadis SH, Chen JK, Koberstein JT, Siegel AF, Sohn JE, Emerson JA (1987) The determination of interfacial tension by video image processing of pendant fluid drops. J Colloid Interface Sci 119:55–66
156. Bashforth S, Adams JC (1882) An attempt to test the theory of capillary action. Cambridge University Press and Deighton, Bell, London
157. Andreas JM, Hauser EA, Tucker WB (1938) J Phys Chem 42:1001–1019
158. Fordham S (1948) On the calculation of surface tension from measurements of pendant drops. Proc R Soc Lond A 194:1–16
159. Stauffer CE (1965) The measurement of surface tension by the pendant drop technique. J Phys Chem 69:1933–1938
160. Huh C, Reed RL (1983) A method for estimating interfacial tensions and contact angles from sessile and pendant drop shapes. J Colloid Interface Sci 91:472–484
161. Boyce JF, Schürch S, Rotenberg Y, Newmann AW (1984) The measurement of surface and interfacial tension by the axisymmetric drop technique. Colloids Surf 9:307–317
162. Girault HH, Schiffrin DJ, Smith BDV (1982) Drop image processing for surface and interfacial tension measurements. J Electroanal Chem 137:207–217
163. Girault HH, Schiffrin DJ, Smith BDV (1984) The measurement of interfacial tension of pendant drops using a video image profile digitizer. J Colloid Interface Sci 101:257–266
164. Demarquette NR, Kamal MR (1994) Interfacial tension in polymer melts. I: An improved pendant drop apparatus. Polym Eng Sci 34:1823–1833
165. Rotenberg Y, Boruvka L, Newmann AW (1983) Determination of surface tension and contact angle from the shapes of axisymmetric fluid interfaces. J Colloid Interface Sci 93:169–183
166. Siegel AF (1982) Robust regression using repeated medians. Biometrika 69:242–244

167. Olshan AF, Siegel AF, Swindler DR (1982) Robust and least-squares orthogonal mapping: methods for the study of cephalofacial form and growth. Am J Phys Anthropol 59:131–137

168. Siegel AF, Benson RH (1982) A robust comparison of biological shapes. Biometrics 38:341–350

169. Bhatia QS, Chen JK, Koberstein JT, Sohn JE, Emerson JA (1985) The measurement of polymer surface tension by drop image processing: application to PDMS and comparison with theory. J Colloid Interface Sci 106:353–359

170. Bhatia QS, Pan DH, Koberstein JT (1988) Preferential surface adsorption in miscible blends of polystyrene and poly(vinyl methyl ether). Macromolecules 21:2166–2175

171. Jalbert C, Koberstein JT, Yilgor I, Gallagher P, Krukonis V (1993). Molecular weight dependence and end-group effects on the surface tension of poly(dimethylsiloxane). Macromolecules 26:3069–3074

172. Fleischer CA, Morales AR, Koberstein JT (1994) Interfacial modification through end group complexation in polymer blends. Macromolecules 27:379–385

173. Anastasiadis SH, Hatzikiriakos SG (1998) The work of adhesion of polymer/wall interfaces and its association with the onset of wall slip. J Rheol 42:795–812

174. Gaines GL Jr (1972) Surface and interfacial tension of polymer liquids – a review. Polym Eng Sci 12:1–11

175. Hata Y, Kasemura T (1979) In: Lee L-H (ed) Polymer engineering and technology, vol 12A. Plenum, New York

176. Wu S (1978) Interfacial energy, structure, and adhesion between polymers. In: Paul DR, Newman S (eds) Polymer blends, vol 1. Academic, New York, p 244

177. Bailey AI, Salem BK, Walsh DJ, Zeytountsian A (1979) The interfacial tension of poly (ethylene oxide) and poly(propylene oxide) oligomers. Colloid Polym Sci 257:948–952

178. LeGrand DG, Gaines GL Jr (1975) Immiscibility and interfacial tension between polymer liquids: dependence on molecular weight. J Colloid Interface Sci 50:272–279

179. Gaines GL Jr, Gaines GL III (1978) The interfacial tension between n-alkanes and poly (ethylene glycols). J Colloid Interface Sci 63:394–398

180. Girifalco LA, Good RJ (1957) A theory for the estimation of surface and interfacial energies. I. Derivation and application to interfacial tension. J Phys Chem 61:904–909

181. Sanchez IC (1983) Bulk and interface thermodynamics of polymer alloys. Ann Rev Mater Sci 13:387–412

182. LeGrand DG, Gaines GL Jr (1971) Surface tension of mixtures of oligomers. J Polym Sci C 34:45–51

183. Gibbs JW (1928) Collected works, vol 1, 2nd edn. Longmans, New York

184. Sancez IC (1984) On the nature of polymer interfaces and interphases. Polym Eng Sci 24:79–86

185. Cahn JW (1978) In: Blakely JJ, Johnson WC (eds) Segregation to interfaces. ASM seminar series. Cleveland, OH

186. Antonow G (1907) Surface tension at the limit of two layers. J Chim Phys 5:372–385

187. Antonoff G (1942) On the validity of Antonoff's rule. J Phys Chem 46:497–499

188. Antonoff G, Chanin M, Hecht M (1942) Equilibria in partially miscible liquids. J Phys Chem 46:492–496

189. Good RJ, Girifalco LA, Kraus G (1958) A theory for estimation of interfacial energies. II. Application to surface thermodynamics of teflon and graphite. J Phys Chem 62:1418–1421

190. Good RJ, Girifalco LA (1960) A theory for estimation of interfacial energies. III. Estimation of surface energies of solids from contact angle data. J Phys Chem 64:561–565

191. Good RJ (1964) In: Fowkes FM (ed) Contact angle, wettability and adhesion. Advances in chemistry series, No. 43. American Chemical Society, Washington, DC, p 74

192. Good RJ, Elbing E (1970) Generalization of theory for estimation of interfacial energies. Ind Eng Chem 62:54–59

193. Wu S (1969) Surface and interfacial tensions of polymer melts: I. Polyethylene, polyisobutylene, and polyvinyl acetate. J Colloid Interface Sci 31:153–161

194. Wu S (1970) Surface and interfacial tensions of polymer melts. II. Poly(methyl methacrylate), poly(n-butyl methacrylate), and polystyrene. J Phys Chem 74:632–638
195. Wu S (1973) Polar and nonpolar interactions in adhesion. J Adhes 5:39–55
196. Fowkes FM (1964) In: Fowkes FM (ed) Contact angle, wettability and adhesion. Advances in chemistry series, No. 43. American Chemical Society, Washington, DC, p 99
197. LeGrand DG, Gaines GL Jr (1969) The molecular weight dependence of polymer surface tension. J Colloid Interface Sci 31:162–167
198. LeGrand DG, Gaines GL Jr (1973) Surface tension of homologous series of liquids J Colloid Interface Sci 42:181–184
199. Helfand E (1975) Theory of inhomogeneous polymers: fundamentals of the Gaussian random-walk model. J Chem Phys 62:999–1005
200. Tagami Y (1980) Effect of compressibility upon polymer interface properties. Ferroelectrics 30:115–116
201. Tagami Y (1980) Effect of compressibility upon polymer interface properties. J Chem Phys 73:5354–5362
202. Helfand E (1974) Theory of inhomogeneous polymers – lattice model of concentrated solution surfaces. Polym Prepr 15:246–247
203. Helfand E (1975) Theory of inhomogeneous polymers: lattice model for polymer–polymer interfaces. J Chem Phys 63:2192–2198
204. Helfand E (1976) Theory of inhomogeneous polymers. Lattice model for solution interfaces. Macromolecules 9:307–310
205. Weber TA, Helfand E (1976) Theory of inhomogeneous polymers. Solutions for the interfaces of the lattice model. Macromolecules 9:311–316
206. Flory PJ (1953) Principles of polymer chemistry. Cornell University Press, Ithaca, NY
207. Roe R-J (1975) Theory of the interface between polymers or polymer solutions. I. Two components system. J Chem Phys 62:490–499
208. Helfand E (1982) Polymer interfaces. In: Solc K (ed) Polymer compatibility and incompatibility: principles and practice. MMI symposium series, vol 2. Michigan Molecular Institute Press, Harwood Academic, New York, p 143
209. Kammer H-W (1977) Surface and interfacial tension of polymer melts – thermodynamic theory of interface between immiscible polymers. Z Phys Chem (Leipzig) 258:1149–1161
210. Guggenheim EA (1940) The thermodynamics of interfaces in systems of several components. Trans Faraday Soc 35:397–412
211. Hong KM, Noolandi J (1981) Theory of inhomogeneous multicomponent polymer systems. Macromolecules 14:727–736
212. Edwards SF (1965) The statistical mechanics of polymers with excluded volume. Proc Phys Soc Lond 85:613–624
213. Freed KF (1972) Functional integrals and polymer statistics. Adv Chem Phys 22:1–128
214. Hong KM, Noolandi J (1981) Theory of interfacial tension in ternary homopolymer-solvent systems. Macromolecules 14:736–742
215. Gaillard P, Ossenbach-Sauter M, Riess G (1982) Polymeric two-phase systems in the presence of block copolymers. In: Solc K (ed) Polymer compatibility and incompatibility: principles and practice. MMI symposium series. Michigan Molecular Institute Press, Harwood Academic, New York, p 289
216. Cahn JW, Hilliard JE (1958) Free energy of a nonuniform system. I. Interfacial free energy. J Chem Phys 28:258–267
217. van der Waals JD (1893) Verh. Konink. Akad. Weten. Amsterdam 1, 8 [translated in English in Rowlinson, J. S. Translation of J. D. van der Waals' "The thermodynamik theory of capillarity under the hypothesis of a continuous variation of density". J Stat Phys 1979, 20, 197–200]
218. van der Waals JD, Kohnstamm Ph (1908) Lehrbuch der Thermodynamik, vol I. Mass and van Suchtelen, Leipzig

Interfacial Tension in Binary Polymer Blends and the Effects of Copolymers

219. Lord Rayleigh (1892) On the theory of surface forces. II. Compressible fluids. Philos Mag 33:209–220
220. Yang AJM, Fleming PD III, Gibbs JH (1976) Molecular theory of surface tension. J Chem Phys 64:3732–3747
221. Fleming PD III, Yang AJM, Gibbs JH (1976) A molecular theory of interfacial phenomena in multicomponent systems. J Chem Phys 65:7–17
222. Adolf D, Tirrell M, Davis HT (1985) Molecular theory of transport in fluid microstructures: diffusion in interfaces and thin films. AlChE J 31:1178–1186
223. Bongiorno V, Davis HT (1975) Modified Van der Waals theory of fluid interfaces. Phys Rev A 12:2213–2224
224. Carey BS, Scriven LE, Davis HT (1978) On gradient theories of fluid interfacial stress and structure. J Chem Phys 69:5040–5049
225. Margenau H, Murphy GM (1943) The mathematics of physics and chemistry. D. Van Nostrand, Princeton, NJ
226. Poser CI, Sanchez IC (1979) Surface tension theory of pure liquids and polymer melts. J Colloid Interface Sci 69:539–548
227. Sanchez IC (1983) Liquids: surface tension, compressibility, and invariants. J Chem Phys 79:405–415
228. Halperin A, Pincus P (1986) Polymers at a liquid–liquid interface. Macromolecules 19:79–84
229. Poser CI, Sachez IC (1981) Interfacial tension theory of low and high molecular weight liquid mixtures. Macromolecules 14:361–370
230. Vrij A (1968) Equation for the interfacial tension between demixed polymer solutions. J Polym Sci A-2 6:1919–1932
231. de Gennes P-G (1980) Dynamics of fluctuations and spinodal decomposition in polymer blends. J Chem Phys 72:4756–4763
232. Ronca G, Russell TP (1985) Thermodynamics of phase separation in polymer mixtures. Macromolecules 18:665–670
233. Pincus P (1981) Dynamics of fluctuations and spinodal decomposition in polymer blends. II. J Chem Phys 75:1996–2000
234. Binder K (1983) Collective diffusion, nucleation, and spinodal decomposition in polymer mixtures. J Chem Phys 79:6387–6409
235. Schichtel TE, Binder K (1987) Kinetics of phase separation in polydisperse polymer mixtures. Macromolecules 20:1671–1681
236. Debye P (1959) Angular dissymmetry of the critical opalescence in liquid mixtures. J Chem Phys 31:680–687
237. Sanchez IC, Lacombe RH (1976) An elementary molecular theory of classical fluids. Pure fluids. J Phys Chem 80:2352–2362
238. Lacombe RH, Sanchez IC (1976) Statistical thermodynamics of fluid mixtures. J Phys Chem 80:2568–2580
239. Sanchez IC, Lacombe RH (1977) An elementary equation of state for polymer liquids. J Polym Sci Polym Lett 15:71–75
240. Sanchez IC, Lacombe RH (1978) Statistical thermodynamics of polymer solutions. Macromolecules 11:1145–1156
241. Joanny JF (1978) These 3érne Cycle. Université Paris 6
242. de Gennes P-G (1979) Scaling concepts in polymer physics. Cornell University Press, Ithaca, NY
243. Roe R-J (1986) SAXS study of micelle formation in mixtures of butadiene homopolymer and styrene-butadiene block copolymer. 3. Comparison with theory. Macromolecules 19:728–731
244. Debye P (1947) Molecular-weight determination by light scattering. J Phys Colloid Chem 51:18–32

245. Owens JN, Gancarz IS, Koberstein JT, Russell TP (1989) Investigation of the microphase separation transition in low-molecular-weight diblock copolymers. Macromolecules 22:3380–3387
246. Joanny JF, Leibler L (1978) Interface in molten polymer mixtures near consolute point. J Phys (Paris) 39:951–953
247. Lifshitz IM (1968) Some problems of the statistical theory of biopolymers. Zh Eksp Teor Fiz 55:2408 [Sov Phys – JETP (1969) 28:1280–1286]
248. Flory PJ, Orwoll RA, Vrij A (1964) Statistical thermodynamics of chain molecule liquids. I. An equation of state for normal paraffin hydrocarbons. J Am Chem Soc 86:3507–3514
249. Nose T (1976) Theory of liquid–liquid interface of polymer systems. Polym J 8:96–113
250. de Gennes P-G (1977) Qualitative features of polymer demixtion. J Phys Lett 38:L441–L443
251. Ouhadi T, Fayt R, Jérôme R, Teyssié Ph (1986) Molecular design of multicomponent polymer systems. 9. Emulsifying effect of poly(alpha-methylstyrene-b-methlyl methacrylate) in poly(vinylidene fluoride)/poly(alpha-methylstyrene) blends. Polym Commun 27:212–215
252. Ouhadi T, Fayt R, Jérôme R, Teyssié Ph (1986) Molecular design of multicomponent polymer systems. X. Emulsifying effect of poly(styrene-b-methyl methacrylate) in poly (vinylidene fluoride)/noryl blends. J Polym Sci Polym Phys 24:973–981
253. Ouhadi T, Fayt R, Jérôme R, Teyssié Ph (1986) Molecular design of multicomponent polymer systems, XI, emulsifying effect of poly(hydrogenated diene-b-methyl methacrylate) in poly(vinylidene fluoride)/polyolefins blends. J Appl Polym Sci 32:5647–5651
254. Ruegg ML, Reynolds BJ, Lin MY, Lohse DJ, Balsara NP (2007) Minimizing the conversion of diblock copolymer needed to organize blends of weakly segregated polymers by tuning attractive and repulsive interactions. Macromolecules 40:1207–1217
255. Meier DJ (1969) Theory of block copolymers. I. Domain formation in A-B block copolymers. J Polym Sci C 26:81–98a
256. Mason JA, Sperling LH (1976) Polymer blends and composites. Plenum, New York
257. Riess G, Jolivet Y (1975) In: Platzer NAJ (ed) Copolymers, polyblends, and composites. Advances in chemistry series, vol 142. American Chemical Society, Washington, DC, p 243
258. Riess G, Kohler J, Tournut C, Banderet A (1967) Über die verträglichkeit von copolymeren mit den entsprechenden homopolymeren. Makromol Chem 101:58–73
259. Fayt R, Jérôme R, Teyssié Ph (1986) Characterization and control of interfaces in emulsified incompatible polymer blends. In: Utracki LA (ed) Polyblends-'86, NRCC/IMRI polymers symposium series. NRCC/IMRI, Montreal, p 1
260. Ouhadi T, Fayt R, Jérôme R, Teyssié Ph (1987) Characterization and control of interfaces in emulsified incompatible polymer blends. Polym Eng Sci 27:328–334
261. Anastasiadis SH, Russell TP, Satija SK, Majkrzak CF (1990) The morphology of symmetric diblock copolymers as revealed by neutron reflectivity. J Chem Phys 92:5677–5691
262. Anastasiadis SH, Russell TP, Satija SK, Majkrzak CF (1989) Neutron reflectivity studies of thin diblock copolymer films. Phys Rev Lett 62:1852–1855
263. Russell TP, Menelle A, Hamilton WA, Smith GS, Satija SK, Majkrzak CF (1991) Width of homopolymer interfaces in the presence of symmetric diblock copolymers. Macromolecules 24:5721–5726
264. Patterson HT, Hu KH, Grindstaff TH (1971) Measurement of interfacial and surface tensions in polymer systems. J Polym Sci C 34:31–43
265. Wilson DJ, Hurtrez G, Riess G (1985) Colloidal behavior and surface activity of block copolymers. In: Walsh DJ, Higgins JS, Maconnachie A (eds) Polymer blends and mixtures. NATO Advanced Science Institute Series. Mortinus Nijhoff, Dordrecht, The Netherlands, p 195
266. Gaillard P, Ossenbach-Sauter M, Riess G (1980) Tensions interfaciales de systèmes polymères biphasiques en présence de copolymères séquencés. Makromol Chem Rapid Commun 1:771–774

267. Noolandi J (1987) Interfacial properties of block copolymers in immiscible homopolymer blends. Polym Prepr 28:46–47
268. Pavlopoulou E, Anastasiadis SH, Iatrou H, Moshakou M, Hadjichristidis N, Portale G, Bras W (2009) The micellization of miktoarm star S_nI_n copolymers in block copolymer/homopolymer blends. Macromolecules 42:5285–5295
269. Gia HB, Jérôme R, Teyssié Ph (1980) Star-shaped block copolymers. III. Surface and interfacial properties. J Polym Sci Polym Phys 18:2391–2400
270. Adamson AW, Gast AP (1997) Physical chemistry of surfaces, 6th edn. Wiley, New York–
271. Piirma I (1992) Polymeric surfactants. Marcel Dekker, New York
272. Nakamura K, Endo R, Takada M (1976) Surface properties of styrene–ethylene oxide block copolymers. J Polym Sci Polym Phys 14:1287–1295
273. Owen MJ, Kendrick TC (1970) Surface activity of polystyrene–polysiloxane–polystyrene ABA block copolymers. Macromolecules 3:458–461
274. Munch MR, Gast AP (1988) Block copolymers at interfaces. 2. Surface adsorption. Macromolecules 21:1366–1372
275. Helfrich W (1973) Elastic properties of lipid bilayers – theory and possible experiments. Z Naturforsch C 28:693–703
276. Stammer A, Wolf BA (1998) Effect of random copolymer additives on the interfacial tension between incompatible polymers. Macromol Rapid Commun 19:123–126
277. Budkowski A, Losch A, Klein J (1995) Diffusion-limited segregation of diblock copolymers to a homopolymer surface. Israel J Chem 35:55–64
278. Semenov AN (1992) Theory of diblock–copolymer segregation to the interface and free surface of a homopolymer layer. Macromolecules 25:4967–4977
279. Morse DC (2007) Diffusion of copolymer surfactant to a polymer/polymer interface. Macromolecules 40:3831–3839
280. Taylor GI (1934) The formation of emulsions in definable fields of flow. Proc R Soc Lond A 146:501–523
281. Noolandi J (1984) Recent advances in the theory of polymeric alloys. Polym Eng Sci 24:70–78
282. Leibler L (1982) Theory of phase equilibria in mixtures of copolymers and homopolymers. 2. Interfaces near the consolute point. Macromolecules 15:1283–1290
283. Noolandi J, Hong KM (1980) Theory of inhomogeneous polymers in presence of solvent. Ferroelectrics 30:117–123
284. Selb J, Marie P, Rameau A, Duplessix R, Gallot Y (1983) Study of the structure of block copolymer–homopolymer blends using small angle neutron scattering. Polym Bull 10:444–451
285. Retsos H, Terzis AF, Anastasiadis SH, Anastassopoulos DL, Toprakcioglu C, Theodorou DN, Smith GS, Menelle A, Gill RE, Hadziioannou G, Gallot Y (2001) Mushrooms and brushes in thin films of diblock copolymer/homopolymer mixtures. Macromolecules 35:1116–1132
286. Shull KR (1993) Interfacial phase transitions in block copolymer/homopolymer blends. Macromolecules 26:2346–2360
287. de Gennes P-G (1980) Conformations of polymers attached to an interface. Macromolecules 13:1069–1075
288. Noolandi J (1991) Interfacial tension in incompatible homopolymer blends with added block copolymer. Die Makromol Chem Rapid Commun 12:517–521
289. Munch MR, Gast AP (1988) Block copolymers at interfaces. 1. Micelle formation. Macromolecules 21:1360–1366

Adv Polym Sci (2011) 238: 271–328
DOI: 10.1007/12_2010_92
© Springer-Verlag Berlin Heidelberg 2010
Published online: 31 July 2010

Theory of Random Copolymer Fractionation in Columns

Sabine Enders

Abstract Random copolymers show polydispersity both with respect to molecular weight and with respect to chemical composition, where the physical and chemical properties depend on both polydispersities. For special applications, the two-dimensional distribution function must adjusted to the application purpose. The adjustment can be achieved by polymer fractionation. From the thermodynamic point of view, the distribution function can be adjusted by the successive establishment of liquid–liquid equilibria (LLE) for suitable solutions of the polymer to be fractionated. The fractionation column is divided into theoretical stages. Assuming an LLE on each theoretical stage, the polymer fractionation can be modeled using phase equilibrium thermodynamics. As examples, simulations of stepwise fractionation in one direction, cross-fractionation in two directions, and two different column fractionations (Baker–Williams fractionation and continuous polymer fractionation) have been investigated. The simulation delivers the distribution according the molecular weight and chemical composition in every obtained fraction, depending on the operative properties, and is able to optimize the fractionation effectively.

Keywords Continuous thermodynamics · Fractionation in column · Theory of copolymer fractionation

---

S. Enders
TU Berlin, Fachgebiet "Thermodynamik und Thermische Verfahrenstechnik", TK 7, Strasse des 17. Juni 135, 10623 Berlin, Germany
e-mail: Sabine.Enders@TU-Berlin.de

Contents

1 Introduction 274
2 Theory 276
   2.1 Liquid–Liquid Phase Equilibrium of Copolymer Solutions 276
   2.2 Stepwise Fractionation Procedure 284
   2.3 Baker–Williams Fractionation 288
   2.4 Continuous Polymer Fractionation 291
3 Results and Discussion 296
   3.1 Liquid–Liquid Phase Equilibrium of Copolymer Solutions 296
   3.2 Stepwise Fractionation Procedure 299
   3.3 Baker–Williams Fractionation 309
   3.4 Continuous Polymer Fractionation 316
4 Summary 324
References 325

Symbols

f Segment-molar activity coefficient
F Fraction
k Reciprocal to the uniformity, see (15)
m Number of stages in the fractionation-column
P Pressure
r Segment number
R Ideal gas constant
T Temperature
U Nonuniformity (30)
v Number of volume increments in the Baker–Williams column
W Intensive two-dimensional distribution function
w Extensive two-dimensional distribution function
X Segment fraction
y Chemical composition of statistical copolymers
Z Segment fraction of the solvent in the solvent + nonsolvent mixture

Abbreviations

CF   Cross-fractionation
CPF   Continuous polymer fractionation
CSF   Continuous spin fractionation
EA   Extraction agent
EVA   Ethylene vinyl acetate copolymer
FD   Feed in continuous polymer fractionation
GPC   Gel permeation chromatograph

Theory of Random Copolymer Fractionation in Columns

SEC   Size exclusion chromatography
SPF   Successive precipitation fractionation
SSF   Successive solution fractionation

Greek Symbols

μ Segment-molar chemical potential
Γ Gamma function
γ Parameter defined in (5)
χ Flory–Huggins interaction parameter
ε Parameter of the Stockmayer distribution function (15) describing the broadness of the chemical heterogeneity
ϕ Quotient of the total amount of segments in phase II and in phase I
ρ Abbreviation, defined in (9) and (11)
η Parameter of the Baker–Williams column in (45)
τ Parameter of the Baker–Williams column in (45)
θ Parameter of the Baker–Williams column in (45)
λ Parameter of the Baker–Williams column in (45)

Subscripts

A   Solvent
E   Excess quantity
M   Mixture
m Number of stages in the fractionation-column
Max   Number of maximal stage in CPF-column
i Component i in the mixture
N   Number-average quantity
n Number of volume increment in the Baker–Williams column
W   Mass-average quantity

Superscripts

I   Phase I
II   Phase II
F   Feed phase

1 Introduction

Synthetic copolymers are always polydisperse, i.e., they consist of a large number of chemically similar species with different molar masses and different chemical compositions. Owing to this polydispersity, characterization of copolymers does usually not provide the number of individual molecules or their mole fraction, mass fraction, etc. but requires the use of continuous distribution functions or their averages. Continuous thermodynamics, developed by Rätzsch and Kehlen [1], can be directly applied to the calculation of thermodynamic properties, including phase equilibria, because this theoretical framework is based completely on continuous distribution functions, which include all the information about these functions and allow an exact mathematical treatment of all related thermodynamic properties. Continuous thermodynamics have been used for calculation of phase equilibria of systems containing two-dimensional distributed copolymers [1–8]. The purpose of this contribution is the application of continuous thermodynamics to copolymer fractionation according to the chemical composition and molecular weight.

Basic research concerning the physical–chemical behavior of polymer solutions is overwhelmingly confined to a few polymers, like polystyrene, that can be polymerized anionically to yield products of narrow molecular weight distribution. One of the reasons for this choice lies in the fact that most polymer properties are not only dependent on the degree of polymerization but are also strongly affected by the broadness of the molecular weight distribution. It is desirable to produce nearly monodisperse polymers. One possibility for doing so is the use of polymer fractionation. Fractionations of polymers are carried out for two different purposes. One purpose is the analytical determination of the molar weight distribution and the other is the preparation of fractions large enough in size to permit study of their properties. In analytical fractionation, the amount of initial polymer is usually small. The fractions do not need to be separated and are often characterized online in automated fractional dissolution procedures. Some column techniques are in use that are based on thermodynamic equilibrium principles and make use of either liquid/liquid or liquid/solid phase separations. In preparative polymer fractionation, scaling-up problems are the main issue, because the necessary amount of initial polymer increases considerably when the purity requirements of the fractions are raised [9–11].

The fractionation of copolymers presents a special problem. For a chemically homogeneous polymer, solubility only depends on molecular weight distribution. In the case of chemically inhomogeneous materials, such as copolymers, solubility is determined by the molecular weight distribution, as well as by chemical composition. In the case of copolymers, both distributions can change during the course of fractionation. The efficiency of any given copolymer fractionation can be estimated from the data on the heterogeneity of fractions in molecular weight (molecular heterogeneity) and in composition (composition heterogeneity).

One of the long-sought "Holy Grails" of polymer characterization has been the simultaneous determination of polymer composition as a function of molecular weight distribution. The combination of a solvent evaporative interface between a gel permeation chromatograph (GPC) and a Fourier transform infrared spectrometer has provided one useful solution to the problem of determining polymer composition as a function of molecular weight for different polymers [12]. Poly (ethylene-*co*-acrylic acid) copolymers were fractionated with supercritical propane, butane, and dimethyl ether [13]. It was possible to carry out the fractionation with respect to the molecular weight using increasing pressure at constant temperature. Additionally, it was possible to fractionate these acid copolymers with respect to chemical composition by first using one of the poor quality solvents (propane or butane) that solubilized the nonpolar ethylene-rich oligomers, and then using dimethyl ether, a very strong solvent for these acid copolymers, to solubilized the acid-rich oligomers. Other examples for the application of supercritical fluids can be found in the literature [14, 15].

One of the most common methods for carrying out analytical polymer fractionation is that of size-exclusion chromatography (SEC), also known as GPC. The polymer solution is passed through a column packed with porous gel beads. The pores have radii comparable in magnitude to the root-mean-square radius of gyration of an average polymer molecule in the sample. The larger molecules are preferentially excluded from the pores, inside which their more extended conformations are forbidden, and are eluted from the column earlier than the smaller molecules. Online detection of the eluent invariably involves refractivity to monitor polymer concentration, and might include light scattering and viscosity. This method was applied to characterize different polymers, like styrene–acrylonitrile copolymers [16–18], α-methylstyrene–acrylonitrile copolymers [19], styrene–methyl methacrylate copolymers [18, 20, 21], styrene–ethyl methacrylate copolymer [22–24], styrene–2 methoyethyl methacrylate copolymer [25], and ethylene terephthalate–tetramethylene ether [26]. Other methods developed for the characterization of copolymers are: fractionation in demixing solvents [27–29], combination of GPC with temperature-programmed column fractionation (TPCF) [30, 31], interaction chromatography [32], column elution method [33], temperature rising elution fractionation (TREF) [34, 35], combination of SEC with precipitation chromatography [36, 37], and crystallization [38, 39].

Within this contribution, we focus our attention on two methods, namely the Baker–Williams fractionation (BW) [40, 41] and continuous polymer fractionation (CPF) [42]. The BW method leads to fractions with a very low nonuniformity and is deemed to be the most effective technique [43, 44]. CPF allows the isolation of fractions on the 100 g scale.

Only a few papers can be found in the literature that deal with theoretical questions on this topic. With the help of continuous thermodynamics, a theory to model stepwise fractionation of homopolymers was developed [45–47]. This theoretical framework could be extended to fractionation in columns [48–50]. The application of the developed theory was able to contribute to the improvement of the fraction technique [49]. Folie [51] studied the fractionation of copolymers. In his theoretical framework, the polymer was described by pseudocomponents with respect to the molecular weight; however, the polydispersity with respect to the

276   S. Enders

chemical composition was neglected. First, Litmanovich and Shtern [52] modeled stepwise copolymer fractionation, where both polydispersities were considered. Later, Ogawa and Inaba [53] also suggested a similar model.

This contribution aims at the development of a theoretical tool for optimization of copolymer fractionation in columns, where both polydispersities are completely taken into account.

2 Theory

2.1 Liquid–Liquid Phase Equilibrium of Copolymer Solutions

Fractionations are usually carried out using a solvent (A), a nonsolvent (B) and the polymer to be fractionated. This means that, for phase equilibrium calculations, a ternary system must be investigated. Due to the very large number of different chemical species, the composition of polydisperse systems is not described by the mole fraction of the individual components, but by a continuous distribution function. In the case of statistical copolymers, a two-dimensional distribution function according the molecular mass and the chemical composition must be used. Usually in polymer thermodynamics, all molecules are imagined to be divided into segments of equal size. With a standard segment defined as the ratio of the van der Waals volume of the considered species and the van der Waals volume of an arbitrary chosen species (for instance one of the solvents or one of the monomers), a segment number r can be defined for each kind of molecule. The introduction of the segment number leads to segment-molar physical quantities. The chemical composition of a copolymer, built up from two different monomers, can be described by the variable y. It is given by the ratio of the segment number of one monomer and the sum of the segment numbers of both monomers, and hence y is related to the amount of one monomer in the copolymer. The intensive distribution function, $W(r, y)$, has to fulfill the normalization condition:

$$\int_0^\infty \int_0^1 W(r,y)\mathrm{d}y\mathrm{d}r = 1. \tag{1}$$

$W(r, y)\mathrm{d}y\mathrm{d}r$ represents the segment fraction of all copolymer species having segment numbers between r and $r + \mathrm{d}r$ and chemical compositions between y and $y + \mathrm{d}y$.

Due to the polydispersity, the demixing behavior becomes much more complicated for a polydisperse polymer in comparison with a monodisperse polymer, as shown in Fig. 1. The binodal curve in this system splits into three kinds of curves: a cloud-point curve, a shadow curve, and an infinite number of coexistence curves. The meaning of these curves becomes clear if one considers the cooling process.

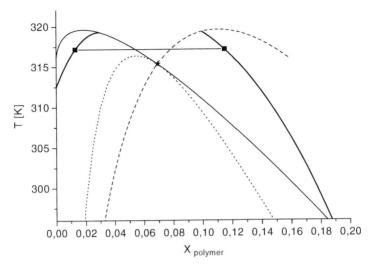

Fig. 1 Schematic liquid–liquid phase diagram for a polydisperse polymer in a solvent: *Solid line* cloud-point curve, *broken line* shadow curve, *dotted line*: spinodal curve, *star* critical point, *thick lines* coexisting curves, *solid line with squares* tie line, $X_{polymer}$ segment fraction of polymer

When reaching the cloud-point curve, at lowering of the temperature, the overall polymer content of the first droplets of the precipitated phase does not correspond to a point on the cloud-point curve but to the corresponding point on the shadow curve. With further lowering of the temperature, the two coexisting phases do not change their overall polymer content according to the cloud-point curve or to the shadow curve but according to the related branches of the coexistence curves. The overall polymer content of the coexisting phases is given by the intersection points of the horizontal line, at the considered temperature, with these branches (tie line in Fig. 1). The coexistence curves are usually not closed curves but are divided into two branches beginning at corresponding points of the cloud-point curve and the shadow curve. Only if the composition of the initial homogeneous phase equals that of the critical point is a closed coexistence curve obtained, whose extremum is the critical point. Moreover, at this point, the cloud-point curve and shadow curve intersect. It can be seen from Fig. 1 that, for solutions of polydisperse polymers, the critical point is not located at the extremum of the cloud-point curve or of the shadow curve. This is in contrast to strictly binary systems where the cloud-point curve, shadow curve, and all coexistence curves become identical.

Homopolymers in coexisting phases show different molar weight distributions, which are also different from that of the initial homogeneous system (Fig. 2). This effect is called the fractionation effect and can be used for the production of tailor-made polymers. The phase with a lower polymer concentration (sol phase) contains virtually only the polymers with a lower molecular weight. Consequently, polymers having a high molecular weight remain in the concentrated phase (gel phase). The cloud-point curve always corresponds to the molecular weight distribution of the

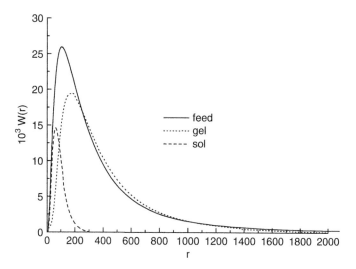

Fig. 2 Fractionation effect for homopolymers

initial polymer, but the first droplets of the formed coexisting new phase never do (with the exception of the critical point) and, hence, they are not located at the cloud-point curve but on the shadow curve. In the case of statistical copolymers, the distribution function according to the chemical composition also differs in the two phases; however, how this distribution changes cannot be predicted a priori. The residence (sol or gel phase) of the molecules with a high value of y depends on the thermodynamic properties of the selected solvent mixture.

To perform phase equilibrium calculations, the starting point is the segment-molar chemical potential, m_i, related to the segment-molar Gibbs free energy of mixing. According to the well-known Flory–Huggins lattice theory [54], the segment-molar chemical potential (μ_i) for the solvents A and B reads:

$$\mu_i = \mu_{i0}(T,P) + RT\left[\frac{1}{r_i}\ln X_i + \frac{1}{r_i} - \frac{1}{r_M}\right] + RT\ln f_i \quad i = A, B, \quad (2)$$

where the first term represents the segment-molar chemical potential of the pure solvents at system temperature T and system pressure P. The second term on the right hand side is the Flory–Huggins contribution (with $f_i = 1$), accounting for the difference in molecular size. In order to describe the deviation from a Flory–Huggins mixture (with $f_i = 1$), the segment-molar activity coefficients, f_i, are introduced. The number-average segment number, r_M, in (2) is for a ternary system, built up from solvent, nonsolvent, and copolymer, and is given by:

$$\frac{1}{r_M} = \frac{X_A}{r_A} + \frac{X_B}{r_B} + \frac{X}{r_N} = \frac{X_A}{r_A} + \frac{X_B}{r_B} + \int_0^\infty\int_0^1 \frac{XW(r,y)}{r}\,dydr, \quad (3)$$

where r_N is the number-average segment number of the copolymer. The segment-molar chemical potential for the copolymer species thus depends on the segment number, r, and the chemical composition, y [4, 6]:

$$\mu(r,y) = \mu_0(T,P,r,y) + RT\left[\frac{1}{r}\ln XW(r,y) + \frac{1}{r} - \frac{1}{r_M}\right] + RT\ln f(r,y). \quad (4)$$

Similar to (2), in (4) the first term is the segment-molar chemical potential of the pure copolymer species with the segment number r and the chemical composition y. The second term displays the Flory–Huggins term and the last term characterizes the deviation from the Flory–Huggins mixture, where the segment-molar activity coefficient can, in principle, depend on the molecular weight and the chemical composition. Whereas the dependence on molecular weight can often be neglected, the dependence on chemical composition plays an important role [3, 6].

Rätzsch et al. [3] suggested the following model for the segment-molar excess Gibbs free energy of mixing (G^E) in order to describe the deviation from the Flory–Huggins mixture:

$$\frac{G^E}{RT} = X_A X \frac{\chi_{AP}}{T}(1 + p_A X)(1 + \gamma_A y_W) + X_B X \frac{\chi_{BP}}{T}(1 + p_B X)(1 + \gamma_B y_W)$$
$$+ X_A X_B \frac{\chi_{AB}}{T}, \quad (5)$$

where y_W is the weight-average chemical composition. This quantity can be calculated using:

$$y_W = \int_0^\infty \int_0^1 yW(r,y)\mathrm{d}y\mathrm{d}r. \quad (6)$$

The influence of the chemical composition in (5) can be derived using a simplified version of Barker's lattice theory [55]. The most important consequence of (5) is the fact that the segment-molar excess Gibbs free energy of mixing and, hence, the activity coefficients depend only on the average value (y_W) of the distribution function, but not on the distribution function itself. In continuous thermodynamics, the phase equilibrium conditions read:

$$\mu_i^I = \mu_i^{II} \quad \mu^I(r,y) = \mu^{II}(r,y) \quad i = A, B. \quad (7)$$

Here, the phase equilibrium condition for the copolymer holds for all polymer species within the total segment number and chemical composition range of the system. This equation is valid for the total interval of the values of the identification variables r and y found in the system. Replacing the segment-molar chemical potentials for the solvents in (7) according to (2) and rearranging results in:

$$X_i^{II} = X_i^I \exp(r_i \rho_i) \quad i = A, B, \quad (8)$$

where the abbreviation ρ_i can be calculated using:

$$\rho_i = \frac{1}{r_M^{II}} - \frac{1}{r_M^I} - \ln f_i^{II} + \ln f_i^I \quad i = A, B. \tag{9}$$

The activity coefficients in (9) can be derived using standard thermodynamics in combination with (5). The replacement of the segment-molar chemical potential of the copolymer species in (7) according (4) leads to:

$$X^{II} W^{II}(r, y) = X^I W^I(r, y) \exp(r\rho(y)), \tag{10}$$

where the abbreviation ρ is given by:

$$\rho(r, y) = \frac{1}{r_M^{II}} - \frac{1}{r_M^I} - \ln f^{II}(y) + \ln f^I(y). \tag{11}$$

Equation (10) is valid for all r and y values found in the system and permits the calculation of an unknown distribution function, $W^{II}(r, y)$. The activity coefficients in (11) can be derived using standard thermodynamics in combination with (5). Integration of (10) and applying the normalization condition (1) results in:

$$X^{II} = \int_0^\infty \int_0^1 X^I W^I(r, y) \exp(r\rho(r, y)) dy dr. \tag{12}$$

To deal with the problem of calculation of the cloud-point curve and the corresponding shadow curve, the temperature of a given phase I is changed at constant pressure until the second phase II is formed. Thus, the unknowns of the problem are the equilibrium temperature, T, the composition of the second phase, X^{II} and X_A^{II}, and the distribution function, $W^{II}(r, y)$. To calculate them, the phase equilibrium conditions (8) and (12) are used. In this system of equations, the unknown distribution function $W^{II}(r, y)$ and the other scalar unknowns T, X^{II}, and X_A^{II} are connected; however, the unknown distribution function $W^{II}(r, y)$ occurs only with the average values r_N^{II} and y_W^{II}. This situation allows a separation of the problem of the unknown distribution function by considering r_N^{II} and y_W^{II} as additional scalar unknowns and their defining equations:

$$\frac{X^{II}}{r_N^{II}} = \int_0^\infty \int_0^1 \frac{X^I W^I(r, y)}{r} \exp(r\rho(r, y)) dy dr \tag{13}$$

and

$$y_W^{II} X^{II} = \int_0^\infty \int_0^1 y X^I W^I(r, y) \exp(r\rho(r, y)) dy \, dr, \tag{14}$$

Theory of Random Copolymer Fractionation in Columns

as additional scalar equations. For calculation of the cloud-point curve and the shadow curve, five equations, namely (8) and (12)–(14), must be solved simultaneously. Furthermore, for numerical calculations, r_N^{II} might be eliminated by means of (8) and (9). The unknown distribution function of the copolymer, $W^{II}(r, y)$, in the shadow phase can be calculated using (10).

Rätzsch et al. [3, 6] could demonstrate that the integrals occurring in (12)–(14) can be solved analytically under certain circumstances, namely if the Stockmayer distribution function [56] is used:

$$W(r, y) = \frac{k^{k+1}}{r_N \Gamma(k+1)} \left(\frac{r}{r_N}\right)^k \exp\left(-k\frac{r}{r_N}\right) \sqrt{\frac{r}{2\pi\varepsilon}} \exp\left(-\frac{r(y - y_W)^2}{2\varepsilon}\right). \quad (15)$$

The first two factors in (15) are a generalized Schulz–Flory distribution with respect to the segment number r. The parameters are k, describing the nonuniformity, and r_N the number-average segment number. Γ is the Γ function. The last two factors are a Gaussian distribution with respect to the chemical composition y, with a standard deviation of $\sqrt{\varepsilon/r}$. The parameters of the Gaussian distribution are the weight-average chemical composition, y_W, and the quantity ε describing the broadness of the chemical heterogeneity. They can be estimated by the kinetic copolymerization parameters because (15) was derived by the kinetics of statistical copolymerization [56]. However, in reality, many copolymers show broad and asymmetric chemical distributions that are not of the Stockmayer type (15). Rätzsch et al. [6] suggested the replacement of the Gaussian distribution function in (15) by the Γ function. The combination of the Γ function for the chemical heterogeneity with the Schulz–Flory distribution for the heterogeneity of the molecular weight reads:

$$W(r, y) = \frac{k^{k+1}}{r_N \Gamma(k+1)} \left(\frac{r}{r_N}\right)^k \exp\left(-k\frac{r}{r_N}\right) \frac{\Gamma(\alpha + \beta + 2)}{\Gamma(\alpha + 1)\Gamma(\beta + 1)} y^\alpha (1 - y)^\beta, \quad (16)$$

where α and β are the parameters describing the distribution of the chemical composition. For the special case $\alpha = \beta$, a symmetrical distribution function results. The value of α indicates the broadness of the distribution, where the limiting case $\alpha \to \infty$ leads to a monodisperse copolymer with respect to the chemical composition. This distribution function displays a large flexibility and allows for the description of asymmetrical distributions. The most important difference between both distribution functions is that in (15) the chemical heterogeneity depends on the segment number, whereas in (16) the chemical composition does not depend on the segment number.

An alternative approach to the calculation of the cloud-point and shadow curve is the application of an equation of state (i.e., [7, 57–60]). The stability conditions in terms of spinodal and critical point are given by Browarzik and Kehlen [61].

To calculate the coexistence curve where a feed phase F splits into the coexisting phases I and II, the mass balance of the copolymer:

$$X^{F}W^{F}(r,y) = (1 - \phi)X^{I}W^{I}(r,y) + \phi X^{II}W^{II}(r,y) \tag{17}$$

must also be applied [3, 6]. The quantity ϕ is the quotient of the total amount of segments in phase II and in feed phase F and equals the fraction of the feed volume that forms phase II. The mass balance for the solvents reads:

$$X_{i}^{F} = (1 - \phi)X_{i}^{I} + \phi X_{i}^{II} \quad i = A, B. \tag{18}$$

Additionally, two balance equations related to the moments of the distribution function can be formulated:

$$\frac{X^{F}}{r_{N}^{F}} = (1 - \phi)\frac{X^{I}}{r_{N}^{I}} + \phi\frac{X^{II}}{r_{N}^{II}} \tag{19}$$

and

$$y_{W}^{F}X^{F} = (1 - \phi)y_{W}^{I}X^{I} + \phi y_{W}^{II}X^{II}. \tag{20}$$

Besides the feed, two of the three variables T, P, and ϕ have to be specified. Starting with the phase equilibrium conditions (8) and (9), the balances (17)–(20) can be used to eliminate the quantities referring to one of the two coexisting phases (for instance phase I), which leads to:

$$X_{i}^{II} = \frac{X_{i}^{F}}{\phi + (1 - \phi)\exp(-r_{i}\rho_{i})} \quad i = A, B. \tag{21}$$

$$X^{II}W^{II}(r,y) = \frac{X^{F}W^{F}(r,y)}{\phi + (1 - \phi)\exp(-r\rho_{B}(r,y))} = \frac{K(r,y)X^{F}W^{F}(r,y)}{\phi}, \tag{22}$$

where the precipitation rate $K(r,y)$ is defined as [46]:

$$K(r,y) = \frac{\phi X^{II}W^{II}(r,y)}{X^{F}W^{F}(r,y)} = \frac{\phi}{\phi + (1 - \phi)\exp(-r\rho(r,y))}. \tag{23}$$

Integration of (22) results in:

$$X^{II} = \int_{r}\int_{y}\frac{K(r,y)X^{F}W^{F}(r,y)}{\phi}dydr. \tag{24}$$

Theory of Random Copolymer Fractionation in Columns

The number-average segment number in the second phase is given by:

$$\frac{X^{\mathrm{II}}}{r_{\mathrm{N}}^{\mathrm{II}}} = \int\limits_{r}\int\limits_{y} \frac{K(r,y)X^{\mathrm{F}}W^{\mathrm{F}}(r,y)}{r\phi}\mathrm{d}y\mathrm{d}r. \tag{25}$$

The weight-average chemical composition in the second phase is given by:

$$y^{\mathrm{II}}X^{\mathrm{II}} = \int\limits_{r}\int\limits_{y} \frac{yK(r,y)X^{\mathrm{F}}W^{\mathrm{F}}(r,y)}{\phi}\mathrm{d}y\mathrm{d}r. \tag{26}$$

In contrast to the cloud-point problem, the integrals occurring in (24)–(26) cannot be solved analytically. In order to calculate the coexisting curves, (21) and (24)–(26) must be solved simultaneously, where the occurring integrals must be estimated using a numerical procedure. If the quantities of feed solution (X^{F}, $W^{\mathrm{F}}(r,y)$, and $X_{\mathrm{A}}^{\mathrm{F}}$) are known, the unknowns are the temperature T (or the quantity ϕ); the polymer concentration in the second phase, X^{II}; the solvent concentration in the second phase, $X_{\mathrm{A}}^{\mathrm{II}}$; the number-average segment number in the second phase, $r_{\mathrm{N}}^{\mathrm{II}}$; and the weight-average chemical composition, $y_{\mathrm{W}}^{\mathrm{II}}$. The corresponding quantities of the first phase can be estimated using the balance equations (17)–(20). If the selected G^{E} model is not dependent on the segment number, a simplification is possible. Inserting the solvent equilibrium conditions [(21) with $i = \mathrm{B}$] the system can be reduced to four equations by eliminating, $r_{\mathrm{N}}^{\mathrm{II}}$. The final equation reads:

$$\begin{aligned} 0 = \int\limits_{r}\int\limits_{y} \frac{K(r,y)X^{\mathrm{F}}W^{\mathrm{F}}(r,y)}{r\phi}\mathrm{d}y\mathrm{d}r - \frac{1}{r_{\mathrm{M}}^{\mathrm{F}}} + \frac{X_{\mathrm{A}}^{\mathrm{II}}}{r_{\mathrm{A}}} + \frac{X_{\mathrm{B}}^{\mathrm{II}}}{r_{\mathrm{B}}} - (1-\phi) \\ \times \left[\frac{1}{r_{\mathrm{B}}}\ln\left(\frac{X_{\mathrm{B}}^{\mathrm{II}}}{X_{\mathrm{B}}^{\mathrm{I}}}\right) + \ln f_{\mathrm{B}}^{\mathrm{II}} - \ln f_{\mathrm{B}}^{\mathrm{I}} \right], \end{aligned} \tag{27}$$

where $p(r,y)$ in (23) is replaced by $p(y)$ and can be calculated by:

$$p(y) = \frac{1}{r_{\mathrm{B}}}\ln\left(\frac{X_{\mathrm{B}}^{\mathrm{II}}}{X_{\mathrm{B}}^{\mathrm{I}}}\right) + \ln f_{\mathrm{B}}^{\mathrm{II}} - \ln f_{\mathrm{B}}^{\mathrm{I}} - \ln f^{\mathrm{II}}(y) + \ln f^{\mathrm{I}}(y) \tag{28}$$

and

$$p_{\mathrm{A}} = \frac{1}{r_{\mathrm{B}}}\ln\left(\frac{X_{\mathrm{B}}^{\mathrm{II}}}{X_{\mathrm{B}}^{\mathrm{I}}}\right) + \ln f_{\mathrm{B}}^{\mathrm{II}} - \ln f_{\mathrm{B}}^{\mathrm{I}} + \ln f_{\mathrm{A}}^{\mathrm{I}} + \ln f_{\mathrm{A}}^{\mathrm{II}}. \tag{29}$$

The coexisting-curve problem is now given by (21) with $i = \mathrm{A}$, (24), (26), and (27), where (28) and (29) can be used to compute $p(y)$ and p_{A}.

Fractionation efficiency is the central feature in the calculation of fractionation, and must be judged by an objective criterion. For fractionation with respect to the molar mass, the uniformity of every fraction, i, can be used. The uniformity (U_i) is defined by:

$$U_i = \frac{M_{W,i}}{M_{N,i}} - 1 = \frac{r_{W,i}}{r_{N,i}} - 1. \tag{30}$$

The more U deviates from zero, the less efficient is the fractionation. The fractionation with respect to the chemical composition is characterized by the distribution function itself.

Thermodynamic principles are relevant to separation processes that make use of the distribution of macromolecules between two phases. These two phases may form a partially miscible system. The diluted phase is called sol phase I, and the polymer-rich phase is the gel phase II. The distribution coefficient depends on molar mass and on chemical composition. Figure 2 depicts the distribution functions in the sol and gel phases. Fractionation can be achieved in a single solvent by a change of temperature, but it is often more practical to vary the solvent quality by using a binary solvent mixture composed of a nonsolvent and a good solvent, usually miscible in all proportions. The solvent composition can be used to fine-tune the solvent quality at constant temperature.

2.2 Stepwise Fractionation Procedure

A classical method for fractionating a polydisperse polymer is to dissolve the polymer completely in a good solvent and then, progressively, to add small amounts of a poor solvent (nonsolvent). In the case of homopolymers, the high molecular weight polymer precipitates first. As more nonsolvent is added, progressively lower molecular weight polymer precipitates.

To obtain quantitative representation of fractionation, a model for the thermodynamic properties of the copolymer + solvent + nonsolvent system and the original two-dimensional distribution function are required. Rätzsch et al. [46] presented the application of continuous thermodynamics to successive homopolymer fractionation procedures based on solubility differences. This method is now applied to copolymer fractionation. The liquid–liquid equilibria (LLE) of polymer solutions forms the thermodynamic background for these procedures. The introduction of the precipitation rate (23) permits calculation of the distribution functions in the sol and gel phases of every fractionation step, i, according to:

$$
\begin{aligned}
X_i^{I} W_i^{I}(r,y) &= \frac{1 - K_i(r,y)}{1 - \phi_i} X_i^{F} W_i^{F}(r,y) \\
X_i^{II} W_i^{II}(r,y) &= \frac{K_i(r,y)}{\phi_i} X_i^{F} W_i^{F}(r,y).
\end{aligned}
\tag{31}
$$

These relations provide the unknown distribution functions $W_i^I(r,y)$ and $W_i^{II}(r,y)$ directly.

Figure 3 presents the schemes for successive precipitation fractionation (SPF) and successive solution fractionation (SSF). In both cases, by lowering the temperature (or adding nonsolvent) a homogeneous polymer solution (called feed phase F) splits into two coexisting phases, a polymer-lean sol phase I and a polymer-rich gel phase II, which are then separated. In SPF (Fig. 3a), the polymer is isolated from phase II as fraction F1. Phase I directly forms the feed phase for the next fractionation step, etc.

In case of SSF (Fig. 3b), fraction F1 is obtained from phase I. Phase II is diluted by adding solvent up to the volume of the original feed phase, corresponding, to a very good approximation, to the same total amount of segments. This phase is used as a feed phase for step 2, etc. In the last fractionation step, the polymer of phase I in the case of SPF, or of phase II in the case of SSF, forms the final polymer fraction. All coexisting pairs of phase I and II are presumed to be in equilibrium. Hence, it is possible to apply all equations introduced above. To indicate the different separation steps 1,2,..., the corresponding number is added as a subscript.

According to the remarks made above, the total number of segments in SSF (Fig. 3b) is the same to a very good approximation in all feed phases. This leads to the following relations:

$$X_{i+1}^F W_{i+1}^F(r,y) = \phi_i X_i^{II} W_i^{II}(r,y) \quad r_{N,i+1}^F = r_{N,i}^{II} \quad y_{W,i+1}^F = y_{W,i}^{II}. \quad (32)$$

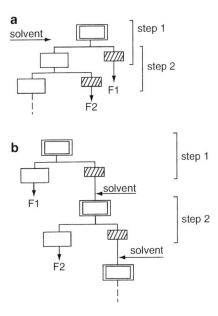

Fig. 3 Schemes of successive fractionation procedures: (**a**) successive precipitation fractionation (SPF), (**b**) successive solution fractionation (SSF). *F1* and *F2* are successive fractions

A constant feed volume can be achieved by adding solvent or solvent mixtures. The necessary amount is given by the amount separated with the polymer-lean phase:

$$n^*_{A,i} + n^*_{B,i} = n^I_i = n^F_i(1 - \phi_i) \quad c_i = \frac{n^*_{A,i}}{n^F_1}. \tag{33}$$

The quantity $n^*_{A,i}$ means the amount of segments of solvent A, which must be added at every fraction step i. This must be done to ensure that the amount of segment keeps constant at every fractionation step. The ratio c characterizes the composition of the solvent mixture added in every fractionation step. Working without a concentration gradient means that this ratio keeps constant during the fractionation procedure. The composition (Z) of the solvent mixture can be expressed by:

$$Z = \frac{X_A}{1 - X}. \tag{34}$$

The feed solvent composition for the fractionation step $(i + 1)$ results from the mass balance (33):

$$Z^F_{i+1} = \frac{Z^{II}_i\left(1 - X^{II}_i\right)\phi_i + c_{i+1}}{1 - X^{II}_i\phi_i}. \tag{35}$$

From (32) as applied to fractionation step i, the distribution function of the i-th polymer fraction $W^I_i(r, y)$ can be derived in a direct and explicit form:

$$X^{II}_i W^{II}_i(r, y) = \frac{1 - K_i(r, y)}{1 - \phi_i} \prod_{j=1}^{i-1} K_j(r, y) X^F_1 W^F_1(r, y). \tag{36}$$

This relation corresponds to the fractionation scheme depicted in Fig. 3. In steps $j = 1, \ldots, i - 1$, the polymer-rich phase II is taken to correspond to the occurrence of the factor $K_j(r, y)$ for $j = 1, \ldots, i - 1$, according to (31). The polymer-lean phase I in step i is taken to correspond to the factor $(1 - K_i(r, y))/(1 - \phi_i)$, according to (31). The unknown quantities, $X^{II}_j, Z^{II}_j, r^{II}_{N,j}, y^{II}_{W,j}$, and ϕ_j (or T_j) for $j = 1, \ldots, i - 1$ can be calculated successively with the help of the equations for LLE discussed above.

In SPF (Fig. 3a), phase I from step i is used directly as the feed phase for step $i + 1$. Hence, the following relations are valid:

$$W^F_{i+1}(r, y) = W^I_i(r, y) \quad r^F_{N,i+1} = r^I_{N,i} \quad y^F_{W,i+1} = y^I_{W,i}. \tag{37}$$

According to the fractionation scheme, the amount of feed segments is not constant. In order to take this effect into account, the quantity λ_i is defined:

$$\lambda_i = \frac{n^F_i}{n^F_1}. \tag{38}$$

For $i = 1$, $\lambda_1 = 1$. In all other cases the material balance could be used to calculated λ_i:

$$\lambda_{i+1} = (1 - \phi_i)\lambda_i + c_{i+1} + d_{i+1} \quad d_i = \frac{n^*_{B,i}}{n^F_1}. \tag{39}$$

The quantity $n^*_{B,i}$ means the amount of segments of solvent B, which must be added at every fraction step i. This must be done to ensure that the amount of segment keeps constant at every fractionation step. The quantity d_i describes the ratio of the added amount of solvent at every fractionation step and the amount of segment in the feed phase for the first fractionation step.

The polymer feed concentration for every fractionation step is given by:

$$X^F_{i+1} = \frac{X^I_i}{1 + \frac{c_{i+1}+d_{i+1}}{(1-\phi_i)\lambda_i}}. \tag{40}$$

If the fractionation is carried out only by lowering the temperature, then the polymer feed concentration in the step $i + 1$ is directly the polymer concentration of sol phase from step i. The composition of the solvent for the fractionation step $i + 1$ reads:

$$Z^F_{i+1} = \frac{Z^I_i(1 - X^I_i)(1 - \phi_i)\lambda_i + c_{i+1}}{(1 - X^I_i)(1 - \phi_i)\lambda_i + c_{i+1} + d_{i+1}}. \tag{41}$$

The fractions are taken from the gel phase. Therefore the distribution function in every fraction step i can be computed using:

$$X^{II}_i W^{II}_i(r,y) = \frac{K_i(r,y)}{\phi_i} \prod_{j=1}^{i-1} \frac{1 - K_j(r,y)}{1 - \phi_j} X^F_1 W^F_1(r,y). \tag{42}$$

This equation permits the direct and explicit calculation of the distribution function of the polymer fraction i from the distribution function of the original polymer. Again, the form of this relation corresponds to the fractionation scheme applied (Fig. 3a). In steps $j = 1,\ldots,i - 1$, the polymer-lean phase I is taken to correspond to the occurrence of the factor $(1 - K_j(r, y))/(1 - \phi_j)$ for $j = 1,\ldots,$ $i - 1$, according to (31). In step i, the polymer-rich phase II is taken to correspond to the factor $K_i(r,y)/\phi_i$, according to (31). The unknown quantities, $X^{II}_j, Z^{II}_j, r^{II}_{N,j}, y^{II}_{W,j}$, and ϕ_j (or T_j) for $j = 1,\ldots,i - 1$ can be calculated successively with the help of the equations for LLE discussed above.

For example, SSF and SPF were applied to styrene–acrylonitrile copolymer in either toluene [62] or a mixture of methyl ethyl ketone and cyclohexane [63] as solvent. These types of fractionation are also called one-direction fractionations.

The cross-fractionation (CF) of copolymers suggested by Rosenthal and White [64] is a combination of several successive precipitation procedures and is also called two-direction fractionation. The original copolymer is fractionated first in a precipitant–solvent system such that the mean content of A units of fractions, say, increases with an increase in their molecular weights. Then, every intermediate fraction is separated in another precipitant–solvent system, where the mean content of A units of fractions diminishes with an increase in their molecular weight. Rosenthal and White [64] assert that the final fractions separated in such a way are highly homogeneous both in molecular weight and in chemical composition.

For example, this method was carried out for various copolymers, namely styrene–methyl methacrylate copolymer [65–67], epoxide resins [68], styrene–acrylic acid copolymer [69], styrene–2-methoxyethyl methacrylate copolymer [70, 71], ethylene–α-olefin copolymer [72], partially modified dextran–ethyl carbonate copolymer [73], vinyl chloride–vinyl acetate copolymer [43], styrene–acrylonitrile copolymer [74], and styrene–butadiene copolymer [75].

The stepwise fractionation procedures (SSF and SPF) are one-direction fractionations and form the basis of cross-fractionation, where first the original polymer is fractionated in intermediate fractions using one solvent system and afterwards each intermediate fraction is further fractionated yielding the final fractions using another solvent system. There are four different possibilities for a fractionation strategy:

(a) SSF/SPF
(b) SSF/SSF
(c) SPF/SSF
(d) SPF/SPF

The theoretical framework introduced above can also be applied to cross-fractionation. This can be achieved by combination of the equations according to the selected fractionation strategy.

2.3 Baker–Williams Fractionation

Precipitation fractionation as developed by Baker and Williams [40] is one of the best-known column fractionation procedures. The fractionation is performed in a glass-bead-filled column with a temperature gradient down the column (Fig. 4). To start the fractionation, the total polymer is precipitated on the glass beads in a section at the entry of the column (or in a separate vessel). In a mixing vessel, a nonsolvent and a solvent are mixed to form a mixture with progressively increasing solvent power through continuous enrichment of the solvent. The polymer is dissolved by adding the solvent mixture. The resulting sol phase moves relatively slowly in the column, and the polymer in a given increment of the liquid sol stream becomes less soluble due to the temperature gradient and precipitates partially on the glass beads as a gel phase. The fractionation is achieved by the repeating exchange of polymer molecules between the stationary gel phase and the mobile sol phase. The superposition of a solvent + nonsolvent gradient and a temperature gradient leads to a very high separation efficiency.

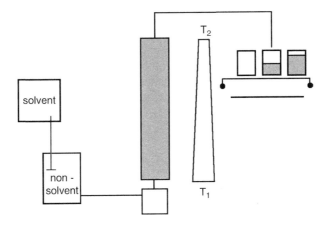

Fig. 4 Schematic of the Baker–Williams column, showing the temperature gradient down the column from higher temperature T_1 to lower temperature T_2

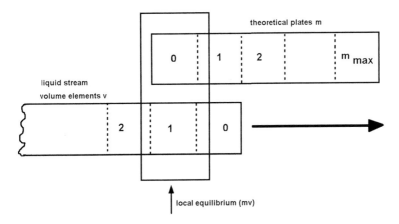

Fig. 5 Theoretical model of the Baker–Williams column

The theoretical treatment is based on a model subdividing the column into stages and the solution stream into parts with equal volumes. Hence, the column fractionation is considered as a combination of many local LLEs and treated in an analogous way as successive fractionation procedures. Rätzsch et al. [50] developed a model in order to simulate the fractionation of homopolymers according to the molar mass in a BW column by a number of local equilibria, similar to the model suggested by Smith [76] and by Mac Lean and White [77].

The column is subdivided into stages, labeled with m, starting with $m = 0$ (Fig. 5). The liquid stream is also subdivided into increments with equal volumes, labeled with v, starting with $v = 0$. At time zero the volume increment $v = 0$ fills stage $m = 0$; at time one the volume increment $v = 0$ occupies stage $m = 1$ and the volume increment $v = 1$ occupies stage $m = 0$, etc. Each volume increment v at each stage m is considered to form liquid–liquid equilibrium (mv) between the sol phase I and the gel phase II.

The gel phase II, which is coated on the surface of the small glass beads, is stationary, i.e., it remains at the same stage m during the progress of time. However, the moving sol phase I always remains in the same volume increment v. Figure 5 depicts this situation. Starting the fractionation, the total polymer is assumed to be precipitated at stage $m = m_P = 0$ or to be distributed evenly among the $m_P + 1$ stages from $m = 0$ to $m = m_P$. The temperature gradient is expressed by [50]:

$$
\begin{aligned}
T_m &= T_0 & m &< m_P \\
T_m &= T_0 - (m - m_P)\Delta T & m &> m_P.
\end{aligned}
\tag{43}
$$

Here T_m is the temperature of stage m and ΔT is the constant temperature difference between neighboring stages. The segment fraction Z of the solvent in the solvent + nonsolvent mixture supplied to the entry (*) of the column, $Z_{v,0}^*$, is assumed to be given by [50]:

$$
Z_{0,v}^* = Z_{0,0}^* + Z^* \left(1 - \exp\left(-\frac{v}{v^*}\right)\right),
\tag{44}
$$

where $Z_{0,0}^*$, ΔZ^*, and v^* are the parameters of this function. The polymer fractions are obtained from the sol phase I of the last stage.

The suggested theory is based on the model described above, which subdivides the column fractionation procedure into many local phase equilibria (Fig. 5). In this way, the phase equilibrium relation presented above can be applied. The considered volume increment v and the considered column stage m are indicated as subscripts of the corresponding quantities. The feed quantities for every LLE can be calculated by applying the above model. However, it has to be taken into account that the feed phase is not a homogenous phase. The feed phase $(m + 1, v + 1)^F$ is the sum of the mobile phase $(m,v + 1)^I$ and the stationary phase $(m + 1, v)^{II}$. After equilibrium, the sol phase $(m + 1, v + 1)^I$ and the gel phase $(m + 1, v + 1)^{II}$ are formed. Therefore, the mass balance for the copolymer reads:

$$
\eta\left[\lambda(1 - \phi)X^I W^I(r, y)\right]_{m,v+1} + \tau\left[\lambda\phi X^{II} W^{II}(r, y)\right]_{m+1,v} + \theta\left[X^F W^F(r, y)\right]_{0,0}
$$
$$
= \left[\lambda X^F W^F(r, y)\right]_{m+1,v+1},
\tag{45}
$$

where the parameters η, τ, and θ are given by:

$$
\begin{aligned}
\eta &= 0 & \text{for} \quad m + 1 = 0 & \quad \eta = 1 & \text{for} \quad m + 1 > 0 \\
\tau &= 0 & \text{for} \quad v + 1 = 0 & \quad \tau = 1 & \text{for} \quad v + 1 > 0 \\
\theta &= 0 & \text{for} \quad m + 1 > m_P & \quad \theta = 0 & \text{for} \quad m + 1 < m_P \quad \text{and} \quad v + 1 = m.
\end{aligned}
\tag{46}
$$

Theory of Random Copolymer Fractionation in Columns

The quantity $\lambda_{m,v}$ in (45) measures the ratio of the segments in the feed phase $(m,v)^F$ and the feed phase $(0,0)^F$. At stage $m = 0$ the condition:

$$\lambda_{0,v} = 1 \tag{47}$$

is realized by the amount of solvent mixture added. In all other cases, λ_{mv} follows from the phases combined, leading to:

$$\left[\lambda X^F\right]_{m+1,v+1} = \eta\left[\lambda(1-\phi)X^I\right]_{m,v+1} + \tau\left[\lambda\phi X^{II}\right]_{m+1,v} + \theta X_{0,0}^F. \tag{48}$$

The ratio of the solvent in the solvent + nonsolvent mixture in the corresponding feed phase is given by:

$$\left[\lambda Z^F\left(1 - X^F\right)\right]_{m+1,v+1} = \eta\left[\lambda(1-\phi)Z^I\left(1 - X^I\right)\right]_{m,v+1}$$
$$+ \tau\left[\lambda\phi Z^{II}\left(1 - X^{II}\right)\right]_{m+1,v} + \sigma\left(1 - \phi_{0,v}\right)Z_{0,v+1}^* \tag{49}$$

with:

$$\sigma = 1 \quad \text{for} \quad m+1 = 0 \qquad\qquad \sigma = 0 \quad \text{for} \quad m+1 > 0. \tag{50}$$

The quantity $Z_{0,v+1}^*$ can be calculated using (44). The combination of (31), as applied to the considered equilibrium, and of (45) interrelates the polymer distributions in the feeds of neighboring equilibria:

$$\left[\lambda X^F W^F(r,y)\right]_{m+1,v+1} = \left[\lambda(1 - K(r,y))X^F W^F(r,y)\right]_{m+1,v}$$
$$+ \left[\lambda K(r,y)X^F W^F(r,y)\right]_{m,v+1}. \tag{51}$$

This equation permits the direct and explicit calculation of the various copolymer distribution functions $W_{m,v}^F(r,y)$ from the distribution function $W_{0,0}^F(r,y)$.

2.4 Continuous Polymer Fractionation

The production of sufficient amounts of narrowly distributed polymer samples, which cannot by synthesized with narrow molecular weight distribution, has been too laborious, except for special cases like the investigation of dilute solutions, for which only small polymer samples are required. This situation was strongly improved by the development of a new technique. Wolf et al. [78] suggested a continuous polymer fractionation method for homopolymers that allows fractions on the 100 g scale.

Fig. 6 Scheme of the CPF in a Gibbs phase triangle

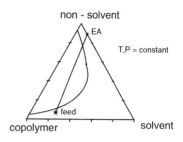

The polymer to be fractionated is dissolved in a solvent + nonsolvent mixture, and this solution (feed) is extracted continuously by a second liquid (extracting agent, EA), which contains the same solvent components as the feed (Fig. 6). The fractionation can be performed with a pulsed counter-current extraction apparatus. Fractionation is achieved by the fact that the molecules are distributed over the counter-current phases according to their chain length. The feed leaves the column as gel and the EA as sol.

The solvent components in the feed and in the EA are chosen such that (a) the entire system formed by the starting polymer and the solvent components exhibits a miscibility gap at the temperature of operation; (b) that, in the Gibb's phase triangle, the composition of the feed corresponds to a point outside of this miscibility gap; and (c) that the EA is composed in such a way that the straight line drawn between feed and EA (working line) intersects the miscibility gap (Fig. 6).

The ratio of flows of feed and EA is chosen such that the working point (average composition of the total content of the apparatus under stationary operating conditions) is located at higher polymer concentration than the intersection of the working line with the branch of low polymer concentration of the demixing curve (Fig. 6). As the original feed comes into first contact with the already polymer-loaded phase originating from the pure EA and moving in the opposite direction, the most easily soluble low molecular weight polymer molecules will be transferred to this dilute phase, which in turn segregates its most sparingly soluble high molecular weight material to the more concentrated phase in order to achieve phase equilibrium. By providing for a high number of such equilibria in the course of the counter-current extraction (proportional to the number of bottoms when a sieve-bottom column is used), the polymer contained in the more concentrated phase will have lost practically all the low molecular weight material contained in the original sample up to a certain characteristic chain length, when it leaves the apparatus after a final extraction by the pure EA. In an analogous manner, the molecular weight distribution of the polymer contained in the less-concentrated phase will narrow as the counter-current extraction proceeds. The feed leaves the column as gel and the EA as sol. The distribution of the polymer on the sol and gel phases can be regulated by the solvent + nonsolvent ratio in the EA and by the ratio of the flux rates of feed and EA. The fractionation efficiency can be increased by additional measures, like pulsation,

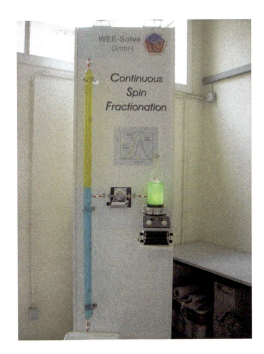

Fig. 7 Photograph of the CSF apparatus

which leads to a higher dispersion of the phases. If the desired separation cannot be obtained in one step, the gel can directly be used as feed, whereas in the case of the sol the polymer must be precipitated before it can be used as feed again. This method was applied to numerous polymers, i.e., polyethylene [79], hydroxyethyl starch [80], polycarbonate [81], polyacrylacid [82], and polyisobutylene [83, 84].

This fractionation method was further developed in 2002 [85, 86] by employing a spinning nozzle (Fig. 7). The new method is called continuous spin fractionation (CSF). The feed phase is pressed through a spinning nozzle to jet threads of the viscous polymer solution, which disintegrate rapidly because of the Rayleigh instability. In this manner, one obtains a large amount of tiny droplets in the desired EA. The droplets with a large ratio of surface to volume facilitate the escape of short chains from entanglements of higher macromolecules due to the short distance of transport. For that reason, CSF not only eliminated the damming-back problem of CPF, but can also be operated successfully at considerably higher polymer concentrations.

The CPF was applied also to copolymers [87, 88]. For example, polycarbonate–siloxane (PC–Si) copolymers are characterized by an outstanding thermal stability, good weathering properties, excellent flame retardancy, and high impact resistance at low temperature [87]. PC–Si materials are used in numerous applications including windows, roofing, contact lenses, and gas-permeable membranes. Depending on the siloxane-block length and domain size, the copolymers are transparent,

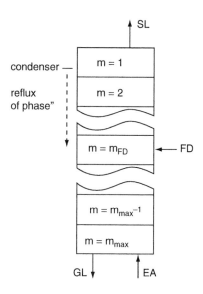

Fig. 8 Model of the improved CPF column. *SL* sol, *GL* gel

translucent, or opaque [89]. The CPF allows fractionation according to the chemical composition and according to the molecular weight and, hence, the tailoring of the copolymer for a certain application purpose.

Rätzsch et al. [48, 49] proposed a theoretical treatment of the CPF similar to that of the BW fractionation. The CPF column is divided into a number of stages m (Fig. 8). The EA enters the column at stage $m = m_{Max}$ and leaves it as sol at stage $m = 1$. The stationary state is calculated by repeating calculation of stages $m = 1, \ldots, m_{Max}$. At start ($i = 0$), the column is filled with EA. For the first set of equilibria ($i = 1$), a certain amount of feed (FD) is added at stage $m = 1$, and the related phase II is transferred downwards to the next stage $m = 2$, filled with EA. When phase II has left the column at m_{Max} as the first, nonstationary gel phase, all phases I are shifted by one stage upwards, and stage m_{Max} is filled with pure EA again. The calculation is repeated for $i = 2$ and $m = 1, \ldots, m_{Max}$ etc. The stationary state is reached when the results for i and $i - 1$ no longer change systematically.

Using this simulation procedure, some improvements could be suggested [48]. The pulsating sieve-bottom column was replaced by a non-pulsating column filled with glass beads. In this manner, the number of theoretical plates could be raised considerably. A further improvement of the fractionation efficiency results from the reflux of part of the polymer contained in the sol phase. In practice, this situation was realized by putting a condenser on the top of the column and introducing the feed somewhere near of the upper third of the column (Fig. 8). These suggestions were verified experimentally using the system dichloromethane/diethylene glycol/bisphenol-A polycarbonate [48]. Except for the lowest molecular weight fraction, one obtains nonuniformities on the order of 0.1.

Theory of Random Copolymer Fractionation in Columns

For conventional CPF, where the FD is introduced on the top, the calculation of stationary states starts ($i = 0$) with the entire column filled with EA. For the first set of equilibria ($i = 1$), a certain amount of FD, determined by the chosen working conditions of CPF, is introduced at m_{FD} (Fig. 6). After the equilibrium has been calculated for this plate, phase II is transferred downwards to the next theoretical plate filled with EA and a new equilibrium on plate $m_{FD} + 1$ is determined. This procedure is repeated until phase II leaves the column as the first (nonstationary) gel phase at m_{Max}. In preparation of the next step of calculation ($i = 2$), all phases I are shifted by one plate upwards, i.e., from m to $m - 1$ so that the first (nonstationary) sol phase leaves the column at m_{FD}. Furthermore, plate m_{Max} is again filled with pure EA. After the addition of another portion of FD, the determination of the next set of equilibria ($i = 2$) proceeds as described for $i = 1$. This treatment is repeated until the stationary state is reached, which means that the results for i and $i - 1$ no longer change systematically.

The mass balance for the polymer transfer is formulated in terms of $w(r, y)$, the extensive segment-molar distributions obtained by multiplying $W(r, y)$ by the overall amount of polymer segments present in a given system. For the transfer between neighboring phases, the following relation can be formulated:

$$w^{I}_{m+1,i-1}(r,y) + w^{II}_{m-1,i}(r,y) = w^{F}_{m,i}(r,y). \qquad (52)$$

For $i = 1$, the first term becomes zero and for $m = m_{FD}$ the additional term $w^{FD}_{i}(r, y)$ has to be added on the left side of the equation. For $m = m_{Max}$ the first and for $m = 1$ the second term vanish. In the case of conventional CPF, m_{FD} is equal to unity. For the mass balance for the subdivision of the polymer among the phases coexisting on one theoretical plate, the corresponding equation reads:

$$w^{F}_{m,i}(r,y) = w^{I}_{m,i}(r,y) + w^{II}_{m,i}(r,y). \qquad (53)$$

In the calculation outlined above, the amount of segments contained in a theoretical plate changes with i within the nonstationary phase. For this reason, a quantity ε is introduced as the ratio of the overall amount of segments present in plate m during step i and during step zero. From the material balance between adjacent theoretical plates, one obtains the following relation:

$$\varepsilon_{m,i} = [\varepsilon(1 - \phi)]_{m+1,i-1} + [\varepsilon\phi]_{m-1,i} \qquad (54)$$

with the limiting conditions $\varepsilon_{m,0} = 0$ for $m = 1$ up to $m = m_{FD} - 1$ (improved CPF), and $\varepsilon_{m,0} = 1$ for $m = m_{FD}$ to $m = m_{Max}$. For $m = m_{FD}$, the extra term ε_{FD} (which is determined by the amount of segments added with the feed, normalized to the amount of EA present on this plate for $i = 0$) has to be added on the right side of the equation. For $m = m_{\text{Max}}$ and $i > 0$, the first term becomes unity and for $m = 1$ the second term vanishes. To obtain the material balances for the polymer, (52) and (53), in terms of the normalized distribution function $W(r, y)$, one divides these relations by the total amount of segments in a plate at $i = 0$. This leads to:

$$\left[\varepsilon(1 - \phi)X^{\text{I}}W^{\text{I}}(r, y)\right]_{m+1, i-1} + \left[\varepsilon\phi X^{\text{II}}W^{\text{II}}(r, y)\right]_{m-1, i} = \left[\varepsilon X^{\text{F}}W^{\text{F}}(r, y)\right]_{m, i}, \quad (55)$$

where the special cases discussed in the context of (52) apply analogously, and to:

$$\left[X^{\text{F}}W^{\text{F}}(r, y)\right]_{m, i} = \left[(1 - \phi)X^{\text{I}}W^{\text{I}}(r, y)\right]_{m, i} + \left[\phi X^{\text{II}}W^{\text{II}}(r, y)\right]_{m, i}. \quad (56)$$

As described above, the stationary state is approached by the stepwise calculation of the composition of the coexisting phase using the equation given in Sect. 3.1, where the information concerning all previous states is required in the actual calculation.

3 Results and Discussion

The suggested fractionation theory is based on the LLE of a copolymer solution; therefore, first the calculation procedure related to the LLE is discussed. Additionally, the calculation results are compared with experimental LLE data for ethylene vinyl acetate copolymer (EVA) in methyl acetate taken from literature [90].

Subsequently, the theory is applied to stepwise fractionation using the cross-fractionation procedure. After some model calculations to study the influence of different operative fractionation parameters on the fractionation efficiency, the theoretical results will be again compared with experimental data for the styrene–butadiene copolymer system in two different solvent systems, namely cyclohexane + isooctane and benzene + methyl ethyl ketone [75].

Finally, the theoretical framework is applied to the simulation of column fractionation according two different methods (BW fractionation and CPF). In both types of fractionation, the influence of operative conditions on the fractionation effect with respect to the molecular weight and the chemical composition is investigated. Because of the lack of experimental data, no comparison with experiments was possible.

3.1 Liquid–Liquid Phase Equilibrium of Copolymer Solutions

The copolymer fractionation aims at the production of fractions having a distribution as narrow as possible. For this reason, this chapter focuses on the distributions in the sol and gel phases. Before any calculations can be carried out, the model parameters must be chosen. The model parameters can be divided into:

Theory of Random Copolymer Fractionation in Columns

(a) Parameters describing the feed polymer
(b) Parameters describing the selected solvent + nonsolvent system
(c) Parameters describing the operative conditions

The feed polymer can be characterized by a two-dimensional distribution function, (15) or (16). Typical values for parameters of these distribution functions are:

$$r_N = 100 \quad k = 1 \quad \varepsilon = 0.25 \quad y_W = 0.5 \quad \alpha = \beta = 4. \tag{57}$$

Setting $\alpha = \beta$ in (16) means that the distribution function with respect to the chemical heterogeneity is symmetrical. The parameters describing the selected solvent + nonsolvent system occur in the G^E model [i.e., (5)]. For the model calculation, the following parameter were chosen:

$$\begin{aligned} r_A &= 1 \quad \chi_{AP} = 150 \text{ K} \quad p_A = 0 \quad \gamma_A = 0.5 \\ r_B &= 1 \quad \chi_{BP} = 250 \text{ K} \quad p_B = 0 \quad \gamma_B = 1 \\ \chi_{AB} &= 500 \text{ K}. \end{aligned} \tag{58}$$

The low-molecular weight component A should act as solvent. The parameter χ_{AP} is selected in a way that no demixing with the copolymer occur. The low molecular weight component B takes over the role of the nonsolvent and, hence, the parameter χ_{BP} leads to a miscibility gap with the polymer. The selected operative conditions are:

$$X^F = 0.01 \quad Z^F = 0.2 \quad T = 350 \text{ K}. \tag{59}$$

The asymmetry of the distribution with respect to the chemical composition has a large impact on the phase equilibria of copolymers [6]. First the influence of the feed distribution (16), especially the symmetry, is studied. At constant mass-average chemical composition, the maxima of the distribution according to the chemical composition are shifted to higher values if the distribution become unsymmetric ($\alpha = 4$ and $\beta = 2$). The distribution $W(100, y)$ in the sol and gel phases are plotted in Fig. 9. The symmetry of the feed copolymer has a large impact on the theoretical fractionation results. Caused by the shift of the maxima in the feed distribution, the maxima in the fractions shift also to higher values. Moreover, in the case of a symmetric feed distribution (solid lines in Fig. 9) more polymer molecules will be in the sol phase in comparison with an unsymmetric feed distribution (broken lines in Fig. 9). If a modified version of (16) is used, namely ignoring the polydispersity with respect to the molecular weight, the ratio between the amounts of sol and gel phases does not change if the parameter β is changed from $\beta = 4$ (symmetrical case) to $\beta = 2$ (unsymmetrical case). This finding indicates the complex interactions of both polydispersities.

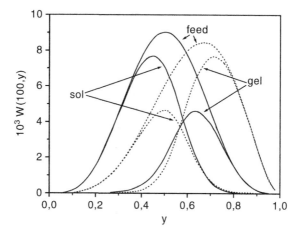

Fig. 9 Distribution of chemical composition at $r = 100$ in the sol and gel phases, where the feed distribution is given by (16): *solid lines* $\alpha = \beta = 4$; *broken lines* $\alpha = 4, \beta = 2$

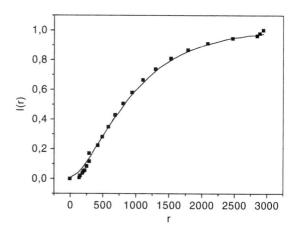

Fig. 10 Fit (*solid line*) of the experimental feed distribution [90] (*squares*) of EVA copolymer

In order to verify the present theoretical framework, the calculation results are compared with experimental data. Schirutschke [90] carried out phase equilibrium experiments (critical point, cloud-point curves) of the system EVA + methyl acetate. The distribution of the copolymer was measured using the successive fractionation procedure and determination of the number-average molecular weight of every obtained fraction. Using this data, the integral distribution function, $I(r)$, can be constructed (Fig. 10) and the parameter of the distribution function with respect to the molecular weight can be estimated. Fitting of the data given in Fig. 10 results in $r_N = 429$ and $k = 0.758$, where the ethylene monomer unit was chosen as standard. The mass-average chemical composition $y_W = 0.375$ was measured

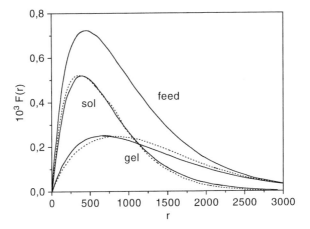

Fig. 11 Comparison of experimental [90] (*solid line*) and calculated (*broken line*) fractionation for EVA in methyl acetate

using elementary analysis. The broadness of the distribution, ε, was estimated using the Stockmayer theory [56], yielding $\varepsilon = 0.25$. The parameters occurring in the G^E model (5) were chosen as:

$$r_A = 2.08 \quad \chi_{AP} = 80.75 \text{ K} \quad p_A = 0.1 \quad \gamma_A = 0.3. \quad (60)$$

Unfortunately, Schirutschke [90] did not give any information about the chemical polydispersity in the experimentally obtained fraction. For this reason, the comparison between the experiment and the modeling results is limited to the fractionation according to the molecular weight. In Fig. 11, the function $F(r)$, which is defined as:

$$F(r) = \int_0^1 W(r,y)\,dy \quad (61)$$

is compared with the experimental data [90], where the experiment was performed at 303.15 K. From Fig. 11, it can be concluded that the proposed theoretical framework is able to describe the polymer distribution in the coexisting phases. Moreover, this comparison also shows the large impact on the LLE of the chemical heterogeneity, even if the broadness of this function is only small.

3.2 Stepwise Fractionation Procedure

Cross-fractionation as a mean of evaluating the molecular weight and chemical-composition distribution of heterogeneous copolymer is composed of two steps.

300   S. Enders

The sample is first fractionated into intermediate fractions in one solvent + non-solvent system (solvent mixture 1). This is followed by further fractionation of these intermediate fractions by another solvent + nonsolvent system (solvent mixture 2).

3.2.1 Influence of the Fractionation Strategy

Using the SSF (Fig. 3b) and SPF (Fig. 3a) techniques, copolymers can be fractionated using the cross-fractionation method by combining the two basic types for fractionation in solvent mixtures 1 and 2. This situation results in four different fractionation strategies: SSF/SPF, SPF/SSF, SPF/SPF, and SSF/SSF, where the solvent mixture is changed for the fractionation of the intermediate fractions. The first question arising in this situation is which strategy should be used in order to optimize the fractionation efficiency. To answer this question, calculation of all four strategies were performed.

In these simulations, it was assumed that both solvent mixtures are made only from one solvent. The applied parameters in the G^E model (5) for the solvent 1 are:

$$r_A = 1 \quad \chi_{AP} = 250 \text{ K} \quad p_A = 0 \quad \gamma_A = 1 \tag{62}$$

and for the solvent 2:

$$r_A = 1 \quad \chi_{AP} = 220 \text{ K} \quad p_A = 0 \quad \gamma_A = 0.5. \tag{63}$$

During the simulation, the original polymer was fractionated into five intermediate fractions using the solvent 1 (62), whereas every intermediate fraction was further divided into five final fractions using the solvent 2 (63). The temperatures for every fractionation step were selected in such a way that in every fraction the same amounts of polymer were present. The polymer feed concentration, expressed in segment fractions, was 0.01. Figure 12 depicts the calculated mass-average chemical composition of every fraction using the four different fractionation strategies. The fractions numbered 10–15 represent the chemical composition of the final fractions, obtained from the first intermediate fraction. The fractions numbered with 20–25 represent the chemical composition of the final fractions, obtained from the second intermediate fraction, and so on. If the fractionation in solvent 1 is carried out using the SPF mechanism (circles and crosses in Fig. 12), the mass-average chemical composition of the intermediate fractions decrease with the number of the fraction. Under this circumstance the fractionation in the solvent 2 can be performed using SPF (circles in Fig. 12) or SSF (crosses in Fig. 12). Except for the final fractions from the first intermediate fraction, where the SPF mechanism leads to a decrease in the y_W values and the SSF to increasing values for this quantity, in all other final fractions the y_W value increases by SSF and decreases by SPF. If the fractionation in the solvent 1 is carried out using the SSF mechanism

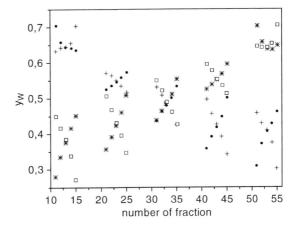

Fig. 12 Mass-average chemical composition for every fraction obtained by cross-fractionation using different fractionation strategies: *circles* SPF/SPF, *crosses* SPF/SSF, *stars* SSF/SPF, and *squares* SSF/SSF

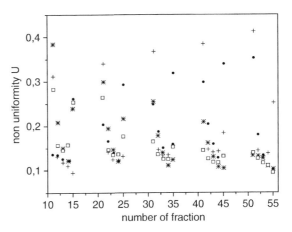

Fig. 13 Nonuniformity for every fraction obtained by cross-fractionation using different fractionation strategies: *circles* SPF/SPF, *crosses* SPF/SSF, *stars* SSF/SPF, and *squares* SSF/SSF

(stars and squares in Fig. 12), the mass-average chemical composition of the intermediate fractions increases with the number of fraction. All fractions, except the fractions obtained from the last intermediate fractions, show a similar behavior. If the fractionation of the intermediate fractions in solvent 2 is carried out using the SPF mechanism (stars in Fig. 12), the mass-average chemical composition increases with the number of fraction, whereas the chemical composition decreases if the second fractionations in the solvent 2 is performed with the SSF method (squares in Fig. 12). Using the data in Fig. 12, it can be concluded the SSF/SSF strategy leads to the fractionation having the highest effectivity in terms of fractionation according to the chemical composition.

In Fig. 13, the fractionation results with respect to the molecular weight are plotted in terms of the nonuniformity of the obtained fractions. Independently of the fractionation strategy applied, all obtained final fractions have a much smaller nonuniformity than the original polymer. Using the SPF/SPF method (circles in Fig. 13), the last fraction of every intermediate fraction always shows the highest value for the nonuniformity. With increasing number of the intermediate fraction this effects is more pronounced. Applying the SPF/SSF mechanism (crosses in Fig. 13), the highest nonuniformity is always observed in the first final fraction of every intermediate fraction, similar to the results for the SSF/SPF mechanism (stars in Fig. 13). Similar to the results discussed above, the SSF/SSF (squares in Fig. 13) leads to the most effective fractionation, also according to the molecular weight, and hence this method can be recommended for the cross-fractionation.

3.2.2 Influence of the Solvent Mixture

The fractionation based on the LLE and the LLE depends strongly on the interactions between the molecules present in the system. In connection with the effectivity of copolymer fractionation, it must be investigated by how much both solvents used in the cross-fractionation procedure should differ in terms of solution power for the copolymer. The quality of solvent can be expressed by the interaction parameter occurring in the G^E model (5). Calculations were done using two different parameter sets, representing a large difference between the interactions of the solvents 1 and 2 with the copolymer:

$$\begin{aligned} \chi_{AP} &= 250 \text{ K} \quad p_A = 0 \quad \gamma_A = 1 \\ \chi_{BP} &= 350 \text{ K} \quad p_B = 0 \quad \gamma_B = 0.15 \end{aligned} \tag{64}$$

and representing only a small difference in these parameters:

$$\begin{aligned} \chi_{AP} &= 250 \text{ K} \quad p_A = 0 \quad \gamma_A = 1 \\ \chi_{BP} &= 220 \text{ K} \quad p_B = 0 \quad \gamma_B = 0.5. \end{aligned} \tag{65}$$

The calculation results are depicted in Fig. 14 in terms of the Breitenbach–Wolf plot with respect to the segment number (Fig. 14a) and with respect to the chemical composition (Fig. 14b). The fractionation effect can be analyzed by the slope of the Breitenbach–Wolf plot. If the two solvents differ strongly, the Breitenbach–Wolf plot with respect to the segment number has a larger slope meaning a higher fractionation effect (Fig. 14a). However, analyzing the data given in Fig. 14b for the fractionation effect according to the chemical composition, the converse can be concluded. The reason for this finding is the difference in the parameter γ_A for the solvent 2. Whereas the Flory–Huggins parameter χ_{AP} has an impact on both fractionations types, the parameter γ_A only has an impact on the fractionation with respect to the chemical heterogeneity. For practical purposes, the χ_{AP} parameters should have a large difference in order to make sure an effective fractionation according the molecular mass, and the γ_A parameter should be always quite

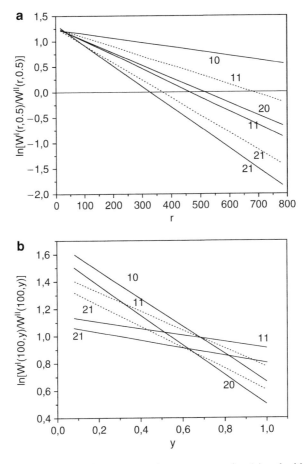

Fig. 14 Breitenbach–Wolf plot with respect to the segment number (**a**) and with respect to the chemical composition (**b**). The *numbers* indicate the final fraction number. The *solid lines* are calculated using the parameter given in (64) and the *broken lines* are obtained using the parameters of (65)

large in order to ensure an effective fractionation according to the chemical composition.

For the production of the intermediate, as for the production of the final fraction, a temperature gradient must also be established. This gradient can be chosen to be linear or nonlinear. The nonlinear temperature gradient is selected in such a way that every final fraction contains the same amount of copolymer segments. The fractionation results are demonstrated in Fig. 15, where the distribution functions with respect to the segment number for the fraction obtained from the third intermediate fractions are shown. It can be recognized clearly that the second approach of keeping the amount of polymer in every final fraction constant is the

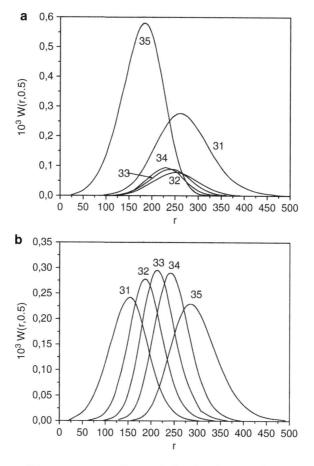

Fig. 15 Influence of the temperature gradient on the fractionation according the segment number for different temperature gradients: (**a**) linear temperature gradient, (**b**) nonlinear temperature gradient. The *numbers* indicate the final fraction number

better choice, because the obtained distribution functions are narrower and the overlapping of the distribution functions is less pronounced.

Additionally, the same calculations were performed in order to investigate whether a more effective fractionation could be achieved using solvent mixtures such as solvent 1 and solvent 2. This situation allows, from the theoretical point of view, a combination of the solvent gradient and a temperature gradient. However, the calculation results make it clear that no improvement could be seen if both gradients were applied. The most important criterion for effective fractionation was equal amounts of polymer in every final fraction. This criterion can be achieved by solvent gradient or temperature gradient, or both. For the solution of a practical fractionation problem, the search for a suitable solvent combination, in particular if

Theory of Random Copolymer Fractionation in Columns

solvent mixtures are to be used, is a very time-consuming task. This task can be abbreviated if only two solvents and not two solvent mixtures are needed. The criterion of equal amounts of polymer in every final fraction can be fulfilled by a suitable nonlinear temperature gradient.

3.2.3 Cross-Fractionation

Teramachi and Kato [75] performed experimentally a cross-fractionation according the SPF/SPF mechanism and two SPFs in different solvent systems of styrene–butadiene copolymer. The copolymer was an industrial product that was polymerized to about 100% conversion with n-butyl lithium. Teramachi and Kato used two solvent systems, namely solvent mixture 1 (cyclohexane + isooctane) and solvent mixture 2 (benzene and methyl ethyl ketone). During all experimental fractionations, the temperature was kept constant at 25°C. In cross-fractionation, the sample was first fractionated into four intermediate fractions in solvent mixture 2 and then each intermediate fraction was fractionated into five fractions in solvent mixture 1. The SPF yield in solvent mixture 1 was 13 fractions, and ten fractions in solvent mixture 2. The original copolymer, as well all obtained fractions, were analyzed using a membrane osmometer for the determination of the number-average molecular weight, and refractive index measurement for the determination of the average chemical composition. In the experiments [75], it was found that the analysis of both one-direction fractionations using different solvent systems led to identical chemical composition distributions for the original polymer; however, both results are not always true. The analysis of the cross-fractionation data gave a broader distribution curve of the chemical heterogeneity and the molecular weight than the one-direction fractionations. The component with low styrene contents (lower than 17 mol%) could only be found by cross-fractionation.

These experimental data give us the possibility to verify the present theory with experiments. From the analysis of the original polymer, two parameters of the Stockmayer distribution function (15), namely $r_N = 884.9$ and $y_W = 0.2799$, are available. The conversion in the polymerization reaction was close to 100% and hence the Stockmayer theory [56] cannot be applied to estimate ε. The values for $\varepsilon = 1.9$ and $k = 2$ were estimated by fitting them to the integral distribution of the original polymer. Next, the parameter of the G^E model (5) must be estimated. Unfortunately, Teramachi and Kato [75] gave no information about the concentration gradient used during the fractionation procedure. Model calculations indicated that the fractionation results can be achieved by a solvent mixture or by a pure solvent. In order to minimize the number of adjustable parameters, both solvent mixtures were simulated by only one "mean" solvent. The fractionation gradient was produced by varying the interaction parameters, γ_A, for both solvent mixtures during the fractionation procedure. This means that these parameters were calculated for every LLE by the solution of the nonlinear system of equations describing the LLE. This is possible because the mass of every fraction was known from the experiments. For the remaining parameters, the following values were assumed for solvent mixture 1:

$$\chi_{AP} = 0.427 \text{ K} \quad p_A = 0.1 \tag{66}$$

and for solvent mixture 2:

$$\chi_{AP} = 0.671 \text{ K} \quad p_A = 0.05. \tag{67}$$

The theoretical frameworks were executed for the calculations of the distributions in the intermediate and in the final fractions. In Fig. 16, the calculated mass-average chemical compositions were compared with the experimental data obtained by the one-direction fractionation [75]. Using the solvent mixture 1, the mass-average chemical composition decreases with increasing number of fraction. Using the solvent mixture 2, this property increases with increasing number of fraction. The reason for this finding is the sign of the parameter, γ_A, which is positive for the solvent mixture 1 and negative for the solvent mixture 2. The agreement between the experimental and calculated data is much better for the fractionation in the solvent mixture 1 than in the solvent mixture 2, especially for the fractions with a high fraction number. During the experiments [75] using the solvent mixture 2, evaporation of the solvent was observed. This effect is not taken into account in the theoretical calculations.

In Fig. 17, the experimental [75] and calculated number-average segment numbers are compared. According to the SPF mechanisms, the copolymers having the highest molecular weight will preferentially be in the first fractions, independent of the chosen solvent mixture. Except for the fractions having a very high molecular weight, the proposed theoretical framework is able to model the experiment

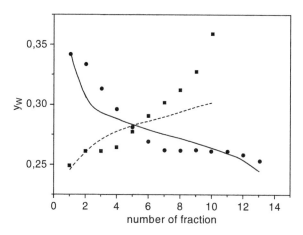

Fig. 16 Comparison of experimental (*symbols*) [75] and calculated (*lines*) mass-average chemical compositions in every fraction after performing a one-direction fractionation using two different solvents: *circles and solid line* solvent mixture 1; *squares and broken line* solvent mixture 2

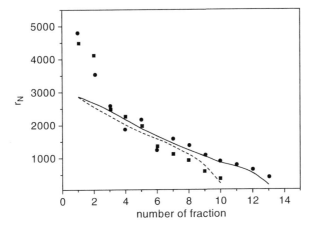

Fig. 17 Comparison of experimental (*symbols*) [75] and calculated (*lines*) number-average segment number in every fraction after performing a one-direction fractionation using two different solvents: *circles and solid line* solvent mixture 1; *squares and broken line* solvent mixture 2

quantitatively. Next, the cross-fractionation should be simulated using the same parameters as for the G^E model. The theoretical results, together with the experimental results [75], are plotted in Fig. 18. Having in mind the approximations introduced in the theoretical framework, it can be notice that the theory is able to model experimental fractionation routines very close to the experimental data.

The effectivity of the fractionation according to the molecular weight can be evaluated by the nonuniformity of the obtained final fractions. This quantity is plotted versus the number of fraction in Fig. 19 for three different fractionation runs, where one is carried out as cross-fractionation and two as one-direction fractionations. Comparing only the one-direction fractionations, it can be recognized that the application of solvent mixture 2 (stars in Fig. 19) leads to a more effective fractionation with respect to the molecular weight, because the fractions have a lower nonuniformity. From the thermodynamic point of view this can be understand by the difference in the Flory–Huggins interaction parameters, χ_{AP} (66, 67). This result demonstrates the important role of the selected solvents. However, the execution of the cross-fraction leads to a strong improvement of the fractionation effectivity, because the nonuniformity is mostly below 0.4. The remarkably high nonuniformity in the last fractions of every intermediate fraction can be explained by the fact that these fractions are the unfractionated remains. In order to improve this situation, more fractionation steps must be carried out.

The enforcement of the copolymer fractionation is motivated by the experimental determination of the two-dimensional distribution function of the original polymer. This is normally done by analysis of the fractionation data (y_{Wi}, r_{Ni}) and construction of the integral distribution function. In Fig. 20, the calculated fractionation data are used for this construction. From Fig. 20a it can be seen that the broadness of the original distribution function with respect to the molecular weight can only be obtained by cross-fractionation. This theoretical result agrees with the

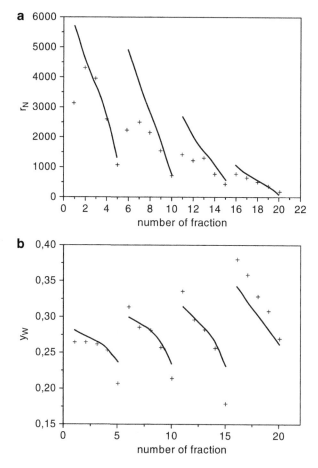

Fig. 18 Comparison of experimental [75] (*symbols*) and calculated (*lines*) cross-fractionations: (**a**) fractionation with respect to the molecular weight, (**b**) fractionation with respect to the chemical composition

experimental observation [75]. Although both one-direction fractionations lead to very close results, they do not find the high molecular weight part of the distribution of the original copolymer. Analyzing the fractionation according to the chemical heterogeneity (Fig. 20b), one can find that the cross-fractionation in comparison to the one-direction fractionation improves the obtained integral distribution function of the original copolymer; however, the complete distribution function could not be recovered. From the thermodynamic point of view, the reason is the selection of the corresponding solvent mixtures. The result could be improved by using a solvent mixture having a larger difference in the parameter γ_A. Again, both one-direction fractionations lead to similar results, but they are not able to reproduce the correct original distribution according to the chemical heterogeneity. If the copolymer

Theory of Random Copolymer Fractionation in Columns

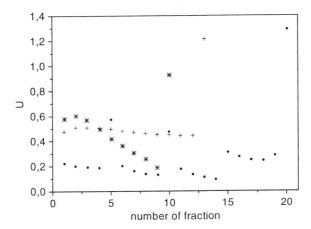

Fig. 19 Comparison of three different fractionation methods: *circles* cross-fractionation, *crosses* successive precipitation fractionation using solvent mixture 1, and *stars* successive precipitation fractionation using solvent mixture 2

fractionation is utilized for analytical purposes, cross-fractionation should be performed, where not only the production of more fractions, but also the application of two different solvent mixtures lead to a significant improvement in the determination of the distribution related to both polydispersities.

In summary, the developed method based on continuous thermodynamics to simulate successive fractionations with respect to the molecular weight and chemical composition is verified by comparison with experimental data, and can be applied for the optimization of a given fractionation problem for analytical and preparative purposes.

3.3 Baker–Williams Fractionation

The most important feature of the BW fractionation method is the high effectivity, especially for analytical purposes. Application of the computer simulation permits the investigation of various effects in the field of column fractionation regarding the effectivity. The simulations were carried out using a copolymer with the following specifications:

$$r_N = 100 \quad k = 1 \quad \varepsilon = 0.3 \quad y_W = 0.5 \quad \alpha = \beta = 4. \tag{68}$$

The selected parameters in the G^E model (5) describing the solvent mixtures are:

$$\begin{aligned} r_A &= 1 & \chi_{AP} &= 150\ \text{K} & p_A &= 0 & \gamma_A &= 0.5 \\ r_B &= 1 & \chi_{BP} &= 250\ \text{K} & p_B &= 0 & \gamma_B &= 1 \\ \chi_{AB} &= 500\ \text{K}. & & & & & & \end{aligned} \tag{69}$$

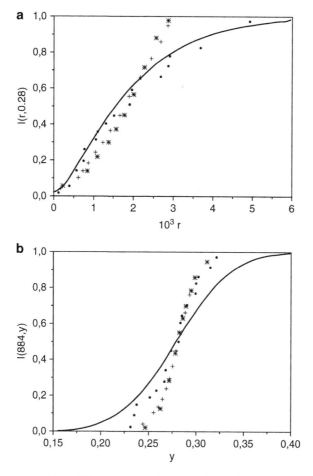

Fig. 20 Comparison of three different fractionation methods: *circles* cross-fractionation, *crosses* successive precipitation fractionation using solvent mixture 1, and *stars* successive precipitation fractionation using solvent mixture 2, with regard to the integral distribution of the original copolymer by (**a**) molecular weight, and (**b**) chemical composition. The *lines* represent the distribution of the original copolymer

For the simulation of the BW column, the operative parameters must also be fixed:

$$X_{0,0}^F = 0.02 \quad m_P = 2 \quad m_{Max} = 8 \quad Z_{0,0}^* = 0.1 \quad v^* = 30. \tag{70}$$

Because the copolymers have a different broadness of the distribution according to the Stockmayer theory (15) or according to (16), different temperature and concentration gradients were established in the column. If the original copolymer

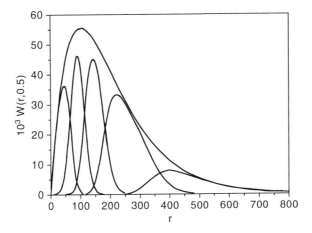

Fig. 21 Simulation results for the Baker–Williams column according to the molecular weight, where the original polymer has a two-dimensional distribution according (16)

is described using the Stockmayer distribution function (15), the operative conditions are:

$$T_0 = 480 \text{ K} \quad \Delta T = 2.5 \quad v_{\text{Max}} = 33 \quad Z^* = 0.25. \tag{71}$$

If the original copolymer is described by the distribution function given in (16), the operative conditions are:

$$T_0 = 400 \text{ K} \quad \Delta T = 5 \quad v_{\text{Max}} = 75 \quad Z^* = 0.4. \tag{72}$$

Figure 21 represents the simulation results for the BW run, where five fractions are formed having an equal amount of the original polymer with a distribution according (16). The obtained fractions show a clearly lower polydispersity than the fractions obtained by successive fractionation methods. The maxima of the distribution functions for the fractions are very close to the original distribution function. This permits an accurate determination of the original distribution function. The last fraction in Fig. 21 has a relative large nonuniformity. However, this can be improved very easily by making more fractions from this material. One advantage of the BW column is the possibility to vary the amount of polymer in the corresponding fraction arbitrarily, without any limits given by the thermodynamics or by the operative parameters of the column. Rätzsch et al. [50] simulated the fractionation of homopolymers having Schulz–Flory distribution functions. They could obtain fractions having a nonuniformity smaller then 0.01. This value could not be reached if copolymers were considered. Usually, the nonuniformities lie between 0.01 and 0.05 for copolymers.

In Fig. 22, the fractionation effect of the BW method with respect to the chemical composition is plotted for a copolymer having a distribution function

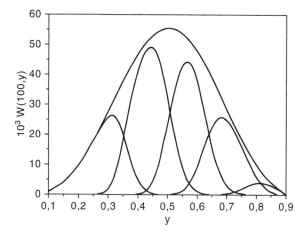

Fig. 22 Simulation results for the Baker–Williams column according to the chemical composition, where the original polymer has a two-dimensional distribution according to (16)

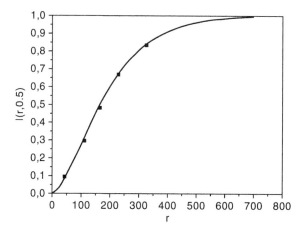

Fig. 23 Simulation results of the Baker–Williams column for a copolymer having a distribution according to (15), where the *symbols* show the fractionation data and the *line* represents the original distribution

given in (16). Again, the application of a BW column leads to a strong improvement in the fractionation effectivity in comparison with the stepwise methods.

The high fractionation effectivity allows correct estimation of the initial copolymer distribution (15) according the molecule weight if only five fractions are formed (Fig. 23). Copolymers distributed according to the Stockmayer distribution function (15) are characterized by a relatively small polydispersity with respect to the chemical composition. In contrast, copolymers showing a distribution given in

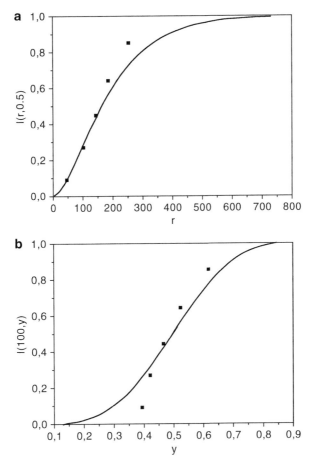

Fig. 24 Simulation results of the Baker–Williams column for a copolymer having a distribution according to (16) with respect to the molecular weight (**a**) and the chemical composition (**b**), where the *symbols* show the fractionation data and the *lines* represent the original distribution

(16) have a broader distribution with respect to the chemical heterogeneity, but the same broadness related to the molecular weight. In Fig. 24, the integral distribution functions constructed from the fractionation data for a copolymer with this distribution function is plotted. From the integral distribution function $I(r, 0.5)$, it can be seen that five fractions are not sufficient to yield the correct original function. Deviation can be found at higher molecular weights (Fig. 24a). The distribution $I(100, y)$ obtained from the fractionation data is also too narrow in comparison with the original distribution (Fig. 24b), where deviations occur at high and at low values of the chemical composition. This result reflects the complex superposition of both kinds of polydispersity.

Rätzsch et al. [50] found by the simulation of the fractionation of homopolymers in the BW column a practical linear relationship between the number of maximal

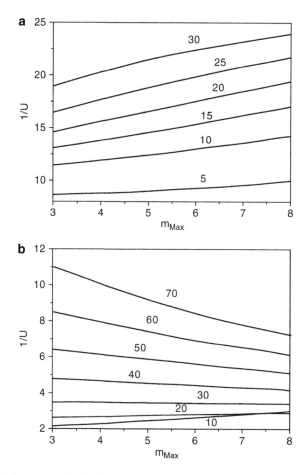

Fig. 25 Simulation results for the Baker–Williams column for different values of the maximal plate number in the column using (15) (**a**) and (16) (**b**) for the feed distribution of the copolymer. The *numbers* are the considered volume element

theoretical plates in the column, m_{Max}, and the reciprocal of the nonuniformity of the obtained fractions. The slope of this relationship was always positive and increased with increasing number of the considered volume element. Figure 25 shows a plot of the reciprocal of the nonuniformity of the copolymer in the corresponding volume element versus the number of theoretical plates established in the column. The results (Fig. 25a) obtained with the copolymer of Stockmayer feed distribution (15) are very similar to those found for the fractionation of homopolymers [50]; however, the slope of the curves are smaller for copolymers than for homopolymers. The number of theoretical plates has a larger impact on the fractionation of homopolymers than on the fractionation of copolymers. Caused by a broader distribution of (16) in comparison with (15), the sign of slope of the studied relation depends, in this case, on the volume element studied (Fig. 25b). A positive slope can only be found at the beginning of the fractionation, meaning lower numbers of the volume element. With increasing liquid stream (increasing number of volume element), the sign of the slope changed to negative, even though the liquid stream was divided into more volume elements.

In order to study this unexpected phenomena further, the distribution functions in the sol phase of a selected volume element ($v = 30$) traveling through the BW column was investigated. For this reason, simulations of a copolymer having a feed distribution given in (16) for columns differing in the number of theoretical plates (m_{Max}) were performed. The simulation results are depicting in Fig. 26. Increasing the length of the column leads to a more effective fractionation of both

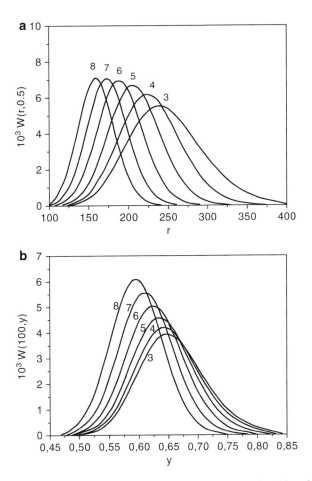

Fig. 26 Distribution of the sol phase with respect to the segment number (**a**) and the chemical composition (**b**) in volume element 30 traveling through the Baker–Williams column with different numbers of maximal theoretical plates m_{Max} (*numbers* in the figure). The original distribution matches (16)

heterogeneities. With increasing length of the column (increasing number m_{Max}) the maxima of the distribution function $W(r, 0.5)$ and the distribution function $W(100, y)$ shift for the observed volume elements to lower values and both distribution functions become narrower. For the volume element $v = 70$ moving through the column, the length of the column has the opposite effect. The practical consequence is the relative large polydispersity in the large fraction.

Increasing the number of theoretical plates, m_{Max}, allows establishing a more flat temperature gradient in the column for the same fractionation results.

In summary, the suggested theoretical model can be applied to answer different questions arising about the efficiency of copolymer fractionation performed in BW columns. This type of column is mostly used for analytical purposes.

3.4 Continuous Polymer Fractionation

CPF has been especially developed to produce large fractions in a relatively short time frame and can be applied for preparative purposes. All traditional procedures, especially the stepwise methods, require a low polymer concentration for good efficiency, and large amounts of solutions must be handled to obtain sufficient material. With this fractionation method, the initial copolymer is divided into two fractions, where these fractions can be used again as feed for the next fractionation run. For homopolymers, this fractionation method is a useful tool for cutting the short molecular weight parts or the extremely high molecular weight parts from the desired product. In the case of copolymers, those with extremely high or low values for the chemical composition can also be removed from the product. The simulation method suggested above is now applied for the optimization of the CPF. This optimization is always done by variation of the parameters describing the column and keeping all others constant. The copolymer (68) used for the simulations and the parameters of the G^{E} model (69) are identical to those used for the simulation of the BW columns. The standard parameter set for the operating conditions is:

$$X^{\text{FD}} = 0.11 \quad Z^{\text{FD}} = 0.5 \quad Z^{\text{EA}} = 0.15 \quad n^{\text{FD}}/n^{\text{EA}} = 0.1$$
$$m_{\text{Max}} = 6 \quad m_{\text{FD}} = 3 \quad T = 520 \text{ K} \quad T_{\text{condenser}} = 500 \text{ K}. \tag{73}$$

For example, the calculated fractionation data for four CPF runs are collected in Table 1, where the initial polymer distribution was a Stockmayer distribution (15). For the fractionation, four CPF runs were simulated in which the obtained gel fractions were directly used as new feed phase.

The data in Table 1 make it clear that no significant fractionation according to the chemical heterogeneity took place. However, the fractionation effect with respect to the molecular weight is characterized by a high effectivity. Except for the first fraction, all other fractions have a nonuniformity lower than 0.06, similar to the results obtained for the simulation of the BW column. The nonuniformities are

Theory of Random Copolymer Fractionation in Columns

Table 1 Calculated fractionation data for CPF, where the original copolymer has a Stockmayer distribution (15)

Fraction no.	T (K)	y_W	r_N	U
1	520	0.49567	74.8	0.673
2	530	0.50651	216.8	0.058
3	540	0.50696	266.0	0.039
4	550	0.50766	311.7	0.037
5	550	0.51159	427.9	0.071

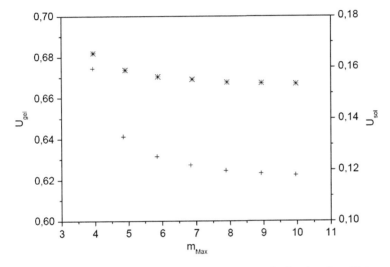

Fig. 27 Influence of the number of theoretical plates, m_{Max}, on the fractionation efficiency with respect to the segment number of CPF, if the original copolymer distribution is given by (15). The *stars* represent the values of the sol fraction and the *crosses* the values from the gel fractions

slightly higher for copolymers in comparison with homopolymers [49], showing the influence of chemical heterogeneity on the fractionation with respect to the molecular weight, even if the polydispersity with respect to the chemical composition is small. The following optimization procedure aims for a much stronger fractionation effect with respect to the chemical composition and, at the same time, to keep the effectivity for the fractionation with respect to the molecular weight. Furthermore, during the optimization it is assumed that the both solvents and the thermodynamic properties, expressed by the parameter of the G^E model, cannot be changed. This means we focus our attention on the optimization of the operative conditions, which can also be changed in practice.

First, the influence of the number of theoretical plates of the CPF column is studied (Fig. 27). The discussion can be done using the nonuniformities in the resulting sol and gel phases. Independently of the isolated phase (sol or gel), the nonuniformity decreases with increasing number of theoretical plates present in the CPF column. However, the decline is only very limited if the column has more

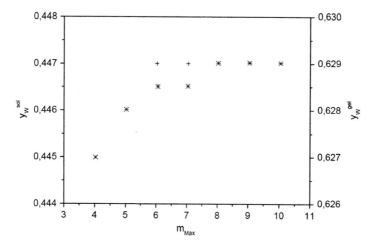

Fig. 28 Influence of the number of theoretical plates, m_{Max}, on the fractionation efficiency with respect to chemical composition of CPF, if the original copolymer distribution is given by (16). The *stars* represent the values of the sol fraction and the *crosses* the values from the gel fractions

than six theoretical plates. This result was also obtained by Rätzsch et al. [49] for the fractionation of homopolymers.

In order to study the influence of the number of theoretical plates on the fractionation according to the chemical composition, simulations using (16) as feed distribution function were performed, because this type of distribution function models a broader distribution. The obtained results are given in Fig. 28. Again, theoretical plates having numbers greater than six do not contribute strongly to the fractionation effect in the CPF column. For practical fractionation, it can recommend to use a column having six theoretical plates.

The next question arising in relation to the CPF column is where the feed phase should be put in the column. The quantity is represented in the theoretical framework by the parameter m_{FD}.

Simulations of the CPF column were carried out for the copolymer with a feed distribution of (16) at different places for the feed input. The results are depicted in Fig. 29 for the fractionation with respect to the segment number and in Fig. 30 for the fractionation with respect to the chemical composition. With increasing m_{FD}, the nonuniformity in the sol as well as in the gel decreases (Fig. 29). Analyzing the fractionation according to the chemical composition leads to the same result (Fig. 30). However, feed input at plate number 4 ($m_{FD} = 4$) cannot be recommended because the polymer concentration on the plate above will be too small. The most effective fractionation can be expected if the feed is added a little above the middle of the columns. If the column with $m_{Max} = 6$ is used, then the feed should be added at $m_{FD} = 3$.

The next operative parameter is the ratio between the amount of feed copolymer solution and the amount of extraction solvent, meaning the working point (Fig. 6).

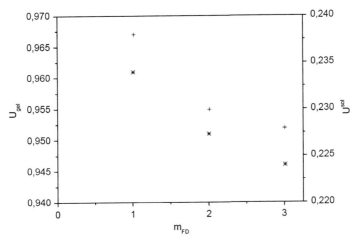

Fig. 29 Estimation of the optimal feed plate for the CPF column for the fractionation according the molecular mass if the original polymer is described by (16). The *stars* represent the values of the sol fraction and the *crosses* the values from the gel fractions

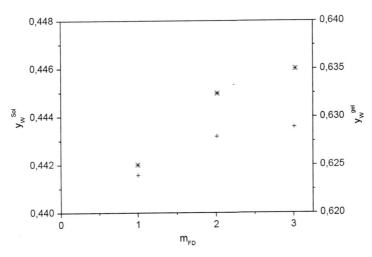

Fig. 30 Estimation of the optimal feed plate for the CPF column for the fractionation according the chemical composition if the original polymer is described by (16). The *stars* represent the values of the sol fraction and the *crosses* the values from the gel fractions

The calculation results obtained by simulation of the fractionation of a copolymer having the initial distribution function given in (16) for different values of n^{FD}/n^{EA} are shown in Fig. 31. The nonuniformity decreases in the gel as well in the sol phase if more extraction agent is used (Fig. 31a). The fractionation according to the chemical composition shows an identical trend. The finding can be explained by

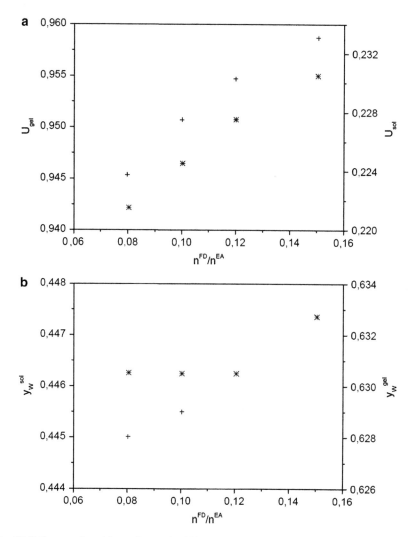

Fig. 31 Influence of working point on the CPF fractionation effect with respect to the molecular weight (**a**) and the chemical composition (**b**) for a copolymer having an initial distribution given by (16). The *stars* represent the values of the sol fraction and the *crosses* the values from the gel fractions

the dilution effect. However, increasing the amount of extraction agent leads to a decrease in the polymer amount present in the column and hence in the produced fraction. In this situation, a compromise between high fractionation effect and the amount of copolymer in the fractions has to be found.

The next operative parameter, which can be optimized by simulation, is the cooling temperature at the condenser. This parameter is also optimized by simulations of the CPF column, where a copolymer is fractionated with a feed distribution

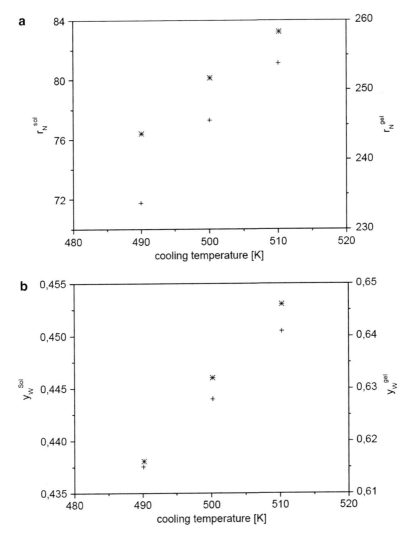

Fig. 32 Influence of cooling temperature in the condenser on the CPF fractionation effect with respect to the molecular weight (**a**) and the chemical composition (**b**) for a copolymer having an initial distribution given by (16). The *stars* represent the values of the sol fraction and the *crosses* the values from the gel fractions

given in (16). The simulation results are demonstrated in Fig. 32 for the fractionation of both polydispersities. From this figure, it can be seen that the optimal temperature in the condenser depends on the phase that forms the final fraction. If the fraction is to be isolated from the sol phase, then the condenser should work at temperatures as low as possible. However, if the fraction is to be taken from the gel phase, the condenser should not be used. These results are of practical importance

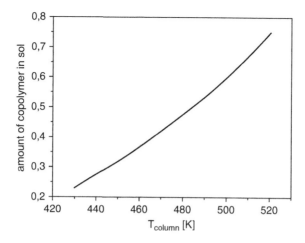

Fig. 33 The relationship of the amount of copolymer present in the sol phase and the temperature in the column

in the case where the CPF is applied for cutting the short or the large molecular weight parts from a synthetic polymer in order to tailor the material properties.

From the calculation results of the stepwise fractionation methods, it is known that the amount of polymer should by equal in all fractions. In CPF, the amount of polymer segments depends on the chosen working point (n^{FD}/n^{EA}) and the temperature in the column. In Fig. 33, the relationship between the temperature in the column and the amount of polymer present in the sol phase is shown.

Using all this knowledge from the simulations above, the following so-called optimized parameters for the CPF column:

$$X^{FD} = 0.11 \quad Z^{FD} = 0.5 \quad Z^{EA} = 0.15 \quad n^{FD}/n^{EA} = 0.08$$
$$m_{Max} = 6 \quad m_{FD} = 3 \quad T_{condenser} = 430 \text{ K} \tag{74}$$

were used to fractionate a copolymer into five fractions by four fractionation runs. The temperature for each fractionation run was selected in such a way that in every fraction there was nearly the same amount of polymer.

The results for this fractionation are presented in Table 2, Fig. 34, and Fig. 35. The differences between the original operating parameters (73) to the optimized parameters (74) are, first, that the n^{FD}/n^{EA} ratio is changed slightly and, second, that the temperature gradient is much more pronounced. The temperature at the condenser is changed from 520 to 430 K, whereby the temperature in the column is raised, especially for the last fractionation run. Similar to the results found for stepwise fractionation, the most important feature for an effective fractionation in column is also that the polymer is equally distributed in the corresponding fractions.

The improvement can be recognized clearly if the fractionation with respect to the chemical composition is analyzed (Table 2 and Fig. 35). The fractionation

Theory of Random Copolymer Fractionation in Columns

Table 2 Calculated fractionation data for CPF, where the original copolymer has a feed distribution given by (16). The operating conditions are given in (74)

Fraction no.	T (K)	y_W	r_N	U
1	430	0.387	43.8	0.992
2	480	0.435	112.8	0.381
3	530	0.502	157.2	0.295
4	580	0.589	222.2	0.214
5	580	0.698	323.1	0.149

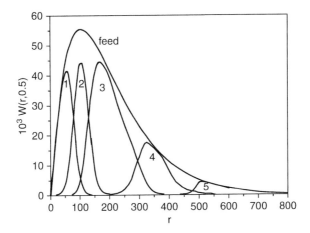

Fig. 34 CPF calculations using optimized operative parameters, where the *numbers* indicate the fraction number

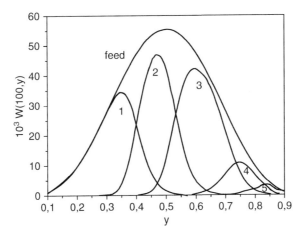

Fig. 35 CPF calculations using optimized operative parameters, where the *numbers* indicate the fraction number

according both polydispersities can be characterized by a high effectivity, which can been seen in Figs. 34 and 35.

In summary, both fractionation methods using columns can be applied for the fractionation of copolymers; however, the separation efficiency for the fractionation with respect to the molecular weight is lower than for the fractionation of homopolymers. For the development of further fractionation methods for copolymers, it is suggested that the CPF column is also used for fractionation in two directions, similar to the cross-fractionation. This can be realized experimentally very simply by changing the solvent mixture in the different runs, necessary to produce different fractions.

4 Summary

For the first time, a theoretical framework for the fractionation of statistical copolymers using successive fractionation methods and columns is introduced, taking into account the polydispersity with respect to molecular weight and chemical composition.

The application of this theoretical framework based on continuous thermodynamics allows the investigation of operating parameters (solvent gradient, temperature gradient, features of the fractionation column, fractionation strategy) on the efficiency of the fractionation, where the two-dimensional distribution of statistical copolymers is completely taken into account. From the thermodynamic point of view, copolymer fractionation is the successive establishing of LLE for suitable solutions of the polymer to be fractionated. Similar to the theoretical description of distillation or extraction columns in chemical engineering, the column is divided into theoretical stages. Assuming an LLE on each theoretical stage, the polymer fractionation can be modeled using phase equilibrium thermodynamics.

From the results of calculations carried out for the successive cross-fractionations, where in principal four different fractionation strategies are possible, it can be concluded that the fractionations in both solvent mixtures should be performed using SSF, meaning that the obtained sol phase should always be taken as a fraction.

During simulation of the fractionation in columns, such as the BW column or the CPF column, the influence of the operative conditions on the fractionation effectivity was investigated. For the simulation of the BW column, the main focus was the analytical purpose and in the simulation of the CPF column, the focus was the preparative purpose. Similar to the results found for the stepwise fractionation, the most important feature for an effective fractionation in column is also that the polymer is equally distributed in the corresponding fractions. This can be achieved by a suitable chosen concentration or temperature gradient, or both.

Acknowledgment Sincere thank is given to Dr. Heike Kahl for support in preparation of the manuscript.

References

1. Rätzsch MT, Kehlen H (1989) Continuous thermodynamics of polymer systems. Prog Polym Sci 14:1–46
2. Wohlfarth C, Rätzsch MT (1990) Continuous thermodynamics of copolymer systems. Adv Polym Sci 98:49–114
3. Rätzsch MT, Kehlen H, Browarzik D (1985) Liquid–liquid equilibrium of polydisperse copolymer solutions. Multivariate distribution functions in continuous thermodynamics. J Macromol Sci Chem A 22:1679–1680
4. Rätzsch MT, Browarzik D, Kehlen H (1989) Refined continuous thermodynamic treatment for the liquid–liquid equilibrium of copolymer solutions. J Macromol Sci Chem A 26:903–920
5. Rätzsch MT, Kehlen H, Browarzik D, Schirutschke M (1986) Cloud-point curves for the system copoly(ethylene-vinyl acetate) plus methylacetate. Measurement and prediction by continuous thermodynamics. J Macromol Sci Chem A 23:1349–1361
6. Rätzsch MT, Browarzik D, Kehlen H (1990) Liquid–liquid equilibrium of copolymer solutions with broad and asymmetric chemical distribution. J Macromol Sci Chem A 27:809–830
7. Browarzik C, Browarzik D, Kehlen H (2001) Phase-equilibrium calculation for solutions of poly(ethylene-co-vinyl acetate) copolymers in supercritical ethylene using a cubic equation of state. J Supercrit Fluids 20:73–88
8. Browarzik D, Rätzsch MT, Wohlfarth C (2003) High pressure phase equilibria in the system ethylene + vinylacetate + (ethylene vinylacetate) copolymer treated by continuous thermo-dynamics. Acta Polymerica 40:457–462
9. Koningsveld R (1970) Preparative and analytical aspects of polymer fractionation. Adv Polym Sci 7:1–69
10. Tung LH (ed) (1977) Fractionation of synthetic polymers. Marcel Dekker, New York
11. Koningveld R, Stockmayer WH, Nies E (2001) Polymer phase diagrams. Oxford University Press, Oxford
12. Provder T, Whited M, Huddleston D, Kuo CY (1997) Characterization of compositional heterogeneity in copolymers and coatings systems by GPC/FTIR. Prog Org Coating 32:155–165
13. Pratt JA, McHugh MA (1996) Supercritical-fluid fractionation of poly(ethylene-co-acrylic acid). J Supercrit Fluids 9:61–66
14. Scholsky KM, O'Connor KM, Weiss CS, Krukonis VJ (1987) Characterization of copolymers fractionated using supercritical fluids. J Appl Polym Sci 33:2925–2934
15. Pratt JA, Lee SH, McHugh MA (1993) Supercritical fluid fractionation of copolymers based on chemical composition and molecular weight. J Appl Polym Sci 49:953–966
16. Glöckner G, Kroschwitz H, Meissner C (1982) HP precipitation chromatography of styrene–acrylonitrile copolymers. Acta Polymerica 33:614–616
17. Glöckner G, van den Berg JHM, Meijerink NLJ, Scholte TG, Koningsveld R (1984) Size exclusion and high-performance precipitation liquid chromatography of styrene–acrylonitrile copolymers. Macromolecules 17:962–967
18. Glöckner G, van den Berg JHM (1987) Copolymer fractionation by gradient high-performance liquid chromatography. J Chromatogr 384:135–144
19. Glöckner G, Ilchmann D (1984) Hochdruck-Fällungschromatographie von α-Methylstyren/ Acrylnitril-Copolymeren. Acta Polymerica 35:680–683
20. Mori S, Uno Y (1987) Operational variables for the separation of styrene–methyl methacrylate copolymers according to chemical composition by liquid adsorption chromatography. Anal Chem 59:90–94
21. Teramachi S, Hasegawa A, Shima Y, Akatsuja M, Nakajima M (1979) Separation of styrene–methyl acrylate copolymer according to chemical composition, using high-speed liquid chromatography. Macromolecules 12:992–996

22. Glöckner G (1987) Quantitative aspects of gradient HPLC of copolymers from styrene and ethyl methacrylate. Chromatographia 23:517–524
23. Glöckner G, Stickler M, Wunderlich W (1987) Investigation of copolymers of styrene and ethyl methacrylate by size exclusion chromatography and gradient HPLC. Fresenius Z Anal Chem 328:67–81
24. Glöckner G, Barth HG (1990) Use of high-performance liquid chromatography for the characterization of synthetic copolymers. J Chromatogr 499:645–654
25. Glöckner G, Stickler M, Wunderlich W (1989) Separation of *stat*-copoly(styrene/2-methoxyethyl methacrylate) samples according to composition by gradient high-performance liquid chromatography. J Appl Polym Sci 37:3147–3161
26. Xu Z, Yuang P, Zhong J, Jiang E, Wu M, Fetters LJ (1989) The characterization of a poly (ethylene terephthalate)–poly(tetramethylene ether) multiblock copolymer via cross fractionation and size exclusion chromatography. J Appl Polym Sci 37:3195–3204
27. Stejskal J, Strakova D, Kratochvil P, Smith SD, McGrath JE (1989) Chemical composition distribution of a graft copolymer prepared from macromonomer: fractionation in demixing solvents. Macromolecules 22:861–865
28. Dong Q, Fan ZQ, Fu ZS, Xu JT (2007) Fractionation and characterization of an ethylene–propylene copolymer produced with a $MgCl_2/SiO_2/TiCl_4$ diester type Ziegler–Natta catalyst. J Appl Polym Sci 107:1301–1309
29. Podešva J, Stejskal J, Prochăzka O, Špaček P, Enders S (1993) Fractionation of a statistical copolymer in a demixing-solvent system: theory and experiment. J Appl Polym Sci 48:1127–1135
30. Kakugo M, Miyatake T, Mizunuma K (1991) Chemical composition distribution of ethylene-1-hexane copolymer prepared with $TiCl_3–Al(C_2H_5)_2Cl$ catalyst. Macromolecules 24:1469–1472
31. van Asten AC, van Dam RJ, Kok WT, Tijssen R, Poppe H (1995) Determination of the compositional heterogeneity by polydisperse polymer samples by the coupling of size-exclusion chromatography and thermal field-flow fractionation. J Chromatogr A 703:245–263
32. Hwang SW, Kim E, Shin C, Kim JH, Ryu DY, Park S, Chang T (2007) Unusual sensitivity of closed-loop phase behavior to chain size and distribution. Macromolecules 40:8066–8070
33. Ogawa T (1990) Fractionation and characterization of polyacetal copolymers by column elution method. J Appl Polym Sci 40:1519–1527
34. Zhang M, Lynch DT, Wanke SE (2000) Characterization of commercial linear low-density polyethylene by TREF-DSC and TREF-SEC cross fractionation. J Appl Polym Sci 75:960–967
35. Kong J, Fab X, Xie Y, Qiao W (2004) Study on molecular chain heterogeneity of linear low-density polyethylene by cross-fractionation of temperature rising elution fractionation and successive self-nucleation/annealing thermal fractionation. J Appl Polym Sci 94:1710–1718
36. van den Glöckner G, Berg JHM, Meijerink NLJ, Scholte TG, Koningsveld R (1984) Two-dimensional analysis of copolymers by size-exclusion chromatography and gradient-elution reversed-phase precipitation chromatography. J Chromatogr 317:615–624
37. Cho KH, Park YH, Jeon SJ, Kim WS, Lee DW (1997) Retention behavior of copolymers in thermal field-flow fractionation and gel permeation chromatography. J Liq Chromatogr Relat Technol 20:2741–2756
38. Nakano S, Goto Y (1981) Development of automatic cross fractionation: combination of crystallizability fractionation and molecular weight fractionation. J Appl Polym Sci 26:4217–4231
39. Springer H, Hengse A, Hinrichsen G (1990) Fractionation and characterization of a 1-butene linear low density polyethylene. J Appl Polym Sci 40:2173–2188
40. Baker CA, Williams RJP (1956) A new chromatographic procedure and its application to high polymers. J Chem Soc:2352–2362
41. Barker PE, Hatt BW, Williams AN (1978) Fractionation of a polymer using a preparative-scale continuous chromatograph. Chromatographia 11:487–493

Theory of Random Copolymer Fractionation in Columns 327

42. Gerrissen H, Roos J, Wolf BA (1985) Continuous fractionation and solution properties of PVC. 1. Continuous fractionation, characterization. Makromol Chemie 186:735–751
43. Kalal J, Marousek V, Svec F (1974) Fraktionierung von Vinylchlorid/Vinylacetat Copolymeren. Angew Makromol Chem 38:45–55
44. Mencer HJ, Kunst B (1978) A modified column method of polymer fractionation. Colloid Polym Sci 256:758–760
45. Wu AH, Prausnitz JM (1990) Fractionation of polydisperse polymer using an antisolvent. Application of continuous thermodynamics. J Appl Polym Sci 39:629–637
46. Rätzsch MT, Kehlen H, Tschersich L (1989) Application of continuous thermodynamics to polymer fractionation. J Macromol Sci Chem A26:921–935
47. Rätzsch MT, Enders S, Tschersich L, Kehlen H (1991) Polymer fractionation calculations using refined free energy relations. J Macromol Sci Chem A28:31–46
48. Rätzsch MT, Kehlen H, Tschersich L, Wolf BA (1991) Application of continuous thermodynamics to polymer fractionation. Pure Appl Chem 63:1511–1518
49. Weinmann K, Wolf BA, Rätzsch MT, Tschersich L (1992) Theory-based improvements of the CPF (continuous polymer fractionation) demonstrated for poly(carbonate). J Appl Polym Sci 45:1265–1279
50. Rätzsch MT, Tschersich L, Kehlen H (1990) Simulation of Baker–Williams fractionation by continuous thermodynamics. J Macromol Sci A27:999–1013
51. Folie B (1996) Single-stage fractionation of poly(ethylene-co-vinyl acetate) in supercritical ethylene with SAFT. AIChE J 42:3466–3476
52. Litmanovich AD, Shtern VY (1967) On the cross-fractionation of copolymers. J Polym Sci C 16:1375–1382
53. Ogawa T, Inaba T (1974) Analysis of the fractionation of ethylene–propylene copolymerization products. J Appl Polym Sci 18:3345–3363
54. Flory PJ (1953) Principles of polymer chemistry. Cornell University Press, Ithaca
55. Barker JA (1952) Cooperative orientation effects in solutions. J Chem Phys 20:1526–1531
56. Stockmayer WH (1945) Distribution of chain lengths and compositions in copolymers. J Chem Phys 13:199–207
57. Becker F, Buback M, Latz H, Sadowski G, Tumakaka F (2004) Cloud-point curves of ethylene-(meth)acrylate copolymers in fluid ethene up to high pressures and temperature – experiment study and PC-SAFT modeling. Fluid Phase Equilib 215:263–282
58. Kleiner M, Tumakaka F, Sadowski G, Latz H, Buback M (2006) Phase equilibria in polydisperse and associating copolymer solutions: poly(ethylene-co-(meth)acrylic acid)–monomer mixture. Fluid Phase Equilib 241:113–123
59. Gross J, Spuhl O, Tumakaka F, Sadowski G (2003) Modeling copolymer systems using the perturbed-chain SAFT equation of state. Ind Eng Chem Res 42:1266–1274
60. Wohlfarth C, Rätzsch MT (1987) Kontinuierliche Thermodynamik des Entmischungsgleichgewichtes: Berechnung der Schattenkurve im System Ethylen + Vinylacetat + (Ethylen-Vinylazetat)-Copolymer. Acta Polymerica 38:156–158
61. Browarzik D, Kehlen H (1996) Stability of polydisperse fluid mixtures. Fluid Phase Equilib 123:17–28
62. Teramachi S, Tomioka H, Sotokawa M (1972) Phase-separation phenomena of copolymer solutions and fractionation of copolymers by chemical composition. J Macromol Sci A6:97–107
63. Teramachi S, Nagasawa M (1968) The fractionation of copolymers by chemical composition. J Macromol Sci A2:1169–1179
64. Rosenthal AJ, White BB (1952) Fractionation of cellulose acetate. Ind Eng Chem 44:2693–2696
65. Teramachi S, Kato Y (1971) Study of the distribution of chemical composition in low-conversion copolymer by cross-fractionation. Macromolecules 4:54–56
66. Teramachi S, Hasegawa A, Hasegawa S, Ishibe T (1981) Determination of chemical composition distribution in a high-conversion copolymer of styrene and methyl acrylate by cross fractionation. Polym J 4:319–323

67. Teramachi S, Hasegawa A, Yoshida S (1982) Comparison of cross fractionation and TLC methods in the determination of the compositional distribution of statistical copolymer. Poly J 14:161–164
68. Spychaj T (1987) Solvent/non-solvent solubility and polarity parameters in fractional separation of oligomers 2. Epoxy resins. Angew Makromol Chem 149:127–138
69. Spychaj T, Hamielec AE (1988) Solvent/non-solvent solubility and polarity parameters in fractional separation of oligomers 2. Styrene–acrylic acid copolymer. Angew Makromol Chem 157:137–151
70. Stejskal J, Kratochvil P (1978) Fractionation of a model random copolymer in various solvent systems. Macromolecules 11:1097–1103
71. Stejskal J, Kratochvil P, Strakova D (1981) Study of the statistical chemical heterogeneity of copolymers by cross fractionation. Macromolecules 14:150–154
72. Bodor G, Dalcolmo HJ, Schröter O (1989) Structural and property correlation of ethylene-α-olefin copolymers. Colloid Polym Sci 267:480–493
73. Arranz F, Sanchez-Chaves M (1985) Chemical heterogeneity of a partially modified dextran with ethyl carbonate groups. Angew Makromol Chem 135:139–149
74. Teramachi S, Fukao T (1974) Cross fractionation of styrene–acrylonitrile copolymer. Polym J 6:532–536
75. Teramachi S, Kato Y (1970) Cross fractionation of styrene–butadiene copolymer. J Macromol Sci A4:1785–1796
76. Smith WV (1970) Precipitation chromatography theory. J Polym Sci A 2(8):207–224
77. Mac Lean DL, White JL (1972) The precipitation chromatographic column: theory, critique and method of design. Polymer 13:124–132
78. Geerissen H, Roos J, Wolf BA (1985) Continuous fractionation and solution properties of PVC, 1. Continuous fractionation, characterization. Makromol Chem 186:735–751
79. Geerissen H, Schützeichel P, Wolf BA (1990) Large scale fractionation of polyethylene by means of the continuous polymer fractionation (CPF) method. Makromol Chem 191:659–670
80. Gosch CI, Haase T, Wolf BA, Kulicke WM (2002) Molar mass distribution and size of hydroxyethyl starch fractions obtained by continuous polymer fractionation. Starch/Stärke 54:375–384
81. Hagenaars AC, Pesce JJ, Bailly C, Wolf BA (2001) Characterization of melt-polymerized polycarbonate: preparative fractionation, branching distribution and simulation. Polymer 42:7653–7661
82. Haberer M, Wolf BA (1995) Continuous fractionation of poly(acrylic acid). Angew Makromol Chem 228:179–184
83. Geerissen H, Roos J, Schützeichel P, Wolf BA (1987) Continuous fractionation and solution properties of PIB. I. Search for the best mixed solvent and first results of the continuous polymer fractionation. J Appl Polym Sci 34:271–285
84. Geerissen H, Schützeichel P, Wolf BA (1987) Continuous fractionation and solution properties of PIB. II. CPF optimization. J Appl Polym Sci 34:287–305
85. Samadi F, Eckelt J, Wolf BA, López-Villanueva FJ, Frey H (2007) Branched versus linear polyisoprene: fractionation and phase behavior. Eur Poly J 43:4236–4243
86. Eckelt J, Haase T, Loske S, Wolf BA (2004) Large scale fractionation of macromolecules. Macromol Mater Eng 289:393–399
87. Xiong X, Eckelt J, Wolf BA, Zhang Z, Zhang L (2006) Continuous spin fractionation and characterization by size-exclusion chromatography for styrene–butadiene block copolymers. J Chromatogr A 1110:53–60
88. Hagenaars AC, Bailly C, Schneider A, Wolf BA (2002) Preparative fractionation and characterization of polycarbonate/eugenol-siloxane copolymers. Polymer 43:2663–2669
89. Schmiedbauer J, Sybert PD (2000) In: Legrand DG, Bendler JT (eds) Handbook of polycarbonate science and technology. Marcel Dekker, New York
90. Schiruschke M (1976) Quellungsmessungen an Hochpolymeren. Diploma thesis, Technical University Merseburg

Adv Polym Sci (2011) 238: 329–387
DOI: 10.1007/12_2010_82
© Springer-Verlag Berlin Heidelberg 2010
Published online: 13 July 2010

Computer Simulations and Coarse-Grained Molecular Models Predicting the Equation of State of Polymer Solutions

Kurt Binder, Bortolo Mognetti, Wolfgang Paul, Peter Virnau, and Leonid Yelash

Abstract Monte Carlo and molecular dynamics simulations are, in principle, powerful tools for carrying out the basic task of statistical thermodynamics, namely the prediction of macroscopic properties of matter from suitable models of effective interactions between atoms and molecules. The state of the art of this approach is reviewed, with an emphasis on solutions of rather short polymer chains (such as alkanes) in various solvents. Several methods of constructing coarse-grained models of the simple bead–spring type will be mentioned, using input either from atomistic models (considering polybutadiene as an example) or from experiment. Also, the need to have corresponding coarse-grained models of the solvent molecules is emphasized, and examples for various dipolar and quadrupolar fluids and their mixtures with short alkanes are given. Finally, we mention even more simplified models, such as the bond fluctuation model on the simple cubic lattice, treating applications like micelle formation in block copolymer solutions or isotropic–nematic phase transitions in solutions of stiff polymers as case studies. Comparisons with pertinent predictions from approximate analytical theories will be briefly mentioned, as well as applications, to understand experimental results.

Keywords Coarse-graining · Molecular dynamics · Monte Carlo · Phase diagrams · Unmixing of solutions

---

K. Binder (✉), W. Paul, P. Virnau, and L. Yelash
Institut für Physik, Johannes Gutenberg-Universität Mainz, Staudinger Weg 7, 55099 Mainz, Germany
e-mail: kurt.binder@uni-mainz.de

B. Mognetti
The Rudolf Peierls Centre for Theoretical Physics, 1 Keble Road, Oxford OX1 3NP, UK
e-mail: b.mognetti@physics.ox.ac.uk

Contents

1	Introduction	331
2	Molecular Models for Polymers and Solvents	333
	2.1 Atomistic Models	333
	2.2 Coarse-Grained Models in the Continuum and on the Lattice	341
	2.3 Mapping Atomistic Models to Coarse-Grained Models	346
3	Basic Aspects of Simulation Methods	355
	3.1 Molecular Dynamics	355
	3.2 Monte Carlo	362
4	Modeling the Phase Behavior of Some Polymer Solutions: Case Studies	366
	4.1 Alkanes in Carbon Dioxide	366
	4.2 Alkanes in Dipolar Solvents	370
	4.3 Solutions of Stiff Polymers and the Isotropic–Nematic Transition	372
	4.4 Solutions of Block Copolymers and Micelle Formation	375
5	Conclusions and Outlook	378
	References	380

Abbreviations and Symbols

AO	Asakura–Oosawa model
CMC	Critical micelle concentration
DPD	Dissipative particle dynamics
EPM	Elementary physical model
EPM2	Second elementary physical model
FENE	Finitely extensible nonlinear elastic
GEMC	Gibbs ensemble Monte Carlo
LJ	Lennard–Jones
MC	Monte Carlo
MD	Molecular dynamics
NIST	National Institute of Standards and Technology
$\mathcal{N}pT$	Constant particle number, constant pressure, and constant temperature ensemble
$\mathcal{N}VE$	Constant particle number, constant volume, and constant energy ensemble (= microcanonical ensemble)
$\mathcal{N}VT$	Constant particle number, constant volume, and constant temperature ensemble (= canonical ensemble)
PC-SAFT	Perturbed-chain statistical associating fluid theory
rRESPA	Reversible reference systems propagator algorithm
SAFT	Statistical associating fluid theory
SAW	Self-avoiding walk
SCFT	Self-consistent field theory
SM	Stockmayer model
SW	Square well

Computer Simulations and Coarse-Grained Molecular Models Predicting 331

TIμVT Thermodynamic integration in the constant chemical potential, constant volume, and constant temperature ensemble

TPT Thermodynamic perturbation theory

TPT1-MSA First-order thermodynamic perturbation theory combined with the mean spherical approximation

1 Introduction

Knowledge of the equation of state of polymeric systems (polymer melts, solutions, and blends) is a central prerequisite for many applications. Theoretical modeling of the phase behavior of polymeric systems on the basis of statistical thermodynamics has been a challenging problem for decades [1–11]. The initially [1,2] proposed lattice model of Flory and Huggins involves many crude approximations, with well-known shortcomings [6,10]; however, in complicated cases (e.g., ternary systems such as a polymer solution in a mixed solvent or a polymer blend in a single solvent) this approach might still be the method of choice [12]. One represents a (flexible) linear macromolecule by a (self-avoiding) random walk on a (typically simple cubic) lattice, such that each bead of the polymer occupies a lattice site (multiple occupancy of sites being forbidden, of course, to model excluded volume interactions between the effective monomers). The chemical bonds between neighboring monomers of the chain molecule then are just the links between neighboring lattice sites. Solvent molecules are often simply represented by vacant sites (V) of the lattice. When one deals with binary blends, two types of monomers (A and B) occur and, apart from the chain lengths N_A, N_B of the macromolecules, which are proportional to their molecular weights, several interaction parameters come into play. Even if we restrict enthalpic forces to nearest-neighbor interactions, three types of (pairwise) interaction parameters ε_{AA}, ε_{AB} and ε_{BB} are introduced, which then can be translated into the well-known Flory–Huggins parameters [6, 9]. Of course, the model can also be generalized to other chain architectures, e.g., block copolymers [6, 9, 11, 13]. However, although the lattice model underlying Flory–Huggins theory [1–8] and its generalizations (e.g., [14]) is an extremely simplified description of any polymeric material, the statistical thermodynamics of this model is rather involved because the analytic treatments require mean field approximations and further uncontrolled approximations [6, 15, 16]. The mean field treatment implies that critical exponents characterizing the singularities of the equation of state near critical points are those [6] of the Landau theory [17]. Studying the Flory–Huggins lattice model by Monte Carlo (MC) simulation methods [18–20], one avoids these approximations and obtains the correct Ising-like critical behavior [6, 15, 16, 21], which has also been established experimentally for the critical points of both polymer solutions and polymer blends (see, e.g., [22, 23]). Also, far from the critical region, the approximate counting of nearest-neighbor contacts between different chains (note that intrachain contacts do not contribute to phase separation) invalidate simple relations between the basic energy parameters

ε_{AA}, ε_{AB} and ε_{BB} and the corresponding Flory–Huggins parameters in the expression describing the free energy of mixing [6, 15, 16]. If the latter is adjusted to describe the simulation "data", a nontrivial dependence of the Flory–Huggins parameter(s) on temperature and volume fractions (ϕ_A, ϕ_B) of the two types of monomers in a blend results [9,16] (note that $\phi_A + \phi_B = 1 - \phi_V$, where ϕ_V is the volume fraction of vacant sites, which may represent "free volume" [24] or solvent [6]).

When one fits the Flory–Huggins theory to experiment [7], nontrivial dependence of Flory–Huggins interaction parameters on temperature and volume fractions also result, but might have other reasons than those noted above: in particular, it is important to take into account the disparity between size and shape of effective monomers in a blend, and also the effects of variable chain stiffness and persistence length [25, 26]). To some extent, such effects can be accounted for by the lattice cluster theories [27–30], but the latter still invokes the mean-field approximations, with the shortcomings noted above. In the present article, we shall focus on another aspect that becomes important for the equation of state for polymer materials containing solvent: pressure is an important control parameter, and for a sufficiently accurate description of the equation of state it clearly does not suffice to treat the solvent molecules as vacant sites of a lattice model. In most cases it would be better to use completely different starting points in terms of off-lattice models.

A basic approach for the description of polymer chains in the continuum is the Gaussian thread model [26, 31]. Treating interactions among monomers in a mean-field-like fashion, one obtains the self-consistent field theory (SCFT) [11, 32–36] which can also be viewed as an extension of the Flory–Huggins theory to spatially inhomogeneous systems (like polymer interfaces in blends, microphase separation in block copolymer systems [11, 13], polymer brushes [37, 38], etc.). However, with respect to the description of the equation of state of polymer solutions and blends in the bulk, it is still on a simple mean-field level, and going beyond mean field to include fluctuations is very difficult [11, 39–42] and outside the scope of this article.

A powerful theory that combines the Gaussian thread model of polymers with liquid-state theory is the polymer reference interaction site model [43, 44]. This approach accounts for the de Gennes [5] "correlation hole" effect, and chemical detail can be incorporated (in the framework of somewhat cumbersome integral equations that are difficult to solve and need various approximations to be tractable). Also, in this theory the critical behavior always has mean field character. The same criticism applies to the various versions of the statistical associating fluid theory (SAFT) [45–54], which rely on thermodynamic perturbation theory (TPT) with respect to the treatment of attractive interactions between molecules (or the beads of polymer chains). Some of those theories [50, 52, 53] seem to perform rather well when one restricts attention to the region far away from critical points, as a comparison with the corresponding MC simulation shows [53, 55, 56]. We shall discuss these comparisons in Sects. 4.1 and 4.2. Other variants of this approach, such as the perturbed chain statistical associating fluid theory (PC-SAFT) approach [51]include additional approximations and therefore sometimes suffer from spurious results, such as artificial multiple critical points in the phase diagram [54]. For this reason, this approach will not be emphasized here.

It should be clear from this brief survey of various attempts to develop theories for the equation of state for polymeric materials that we are still far from a fully satisfactory solution for this difficult problem. In this situation, computer simulations of molecular models are an attractive alternative. However, this approach is also plagued with some problems: although a small-molecule fluid, apart from the critical region, does not develop spatial correlations on scales larger than a few nanometers [57], a single macromolecular coil exhibits a nontrivial structure from the length of chemical bonds (0.1 nm) over the persistence length (1 nm) to the gyration radius (10 nm) [3, 5, 8, 25, 26, 31, 58]. As a consequence, simulational modeling must either restrict attention to relatively short polymer chains [10, 53, 55, 56, 59–74], or consider coarse-grained models [6, 9, 10, 15, 16, 21, 53, 55, 56, 75–83]. Even for both small-molecule systems and for coarse-grained models it is essential that one considers temperatures far above a possible glass transition temperature [71, 81–86], particularly if one applies molecular dynamics (MD) simulation methods [87–91]. We note that MC methods for polymers have been devised where moves occur that involve bond crossing or bond breaking, etc., [92–95], allowing equilibration of dense melts for very long chains. Since we are not aware that such algorithms have been broadly used for the study of thermodynamic properties of polymer blends, we shall not address these advanced algorithms (as well as other specialized MC algorithms for lattice models of polymers [96]) in the present article.

Finally, we mention the very promising idea of mapping atomistic models to coarse-grained ones, thus putting some information on chemical details into the effective parameters of a coarse-grained model in a systematic way [97–120]. In Sect. 2, we shall briefly review both atomistic and coarse-grained models, as well as mention some aspects of this systematic coarse-graining approach.

The outline of this article is as follows: after a short discussion of some of the models (Sect. 2) we recall the basic aspects of MD and MC methods (Sect. 3). Results of simulations of chemically detailed atomistic models for short alkanes, polyethylene melts, and polybutadiene melts are mentioned. Section 4 is devoted to a discussion of coarse-grained models for the description of the phase behavior of alkanes in various solvents (Sects. 4.1 and 4.2). Also, qualitative models for semiflexible polymers that exhibit nematically ordered phases [121–123] and for block copolymer solutions that exhibit micelle formation [124, 125] will be discussed. Section 5 presents our conclusions.

2 Molecular Models for Polymers and Solvents

2.1 Atomistic Models

In this article, we confine our attention to the modeling of polymeric systems in the framework of classical statistical mechanics. Processes where electronic degrees of freedom are involved (such as chemical reactions) are outside the scope of the present review. Also, we will not consider quantum fluctuations due to the nuclei (e.g., in orthorhombic crystalline polyethylene the quantum-mechanical zero-point motion of the atoms does affect the thermal expansion and elastic constants of the material [126]).

Thus, the starting point of both MD [87–91] and MC methods [18–20] is a classical potential $U(\{\vec{r}_i\})$ (often it is referred to as force field [71, 81, 82, 127]), which contains only the positions of all the atoms $\{\vec{r}_i\}$ as variables. Typically, $U(\{\vec{r}_i\})$ is decomposed into contributions describing intramolecular forces along a polymer chain, which are described by bond stretching potentials $U_\ell(|\vec{r}_{ij}|)$, $\vec{r}_{ij} = \vec{r}_i - \vec{r}_j$ (where ℓ is the bond length), bond angle potentials $U_{\text{bend}}(\Theta_{ijk})$ (describing local bond angles), torsional potentials $U_{\text{tors}}(\phi_{ijk\ell})$, and last but not least nonbonded interactions $U_{\text{nb}}(\vec{r}_{ij})$. The latter are typically assumed to be pairwise additive. For example, the bond length potential is often assumed to have a simple harmonic form:

$$U_\ell(|r_{ij}|) = \frac{1}{2} k_\ell \left(|\vec{r}_{ij}| - \ell_0 \right)^2 , \tag{1}$$

where k_ℓ is a "spring constant" for the chemical bond between the two neighboring atoms in a polymer chain, and ℓ_0 is their (classical) ground-state distance. Also, the bending potential often is assumed to be harmonic in the angle Θ_{ijk} formed between two successive bonds $\vec{r}_{ij}$ and $\vec{r}_{jk}$ along a chain:

$$U_{\text{bend}}(\Theta_{ijk}) = \frac{1}{2} k_\Theta (\Theta_{ijk} - \Theta_0)^2 , \tag{2}$$

where again k_Θ is a spring constant but now for chain bending, and Θ_0 the classical ground state value for the bond angle. Finally, the torsional potential (defined in terms of the angle $\Theta_{ijk\ell}$ that the bond $\vec{r}_{k\ell}$ makes with its projection into the plane formed by the bonds $\vec{r}_{ij}$ and $\vec{r}_{jk}$) can be parameterized as:

$$U_{\text{tors}}(\phi_{ijk\ell}) = \frac{1}{2} \sum_{n=1}^{n_{\max}} k_n \left[1 - \cos(n\phi_{ijk\ell}) \right], \tag{3}$$

where further constants $\{k_n\}$, and $n_{\max}$ enter. For neutral polymers, for which Coulomb interactions can be disregarded, the nonbonded interactions are often assumed to have the simple Lennard–Jones (LJ) form:

$$U_{\text{LJ}}(r) = 4\varepsilon \left[\left(\frac{\sigma}{r} \right)^{12} - \left(\frac{\sigma}{r} \right)^6 \right], \tag{4}$$

with ε describing the strength and σ the range of this potential. Note that $U_{\text{LJ}}(r)$ acts both between monomers of different chains and between monomers of the same chain if they are neither nearest, nor next-nearest, nor third-nearest neighbors along the chain (so that none of the interactions in (1)–(3) would apply).

Note that (1)–(4) describe the simplest case of an atomistic model of polymer chains, and a model of this type is in fact useful for describing polyethylene melts. In fact, a further approximation is commonly is invoked, namely the "united atom" approximation: rather than treating all carbon and hydrogen atoms of the polymer C_nH_{2n+2} explicitly, one lumps the hydrogens and carbons of the CH_2 groups along the chain interior as well as the CH_3 groups at the chain ends together into "superatoms". This reduction of the degrees of freedom can be viewed as a first step of "coarse-graining". Just as in a fully atomistic model (where C atoms and H atoms are individually considered), one must distinguish in the parameters ε, σ of the LJ interaction (4) whether one considers a C–C, C–H or H–H pair of atoms; on the united atom level one can distinguish whether one considers CH_2–CH_2, CH_2–CH_3 or CH_3–CH_3 pairs.

We emphasize that the intrachain potentials written in (1)–(3) should only be taken as simple generic examples. In cases of practical interest, the potentials are often more complicated, and it is necessary to consider cross-couplings between different degrees of freedom, i.e., coupling terms between bond stretching and bond angles, or coupling between bond angles and torsional angles, etc. Although parts of these potentials can be obtained from ab initio quantum chemistry methods [127, 128], in some cases a group of four successive monomers (i, j, k, l) already contain too many degrees of freedom to allow a highly accurate quantum-chemical treatment, and additional simplifications need to be introduced, which must be carefully tested against suitable experimental data. Of course, potentials such as the non-bonded LJ interaction (4) and its parameters are purely empirical. Thus, there is still ongoing research on the construction and further improvement of suitable force fields [127, 128]. Another important aspect is that the level of detail that is desirable for a force field also depends on the applications that one wants to use it for. For example, for a study of polyethylene in the melt [60, 61], it was clearly admissible to simplify the problem by using a united atom model, as described above. However, for a computational modeling study of crystalline orthorhombic polyethylene, an all-atom description was obviously required [126]. Despite the fact that this polymer is the chemically simplest macromolecule, the potential that was used [126] contained no less than 36 parameters [129], and it was found that for some properties the accuracy was still unsatisfactory [126].

In the present article, we shall not discuss crystalline polymers further, and thus potentials of the type described by (1)–(4) suffice. Even then, for dense melts containing long chains, equilibration by either MD or MC methods is very difficult. In cases such as polybutadiene, where the different conformational states are almost isoenergetic, one can simplify the problem further by setting the torsional potential to zero [70, 71], which leads to a considerable speed-up of the dynamics at low temperature. It has been checked that such a bead–spring model plus bond-angle potential is still able to reproduce both single chain structure factors and the structure factor describing the collective scattering from the melt [70].

When one deals with an atomistically detailed description of a polymer solution, one must also pay appropriate attention to the model used to describe the solvent molecules [55, 56]. For example, when we describe polyethylene in terms of an united atom model [56, 61] and use as a generic case methane (CH_4) as solvent, it would not make sense to treat the hydrogen atoms in the model of CH_4 explicitly while the CH_2 and CH_3-groups of polyethylene are treated as superatoms. Thus, when CH_4 is treated as a point particle as well, the only interaction between the solvent molecules that is left is also of the LJ type (4); remember that CH_4 is a neutral molecule that possesses neither a dipole moment nor a quadrupole moment. If one adjusts the LJ parameters $\varepsilon_{SS}, \sigma_{SS}$ for CH_4 such that the experimental critical temperature T_c and experimental density ρ_c are reproduced by the model, $T_c = 190.6$ K and $\rho_c = 10.1$ mol L^{-1} [130], then the vapor–liquid coexistence data both in the temperature–density plane (Fig. 1a) and in the pressure–temperature plane (Fig. 1b) are indeed reproduced over a wide temperature regime [56], as is the temperature dependence of the interfacial tension (Fig. 1c). For the sake of computational efficiency of the MC simulations needed to establish the phase diagram of the considered LJ model with sufficient accuracy, it was decided [131] to simulate a LJ model with truncated and shifted interactions rather than using the full potential (4):

$$U_{ij}(r) = U_{LJ}(r) + 127\varepsilon_{SS}/4096, \quad R \leq r_c = 2^{7/6}\sigma_{ss} , \tag{5}$$

whereas $U_{ij}(r > r_c) \equiv 0$. Note that the additive constant in (5) is chosen such that the potential $U_{ij}(r)$ between particles i and j is continuous at $r = r_c$. Figure 1 shows that the united atom approximation for methane does reproduce the liquid–vapor coexistence of this fluid with very good accuracy. When one uses the LJ parameters $\varepsilon_{SS}, \sigma_{SS}$ for this solvent (determined by the fit of the critical point as obtained from MC simulation to the experimental critical point data) as an input for approximate equation of state theories such as first-order TPT combined with the mean spherical approximation (TPT1-MSA) [50, 52], one obtains [56] a reasonable agreement with experimental data for $T \leq 170$ K. At higher temperatures deviations appear; in particular, TPT1-MSA overestimates the critical temperature significantly, and predicts a parabolic shape of the coexistence curve in the critical region (the difference between the coexisting liquid and vapor densities scales as $\rho_l - \rho_v \propto \sqrt{1 - T/T_c}$, while the actual coexistence curve is flatter, $\rho_l - \rho_v \propto (1 - T/T_c)^{\beta}$ with a non-mean-field exponent $\beta \approx 0.326$ [132, 133]). This discrepancy between TPT1-MSA and experiment (and simulation results) illustrates a general shortcoming of all mean-field-type equation of state descriptions in the critical region, as emphasized already in the Introduction.

The accuracy of the united atom description for polyethylene has also been carefully tested in the literature [60, 71], both by comparison of simulation results with experiments and with simulations dealing with an all-atom model where hydrogen atoms are explicitly considered. Of course, for polyethylene the vapor liquid critical point would occur at very high temperatures, where the macromolecule would no longer be chemically stable, and is of no physical interest; thus one uses data for single chain and collective structure factors to gauge the accuracy of the simulation models in this case.

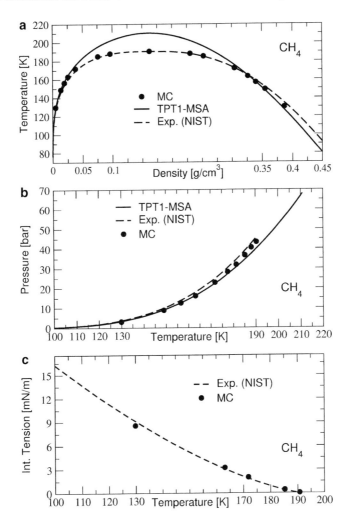

Fig. 1 Coexistence curve for CH$_4$ in the temperature–density plane (**a**), vapor pressure at coexistence (**b**), and surface tension versus temperature (**c**). *Dashed curves* are experimental results [130], *circles* show the MC results [56], while the *solid curves* in parts (**a**) and (**b**) show the results of the TPT1-MSA theory using the same interaction potential (with $\varepsilon_{ss} = 2.636 \times 10^{-21}$ J, $\sigma_{ss} = 3.758$ Å, in (5)) as the MC simulation. From Mognetti et al. [56]

United atoms models become more involved when chemically more complicated polymers are considered. Sometimes additional simplifications are also introduced, e.g., one may take bond lengths $|\vec{r}_{ij}|$ between neighboring bonds to be rigid, rather than allowing variation according to a harmonic potential. For example, for 1,4-polybutadiene, a force field with rigid bonds was proposed [134] that has three different rigid bond lengths, appropriate for the three distinct bonds occurring along the backbone of the chain: $\ell_0 = 1.53$ Å for the CH$_2$–CH$_2$ bond, $\ell_0 = 1.50$ Å for the

CH$_2$–CH bond, and $\ell_0 = 1.34$Å for the CH–CH bond. In this polymer, one also needs two different bond-angle potentials (CH–CH–CH$_2$ and CH–CH$_2$–CH$_2$), and four different torsional potentials (at the double bond differentiating *cis* from *trans*, at the allyl bond next to the double bond in the *cis* and in the *trans* group, and at the alkyl bond linking the monomers together).

Finally, we stress that the proper choice of atomistic model for the solvent molecules could be a tricky problem. Consider, e.g., the case of (supercritical) carbon dioxide (CO$_2$), which plays an important role in chemical technology [135–139], e.g., as a blowing agent in the production of polymeric foams [137–139]. Despite longstanding efforts, there is no consensus in the literature on the "best" effective potential describing the interaction between CO$_2$ molecules [131]. Figure 2 presents data for liquid–vapor coexistence [130] and the interfacial tension [130] of CO$_2$, and compares them with various pertinent theoretical predictions (adapted from [131]). There have been many proposals on how to fully parameterize all-atom potentials of this linear molecule [140–152], and coarse-grained models have also been proposed [53, 131, 153–155]. Figure 2 presents a counterpart to Fig. 1, where experimental data for CO$_2$ are compared to various theoretical predictions obtained from the computer simulation of such models [53, 131, 156]. It is clearly seen that there still occur significant disagreements between most of these computations and the experimental data (and there is also disagreement between the theoretical models themselves). It is also clear that for molecules such as CO$_2$, which carry sizable quadrupole moment, a reduction to ordinary point particles interacting with LJ forces and nothing else (as attempted in [53]) is not a good choice, while amending this simple model with a (spherically averaged) quadrupolar interaction [131] yields very satisfactory results. When one takes the full quadrupolar interaction into account [157, 158], no significant improvement in the description of the equation of state is achieved, although structural properties (such as orientational correlations among molecules) can be accounted for more accurately [158]. Alternatively, one can get very accurate description of equation of state data from computer simulation of a two-center LJ model [153–155], but an atomistic interpretation of such a description is also lacking.

For solvents such as ammonia (NH$_3$) or hydrogen sulfide (H$_2$S), it is important to realize that such molecules carry a dipole moment μ. If one uses an all-atom model, it amounts to work with suitable partial charges on the sites of the atoms, and to deal with Coulomb interactions between the atoms of different molecules. If one wants to integrate hydrogen atoms in NH$_3$ or H$_2$S into an united atom, as done for CH$_2$ or CH$_3$ groups or methane, one can work with the Stockmayer model (SM) where the molecules are treated as point particles interacting with LJ plus dipolar forces, $r = |\vec{r}_i - \vec{r}_j|$, $\vec{\mu}_i$ being an unit vector in the direction of the dipole moment (which has the strength μ):

$$V_{SM}(\vec{r}) = 4\varepsilon_{SS}\left[\left(\frac{\sigma_{ss}}{r}\right)^{12} - \left(\frac{\sigma_{ss}}{r}\right)^6\right] + \frac{\mu^2}{r^3}\left[\vec{\mu}_i \cdot \vec{\mu}_j - \frac{3}{r^2}(\vec{\mu}_i \cdot \vec{r})(\vec{\mu}_j \cdot \vec{r})\right]. \qquad (6)$$

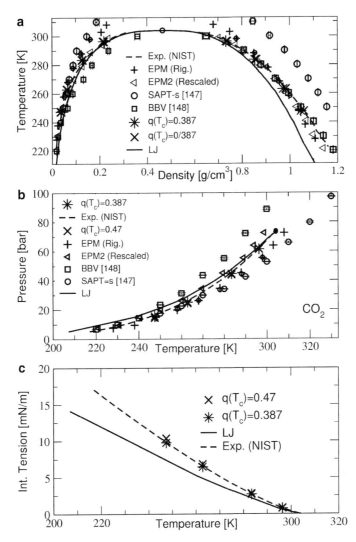

Fig. 2 Coexistence curve for CO_2 in the temperature–density plane (**a**), vapor pressure at coexistence (**b**), and surface tension versus temperature (**c**). *Dashed curves* are the experimental data, while *solid curves* describe the prediction of the simple (truncated and shifted) LJ model (5), where the critical temperature and density are adjusted to coincide with experiment to fix the two parameters ε_{ss} and σ_{ss}, as for Fig. 1. *Stars* and *crosses* denote the results of [131] for the parameter $q(T_c)$ that controls the strength of the quadrupolar interaction being chosen as $q(T_c) = 0.387$ or $q(T_c) = 0.47$, respectively (see Sect. 2.2). *Plus symbols* and *triangles* are the result of atomistic models called EPM and EPM2 [146]. *Small circles near the pluses* are the results for flexible monomers [146], which give essentially the same results for the thermodynamic properties as the model for rigid molecules. *Big circles* and *squares* are simulation results [156] for two ab initio potentials [146, 150]. Note that the interaction parameters of the EPM2 models have similarly been rescaled to fit the critical density and temperature of the experiment as done in Fig. 1, and that no prediction for the liquid–vapor surface tension from the atomistic models is available. From Mognetti et al. [131]

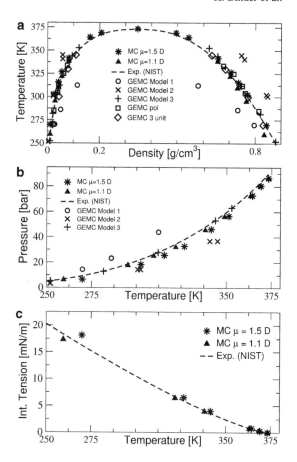

Fig. 3 Coexistence densities, coexistence pressure, and interface tension of hydrogen sulfide. The *dashed lines* are experimental results [130]. *Symbols* are MC results of a coarse-grained model (see Sect. 2.3) using two different values for the dipole moment μ, $\mu = 1.5$ D and $\mu = 1.1$ D, respectively, and results of various GEMC simulations of atomistic models [159–163]. *Model 1* depicts results tabulated in [159] for a three-atom model of H_2S [160]. *Model 2* shows results where H_2S is represented by a model with four "atoms" [161]. *Model 3* is a reparametrization of model 2 in [159] and *pol* is a model [162] that includes a polarizable site (for a total of five interacting "atoms" per molecule). Finally, *3 unit* is a recent [163] reparametrization of a three-site model, which also includes a three-center LJ interaction. From Mognetti et al. [55]

However, the slow decay of the dipole–dipole interaction in (6) (proportional to r^{-3}) makes the use of Ewald summation techniques necessary when one uses (6) in a simulation. Thus, the computational effort of working with the full Stockmayer potential (6) is comparable to the effort needed for all-atom models with partial charges at the atoms.

As an example for the problems in obtaining an accurate description of the thermodynamic properties of dipolar fluids from the computer simulation of atomistic models, Fig. 3 presents a comparison [55] of experimental data for H_2S [130] with a coarse-grained model [55] that we shall explain in Sect. 2.3 and various simulations of atomistic models [159–163]. Again, one concludes that simple versions of atomistic models do a rather poor job, whereas the more complicated recent versions can provide a reasonable description of the coexistence data, although (due to the use of Gibbs ensemble MC, GEMC, data) they neither yield results close to criticality nor provide any information on the surface tension. The coarse-grained model proposed in [55], based on a mapping of the SM (6) on an effective LJ model (see Sect. 2.3) overcomes both of these limitations if the effective dipole moment μ is adequately adjusted.

2.2 Coarse-Grained Models in the Continuum and on the Lattice

For long flexible polymer chains it has been customary for a long time [1, 2] to reduce the theoretical description to the basic aspects such as chain connectivity and to excluded volume interactions between monomers, features that are already present when a macromolecule is described by a self-avoiding walk (SAW) on a lattice [3]. The first MC algorithms for SAW on cubic lattices were proposed in 1955 [164], and the further development of algorithms for the simulation of this simple model has continued to be an active area of research [77, 96, 165–169]. Dynamic MC algorithms for multichain systems on the lattice have also been extended to the simulation of symmetric binary blends [15, 16]; comprehensive reviews of this work can be found in the literature [6, 81, 82]. It turns out, however, that for the simulation both of polymer blends [6, 9, 21, 82, 170, 171] and of solutions of semiflexible polymers [121–123], the bond fluctuation model [76, 79, 80] has a number of advantages, and hence we shall focus attention only on this lattice model.

Using the lattice spacing of the simple cubic lattice as the unit of length, $a = 1$, each coarse-grained macromolecule is represented as a chain of effective monomers connected by bond vectors, which can be taken from the set $\{(\pm 2, 0, 0), (\pm 2, \pm 1, 0), (\pm 2, \pm 1, \pm 1), (\pm 2, \pm 2, \pm 1), (\pm 3, 0, 0)(\pm 3, \pm 1, 0)\}$, including also all permutations between these coordinates. Altogether 108 different bond vectors occur, which lead to 87 different angles between successive bonds. Each effective monomeric unit is represented by an elementary cube of the lattice, blocking all eight sites at the corners of this cube from further occupation, thus realizing the excluded volume interaction between the monomers. Allowing for two types of polymers (A and B) in the system, it then is natural to also allow for (attractive) interactions of somewhat longer range between any pair of monomers (α, β). These interactions in most cases were assumed to have the simple square well (SW) form:

$$U_{\text{SW}}^{\alpha\beta}(r) = \begin{cases} -\varepsilon_{\alpha\beta} & 2 \leq r \leq r_{\text{c}}, \\ 0 & r > r_{\text{c}}. \end{cases} \tag{7}$$

In most cases, $r_{\text{c}} = \sqrt{6}$ was used (so all neighbors in the first-neighbor shell in a dense melt, defined from the first peak position in the radial pair distribution function $g(r)$ between monomers, are included [170–172]). The extremely short-range case $r_{\text{c}} = 2$ was also used [21]; then monomers attract each other only when they are nearest neighbors on the lattice. Of course, (7) also includes, as a special case, the case of a polymer solution $(\alpha = \beta)$ where only a simple species of polymer is present [173].

In this model, one can also introduce effective potentials that depend on the bond length b such as [121]:

$$U_b(b) = \varepsilon_0(b - b_0)^2, \tag{8}$$

and on the bond angle Θ between successive bonds, e.g., [121, 123]:

$$U_{\text{bending}}(\Theta) = -f\cos\Theta(1 + c\cos\Theta). \tag{9}$$

In [121–123], the constants ε_0, b_0, f, and c were chosen quite arbitrarily as $\varepsilon_0 = 4, b_0 = 0.86$ and $c = 0.03$ (ε_0 and f are quoted in units of absolute temperature $k_B T$, k_B being Boltzmann's constant). On the other hand, if one chooses a bond length potential [174, 175] defined as $U_b(b) = 0$ if $b = \sqrt{10}$ and $U_b(b) = 1$ otherwise, a rather good model for the glass transition of polymers is obtained [84] due to the resulting "geometric frustration" [174, 175]. Finally, we mention that occasionally one finds in the literature (e.g., [176–179]) another version of the bond fluctuation model, in which monomers take a single lattice site only and the bond vectors are allowed to be $\{(\pm 1, 0, 0), (\pm 1, \pm 1, 0)$ and sometimes also [178] $(\pm 1, \pm 1, \pm 1)\}$. All permutations between these coordinates are included, but this model will not be discussed further here because it has mostly only been applied to study mesophase ordering of block copolymers. We stress that the advantage of the bond fluctuation model [76, 79, 80] as described above is that at a volume fraction of $1\Phi = 0.5$ of occupied lattice sites, one reproduces both the single chain structure factor (as described by the Debye function [8]) of polymer chains as well as the collective structure factor of dense melts [175] qualitatively in a reasonable way. If one uses a dynamic algorithm in which monomers are chosen at random, and a lattice direction $(\pm x, \pm y, \pm z)$ is chosen at random, and a move of the monomer by one lattice unit is attempted as a trial move according to the Metropolis MC algorithm [18–20], then a qualitatively reasonable description of the polymer dynamics is also obtained [80, 82, 175]. For short chains, the dynamics correspond to the Rouse model [5, 8, 31] whereas, for long chains, reptation [5, 8, 31] is observed since for the chosen bond lengths no bond crossing is possible [79, 80]. On the other hand, in order to allow for a fast equilibration, moves can be introduced (such as the slithering snake algorithm [82, 83] or monomeric jumps over larger distances that allow for bond crossing [95]) that have no counterpart in the real dynamics of polymers, but do not alter the static properties of the model. Therefore, the bond fluctuation model has also been broadly used (e.g., [9, 180]) to simulate the dynamics of spinodal decomposition [181, 182] of polymer blends.

We now turn to coarse-grained off-lattice models. One strategy is to stay as close to the atomistic model as possible but to eliminate many degrees of freedom, e.g., for modeling alkane chains [183–185], both the bond length ℓ of C–C bonds and the bond angle Θ is fixed ($\ell = 1.54\text{Å}, \Theta = 112°$), but the torsional angle $\phi_{ijk\ell}$ (3) is kept as a variable. The advantage of such a model (with suitable choices of the torsional and nonbonded potentials [185]) is that one can still make a direct connection with polyethylene melts. Both local MC moves (where two subsequent

Computer Simulations and Coarse-Grained Molecular Models Predicting 343

monomers are rotated together to new positions, thus restricting the torsional angle to $\pm 60°$ and $180°$) and nonlocal ones (slithering-snake moves or "pivot rotations" [165]) have been implemented [185]. However, for the application of MD techniques (one of the principle advantages of off-lattice models in comparison with lattice models of polymers is [81–84] that MD accounts better for dynamic properties) models without such constraints for bond lengths and angles are more convenient. To avoid the small MD time step that the (rather stiff) potentials for bond lengths and bond angles [(1) and (2)] necessitate, one uses coarse-grained bead–spring models with rather soft "springs". The most commonly used "spring potential" is the finitely extensible nonlinear elastic (FENE) potential [75, 78, 186, 187]:

$$U_{\text{FENE}}(r) = -\frac{k}{2}R_0^2 ln\left(1 - r^2/R_0^2\right), \tag{10}$$

where the parameters k, R_0 can be chosen as $k = 7$, $R_0 = 2$ [186] or as $k = 30$, $R_0 = 1.5$ [84], for instance when one chooses a (truncated and shifted) LJ potential [such as (4) and (5)] and measures lengths in units of σ and energies in units of ε_{pp} (we use here indices "pp" to distinguish these interactions from those of the solvent). Note that in this model the LJ potential [(4) and (5)] acts between any pair of monomers, including nearest neighbors along a chain; thus the total potential for the length of an effective bond is in fact the sum of (5) and (10), $U_{\text{bond–length}}(r) = U_{ij} + U_{\text{FENE}}(r)$, whereas between nonbonded pairs only (5) acts. Although the minimum of (5) occurs at $U_{ij}(r = r_{\min})$ with $r_{\min} = 2^{1/6}\sigma_{\text{pp}}$, the minimum of the bond potential occurs at [84] $U_{\text{bond–length}}(r = r'_{\min})$ with $r'_{\min} \approx 0.96\sigma_{\text{pp}}$. The fact that $r_{\min} \neq r'_{\min}$ and that the ratio $r_{\min}/r'_{\min}$ does not fit to any simple crystal structure is responsible for the occurrence of glass-like freezing-in of this bead–spring model at low temperatures. At densities $\rho\sigma_{\text{pp}}^3 = 1$, the glass transition occurs roughly at $k_B T_g/\varepsilon_{\text{pp}} \approx 0.4$ [84], whereas the Θ-temperature (i.e., the temperature at which in the dilute limit very long bead–spring chains collapse into a dense globule) is much higher, namely $k_B\Theta/\varepsilon_{\text{pp}} \approx 3.3$ [10]. Thus, for many applications of the bead–spring model based on (4), (5) and (10), the glass-like behavior at low temperatures does not restrict its use in computer simulations. It has the advantage that both MC and MD methods are readily applicable for its study [10, 84].

This bead–spring model is an appropriate description for a very flexible chain, but an analog of the bond angle potential [(2) in the atomistic model or (9) for the bond fluctuation model], is not included here for simplicity. However, when one adjusts $\varepsilon_{\text{pp}}, \sigma_{\text{pp}}$ to the vapor–liquid critical temperatures and densities of short alkanes, as done for methane (Fig. 1), one obtains a rather good description of vapor–liquid coexistence data and the interfacial tension over a broad temperature range [56] (Fig. 4). Although it is known that alkanes do have a bond angle potential for the C–C bonds, it is ignored here because the simple bead–spring model based on (10) makes sense only if the effective monomers correspond to larger units formed by integrating several (e.g., about $n = 3$) carbon atoms in one unit. Thus, the bond potential $U_{\text{bond–length}}(r)$ defined above does not represent a single (stiff!)

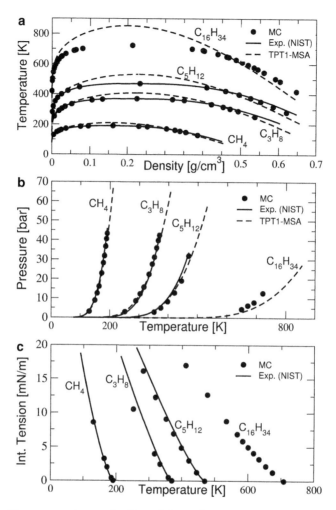

Fig. 4 Liquid–vapor coexistence densities of short alkanes in the temperature–density plane (**a**), corresponding coexistence pressure (**b**), and interfacial tension (**c**). *Symbols* are MC results, *solid curves* are experimental results [130], and *dashed curves* are predictions using the TPT1-MSA, employing the same choices of $\varepsilon_{pp}, \sigma_{pp}$ as an input as for the MC simulations. From Mognetti et al. [56]

C–C (chemical) bond, but rather represents a (softer!) effective bond between these effective units (Fig. 5). In this spirit, the (elongated) molecule C_3H_8 is still described by an effective point particle, whereas $C_{16}H_{34}$ is represented by a pentamer (Fig. 5). Thus, Fig. 5 gives a motivation for the use of these very simple coarse-grained bead–spring models, which are computationally orders of magnitude faster to simulate than a fully atomistic model. Note that one gains a factor of ten in the number of atoms when going from $C_{16}H_{34}$ to a pentamer, and also that for

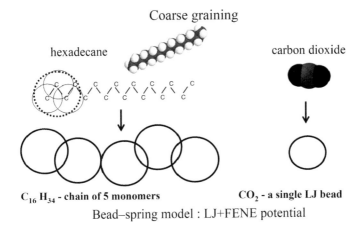

Fig. 5 Illustration of the interpretation of the coarse-grained models for polymers and solvent: In the case of short alkanes, typically three C–C bonds are taken together in one effective unit (*dotted circle*). The oligomer $C_{16}H_{34}$, containing 50 atoms or 16 united atoms, is thus reduced to an effective chain of five beads. Neighboring beads along a chain interact with a combination of LJ and FENE potentials. Nonbonded beads only interact with a single LJ potential. The solvent molecule (CO_2 in the present case) is represented by a point particle (in the case of CO_2 or C_6H_6, it carries a quadrupole moment; in the case of NH_3 or H_2S, it carries a dipole moment). From Yelash et al. [234]

the softer potential $U_{bond-length}(r)$ a significantly larger MD time step can be applied than when one uses (2), (3), etc.

Of course, when one deals with cooperative phenomena in a system containing a great number of very long chains, even the use of the bond fluctuation model or the bead–spring model in simulations is a big effort. So the question arises: is even a much coarser view of polymers useful? In the extreme case, a whole polymer chain is represented by a (very soft) effective particle. Murat and Kremer [188] suggested replacing the chains by soft ellipsoidal particles that can overlap strongly in the melt to take into account the fact that in the volume V taken by one chain of length N and radius $R_g \propto N^{1/2}$, $V \propto N^{3/2}$, there is space for a large number ($\propto N^{1/2}$) of other chains because the monomer density of the considered chain scales as $\rho \propto N/V \propto N^{-1/2}$.

The idea to coarse-grain the description of a system containing a very large number of polymer chains such that each chain is represented by a single effective particle dates back to the Asakura–Oosawa model [189–191] of polymer–colloid mixtures. In this model, the colloidal particles (e.g., cross-linked polystyrene spheres with radii in the size range 100 nm $< R_c <$ 1 μm) are represented as hard spheres, which have no other interactions than excluded volume interactions between themselves and with the polymers. The polymers are taken as soft spheres of radius R_p (which is thought to be of the order of the gyration radius of the chains) and are treated like particles of an ideal gas (i.e., they may overlap with no energy cost). The solvent molecules of these colloidal dispersions are not considered explicitly. This model is extremely popular in colloid science (e.g., [192–194]) because it predicts the phase diagram of these systems in qualitative agreement with experiment (e.g., [194, 195]). As is well known, the depletion attraction between the colloidal particles caused by the polymers can cause (entropically driven) phase separation between polymers and colloids. Due to the easy observability of the large colloidal particles, such systems have become model systems for the study of phase separation and interfacial phenomena (see, e.g., [194–198]).

Of course, the assumption that polymer coils can interpenetrate each other in solution with no free energy cost is approximately true at best in a solvent under theta conditions [3–8], but not in a good solvent. Thus, there have been numerous attempts to include the resulting soft repulsive interaction between the effective spheres representing polymers in a good solvent [199–204]. For example, Zausch et al. [204] made a simple choice for the polymer–polymer potential that was very convenient from the computational point of view, namely:

$$U_{ij}^{\mathrm{pp}}(r) = \varepsilon_{\mathrm{pp}}\left[1 - 10(r/r_{\mathrm{c}})^{3} + 15(r/r_{\mathrm{c}})^{4} - 6(r/r_{\mathrm{c}})^{5}\right], \tag{11}$$

which vanishes at $r = r_{\mathrm{c}}$; also, the force is continuous there. This potential is essentially a polynomial expansion of a cosine function. For the colloid–colloid and colloid–polymer interactions, standard Weeks–Chandler–Anderson potentials [57, 87] (i.e., LJ potentials cut at the minimum and shifted to zero there) were used. Figure 6 shows the phase diagram of this model and compares it to the corresponding phase diagrams of the simple Asakura–Oosawa model [204], for the case of $R_{\mathrm{p}}/R_{\mathrm{c}} = 0.8$. One notes that both phase diagrams are very similar to each other (and to corresponding experimental data, e.g., [195]). Thus, one sees from this example once more that the phase behavior can be insensitive to structural details; what matters is a sufficiently accurate description of effective potentials.

2.3 Mapping Atomistic Models to Coarse-Grained Models

In Sect. 2.1, it was argued that the use of atomistic models employing full chemical detail might need enormous computer resources in many cases and hence would not often be economical. In Sect. 2.2, we have seen that coarse-grained models, which are by far more economical for use in computer simulations, can yield useful information on the phase behavior of various systems, if one has good enough effective potentials. In Sect. 2.2, it was also argued that a good description of the phase behavior over a wide range of densities and temperatures is obtained if the effective potentials are chosen such that the experimental critical temperature and density (if known) are correctly reproduced. Although critical point data are available [130] for small molecules and oligomers, no such information exists for long macromolecules. Thus, it is very desirable to predict accurate effective interactions by other means: one very popular approach [97–119] attempts to construct coarse-grained models systematically from atomistic ones by integrating-out local

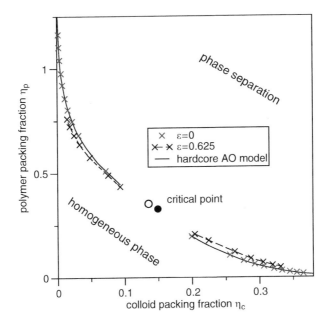

Fig. 6 Phase diagram of colloid–polymer mixture models in the plane of variables colloid volume fraction $\eta_c = \rho_c(4\pi R_c^3/3)$, where ρ_c is the density of colloidal particles, and polymer volume fraction $\eta_p\{\eta_p = \rho_p(4\pi R_p^3/3)\}$, where ρ_p is the density of polymer chains, for the Asakura–Oosawa model. In the case of soft potentials, such as (10), effective radii are obtained from the Barker–Henderson approach, see [204]. *Open circle* shows the critical point of the Asakura–Oosawa model, and the *closed circle* is the critical point of the model with $\varepsilon = 0.625$. Adapted from Zausch et al. [204]

degrees of freedom but keeping suitable degrees of freedom fixed at the scale of the coarse-grained model.

This very promising approach is still under development [104–119]; for this reason we shall not attempt to give an exhaustive review, but rather confine ourselves to the flavor of the approach, using as a specific example the case of 1,4-polybutadiene, for which well-established atomistic potentials exist [134].

In most cases [97–119], it has been useful to define the coarse-grained repeat units such that they comprise $n = 3 - 5$ C–C bonds along the backbone of the chain. This approach has also been used here (Fig. 7), such that one coarse-grained unit represents one butadiene monomer, hence containing four united atoms. The thickness of a polybutadiene chain is roughly 4.5 Å, which agrees with the size of the polybutadiene monomer [111], and hence this mapping is plausible from simple geometric considerations. In this mapping, no distinction is made between *cis* and *trans* units on the coarse-grained scale, so that on that scale a homopolymer model results.

One needs effective potentials for the degrees of freedom on the coarse-grained scale: the length L of the bond between two coarse-grained monomers, the angle Θ between two subsequent bonds, and the nonbonded interaction $U_{ij}^{cg}(r)$ between the

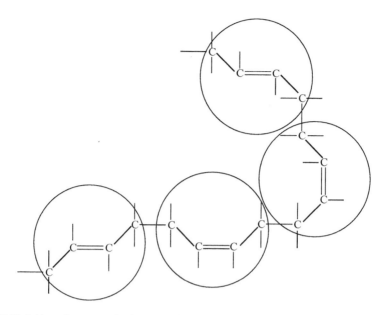

Fig. 7 Definition of coarse-grained repeat units (symbolized by *circles*) for the 1,4-polybutadiene chain. The atomistic model is an united atom model of a random copolymer of 45% *cis* and 55% *trans* content, without vinyl units. Bond lengths are constrained ($\ell = 1.53$ Å for the CH$_2$–CH$_2$ bond, $\ell = 1.5$ Å for the CH$_2$–CH bond and $\ell = 1.34$ Å for the CH–CH bond). From Strauch et al. [117]

coarse-grained (cg) effective monomers i,j. The intramolecular potentials $U_{\text{bond}}(L)$ and $U_{\text{angle}}(cos\Theta)$ can be found by the "Boltzmann inversion" [105–110] from the probability distribution $P(L), P(cos\Theta)$ of these degrees of freedom sampled from a simulation of the full atomistic model. For this simulation of the full atomistic model, a relatively short run of a relatively small system might suffice, at least in favorable cases. Thus [117]:

$$U_{\text{bond}}(L) = -k_B T ln[P(L)], \quad (12)$$

$$U_{\text{angle}}(cos\Theta) = -k_B T ln[P(cos\Theta)]. \quad (13)$$

Figure 8 shows the resulting potentials at three temperatures. In principle, effective potentials obtained from the Boltzmann inversion [(12) and (13)] will contain entropic contributions and, hence, in general must be state-point dependent; however, Fig. 8 suggests that in favorable cases this dependence is small.

However, it is difficult to obtain the nonbonded interaction from the same route: in addition, atomistic potentials for nonbonded interactions are often rather unreliable. Thus, the form of the nonbonded interaction was chosen to be a priori fixed to be of the LJ type:

$$U_{\text{LJ}}^{\text{cg}}(r) = 4\varepsilon_{\text{pp}}\left[(\sigma_{\text{pp}}/r)^n - (\sigma_{\text{pp}}/r)^m\right] \quad (14)$$

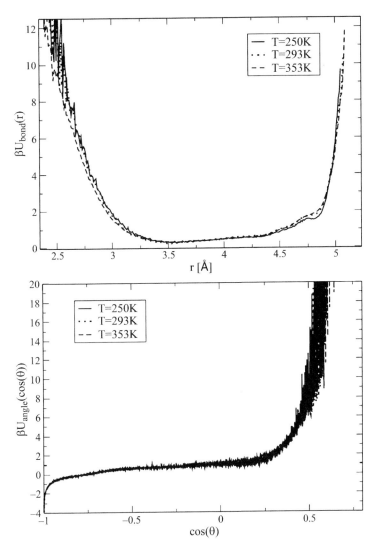

Fig. 8 Effective bond length potential $U_{bond}(r)/k_BT$ of the coarse-grained model for 1,4-polybutadiene (*upper graph*) and effective bond angle potential $U_{angle}(cos\Theta)/k_BT$ (*lower graph*). Three temperatures are included, as indicated. From Strauch et al. [117]

but it is no longer required that $n = 12$ and $m = 6$. Rather, one can argue that on the coarse-grained level, the nonbonded interaction should also be somewhat softer, and hence $r = 7, m = 4$ is a better choice [205]. In order to determine the parameters $\varepsilon_{pp}, \sigma_{pp}$ of (14), Strauch et al. [117] referred to a simulation at zero pressure of the full atomistic model to record the temperature dependence of the density (Fig. 9a) [206]. Simulating also the coarse-grained model by *NpT* MC methods for

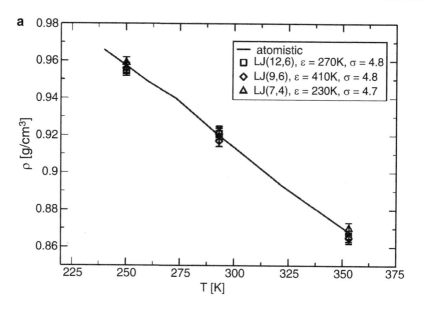

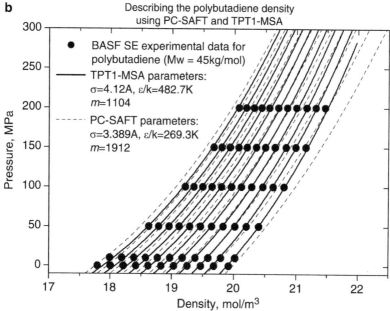

Fig. 9 (a) Zero pressure isobar in the 1,4-polybutadiene melt. The *line* shows MD results from the chemically realistic model. The *symbols* show average densities in the NpT MC simulation for the optimal choices of parameters for different versions of the LJ-type interaction. From Strauch et al. [117]. (b) Comparison between experimental data for polybutadiene melts in the temperature range from 299 to 461 K (*symbols*) and calculations using PC-SAFT (*dashed curves*) or TPT1-MSA (*solid curves*) models. Parameters of the fits are quoted in the figure, where m is the effective degree of polymerization, which is also treated as a fit parameter; and σ and ε refer to a nonbonded LJ (12,6) potential). The bond length potential is the FENE + LJ potential of Sect. 2.2, and no bond angle potential is used. Adapted from Binder et al. [120]

Computer Simulations and Coarse-Grained Molecular Models Predicting

various choices of $\varepsilon_{pp}, \sigma_{pp}$, one finds the optimal choices of parameters for different versions (i.e., choices of n and m) of the LJ interaction. In all cases, the values obtained for σ and ε are completely reasonable, σ being 4.7 or 4.8 Å, and ε being between 230 and 410 K.

Alternatively, one can try to obtain parameters for ε and σ through fitting data for the density $\rho = \rho(p, T)$ taken from experiment to a model calculation (Fig. 9b). One notes that the TPT1-MSA equation of state, which does not include a bond-bending potential and uses the (12,6) LJ potential rather than the (7,4) LJ potential of Strauch et al. [117], can fit a whole family of curves over a wide range of temperature and pressure. Since for the large molecular weight of 1,4-polybutadiene used in the experimental study, all the data fall far below the (unaccessible!) vapor–liquid critical point of the polymer, the inaccuracy of TPT1-MSA near the critical point (noted in Figs. 1 and 4) is irrelevant for the present purposes. For the large chain length ($N \approx 1100$) TPT1-MSA is much more convenient than using computer simulation. The LJ parameters obtained from this treatment ($\sigma = 4.12$ Å, $\varepsilon = 462.7$ K) are in the "same ball park" as those from systematic coarse-graining (due to the use of the FENE potential as a bond length potential in TPT1-MSA and the lack of a bond-bending potential, we cannot expect to find precisely the same parameters, of course!).

Figure 9b also demonstrates that another approximate equation of state widely used in the literature, namely PC-SAFT [51], provides a less satisfactory fit of the experimental data and yields less plausible interaction parameters [120]. It has been shown [54] that, at high pressures, the PC-SAFT calculation predicts too large a density as a result of a spurious liquid–liquid-type phase separation inherent in this equation of state model.

In any case, Figs. 4 and 9 give some evidence that simple coarse-grained models can describe the equation of state of both oligomers and polymer melts reasonably well. Being interested in polymer-plus-solvent systems, we also need (as already discussed) a good coarse-grained model for the solvent. Figures 1–3 show that such models exist and can describe the experimental data over a wide range of temperatures and pressures rather well. In the case of methane (CH_4), modeled as a simple point particle (Fig. 1) with LJ interactions, the choice of parameters for the coarse-grained model is a nontrivial matter when dipolar or quadrupolar interactions also are present, as in the case of H_2S (Fig. 3) or CO_2 (Fig. 2). In the dipolar case, a popular coarse-grained model is the SM (6), and an analogous model with quadrupole–quadrupole interactions is [131, 157, 158]:

$$U_{QQ}(r) = 4\varepsilon_{ss}\left[\left(\frac{\sigma_{ss}}{r}\right)^{12} - \left(\frac{\sigma_{ss}}{r}\right)^{6} - \frac{3}{16}q_F\left(\frac{\sigma_{ss}}{r}\right)^{5}f_{QQ}\left(\Theta_i, \Theta_j, \phi_i, \phi_j\right)\right], \qquad (15)$$

where $q_F = Q^2/\varepsilon_{ss}\sigma_{ss}^5$ is related to the quadrupole moment Q of the molecules, and the function f_{QQ} depends on the polar angles (Θ_i, ϕ_i) and (Θ_j, ϕ_j) of both molecules (taking the direction of $\mathbf{r}_{ij} = \mathbf{r}_i, -\mathbf{r}_j$ as the z-axis). Specifically, the function $f_{QQ}(\Theta_i, \Theta_j, \phi_i, \phi_j)$ can be derived to be [207]:

$$f_{\mathrm{QQ}}(\Theta_i, \Theta_j, \phi_i, \phi_j) = 1 - 5\cos^2\Theta_i - 5\cos^2\Theta_j + 2\sin^2\Theta_j\cos^2(\phi_i - \phi_j)$$
$$- 16\sin\Theta_i\cos\Theta_i\sin\Theta_j\cos\Theta_j\cos(\phi_i - \phi_j) . \tag{16}$$

However, since for many cases of practical interest the absolute strength of the multipolar interactions at typical nearest and next-nearest neighbor distances in the fluid is much weaker than the LJ interactions, one can follow the idea of Müller and Gelb [208] to treat the multipolar interaction only in spherically averaged approximation:

$$U^{\mathrm{eff}}(r) = -(k_{\mathrm{B}}T)ln\langle exp[-U(r, \{\Theta_i, \phi_i\})/k_{\mathrm{B}}T]\rangle_{\{\Theta_i, \phi_i\}} . \tag{17}$$

The isotropically averaged dipolar interaction can then be cast into the form:

$$U_{\mathrm{D}}^{\mathrm{eff}} = 4\varepsilon_{\mathrm{ss}}\left[\left(\frac{\sigma_{\mathrm{ss}}}{r}\right)^{12} - (1 + \lambda)\left(\frac{\sigma_{\mathrm{ss}}}{r}\right)^6\right], \tag{18}$$

with:

$$\lambda = \frac{1}{12}\frac{\mu^4}{\varepsilon_{\mathrm{ss}}\sigma_{\mathrm{ss}}^6 k_{\mathrm{B}}T} \equiv \lambda_{\mathrm{c}}T_{\mathrm{c}}/T , \tag{19}$$

where μ is the dipole moment (cf. (6)). Similarly, the isotropically averaged quadrupolar interaction becomes:

$$U_Q^{\mathrm{eff}} = 4\varepsilon_{\mathrm{ss}}\left[\left(\frac{\sigma_{\mathrm{ss}}}{r}\right)^{12} - \left(\frac{\sigma_{\mathrm{ss}}}{r}\right)^6 - \frac{7}{20}q\left(\frac{\sigma_{\mathrm{ss}}}{r}\right)^{10}\right], \tag{20}$$

where:

$$q = Q^4/\left[\varepsilon_{\mathrm{ss}}\sigma_{\mathrm{ss}}^{10}k_{\mathrm{B}}T\right] = q_{\mathrm{c}}(T_{\mathrm{c}}/T) . \tag{21}$$

Note that (18)–(21) can also be justified in terms of a perturbation expansion of the dipole–dipole or quadrupole–quadrupole part of the interaction in second order in inverse temperature.

Obviously (18) can be interpreted as a LJ potential with renormalized parameters:

$$U_{\mathrm{D}}^{\mathrm{eff}} = 4\tilde{\varepsilon}\left[(\tilde{\sigma}/r)^{12} - (\tilde{\sigma}/r)^6\right], \tag{22}$$

with $\tilde{\varepsilon} = \varepsilon_{\mathrm{ss}}(1 + \lambda)^2$, $\tilde{\sigma}^6 = \sigma_{\mathrm{ss}}^6/(1 - \lambda)$. Notice that λ is proportional to inverse temperature and hence $\tilde{\varepsilon}$ and $\tilde{\sigma}$ are temperature-dependent. Using the knowledge of critical properties for the standard LJ model in three dimensions,

Computer Simulations and Coarse-Grained Molecular Models Predicting

$T_c^* = k_B T_c / \tilde{\varepsilon}(T_c) = 1.312$, $\rho_c^* = \rho_c [\tilde{\sigma}(T_c)]^3 = 0.316$ [209], we can obtain immediately $\tilde{\varepsilon}(T_c)$ and $\tilde{\sigma}(T_c)$ by requiring that $T_c = T_c^{exp}$, $\rho_c = \rho_c^{exp}$:

$$\tilde{\varepsilon}(T_c) = k_B T_c^{exp} / T_c^*, \quad \tilde{\sigma}(T_c) = (M_{mol}\rho_c^* / N_A \rho_c^{exp})^{1/3} . \qquad (23)$$

Here, a factor M_{mol}/N_A was introduced, M_{mol} being the molar mass of the material and N_A Avogadro's number, to convert to the units normally used. With a little algebra, one finds from (19), (22), and (23) an equation for λ_c in terms of experimental properties [55]:

$$\lambda_c = \lambda_{c0}/(1 - \lambda_{c0}), \quad \lambda_{c0} = \mu^4 / \left[12\tilde{\varepsilon}(T_c)\tilde{\sigma}(T_c)^6 k_B T_c^{exp}\right] \qquad (24)$$

together with:

$$\varepsilon_{ss} = \tilde{\varepsilon}(T_c)(1 - \lambda_{c0})^2, \quad \sigma_{ss}^6 = [\tilde{\sigma}(T_c)]^6 / (1 - \lambda_{c0}) . \qquad (25)$$

Of course, this spherical average of the dipolar interaction makes sense only if the dipolar interaction is small enough in comparison with the LJ interaction: with (25) we can now quantify this condition as $\lambda_{c0} \ll 1$. Figure 10a, shows a plot of λ_c versus λ_{c0}, including a number of dipolar fluids using in most cases two values for μ, one being the real dipole moment of the molecule and the other value being based on a (larger) effective dipole moment. This yields a reasonable fit for the equation of state. The need to "renormalize" the dipole moment to be used in the Stockmayer potential (6), indicates the inaccuracy of the latter because of additional steric interactions between the molecules, polarization effects, etc. As shown already in Fig. 3, a modeling based on the simple equations (18), (19), and (22)–(25) yields a reasonably accurate description of the equation of state of fluids with (weak) dipolar interactions.

In the case of quadrupole–quadrupole interactions, the counterpart of (18) is:

$$U_Q^{eff} = 4\varepsilon_{ss}\left[(\sigma_{ss}/r)^{12} - (\sigma_{ss}/r)^6 - (7/20)q(\sigma_{ss}/r)^{10}\right], \qquad (26)$$

with:

$$q = Q^4 / \left[\varepsilon_{ss}\sigma_{ss}^{10} k_B T\right] = q_c T_c / T. \qquad (27)$$

Obviously, in the quadrupolar case the effective potential is not simply a renormalized LJ potential, as in the dipolar case, and hence the estimation of the phase diagram cannot be reduced to the standard LJ problem. However, Mognetti et al. [131] solved this problem by treating q_c as an arbitrary additional parameter in the Hamiltonian, and computing by MC techniques "master curves" $T_c(q_c)/T_c(0)$ versus q_c and $\rho_c(q_c)/\rho_c(0)$ versus q_c, respectively, The conditions that match the

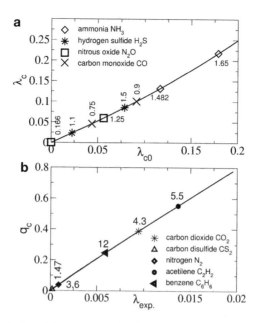

Fig. 10 (a) Universal plot of the parameter λ_c of the isotropically averaged dipolar interaction versus the material parameter λ_{c0} (24). For a few molecules, the parameters λ_c and λ_{c0} are indicated by *symbols*, using both the value of λ_{c0} based on the experimentally reported dipole moments [130] (NH$_3$: $\mu_{exp} = 1.482$ D; H$_2$S: $\mu_{exp} = 1.1$ D; N$_2$O: $\mu_{exp} = 0.166$ D; CO: $\mu_{exp} = 0.75$ D) as well as the effective dipole moment proposed in [55]. All dipole moments are in Debye units. (b) Universal plot of the model parameter q_c of the isotropically averaged quadrupolar interaction versus the material parameter λ_{exp} (27)–(29). As an example, the estimates of q_c are shown by *symbols*, using the experimentally measured quadrupole moment Q of these materials to estimate λ_{exp}. The quadrupole moments quoted in the figure are given in units of D Å. From Mognetti et al. [55, 131]

simulation critical point with the experimental one are fully analogous to (23), namely:

$$\varepsilon_{ss}(q_c) = k_B T_c^{exp}/T_c^*(q_c) , \quad \sigma_{ss}^3(q_c) = \rho_c^*(q_c) M_{mol}/N_A \rho_c^{exp}. \tag{28}$$

The self-consistent solution of (27) and (28) can again be cast in the form of a master curve, where q_c is described as a function of a parameter λ_{exp} (Fig. 10b), with:

$$\lambda_{exp} = Q^4 \left[\rho_c^{exp} N_A/M_{mol}\right]^{10/3}/(k_B T_c^{exp})^2 . \tag{29}$$

In this case, the experimental values of Q typically yield a rather good description of the equation of state of the pure materials. However, when one considers mixtures, one finds that a slight enhancement of Q (again possibly due to

Computer Simulations and Coarse-Grained Molecular Models Predicting | 355

polarization effects, for instance) yields better results. For example, Mognetti et al. [56] recommended for CO_2 the use of $q_c = 0.470$ instead of $q_c = 0.387$ (resulting from the experimental value of Q [130]); for benzene (C_6H_6) the best choice is $q_c = 0.38$ instead of $q_c = 0.247$ (resulting from the experimental value [130]).

At the end of this section, we mention the estimation of interaction parameters between polymer and solvent, or (more generally) between two species A and B in a binary mixture. The simplest possibility is to use the standard Lorentz–Berthelot combining rules [210]:

$$\sigma_{AB} = (\sigma_{AA} + \sigma_{BB})/2 , \quad \varepsilon_{AB} = \sqrt{\varepsilon_{AA}\varepsilon_{BB}}. \tag{30}$$

Of course, (30) is somewhat arbitrary and ad hoc, and many recipes to amend (30) by correction factors can be found in the literature. For example, one may modify the square root rule in (30) by a correction factor ξ that is adjusted in order to improve the agreement with experiment [10, 53, 210]:

$$\sigma_{AB} = (\sigma_{AA} + \sigma_{BB})/2, \quad \varepsilon_{AB} = \xi\sqrt{\varepsilon_{AA}\varepsilon_{BB}}. \tag{31}$$

Of course, other choices of combining rules are considered in the literature [210–214] but are not considerated here. In view of the fact that there are good reasons for also including three-body terms into the description of intermolecular interaction (see, e.g., [215, 216]), using simplified pair potentials of the type described in this section should only be considered as a reasonable approximate first step on the way towards a more rigorous modeling of interactions in real materials.

3 Basic Aspects of Simulation Methods

3.1 Molecular Dynamics

In principle, the idea of MD simulations is very simple: one solves Newton's equations of motion for the interacting many-body system numerically on a computer. Thus, if $U(\{\vec{r}_i\})$ is the total potential acting on particle i (with mass m_i) and position $\vec{r}_i(t)$ at time t, one has to solve:

$$m_i \frac{d^2}{dt^2}\vec{r}_i(t) = -\nabla_i U(\{\vec{r}_i\}) , \quad i = 1,\ldots,\mathcal{N} , \tag{32}$$

where $\mathcal{N}$ is the number of particles (atoms or "pseudo-atoms" such as CH_2 beads, etc.) in the system in the volume V (typically a $L \times L \times L$ box with periodic boundary conditions). Equation (32) is a description in terms of classical mechanics but, invoking the ergodicity hypothesis of statistical mechanics [17], one expects that time averages:

$$\bar{A} = (1/t_{\text{obs}}) \int_0^{t_{\text{obs}}} A(\{\vec{r}_i(t)\}) dt, \quad t_{\text{obs}} \to \infty, \tag{33}$$

of observables $A(\{\vec{r}_k(t)\})$ in the system are equivalent to ensemble averages in the microcanonical (NVE) ensemble [17], where E is the total internal energy of the system:

$$\bar{A} = \langle A(\{\vec{r}_i\}) \rangle_{NVE} . \tag{34}$$

The fact that the microcanonical ensemble average appears here is, of course, due to the fact that the total energy E is conserved for (32). In practice, however, the numerical integration of (32) is not exact and one has to discretize the time axis in terms of finite time steps Δt. Thus, errors may accumulate that violate the conservation law for energy in an undesirable way. These cumulative errors cannot be suppressed entirely, but minimized using symplectic integration schemes [217], such as the Verlet algorithm [81–83, 87], in which the system coordinates $\{\vec{r}_i(t)\}$ are propagated as follows:

$$\vec{r}_i(t + \Delta t) = 2\vec{r}_i(t) - \vec{r}_i(t - \Delta t) + \vec{a}_i(t)(\Delta t)^2 + \mathcal{O}((\Delta t)^4) , \tag{35}$$

where $\vec{a}_i(t) = -\nabla U(\{\vec{r}_i(t)\})/m_i$ denotes the acceleration that acts at the ith particle at time t. Of course, Δt in (35) has to be kept small enough to reach sufficient accuracy (for an atomistic model, "small enough" means a Δt in the range of 1–2 fs, i.e., 10^{-15} s!).

A useful modification of (35) is the so-called Velocity–Verlet algorithm. It explicitly incorporates the velocity $\vec{v}_i(t)$ of the particle:

$$\vec{r}_i(t + \Delta t) = \vec{r}_i(t) + \vec{v}_i(t)\Delta t + \frac{1}{2}\vec{a}_i(t)(\Delta t)^2 , \tag{36}$$

$$\vec{v}_i(t + \Delta t) = \vec{v}_i(t) + \frac{1}{2}[\vec{a}_i(t) + \vec{a}_i(t + \Delta t)]\Delta t; . \tag{37}$$

This algorithm produces integration errors of the same order as the original Verlet algorithm. Its advantage lies in symmetric coordinates for "past" and "future", and it also conserves the phase space volume; i.e., Liouville´s theorem [17] is obeyed. Although energy is not conserved perfectly on a short time scale, there are no systematic energy drifts for large time scales. There exist further suggestions (e.g., the "leapfrog method" [218]) or other algorithms such as predictor–corrector methods [219] that are more accurate at short times but that violate Liouville´s theorem. As these methods are not symplectic, they are less in use today. We also note that rigid constraints (rigid bond lengths, rigid bond angles, etc.) also require different algorithms [87, 218], but this topic is not considered here.

We rather focus on another aspect, namely the desirable choice of statistical ensemble. Although statistical mechanics [17] asserts that in the thermodynamic

Computer Simulations and Coarse-Grained Molecular Models Predicting   357

limit ($\mathcal{N} \to \infty$) all statistical ensembles are equivalent and can be transformed into each other via Legendre transformations, for finite $\mathcal{N}$ these ensembles are not equivalent [220]. If one wishes to study phase transition and phase coexistence, the use of the microcanonical ensemble is somewhat cumbersome [221].

There have been many methods suggested to carry out MD simulations at given temperatures T rather than at given energy E. The first approach that was used was based on velocity rescaling, e.g., the velocities were changed until all velocities satisfied the relation following from the Maxwell–Boltzmann distribution, $m_i\langle\vec{v}_i^2\rangle = 3k_BT/2$. Of course, such a velocity rescaling simulation destroys one of the advantages of MD, namely the possibility to get detailed accurate information on time-displaced correlation functions $\langle A(\{\vec{r}_i(t)\})A(\{\vec{r}_k(t+t')\})\rangle$ of the variables; moreover this technique does not lead to a distribution of variables according to the canonical $\mathcal{N}VT$ ensemble of statistical mechanics [87]. Alternatively, one can couple the system to "thermostats" [87, 218]. Although the popular Berendsen thermostat [222] does not correspond strictly to the $\mathcal{N}VT$ ensemble, and hence we do not recommend its use, the correct $\mathcal{N}VT$ ensemble is obtained implementing the Nose–Hoover thermostat [223, 224]. In this technique, the model system is coupled to a heat bath, which represents an additional degree of freedom represented by the variable $\zeta(t)$. The equation of motion then becomes:

$$d\vec{r}_i/dt = \vec{v}_i(t), \quad m_i d\vec{v}_i(t)/dt = -\nabla_i U(\{\vec{r}_j\}) - \zeta(t)m_i\vec{v}_i(t) , \tag{38}$$

so this coupling enters like a friction force. However, $\zeta(t)$ can change sign because it evolves according to the equation:

$$d\zeta(t)/dt = (2M_b)^{-1}\left(\sum_{i=1}^{\mathcal{N}} m_i\vec{v}_i^2 - 3\mathcal{N}k_BT\right) . \tag{39}$$

M_b is interpreted as the "mass of the heat bath". For appropriate choices of M_b, the kinetic energy of the particles does indeed follow the Maxwell–Boltzmann distribution, and other variables follow the canonical distribution, as it should be for the $\mathcal{N}VT$ ensemble. Note, however, that for some conditions the dynamic correlations of observables clearly must be disturbed somewhat, due to the additional terms in the equation of motion [(38) and (39)] in comparison with (35). The same problem (that the dynamics is disturbed) occurs for the Langevin thermostat, where one adds both a friction term and a random noise term (coupled by a fluctuation–dissipation relation) [75, 78]:

$$m_i \frac{d^2\vec{r}_i(t)}{dt^2} = -\nabla_i U(\{\vec{r}_i(t)\}) - \zeta\frac{d\vec{r}_i}{dt} + \vec{W}_i(t) , \tag{40}$$

$$\langle\vec{W}_i(t) \cdot \vec{W}_j(t')\rangle = \delta_{ij}\delta(t-t')6k_BT\zeta . \tag{41}$$

Equation (41) ensures that the time averages resulting from (40) are equivalent to the canonical ensemble averages. The algorithm is rather robust, and very useful if one is just interested in static averages of the model. However, when one considers the dynamics of polymer solutions, one must be aware that the additional terms in (40) seriously disturb the hydrodynamic interactions, for instance. This latter problem can be avoided by using a more complicated form of friction plus random force, the so-called dissipative particle dynamics (DPD) thermostat [225–227].

However, in order to obtain the strictly correct dynamics of the system as it is described by Newton´s equations, (32) in the canonical ensemble, the proper procedure is to generate a number of initial states in the presence of a (correct) thermostat (such as Nose–Hoover, Langevin, or DPD thermostats) and use these states as initial states of trajectories generated in strictly microcanonical MD runs. Alternatively, one also can generate a number of initial states in the $\mathcal{N}VT$ ensemble (typically it suffices to average over ten such independent states) by MC methods [204, 228, 229]. Although comprehensive studies of the static and dynamic properties of binary mixtures undergoing phase separation have been done for binary LJ mixtures [228, 229], and a soft variant of the Asakura–Oosawa model for colloid–polymer mixtures [204], we are only aware of a single study of the dynamics of chain molecules in binary liquid n-alkane mixtures [69]. Well-relaxed atomistic configurations of binary mixtures of united atom models were produced with a specialized MC algorithm including scission and fusion moves [230], and self-diffusion then was studied using MD runs based on the Velocity–Verlet algorithm with a multiple time step method (the reversible reference systems propagator algorithm, rRESPA [231, 232]). However, the considered mixtures are fully miscible at the temperatures of this study [intermolecular interactions were modeled by (30)], and hence this work is somewhat out of the scope of the present review, which focuses on the study of the equation of state and phase diagrams of polymer-containing systems.

It should be noted that a study of phase behavior by MD methods in the $\mathcal{N}VT$ ensemble is not straightforward, due to various limitations of computer simulation methods in general [233]. If one deals with relatively small molecules and the interactions are short range, it is nowadays possible to run a MD simulation in the two-phase coexistence region long enough until "macroscopic" phase separation is achieved, i.e., phase separation on the scale L of the simulation box, so that a slab-like configuration results, see Figs. 11–13. Due to the effect of the periodic boundary conditions of a cubic $L \times L \times L$ box, in the final equilibrium configuration the interfaces are (on average) planar and parallel to an $L \times L$ surface of the simulation box, provided the volume fractions of both coexisting phases are approximately equal, as was the case for the quench in Fig. 12. If such an equilibrium state of coexisting bulk phases can be achieved, it is possible to record the density profiles of both constituents across the interfaces (Fig. 13) and, therefore, obtain information both on the properties of the bulk coexisting phases (which in this case have already been determined by MC methods [10, 53], see Fig. 12) and on interfacial properties. For example, by recording the pressure tensor across the interface and using the virial theorem [87, 88], it is possible to estimate the interfacial tension from the anisotropy of the pressure tensor [235]. This approach

Fig. 11 Snapshots (**a–e**) of the configuration of a mixture resulting from simulation of a pressure quenching experiment, see Fig. 12, for the model of hexadecane dissolved in carbon dioxide, cf. Fig. 5. In this model, the LJ parameters of both $C_{16}H_{34}$ and CO_2 are fitted to the respective critical temperatures and densities, while intermolecular interactions were described by (31) with $\xi = 0.886$ [10]. The chosen temperature was $T = 486$ K, and the pressure quench was realized by a volume increase, so that the density decreased from a value $\rho^* = 0.8$ (in LJ units using σ_{pp} as length scale) to $\rho^* = 0.45$ at time $t = 0$ (starting from a well-equilibrated state in the one-phase region). The snapshots were taken at times 0 (the initial state before the quench)(**a**), 10 (**b**), 100 (**c**), 1000 (**d**), and 4000 τ (**e**) after the quench. τ is the MD time unit ($\tau = \sigma_{pp}(m/\varepsilon_{pp})^{1/2}$. Masses of CO_2 and effective monomers are both taken as $m = 1$). The quench refers to a mole fraction of CO_2 of $x = 0.6$, and the simulation box contained in total $\mathcal{N} = 435{,}000$ particles. The *inset* in (**c**) shows an enlarged region of size $20 \times 20 \times 5\sigma^3$, marked by the *rectangle in the left-bottom corner*. *Gray spheres* represent the solvent molecules (CO_2) and *dark spheres* the effective beads of $C_{16}H_{34}$ (for clarity no bonds connecting these beads are shown). From Yelash et al. [234]

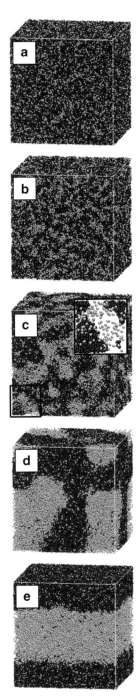

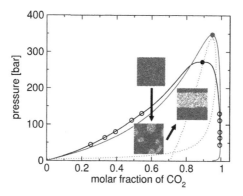

Fig. 12 Isothermal slice through the phase diagram at $T = 486$ K in the pressure–mole fraction plane of the model for the $C_{16}H_{34}$–CO_2 mixture as described in the caption of Fig. 11. The coexistence curve between the polymer-rich phase (*left*) and CO_2 supercritical vapor (near $x = 1$, *right*) has been estimated both by MC [10, 53] (*curve connecting the open circles*) and by TPT1-MSA [10, 53]. *Dotted curves* are the spinodal curves predicted by TPT1-MSA. The closed black circle shows the critical point obtained by MC; the gray circle shows the critical point obtained by TPT1-MSA. Note that TPT1-MSA significantly overestimates the unmixing tendency near the critical pressure, but is reasonably accurate for pressures $p < 200$ bar. The *snapshots* of slices through the simulation box connected by *arrows* indicate the quenching experiment and the resulting structure evolution in the system. From Yelash et al. [234]

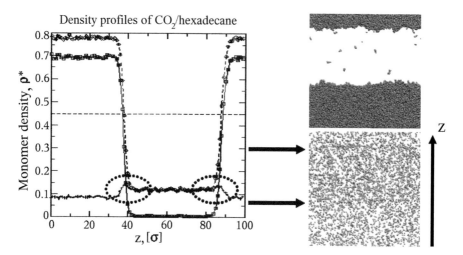

Fig. 13 *Left*: Density profiles of carbon dioxide and hexadecane across the two interfaces in a box of linear dimension $L = 98.88\sigma_{pp}$. Densities are quoted in LJ units ($\rho^* = \rho\sigma_{pp}$). The *dotted ellipses* highlight the interfacial adsorption of CO_2 at the polymer–CO_2 vapor interface. *Right*: The *snapshots* show $L \times L \times 5\sigma$ slices where the positions of the CO_2 molecules (*lower image*) and the polymer (*upper image*) are shown separately. Note that a few hexadecane molecules are dissolved in the vapor phase as well (this finite solubility in the gas decreases rapidly with increasing polymer chain length)

of getting both bulk densities and the interfacial tension of coexisting liquid and vapor phases has demonstrated its power for simple liquids [236], although we are not yet aware of applications to polymer-containing systems. However, alternative methods for estimating the interfacial tension and for analyzing the capillary wave spectrum [237–240] of interfacial fluctuations, and the system size dependence of interfacial widths [237–240], have been used both for the study of interfaces in lattice models of symmetric polymer mixtures [237, 238] and for off-lattice models of polymer solutions [239] and colloid–polymer mixtures [240]. These latter techniques are preferable when the interfacial tension is rather small.

The technique to study phase coexistence via MD by simulating phase separation kinetics in the $\mathcal{N}VT$ ensemble until equilibrium is established [234] becomes cumbersome near critical points, and in any case it requires the simulation of very large systems over a large simulation time. In addition, this method is hardly feasible when the model systems contains long polymers – their diffusion simply is too slow [5, 6, 8, 9, 31]. Experience with the simulation of spinodal decomposition in lattice models of polymer mixtures [9, 180, 241] shows that only the early stages of phase separation are accessible, meaning that the method is unsuitable for studying the equilibrium states of well phase-separated systems.

An alternative technique to study phase coexistence in the $\mathcal{N}VT$ ensemble via MD was applied by Bartke and Hentschke [242], who simulated in rather small systems the van der Waals-like loop in the pressure versus volume isotherm and estimated the coexisting phases from the Maxwell construction, exemplifying the technique for the Stockmayer fluid (6). As usual, the pressure in $\mathcal{N}VT$ simulations is accessible from the virial theorem [87, 88]. It should be noted, of course, that this "loop" in the isotherm, for systems with short-range interactions, has nothing to do with the loop resulting in the van der Waals equation of state from a mean-field-type approximation, but rather reflects finite size effects on phase coexistence [243–245], if equilibrium in the simulation box is reached. Then, a careful analysis of finite size effects is mandatory to avoid misleading conclusions. For polymer-containing systems, the extent to which equilibrium is reached is rather doubtful, particularly at high densities (smaller volumes), and this problem could invalidate the approach in such cases.

Finally, we mention that it is also possible to carry out MD simulations in the $\mathcal{N}pT$ ensemble: then, the pressure p is a given externally controlled variable, and the volume V of the system is a fluctuating variable that is sampled. The dynamics of these volume fluctuations is controlled by coupling to a "barostat" [246]. Although the dynamic correlations between observables are clearly no longer faithfully represented in this approach due to the density fluctuations, and hence a major advantage of MD is lost, MD in the $\mathcal{N}pT$ ensemble may nevertheless have significant advantages for studying phase equilibria because one can restrict attention to the two coexisting phases separately. Thermodynamics tells us that two phases I and II coexist if, apart from having the same temperature and pressure, the chemical potential μ (in a one-component system) is also equal:

$$\mu_{\mathrm{I}}(p,T) = \mu_{\mathrm{II}}(p,T) \,. \tag{42}$$

The chemical potentials in both phases can, in principle, be "measured" in a simulation by the Widom virtual particle insertion/deletion technique [247]. Determining, for both liquid and vapor at a chosen temperature, the chemical potentials as function of pressure, one finds the coexistence pressure $p_{coex}(T)$ from the intersection of both curves. This approach is readily generalized to more-component systems. This technique was first demonstrated for simple models of pure fluids [248, 249] and then extended to more complicated models of molecules [153, 250] describing quadrupolar fluids, and to various mixtures [154]. Again, this method is problematic near critical points. The angle under which the two curves $\mu_I(p, T)$ and $\mu_{II}(p, T)$ cross at $p = p_{coex}$ becomes very small when T is only slightly below T_c, and one has to deal with critical slowing down [229], finite size effects, etc. It is also problematic for large molecules, for which the acceptance of particle insertions becomes too low.

For fully atomistic all-atom models, it is often difficult to find efficient MC moves to relax their configurations, and then MD is normally the method of choice. We note, however, that for chemically realistic models of polymer blends equilibration by MD methods is extremely difficult, if at all possible. Dealing with such systems is still an unsolved challenge.

3.2 Monte Carlo

MC simulations aim to realize the probability distributions considered in statistical thermodynamics numerically using random numbers and to calculate the desired averages of various observables in the system using these distributions [18–20]. There exist numerous extensive reviews describing the specific aspects of MC methods for polymers [77, 82, 84, 90, 96], and thus we focus here only on some salient features that are most relevant when one addresses the estimation of the equation of state and phase equilibria of systems containing many polymers.

These MC methods then are based on the Metropolis algorithm [251], by which one constructs a stochastic trajectory through the configuration space $(\mathbf{X})$ of the system, performing transitions $W(\mathbf{X} \to \mathbf{X}')$. The transition probability must be chosen such that it satisfies the detailed balance principle with the probability distribution that one wishes to study. For example, for classical statistical mechanics, the canonical ensemble distribution is given in terms of the total potential energy $U(\mathbf{X})$, where $\mathbf{X} \equiv (\vec{r}_1, \vec{r}_2, \ldots, \vec{r}_N)$ stands symbolically for a point in configuration space [17]:

$$P_{NVT}(\mathbf{X}) = Z^{-1} exp[-U(\mathbf{X})/k_B T], \qquad (43)$$

Z being the partition function (remember that the free energy F then is [17] $F(N, V, T) = -k_B T ln Z$). The detailed balance principle then requires that:

$$P(\mathbf{X})W(\mathbf{X} \to \mathbf{X}') = P(\mathbf{X}')W(\mathbf{X}' \to \mathbf{X}) . \qquad (44)$$

Computer Simulations and Coarse-Grained Molecular Models Predicting 363

If it is possible to generate a Markov chain of transitions $\mathbf{X} \to \mathbf{X}' \to \mathbf{X}'' \to \ldots$, one can show that, in the limit when the number $\mathcal{M}$ of configurations generated is very large, i.e., $\mathcal{M} \to \infty$, the canonical average of some observable $A(\mathbf{X})$:

$$\langle A(\mathbf{X}) \rangle_{\mathcal{N}VT} = Z^{-1} \int d\mathbf{X} A(\mathbf{X}) exp[-U(\mathbf{X})/k_B T] \qquad (45)$$

can be approximated by a simple arithmetic average over the $\mathcal{M}$ configurations generated:

$$\bar{A} = \mathcal{M}^{-1} \sum_{i=1}^{\mathcal{M}} A(\mathbf{X}_i) . \qquad (46)$$

One of the big advantages of MC methods is that they can be readily generalized to all statistical ensembles of interest. For example, in the grand canonical ensemble of a single component system it is not the particle number $\mathcal{N}$ that is fixed, but rather the chemical potential μ. Thus, in order to realize the distribution $P_{\mathcal{N}VT}$ in the grand canonical ensemble, the moves $\mathbf{X} \to \mathbf{X}'$ must include insertion and deletion of particles (in practice this is easily realizable for small molecules, such as solvent molecules, but becomes difficult for short polymers, and impossible for long polymers because the acceptance rate of such "MC moves" becomes too small). For polymer blends, a particularly useful ensemble is the semigrand canonical ensemble. Suppose we have two types of polymers, A and B, having the same chain length, $N_A = N_B$ (the extension to different chain lengths is discussed in [170, 171]). Then, it is possible to consider a MC move where an A chain is replaced by a B chain (with identical configuration) or vice versa, taking the chemical potential difference $\Delta\mu = \mu_A - \mu_B$ properly into account in the transition probability [6, 15, 16, 21, 82, 170–172]. An example for such an application, extending the method to a mixture of homopolymers and block copolymers, will be presented as a case study in Sect. 4.4. We emphasize, that neither the grand canonical nor the semi-grand canonical ensemble can be used in MD simulations.

The random numbers (actually no strictly random numbers are used, but rather only pseudorandom numbers, generated on the computer by a suitable algorithm [20]) are then used for two purposes: first a trial MC move $\mathbf{X} \to \mathbf{X}'$ is attempted. For example, in a simulation of a polymer-plus-solvent system in the grand canonical ensemble, coordinates of a point in space are chosen at random, and there one attempts to insert an additional solvent particle; or one chooses a randomly selected bead of a polymer chain and attempts to move it to a randomly chosen neighboring position in a small volume region δV around its previous position; etc. Then, one needs to expose this trial configuration to the Metropolis acceptance test. In the canonical ensemble, one simply needs to compute the change in total potential energy ΔU caused by the trial move: if $\Delta U < 0$, the trial move is accepted; if $\Delta U > 0$, one compares $\Delta W \equiv exp[-\Delta U/k_B T]$ with a random number ξ uniformly distributed in the unit interval $\{0, 1\}$. If $\Delta W \geq \xi$, the trial move is accepted, and $\mathbf{X}'$

is taken as the next configuration; otherwise, $\mathbf{X}'$ is rejected, the old configuration $\mathbf{X}$ is counted once more for the average, and a new trial move is attempted.

Although the basic step of MC algorithms is very simple, much know-how is needed to carry out successful MC simulations of dense polymeric systems in practice. One must realize that the subsequently generated configurations $\mathbf{X} \to \mathbf{X}' \to \mathbf{X}'' \to \ldots$ of such a stochastic trajectory in phase space are not statistically independent of each other, but in most cases strongly correlated. In fact, one can give MC sampling a dynamic interpretation: one numerically realizes a master equation for the probability $P(\mathbf{X}, t)$ that a state $\mathbf{X}$ is found at the "time" t of the sampling process [18–20]. So, if in a multichain system in the canonical ensemble the attempted MC moves just consist of small random displacements of the effective monomers, one generates a chain dynamics consistent with the simple Rouse model of polymer dynamics (or reptation model, if the chains are entangled) [5, 8, 31]. Of course, the time scale of the MC sampling process has no a priori interpretation in terms of physical time units and one traditionally uses dimensionless "time" units such as Monte Carlo step (MCS) per monomer. When one wishes to connect this "time" to physical time, one needs to use extra information (e.g., from the energy barriers of torsional potentials, etc.) to map the MC time onto the physical time via a "time rescaling factor" (which depends on temperature and density [101, 104]).

Thus, although MC applications to study the dynamics of polymers (e.g., near their glass transition [82, 86]) exist, an important advantage of MC is that one can abandon the possibility of studying polymer dynamics in favor of a speedup of the sampling by using MC moves that look artificial from the point of view of the real dynamics of polymers in the laboratory, but which are perfectly permissible as a means of creating a trajectory through the configuration space to sample probabilities such as (43). For example, one may allow for moves of monomers over such large distances that the covalent bonds connecting neighboring monomers along a chain are crossed during the move [95]. Such moves do not occur in real polymer melts, where chains can never cross each other, but in MC simulations such moves can be implemented such that they satisfy (44) and hence are perfectly valid to study static equilibrium behavior. This is also true for a large variety of other "artificial" moves, such as the "slithering snake" algorithm [82] (one chooses a chain end of one of the chains at random, and tries to remove the end monomer and attach it to the other chain end in a randomly chosen direction), or algorithms involving chain fission and fusion [95, 96, 230]. However, we shall not describe these algorithms here, but rather direct the reader to the literature [82, 83, 88, 90, 96].

Similarly, "tricks of the trade" are also needed when one wishes to realize the grand canonical ensemble: inserting a polymer chain of moderate length even in a semidilute polymer solution has such a low acceptance probability that a straightforward simulation would never work. This problem can be overcome to some extent by the configurational bias algorithm [88]; for the bond fluctuation model [76, 79, 80], this algorithm has allowed a successful study of the phase diagram of polymer solutions up to a chain length of $N = 60$ [173]. The configurational bias

MC method can also be implemented for off-lattice models, e.g., united atom models for alkanes [59, 252] have been studied up to $C_{70}H_{142}$. Many such studies of liquid–vapor-type phase equilibria, however, do not use the grand canonical ensemble but rather apply the Gibbs ensemble [253]. This method considers two simulation boxes with volumes V_1, V_2 and particle numbers $\mathcal{N}_1$, $\mathcal{N}_2$ such that $V_1 + V_2 = \text{const}$ and $\mathcal{N}_1 + \mathcal{N}_2 = \text{const}$, while both particles and volume can be exchanged between the boxes. In this way, it is ensured that both boxes are not only at the same temperature, but also at the same pressure and the same chemical potential. Thus, this method has been very popular for the estimation of vapor–liquid coexistence curves, both for small molecules [253–257] and for the alkanes [59, 252]. However, most of this work has yielded rather inaccurate data near the critical point, due to finite size effects. If one combines grand canonical simulations with histogram reweighting [258, 259] and umbrella sampling [260, 261] or multi-canonical MC [262], one can obtain precise results including in the region of the critical point (see, e.g., [263–265]). If one uses such simulation data in a finite size scaling analysis [18–20], one can obtain both the coexistence curve and interfacial tension near the critical point very precisely, as has been demonstrated for many systems (e.g., [52, 53, 55, 56, 131, 170–173, 204, 266, 267]). Since a detailed review of these techniques can be found in the literature [10], we refer the reader to this source for technical details on these methods. We mention, however, that in some cases special algorithms are needed to allow the use of grand canonical simulations. For example, for the Asakura–Oosawa model and related models of colloid–polymer mixtures [204], in the regime of interest the density of the polymer coils (that may overlap each other strongly with no or little energy cost) can be so high that it is almost impossible ever to successfully insert a colloidal particle, which must not overlap any polymer or any other colloidal particle. To allow nevertheless successful colloid insertions, an attempted "cluster move" [268] needs to be implemented. In this move, in a spherical region a number n of polymers is removed and a colloidal particle inserted, or the reverse move is attempted, and transition probabilities are defined such that detailed balance (44) is obeyed.

A completely different nonstandard technique to obtain a first overview of the equation of state was recently proposed by Addison et al. [269], whereby a gravitation-like potential is applied to the system, and the equilibrium density profile and the concentration profile of the center of mass of the polymers is computed to obtain the osmotic equation of state. In this "sedimentation equilibrium" method one hence considers a system in the canonical $\mathcal{N}VT$ ensemble using a box of linear dimensions $L \times L \times H$, with periodic boundary conditions in x and y directions only, while hard walls are used at $z = 0$ and at $z = H$. An external potential is applied everywhere in the system:

$$U_{\text{external}}(z) = -mgz = -(k_B T/a)\lambda_g z. \tag{47}$$

Here, m is the mass of a monomer, g is the acceleration due to the gravity-like potential, and a is the length unit (equal to lattice spacing if a lattice model is used).

The dimensionless constant $\lambda_g = amg/k_B T$ characterizes the strength of this potential. For large z, the density profile of an ideal gas of monomers at the lattice would follow the standard barometric height formula; for the monomer density $\rho_m(z)$ or center of mass density $\rho(z)$:

$$\rho_m(z) \propto exp(-mgz/k_B T) = exp(-z/\xi_m), \quad \rho_m(z) = N\rho(z), \qquad (48)$$

$\xi_m = a/\lambda_g$ being a characteristic gravitational length. The variation of the density profile for large z, where the system is very dilute and the ideal gas behavior holds, is hence trivially known, but this knowledge serves as a consistency check of the method. For smaller z, the density profile is nontrivial, however, and from this profile the osmotic equation of state can be estimated if one invokes the local density approximation, $\rho(z)$:

$$\frac{dp(z)}{dz} = -Nmg\rho(z) , \qquad (49)$$

$p(z)$ being the local osmotic pressure at altitude z. Integration of (49) yields:

$$\pi(z) \equiv p(z)/k_B T = N\xi_m^{-1} \int_z^\infty \rho(z')dz' \approx \xi_m^{-1} \int_z^\infty \rho_m(z')dz' , \qquad (50)$$

where the last step again rests on the validity of the local density approximation. Equations (49) and (50) are plausible if ξ_m is large enough, so that $R_g/\xi_m \ll 1$, R_g being the gyration radius of the polymers. The validity of (49) and (50) becomes doubtful when phase coexistence occurs, however, because a rapid variation of $\rho(z)$ may occur across the interface.

This approach has been tested by Ivanov et al. [123], both for fully flexible and for semiflexible chains. Typical data are presented in Fig. 14, for the bond fluctuation model with the bending potential of (9), for $f = 8.0$, $N = 20$ and $\lambda_g = 0.01$. The conclusion from this study is that this approach quickly gives a reliable equation of state in the one-phase regions, and that the gravitation-like potential induces an additional rounding at the first-order transition.

4 Modeling the Phase Behavior of Some Polymer Solutions: Case Studies

4.1 Alkanes in Carbon Dioxide

In Sects. 2.1–2.3 we have shown that the equation of state and the phase diagram of important solvents such as supercritical carbon dioxide can be modeled rather well by a coarse-grained model, where the molecule is described as a point particle

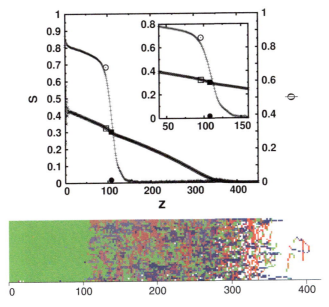

Fig. 14 *Top*: Profiles of the orientational order parameter $S(z)$ (*s-shaped curves*), and of the volume fraction of lattice sites taken by monomeric units, $\phi(z)$ (*thick lines*), for a bond fluctuation model (see Sect. 2.2) on the $80 \times 80 \times 1000$ simple cubic lattice, and chain length $N = 20$, with $\mathcal{N} = 1,600$ chains in the system, choosing parameters $f = 8.0\,(9)$ and $\lambda_g = 0.01$. *Squares* indicate the density values at coexistence and *open circles* indicate the order parameter at the transition. These values were extracted from a bulk grand canonical simulation [123]. The *inset* shows an enlarged plot of the transition region. *Bottom*: Two-dimensional *xz*-map of the coarse-grained order parameter profile for a system snapshot corresponding to the graph. From Ivanov et al. [123]

carrying a quadrupole moment (Fig. 2). It has also been shown that the angular dependence of the quadrupole–quadrupole interaction can be averaged, with no significant loss in accuracy, as far as the phase behavior is concerned. Similarly, the phase diagrams of the pure alkanes are well accounted for by a bead–spring model (Figs. 4 and 5), ignoring bond angle and torsional potentials.

In this section, we ask the question: to what extent can one obtain a reasonable description of the phase behavior of mixtures, when one has a accurate description of the pure material? For a binary mixture (A,B), some information on the interactions between chemically distinct species is indispensable, and we use the simplest assumption for this purpose, namely the Lorentz–Berthelot combining rule (30). This means that we wish to *predict* the phase behavior of the mixture, given some knowledge of the pure components. The question to what extent this works is highly nontrivial: of course, if one allows for sufficiently many additional parameters, an accurate "fitting" of experimental vapor–liquid coexistence data clearly is achievable, but such an approach is rather ad hoc and has little predictive power, and hence is unsatisfactory.

As an example, Fig. 15 compares simulations [56] of this model for the mixture of carbon dioxide and pentane to pertinent experimental data [270]. Two isothermal

Fig. 15 Isothermal slices through the phase diagram of the $CO_2 + C_5H_{12}$ system at $T = 423.48$ K (**a**) and $T = 344.34$ K (**b**). *Closed circles* represent experimental data [270], *asterisks* MC results for the coarse-grained model, and the *solid curve* the corresponding TPT1-MSA prediction. The *dashed curve* shows, for comparison, the TPT1-MSA prediction for a CO_2 model with no quadrupole moment ($q_c = 0$). The *triangle* indicates MC results for the critical point. From Mognetti et al. [56]

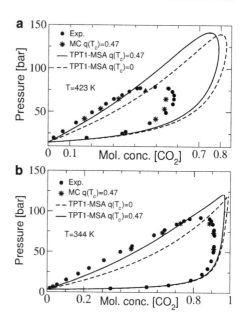

slices in the plane of variables pressure versus molar concentration of CO_2 are shown, and one can see that the two-phase coexistence regions show up as loops extending from pure pentane to rather large CO_2 content, but not reaching pure CO_2 since at these temperatures CO_2 is supercritical. Remarkably, the MC results agree better with experiment than the TPT1-MSA calculation at all molar concentrations. Although one expects that TPT1-MSA overestimates the critical pressure p_c somewhat, for $T = 423.48$ K this overestimation occurs by a factor of about two! It is also interesting to note that TPT1-MSA is also inaccurate for the high pressure branch of the two-phase coexistence loop, although for small CO_2 content the data are far away from any critical region. Since TPT1-MSA here is based on exactly the same interaction parameters as the MC simulation, this discrepancy indicates some shortcoming of TPT1-MSA beyond its inability to accurately describe the critical region.

It also is obvious that ignoring the quadrupolar interaction among CO_2 molecules yields less accurate results, as expected from the experience with pure CO_2.

Figure 16 now considers the behavior of the mixtures of CO_2 and hexadecane, which was already used as a generic system for testing simulation methodologies [10,53]. However, in that work the quadrupolar interactions were ignored, and an ad hoc correction factor $\xi \approx 0.886$ for the Lorentz–Berthelot combining rule was used in order to get qualitatively reasonable results that agreed almost quantitatively with experiment. Including the quadrupolar interactions ($q_c = 0.47$) but leaving $\xi = 1$ has about the same effect as choosing $\xi = 0.9$ in the model without quadrupolar effects. A rather small deviation of ξ from unity would clearly bring the data for $q_c = 0.47$ further upward, and hence create agreement with the experimental data. Of course, one cannot expect that the simple Lorentz–Berthelot combining rule (30)

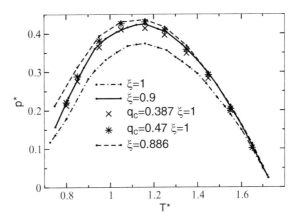

Fig. 16 Critical line of the mixture of the system CO_2 plus hexadecane, projected onto the p^*, T^* plane (pressure p and temperature T rescaled with the LJ parameters of the effective monomers of hexadecane as usual, $p^* = p\varepsilon/\sigma^3$, $T^* = k_B T/\varepsilon$). Data are shown for $q_c = 0$, $\xi = 0.886$ (*top curve*), $\xi = 0.9$ (*middle curve*), and $\xi = 1$ (*bottom curve*). Symbols show the simulation results for $q_c = 0.387$ and $q_c = 0.47$, applying $\xi = 1$ in both cases. From Mognetti et al. [56]

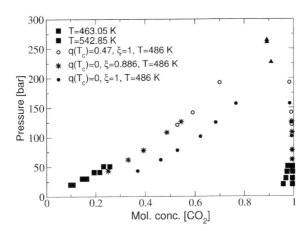

Fig. 17 Isothermal slice through the phase diagram of the $CO_2 + C_{16}H_{34}$ system at $T = 486$ K showing MC results for the model with $q_c = 0.47$ (*open circles*) and comparing them to the results of the model with $q_c = 0$, showing both the choice $\xi = 1$ (*closed circles*) and the choice $\xi = 0.886$ (*asterisks*). *Squares* show two sets of experimental data [271] at two temperatures that bracket the temperature used in the simulation. *Triangles* are MC estimates for the location of the critical point. From Mognetti et al. [56]

works exactly for our grossly simplified coarse-grained model, and thus correction factors ξ that deviate from unity by 1–2% are physically reasonable.

This consideration is underlined by the analysis of isothermal slices through the miscibility gap in the plane of variables pressure versus molar concentration of this system (Fig. 17). It is seen that the MC results that include the quadrupolar interactions ($q_c = 0.47$) and respect the Lorentz–Berthelot rule ($\xi = 1$) fit to the available experimental data [271] about as well as the model predictions using $q_c = 0$ but requiring a large violation of this rule ($\xi = 0.886$).

4.2 Alkanes in Dipolar Solvents

Recently, some model calculations were performed for short alkane chains dissolved in ammonia (NH$_3$) [55]. This solvent can be treated in the spirit of Sect. 2.3: starting from the Stockmayer potential (6), one may average the angular dependence of the dipole–dipole interaction to derive an effective LJ interaction (22) with temperature-dependent LJ parameters. The physical dipole moment of ammonia ($\mu = 1.482$ D) [130] already gives a reasonable first guess of the phase behavior, although the density on the liquid branch of the coexistence curve is somewhat underestimated. Using an enhanced value $\mu = 1.65$ D, based on a similar reasoning as in the case of H$_2$S (see Sect. 2.3 and Fig. 3), a better description of the liquid branch of the coexistence curve results, but the accuracy with which the vapor branch can be described slightly deteriorates. Nevertheless, this choice has a clear advantage for the computation of the critical line for mixtures of ammonia plus alkane (Fig. 18), when we compare both choices with the available experimental data [272–274]. We emphasize again that our model for alkanes is the simple bead–spring model (Fig. 5) where nonane is a trimer of effective beads and hexadecane a fivemer, and the LJ-interactions between ammonia and the effective beads are chosen from the simple Lorentz–Berthelot combining rule (30). In view of the obvious crudeness of the model, we find the rough agreement between model

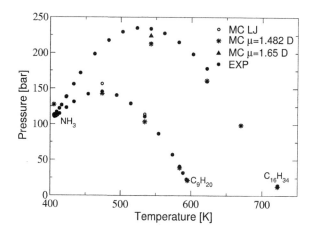

Fig. 18 Critical lines for two alkanes dissolved in ammonia, namely nonane (C$_9$H$_{20}$, *lower set of points*) and hexadecane (C$_{16}$H$_{34}$, *upper set of points*). MC predictions for two choices for the effective dipole moment strength of NH$_3$ are included: $\mu = 1.482$ D is the experimental value [130] of the true dipole moment, $\mu = 1.65$ D is an enhanced value, to be used in the Stockmayer potential (6). *Solid circles* are experimental data [272–274]. From Mognetti et al. [55]

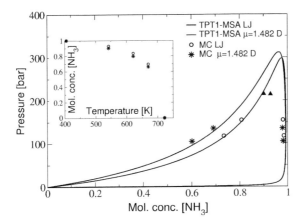

Fig. 19 Isothermal slice through the phase diagram of $NH_3 + C_{16}H_{34}$ at $T = 542.6$ K for the case where the physical value $\mu = 1.482$ D was used for ammonia, and for an even simpler model where the dipole moment of this molecule was completely ignored (denote as MC LJ). *Triangles* denote MC estimates for the critical point in the plane of variables pressure p and molar concentration x of NH_3; *asterisks* and *circles* are selected MC data for the miscibility gap of both models. *Curves* are predictions derived from TPT1-MSA, for the same choice of interaction parameters as in MC. *Inset* shows the critical molar concentration of NH_3 for both models versus temperature. From Mognetti et al. [55]

prediction and experiment that one notes from Fig. 18 quite satisfactory, thus the model should be useful to provide the experimentalist with a rough prediction about the miscibility behavior in such solutions. We also note that the mixtures shown in Fig. 17 belong to "type I" phase diagrams in the Scott–Konynenburg classification [275], i.e., an uninterrupted line of critical points connects the critical point of pure $C_{16}H_{34}$ (or C_9H_{20}) to the critical point of the solvent, unlike the case of $C_{16}H_{34} + CO_2$, which has a "type III" phase diagram.

These types of coarse-grained models for alkanes in dipolar solvents can again be used to make predictions for more specific properties, either by MC simulation or by equation of state calculation based on approximate theories such as TPT1-MSA (Fig. 19). As always, the latter approach overestimates the miscibility gap near the critical pressure, but at lower pressures fair agreement with the MC simulation results is found.

Thus, we feel that Figs. 18 and 19 demonstrate that the coarse-graining approach described in Sect. 2.3 (where the effective intermolecular potentials of both solvent and polymer are taken from fitting suitable experimental input on the equation of state of the respective material, and intermolecular interactions between unlike species are simply estimated by the Lorentz–Berthelot combining rule) is a useful first step in obtaining a rough orientation on phase equilibria of polymer solutions, for polar solvents such as CO_2 and NH_3. As a caveat, however, we mention that cases have been found [55] where the discrepancy between the model prediction and the experimental data is relatively large, e.g., in the case of the $H_2S + C_5H_{12}$ mixture [55], although rather good models for both solvent (Fig. 2) and the alkane

372   K. Binder et al.

oligomer (Fig. 4) have been found. Clearly, more research is necessary to understand better the conditions under which this simple modeling approach is accurate. As a final warning, we emphasize that the present approach is unsuitable for the modeling of aqueous solutions: the strong clustering of water molecules induced by hydrogen bonds renders a united atom description of H_2O clearly unreliable, and also polyelectrolytes will require a quite different approach.

4.3 Solutions of Stiff Polymers and the Isotropic–Nematic Transition

In this subsection, we no longer address a specifically chosen polymer or a specifically chosen solvent, but address the generic problem of the statistical thermodynamics of semiflexible or stiff polymers under good solvent conditions. With increasing concentration of the polymer, a transition from an isotropic solution to a nematic solution is expected [121–123, 276–287]. Even more interesting is the case of semiflexible polymers in concentrated solutions under bad solvent conditions, where the tendency to phase separate and the tendency for nematic order act together [122, 288, 289]. Because only rather qualitative simulation results are available for the latter problem [122], it will not be considered further.

We return to the bond fluctuation model on the simple cubic lattice (Sect. 2.2), with a potential for the bond angle (9), and consider [123] the specific choice of parameters $N = 20, f = 8.0$. For such stiff chains, one is also interested in the global nematic order in the system (in the simulations, $L \times L \times L$ boxes with $L = 90$ lattice spacings were used [123]), which is described by the 3×3 order parameter tensor ($\delta_{\alpha\beta} = $ Kronecker delta):

$$Q_{\alpha\beta} = \frac{1}{\mathcal{N}(N-1)} \sum_{i=1}^{\mathcal{N}(N-1)} \frac{1}{2}\left(3e_i^{\alpha}e_i^{\beta} - \delta_{\alpha\beta}\right), \qquad (51)$$

where e_i^{α} is the αth component of the unit vector connecting monomers i and $i+1$ of a chain (the sum in (51) extends over all bonds of all chains in the system). The largest eigenvalue of $Q_{\alpha\beta}$ can be taken as the nematic order parameter, which we henceforth denote as S.

In the MC simulations using the grand canonical ensemble, two starting configurations need to be used: one box is completely empty, since the solvent molecules in this model are not explicitly considered at all, and the other starting condition is a box maximally dense packed, with chains placed along one coordinate axis (perfectly oriented) having all bond lengths b equal to $b = 2$. For the second case, it is possible to fill the box with a volume fraction ϕ up to $\phi = 1$, whereas this is impossible for the box that is initially empty and then filled with chains according to the configurational bias algorithm and the chosen value of the chemical potential (μ). As expected from experience with simulations of first-order phase transitions in

Fig. 20 Nematic order parameter (**a**) and volume fraction ϕ of lattice sites taken by monomers (**b**) plotted versus the normalized chemical potential $\mu(k_BT \equiv 1$ is chosen here), for the bond fluctuation model on the simple cubic lattice, for chains with $N = 20$ and the bond-angle potential (9) with $f = 8$. The *vertical lines* show the value of the chemical potential at the transition point, $\mu = -166 \pm 0.5$, estimated from the analysis of the osmotic pressure of this polymer solution (see Fig. 21). *Triangles* refer to results obtained from the densely packed starting configuration, while *squares* correspond to the dilute isotropic starting conformation. These data were omitted in (**a**) because the system relaxes into a metastable nematic multidomain rather than into the stable ordered monodomain configuration. From Ivanov et al. [123]

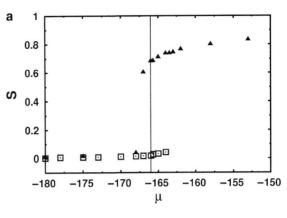

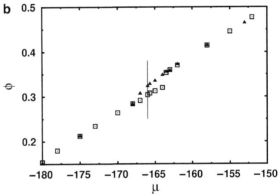

general [18–20], the nematic–isotropic transition shows up via pronounced hysteresis in the S versus μ curve (Fig. 20a) and a (weaker) hysteresis in the corresponding density variation (Fig. 20b). In order to be able to locate the transition point from the isotropic to the nematic solution precisely, a thermodynamic integration method in the grand canonical ensemble (TIμVT method) was used. Denoting the chemical potential of a ideal gas of chains as μ_{id}, and defining $\mu_{ex} = \mu - \mu_{id}$, the osmotic pressure π of the solution becomes ($k_BT \equiv 1$ here) [290–293]:

$$\pi = \rho(1 + \mu_{ex}) - \int_0^\rho \mu^{ex}(\rho')d\rho' , \qquad (52)$$

where $\rho = \mathcal{N}/V$ is the density of polymer chains in the system (note that $\rho = \phi/(8N)$ in our model). Of course, in practice the integral in (52) is discretized, but for a very good accuracy clearly a large number of state points (μ_i, T, V) need to be simulated to render the discretization error negligible. Despite this disadvantage, this old [290] method is still superior in accuracy to any other approach [123]. The

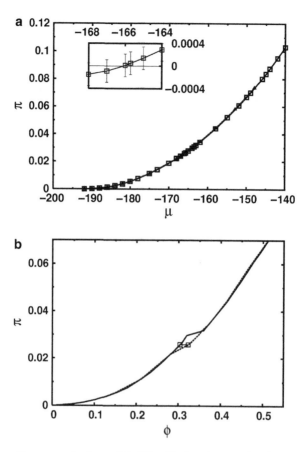

Fig. 21 (a) Osmotic pressure for the model of Fig. 20 plotted versus chemical potential, obtained from the T$I\mu VT$ method. The *inset* shows the difference $\pi_{\text{nematic}} - \pi_{\text{isotropic}}$ in the region close to the isotropic–nematic transition on enlarged scales. *Triangles* correspond to a densely packed starting configuration, while *squares* correspond to a dilute isotropic starting configuration. (b) Equation of state for the bond fluctuation model with $N = 20$, $f = 8$, as a plot of π versus ϕ. *Solid curve* refers to the dilute and *dotted curve* to the densely packed starting configuration, respectively. Two *squares* indicate the densities in the coexisting phases. Note that the hysteresis region around the transition in the simulation is controlled by the kinetics of the algorithm. It has nothing to do with a van der Waals-like loop, and hence the hysteresis region is rather asymmetric. From Ivanov et al. [123]

construction from (52) then should yield two branches in a π versus μ plot: one for the isotropic phase, the other for the nematic phase, intersecting at the transition point $\mu = \mu_t$ under some angle. However, since the transition is only rather weakly of first order, on a large scale for π this intersection is not seen, and one needs to analyze the difference $\pi_{\text{nematic}} - \pi_{\text{isotropic}}$ on a magnified scale to be able to clearly identify the transition (Fig. 21). Nevertheless, Figs. 20 and 21 demonstrate that definite results on both the location of the transition and the associated jumps in the monomer volume fraction and the nematic order parameter can be obtained.

Computer Simulations and Coarse-Grained Molecular Models Predicting 375

However, if one is interested only in a rough overview of the behavior of the model, without precise characterization of these jumps and the location of the transition, the "sedimentation equilibrium" method (Sect. 3.2) is a conceptually simple and straightforward alternative (Fig. 14).

It is still an open problem to extend the above analysis to models (such as studied by Ivanov et al. [122]) where an attractive interaction between the effective monomers is also present, so that variable solvent quality is implicitly modeled. Clearly it will require a major effort to extend the techniques described for short alkanes (Sects. 4.1 and 4.2) to coarse-grained off-lattice models for stiff chains in explicit solvent.

4.4 Solutions of Block Copolymers and Micelle Formation

In this subsection, we return to schematic models of flexible chains again, but consider the extension from (monodisperse) homopolymers to diblock copolymers, i.e., we have a block of A-type monomers (chain length N_A) covalently linked to a block of B-type monomers (chain length N_B), such that the total block copolymer has the composition $f = N_A/N$ where $N = N_A + N_B$. When such block copolymers occur in a solvent, it is natural to assume that the solubility for the two blocks is different. Of particular interest is the case in which the shorter block (say, the A-block, so $f < 1/2$) is under bad solvent conditions, while the solvent is still a good solvent for the B-block. If we then have isolated single block copolymers, the configuration of the chain should then be a collapsed spherical A-globule, with the A–B junction on the surface, so that the B polymer is outside the globule, in a mushroom-like configuration. However, when one considers a (dilute) solution containing many such block copolymers in a selective solvent, one may encounter a transition from an (almost) ideal "gas" of single block copolymer chains to a "gas" of so-called micelles, where in each micelle a number n_{AB} of chains cluster together such that the A-parts form a common "core" (of radius R) while the B-parts form the "corona" of radius S, see Fig. 22).

The theory of micelle formation of such block copolymers (and the related case of smaller surfactant molecules) in solution, within the framework of statistical thermodynamics, is a longstanding and challenging problem, which is still incompletely understood (see, e.g., [124, 294–303]). One wants to predict how the critical micelle concentration (CMC) and the number n_{AB} of chains forming a micelle (and also geometric properties of the micelles, such as the radii R and S, Fig. 22) depend on the parameters of the problem (f, N, interaction parameters $\varepsilon_{AA}, \varepsilon_{AB}$ and ε_{BB}, chain stiffness, etc.). Depending on these parameters, the solvent is partially or completely expelled from the micellar core, and the A–B interface between core and corona may be sharp (as hypothesized in Fig. 22) or diffuse, etc. Thus, a variety of scaling-type predictions exist (see [124] for a brief review), but it is very hard to test them because simulations need to equilibrate large enough systems where many micelles occur and are in equilibrium with a surrounding solution that still contains

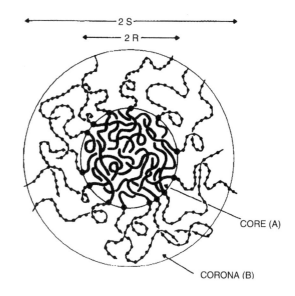

Fig. 22 Spherical block copolymer micelle, assuming strong segregation between the A-blocks (*thick lines*) that form the micellar core and the B-blocks (*chain lines*) that form the corona. The A–B junctions (highlighted by *large dots*) are localized at the surface of the core, which forms a sphere of radius R, while the total micelle forms a sphere of radius S. From Milchev et al. [124]

many block copolymers as single chains. Equilibrium then is established via diffusion and condensation (evaporation) of chains in (from) the micelles, such that all the micelles in the system and the remaining solution have the same chemical potential. Such a chemical equilibrium between the micelles and the solution can, in practice, be established only for rather small N, where the scaling concepts on the micelles are not yet applicable [124]. Many of the simulations of single micelles can be found in the literature (see [124] for some further references), but such work that considers a constrained equilibrium where some value of n_{AB} is a priori imposed cannot answer the questions asked above, which address the full equilibrium aspects of micelle formation from solution.

However, there is one special case where simulations of micelle formation for a model containing reasonably long chains has turned out to be feasible, and this is the case where one uses as a solvent for the $A_f B_{1-f}$ block copolymers B-homopolymers of the same chain length N rather than small molecules [125]. In this case, equilibration is achieved by working in an extension of the semigrand canonical ensemble, using the chemical potential difference between the block copolymers (which we shall denote as species C in the following) and the B-chains acting as a solvent as the external control variable, $\delta\mu = \mu_C - \mu_B$. Choosing $N_C = N_B$, trial moves can be attempted where a block copolymer turns into a homopolymer, $C \to B$, or vice versa, $B \to C$. At fixed chain configuration, just a fraction f of monomers needs to be relabeled as A or B in such a move. The chemical potential difference $\delta\mu$ enters the transition probability of these exchange moves in much the same way as for the semigrand canonical algorithm for ordinary polymer blends [6, 82, 170, 171].

Now we will discuss a few characteristic results obtained in the MC study of Cavallo et al. [125], using the bond fluctuation model for the special case $f = 1/8$

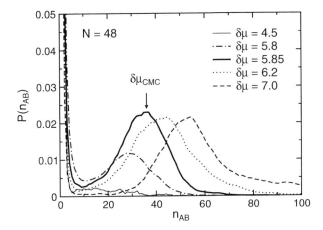

Fig. 23 Cluster size distribution $P(n_{AB})$ versus n_{AB} for the bond fluctuation model of a mixture of homopolymer B and block copolymer $(A_f B_{1-f})$, for $N = 48, f = 1/8$, choosing a $L \times L \times L$ lattice with $L = 96$, a volume fraction $\phi = 0.5$ of occupied lattice sites, and $\varepsilon N = 19.2$ (corresponding to a Flory–Huggins parameter $\chi N \approx 100$). Several values of the exchange chemical potential $\delta\mu$ are shown; $k_B T \equiv 1$ throughout. The *thick curve* denotes the CMC; the weight of the peak for small n_{AB}, corresponding to the solution of block copolymers without micelles, and the weight of the peak near $n_{AB} \approx 35$, due to the micelles, are equal. From Cavallo et al. [125]

and chain lengths in the range $48 \leq N \leq 128$, and the interaction of the type (7), choosing the most symmetric case $\varepsilon_{AA} = \varepsilon_{BB} = -\varepsilon_{AB} = \varepsilon$. If A-monomers from different chains are within the interaction range r_c, these two chains are counted as members of the same micelle. In this way, the number of micelles for each value of n_{AB} can be counted in each system configuration that is analyzed, and the probability distribution $P(N_{AB})$ of the micellar sizes N_{AB} is sampled (Fig. 23). As illustrated in Fig. 23, the CMC is estimated from the equal weight rule applied to the distribution $P(N_{AB})$ for the region of $\delta\mu$ where the distribution is bimodal. Note that this "equal weight rule" is familiar from the MC study of ordinary first-order phase transitions [18–20]. In such cases, however, the two peaks for $L \to \infty$ would turn into two delta functions, but this is not the case here: for finite N, n_{AB} for micelles at the CMC is also finite, so the finite size rounding of $P(n_{AB})$ is due to the finite size of the micelles and not due to the finite simulation volume. Only when $N \to \infty$ can the CMC turn into a sharp thermodynamic phase transition in the standard sense.

The most advanced theoretical studies of micelle formation are based on numerical versions of the SCFT [32–35, 297] of polymers, and such an approach has been worked out by Cavallo et al. [125]. Since the same approach has previously been used to study interfacial properties in polymer blends and compared to corresponding bond fluctuation model simulations, the "translation factor" from the interaction parameter ε of the simulation to the Flory–Huggins χ parameter of the SCFT is already known, and is not an adjustable parameter. SCFT predicts that for large enough N, the distribution $P(n_{AB})$ at the CMC should only depend on the ratio $n_{AB}/N^{1/2}$. Figure 24 demonstrates that, at least in the range of sizes $48 \leq N \leq 128$,

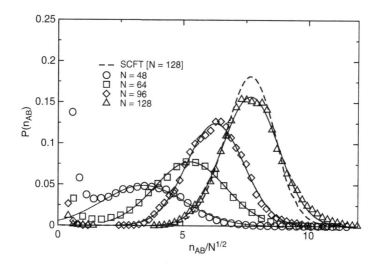

Fig. 24 Cluster size distribution $P(n_{AB})$ at the CMC plotted versus the scaled aggregation number, $n_{AB}/\sqrt{N}$. The *symbols* represent the simulation data (all taken for $\varepsilon N = 19.2$) for different choices of N, as indicated. The *solid lines* show fits with a Gaussian function. For the largest chain length, $N = 128$, the SCFT result (using the equivalent choice of incompatibility between A and B, $\chi N = 100$) is shown as a *dashed line*. The normalization of the SCFT result was chosen such that it agrees with the micellar peak of the MC simulation. From Cavallo et al. [125]

this scaling does not yet hold. One sees that the SCFT distribution is very close to the MC result for $N = 128$, but this is probably a coincidence with no deep meaning (when one studies radial concentration distributions of micelles at the CMC, the data for $N = 128$ have not yet converged to the SCFT result). The reasons for the strong deviations from this simple scaling implied by SCFT are not really clear [125]. Thus, we conclude that the equation of state of block copolymer solutions at the CMC, where micelles form, is far from being understood. Of course, there are many other interesting questions to ask. For example, when one goes beyond the CMC, the micelles can either grow, cluster together in cylindrical objects (some simulation evidence for cylindrical micelles has been seen, e.g., by Viduna et al. [303]), or form micellar mesophase lattices. The non-Gaussian character of $P(n_{AB})$ for $\delta\mu = 7.0$ in Fig. 23 (e.g., the tail extending to $n_{AB} = 100$) is also an indication of some very large, nonspherical micelles in the system at these conditions.

5 Conclusions and Outlook

This article gives a brief review of the state of the art of the modeling of the equation of state of the solutions of (short) polymer chains by MC and MD simulations. Emphasis has been on the type of results that can be obtained, although for more detail on the (already rather elaborate!) technical aspects of the available methods, the interested reader is directed to the available books on simulation methods and to the original articles describing the work that is reviewed here.

A general conclusion is that the study of phase equilibria in polymer solutions by simulation methods is still in its infancy. Due to the large scales of length and time that need to be bridged when one wishes to simulate macromolecules with high molecular weight, the simulation of polymers in general is a challenge if one insists that the simulation reaches complete thermal equilibrium. Although special techniques exist to deal with this problem, both for single-chain problems and for dense melts, the case of solutions seems to have found somewhat less attention in the literature. We also note that in a semidilute solution of very long macromolecules, there exists an intermediate large length scale, the concentration correlation length (sometimes also referred to as the radius of "concentration blobs", i.e., the size of regions over which, in the good solvent limit, excluded-volume interactions are not yet screened out, unlike dense melts where the screening length of excluded-volume interactions is only on the order of a few molecular diameters). This incomplete screening of excluded volume in polymer solution has then an interesting interplay with thermal effects when the solvent quality deteriorates, and ultimately the polymer solution separates into a diluted solution of collapsed globules and a concentrated solution of strongly overlapping chains. The critical point of this phase separation moves towards the theta temperature of the solution (and the critical concentration tends to zero) when the polymer chain length tends to infinity. However, the precise character of this crossover in critical behavior, which is associated with this limit according to theoretical predictions, cannot yet be reliably assessed even by the study of strongly coarse-grained, qualitative models that consider the solvent only implicitly (by postulating suitable weak effective attractions between the effective monomeric units formed from groups of successive chemical monomers along the chain) rather than explicitly. In view of the lack of recent progress with this interesting but difficult problem, we have not reviewed it here, but rather focused only on the phase behavior of solutions of very short chains, considering flexible homopolymers almost exclusively. As an example of the rich science that emerges when this restriction is relaxed, we have pointed towards the possibility of orientational ordering in solutions of stiff polymer chains, and on micelle formation in solutions containing diblock copolymers. Of course, many interesting phenomena exist in solutions containing polymers with more complex architecture (star polymers, comb polymers and bottle brushes, multiblock copolymers, etc.), giving rise to many possibilities of mesophase formation that are outside the scope of our article.

One topic we have addressed in detail is the explicit modeling of solvent molecules, which are described in a simplified, coarse-grained fashion in view of the fact that even the most "atomistic" models are based on united-atom-type approximations for the description of polymer chains, or use even coarser models so that an all-atom modeling of solvent molecules is not warranted. We have also emphasized that in the description of solvent–oligomer phase equilibria, the pressure (in addition to the composition of the solution and its temperature) is an important control variable. We have shown that simulations with predictive power are already possible, at least in favorable cases, when the effective interaction parameters of the pure constituents (solvent, oligomer) are adjusted such that their vapor–liquid equilibria are reasonably well described. Interactions between unlike particles are then described by the Lorentz–Berthelot combining rule. We have shown that dipolar solvents (such as NH_3) and quadrupolar solvents (such as supercritical CO_2) can be described accurately enough by very simple potentials. However, the described modeling of solvents does not include water, nor does the described coarse-graining of the macromolecules in terms of very simple bead–spring models apply to proteins or other biopolymers. Also, solutions of synthetic polyelectrolytes (where one needs to address the effects of counterions and/or salt ions dissolved in the solution) are completely outside the scope of this article.

Finally, we have addressed only the phase behavior of solutions, and have not addressed either the structural properties (e.g., as described by the various pair distribution functions or the size of polymer coils) or the interfacial structure in phase-separated solutions (though we did pay attention to prediction of the interfacial tension between coexisting phases). Also, the kinetics of phase separation (via nucleation and growth or spinodal decomposition) has not been discussed. Thus, we emphasize that, although a few first and promising steps towards the computational modeling of polymer solutions via computer simulations have been taken, many further studies are still necessary to obtain a more complete theoretical understanding of polymer solutions and their properties.

Acknowledgement Part of the work reviewed here has been carried out in a fruitful and productive collaboration with A. Cavallo, V. Ivanov, L. G. MacDowell, M. Müller, M. Oettel, and T. Strauch; it is a pleasure to thank them. Stimulating discussions with H. Weiss and F. Heilmann are acknowledged as well. Financial support has been provided by the BASF SE and by the Deutsche Forschungsgemeinschaft (DFG) in the framework of projects DFG 436 RUS 113/791, PA473/7-1 and PA473/8-1, and Sonderforschungsbereich 625. Computer time grants were provided by John von Neumann Institute for Computing (NIC) and the Network of Excellence SOFTCOMP.

References

1. Huggins MJ (1941) J Chem Phys 9:440
2. Flory PJ (1941) J Chem Phys 9:660
3. Flory PJ (1953) Principles of polymer chemistry. Cornell University Press, Ithaca
4. Tompa H (1956) Polymer solutions. Butterworths, London
5. de Gennes PG (1979) Scaling concepts in polymer physics. Cornell University Press, Ithaca
6. Binder K (1994) Adv Polym Sci 112:181
7. Koningsveld R, Stockmayer W-H, Nies E (2001) Polymer phase diagrams. Oxford University Press, Oxford
8. Rubinstein M, Colby RH (2003) Polymer physics. Oxford University Press, Oxford
9. Ho WH, Yang JS (2005) Adv Polym Sci 156:1
10. Binder K, Müller M, Virnau P, MacDowell LG (2005) Adv Polym Sci 173:1
11. Fredrickson GH (2006) The equilibrium theory of inhomogeneous polymers. Oxford University Press, Oxford

12. Wolf BA (2009) Macromol Theory Simul 18:30
13. Hamley IW (1998) The physics of block copolymers. Oxford University Press, Oxford
14. Sanchez IC, Lacombe RH (1976) J Phys Chem 80:2352
15. Sariban A, Binder K (1987) J Chem Phys 86:5859
16. Sariban A, Binder K (1988) Macromolecules 21:711
17. Landau LD, Lifshitz EM (1958) Statistical physics. Pergamon, Oxford
18. Binder K (ed) (1986) Monte Carlo methods in statistical physics, 2nd edn. Springer, Berlin
19. Binder K, Heermann DW (2002) Monte Carlo simulation in statistical physics: an introduction, 4th edn. Springer, Berlin
20. Landau DP, Binder K (2009) A guide to Monte Carlo simulation in statistical physics, 3rd edn. Cambridge University Press, Cambridge
21. Deutsch HP, Binder K (1993) J Phys (Paris) II 3:1049
22. Melnichenko YB, Anisimov MA, Povodyrev AA, Wignall GD, Sengers JV, Van Hook WA (1997) Phys Rev Lett 79:5266
23. Schwahn D (2005) Adv Polym Sci 183:1
24. Robertson RE (1992) In: Bicerano J (ed) Computational modelling of polymers. Marcel Dekker, New York, p 297
25. Flory PJ (1969) Statistical mechanics of chain molecules. Interscience, New York
26. Grosberg AYu, Khokhlov AR (1994) Statistical physics of macromolecules. AIP, New York
27. Bawendi MG, Freed KF (1988) J Chem Phys 88:2741
28. Freed KF, Dudowicz J (1995) Trends Polym Sci 3:248
29. Foreman KW, Freed KF (1998) Adv Chem Phys 103:335
30. Freed KF, Dudowicz J (2005) Adv Polym Sci 183:63
31. Doi M, Edwards SF (1986) Theory of polymer dynamics. Clarendon, Oxford
32. Helfand E, Tagami Y (1971) J Polym Sci Polym Lett 9:741
33. Schmid F (1998) J Phys Condens Matter 10:8105
34. Matsen MW (2002) J Phys Condens Matter 14:R21
35. Müller M, Schmid W (2005) Adv Polym Sci 185:1
36. Müller M, MacDowell LG (2003) J Phys Condens Matter 15:R609
37. Halperin A, Tirrell M, Lodge TP (1991) Adv Polym Sci 100:31
38. Advincula RC, Brittain WJ, Carter KC, Ruehe J (eds) (2004) Polymer brushes. Wiley-VCH, Weinheim
39. Fredrickson GH, Ganesan V, Drolet F (2002) Macromolecules 35:16
40. Moreira AG, Baeurle SA, Fredrickson GH (2003) Phys Rev Lett 91:150201
41. Baeurle SA, Efimov GV, Nogovitsin EA (2006) Europhys Lett 75:378
42. Baeurle SA (1999) J Math Chem 46:363
43. Schweizer KS, Curro JG (1994) Adv Polym Sci 116:319
44. Schweizer KS, Curro JG (1997) Adv Chem Phys 98:1
45. Wertheim MS (1984) J Stat Phys 35:19
46. Wertheim MS (1984) J Chem Phys 85:2929
47. Chapman WG, Gubbins KE, Jackson G, Radosz M (1989) Fluid Phase Equilib 52:31
48. Blas FJ, Vega LF (1997) Mol Phys 92:135
49. Müller EA, Gubbins KE (2001) Ind Eng Chem Res 40:2198
50. MacDowell LG, Müller M, Vega C, Binder K (2000) J Chem Phys 113:419
51. Gross J, Sadowski G (2001) Ind Eng Chem Res 40:1244
52. MacDowell LG, Virnau P, Müller M, Binder K (2002) J Chem Phys 117:6360
53. Virnau P, Müller M, MacDowell LG, Binder K (2004) J Chem Phys 121:2169
54. Yelash L, Müller M, Paul W, Binder K (2005) J Chem Phys 123:014908
55. Mognetti BM, Virnau P, Yelash L, Paul W, Binder K, Müller M, MacDowell LG (2009) Phys Chem Chem Phys 11:1923
56. Mognetti BM, Virnau P, Yelash L, Paul W, Binder K, Müller M, MacDowell LG (2009) J Chem Phys 130:044101
57. Hansen JP, McDonald IR (1986) Theory of simple liquids. Academic, New York

58. Virnau P, Kantor Y, Kardar M (2005) J Am Chem Soc 127:15102
59. Smit B, Siepmann JL, Karaborni S (1995) J Chem Phys 102:2126
60. Paul W, Yoon DY, Smith GD (1995) J Chem Phys 103:1702
61. Paul W, Smith GD, Yoon DY (1997) Macromolecules 30:7772
62. Smith GD, Paul W, Yoon DY, Zirkel A, Hendricks J, Richter D, Schober H (1997) J Chem Phys 107:4751
63. Paul W, Smith GD, Yoon DY, Farago B, Rathgeber S, Zirkel A, Willner L, Richter D (1998) Phys Rev Lett 80:2346
64. Smith GD, Paul W, Monkenbusch M, Willner L, Richter D, Qui XH, Ediger MD (1999) Macromolecules 32:8857
65. Smith GD, Paul W, Monkenbusch M, Richter D (2000) Chem Phys 261:61
66. Smith GD, Borodin O, Bedrov D, Paul W, Qiu XH, Ediger MD (2001) Macromolecules 34:5192
67. Smith GD, Paul W, Monkenbusch M, Richter D (2001) J Chem Phys 114:4285
68. Smith GD, Borodin O, Paul W (2002) J Chem Phys 117:10350
69. Harmandaris VA, Angelopoulou D, Mavrantzas VG, Theodorou DN (2002) J Chem Phys 116:7656
70. Krushev S, Paul W (2003) Phys Rev E 67:021806
71. Paul W, Smith GD (2004) Rep Progr Phys 67:1117
72. Capaldi F, Rutledge GC, Boyce MC (2005) Macromolecules 38:6700
73. Genix AC, Arbe A, Alvarez F, Colmenero J, Willner L, Richter D (2005) Phys Rev E 72:031808
74. Makrodimitri ZA, Dohrn R, Economou IG (2007) Macromolecules 40:1720
75. Grest GS, Kremer K (1986) Phys Rev A 33:3628
76. Carmesin I, Kremer K (1988) Macromolecules 21:2819
77. Kremer K, Binder K (1988) Comput Phys Rep 7:259
78. Kremer K, Grest GS (1990) J Chem Phys 92:5057
79. Deutsch H-P, Binder K (1991) J Chem Phys 94:2294
80. Paul W, Binder K, Heermann DW, Kremer K (1991) J Phys II 1:37
81. Binder K (1992) In: Bicerano J (ed) Computational modelling of polymers. Marcel Dekker, New York, p 221
82. Binder K (ed) (1995) Monte Carlo and molecular dynamics simulations in polymer science. Oxford University Press, New York
83. Kotelanskii M, Theodorou DN (eds) (2004) Simulation methods for polymers. Marcel Dekker, New York
84. Binder K, Baschnagel J, Paul W (2003) Progr Polym Sci 28:115
85. Binder K, Kob W (2005) Glassy materials and disordered solids: an introduction to their statitsical mechanics. World Scientific, Singapore
86. Baschnagel J, Varnik F (2005) J Phys Condens Matter 17:R851
87. Allen MP, Tildesley DJ (1987) Computer simulation of liquids. Clarendon, Oxford
88. Frenkel D, Smit B (1996) Understanding molecular simulation: from algorithms to applications. Academic, San Diego
89. Leach AR (1996) Molecular modelling. Longman, Harlow
90. Attig N, Binder K, Grubmüller H, Kremer K (eds) (2004) Computational soft matter: from synthetic polymers to proteins. John von Neumann Institute for Computing (NIC), Juelich
91. Ferrario M, Ciccotti G, Binder K (eds) (2006) Computer simulations in condensed matter: from materials to chemical biology, vols 1, 2. Springer, Berlin
92. Karayiannis N, Mavrantzas V, Theodorou D (2002) Phys Rev Lett 88:105503
93. Karayiannis N, Gianousaki A, Mavrantzas V, Theodorou D (2002) J Chem Phys 117:5465
94. Banaszak BJ, de Pablo JJ (2003) J Chem Phys 119:2456
95. Wittmer JP, Beckrich P, Meyer H, Cavallo A, Johner A, Baschnagel J (2007) Phys Rev E 76:011803
96. Binder K, Paul W (2008) Macromolecules 41:4537

Computer Simulations and Coarse-Grained Molecular Models Predicting 383

97. Paul W, Binder K, Kremer K, Heermann DW (1991) Macromolecules 24:6332
98. Baschnagel J, Binder K, Paul W, Laso M, Suter U, Batoulis I, Jilge W, Bürger T (1991) J Chem Phys 95:6014
99. Baschnagel J, Qin K, Paul W, Binder K (1992) Macromoleucles 25:3117
100. Paul W, Pistoor N (1994) Macromolecules 27:1249
101. Tries V, Paul W, Baschnagel J, Binder K (1997) J Chem Phys 106:738
102. Tschöp W, Kremer K, Batoulis J, Bürger T, Hahn O (1998) Acta Polym 49:61
103. Tschöp W, Kremer K, Batoulis J, Bürger T, Hahn O (1998) Acta Polym 49:75
104. Baschnagel J, Binder K, Duruker P, Gusev AA, Hahn O, Kremer K, Mattice WL, Müller-Plathe F, Murat M, Paul R, Santos S, Suter UW, Tries V (2000) Adv Polym Sci 152:41
105. Reith D, Meyer H, Müller-Plathe F (2001) Macromolecules 34:2335
106. Müller-Plathe F (2002) Chem Phys Chem 3:754
107. Müller-Plathe F (2003) Soft Matter 1:1
108. Reith D, Pütz M, Müller-Plathe F (2003) J Comput Chem 24:1624
109. Faller R (2004) Polymer 45:3869
110. Milano G, Müller-Plathe F (2005) J Phys Chem 24:1624
111. Yelash L, Müller M, Paul W, Binder K (2006) J Chem Theory Comput 2:588
112. Sun Q, Faller R (2006) J Chem Theory Comput 2:607
113. Harmandaris VA, Athikari NP, Van der Vegt NFA, Kremer K (2006) Macromolecules 39:6708
114. Sun Q, Pon FR, Faller R (2007) Fluid Phase Equilib 261:35
115. Ghosh J, Faller R (2007) Mol Simul 33:759
116. Prapotnik M, Delle Site L, Kremer K (2008) Ann Rev Phys Chem 59:545
117. Strauch T, Yelash L, Paul W (2009) Phys Chem Chem Phys 11:1942
118. Noid WG, Chu JW, Ayton GS, Krishna V, Izekov S, Voth GA, Das A, Andersen HC (2008) J Chem Phys 128:244114
119. Voth G (ed) (2008) Coarse-graining of condensed and biomolecular systems. Taylor & Francis, Boca Raton
120. Binder K, Paul W, Virnau P, Yelash L, Müller M, MacDowell LG (2008) In: Voth G (ed) Coarse-graining of condensed and biomolecular systems. Taylor & Francis, Boca Raton, p 399
121. Weber H, Paul W, Binder K (1999) Phys Rev E 59:2168
122. Ivanov VA, Stukan MR, Müller M, Paul W, Binder K (2003) J Chem Phys 118:10333
123. Ivanov VA, An EA, Spirin LA, Stukan MR, Müller M, Paul W, Binder K (2007) Phys Rev E 76:026702
124. Milchev A, Bhattacharya A, Binder K (2001) Macromolecules 34:1881
125. Cavallo A, Müller M, Binder K (2006) Macromolecules 39:9539
126. Martonak R, Paul W, Binder K (1998) Phys Rev E 58:1998
127. Smith GD (2005) In: Yip S (ed) Handbook of materials modeling. Springer, Berlin, p 2561
128. Borodin O, Smith GD (2003) J Phys Chem B 108:6801
129. Sorensen RA, Liau WB, Kesner L, Boyd RH (1988) Macromolecules 21:200
130. National Institute of Standards and Technology (2008) NIST Chemistry WebBook. http://webbook.nist.gov/chemistry/
131. Mognetti BM, Yelash L, Virnau P, Paul W, Binder K, Müller M, MacDowell LG (2008) J Chem Phys 128:104501
132. Zinn-Justin J (2001) Phys Rep 344:159
133. Binder K, Luijten E (2001) Phys Rep 344:179
134. Smith GD, Paul W (1998) J Phys Chem A 102:1200
135. Kiran E, Levelt-Sengers JMH (eds) (1994) Supercritical fluids. Kluwer, Dordrecht
136. Kemmere MF, Meyer Th (eds) (2005) Supercritical carbon dioxide in polymer reaction engineering. Wiley-VCH, Weinheim
137. Hikmet RM, Callister S, Keller A (1988) Polymer 29:1378
138. Han JH, Han CD (1990) J Polym Sci B Polym Phys 28:711

139. Krause B, Sijbesma HJP, Münüklü P, van der Vegt NFA, Wessburg W (2001) Macromolecules 34:8792
140. Murthy CS, Singer K, McDonald IR (1981) Mol Phys 44:135
141. Böhm HJ, Meissner C, Ahlrichs R (1984) Mol Phys 53:651
142. Böhm HJ, Ahlrichs R (1985) Mol Phys 55:445
143. Zhu SB, Robinson GW (1989) Comp Phys Commun 52:317
144. Etters RD, Kuchta B (1989) J Phys Chem 90:4537
145. Geiger LC, Ladanyi BM, Chapin ME (1990) J Chem Phys 93:4533
146. Harris JG, Yung KH (1993) J Phys Chem 99:12021
147. Martin MG, Siepmann JI (1998) J Phys Chem B 102:2569
148. Martin MG, Siepmann JI (1999) J Phys Chem B 103:4508
149. Bukowski R, Sadlej J, Jeziorski B, Jankovski P, Szalewicz K, Kucharski SA, Williams HL, Rice BM (1999) J Chem Phys 110:3785
150. Bock S, Bich E, Vogel E (2000) Chem Phys 257:147
151. Vorholz J, Harismiadis VI, Rumpf B, Panagiotopoulos AZ, Maurer G (2000) Fluid Phase Equilib 170:203
152. Zhang Z, Duan Z (2005) J Chem Phys 122:214507
153. Vrabec J, Stoll J, Hasse H (2001) J Phys Chem B 105:12126
154. Stoll J, Vrabec J, Hasse H (2003) AIChE J 49:2187
155. Vrabec J, Stoll J, Hasse H (2005) Mol Simul 31:215
156. Bratschi C, Huber H, Searles DJ (2007) J Chem Phys 126:164105
157. Mognetti BM, Oettel M, Yelash L, Virnau P, Paul W, Binder K (2008) Phys Rev E 77:041506
158. Mognetti BM, Oettel M, Virnau P, Yelash L, Binder K (2009) Mol Phys 107:331
159. Kristof T, Liszi J (1997) J Phys Chem B 101:5480
160. Jorgensen WL (1986) J Phys Chem 90:6379
161. Forester TR, MacDonald IR, Klein ML (1989) Chem Phys 129:225
162. Delhommelle J, Milllie P, Fuchs AH (2000) Mol Phys 98:1895
163. Nath SK (2003) J Phys Chem B 107:9498
164. Rosenbluth MN, Rosenbluth AW (1955) J Chem Phys 23:356
165. Sokal AD (1995) In: Binder K (ed) Monte Carlo and molecular dynamics simulations in polymer science. Oxford University Press, New York, p 47
166. Grassberger P (1997) Phys Rev E 56:3682
167. Hsu H-P, Grassberger P (2003) Eur Phys J B 36:209
168. Prellberg T, Krawczyk J (2004) Phys Rev Lett 92:12062
169. Hsu H-P, Binder K, Klushin LI, Skvortsov AM (2008) Phys Rev E 78:041803
170. Müller M, Binder K (1995) Macromolecules 28:1825
171. Müller M (1999) Macromol Theory Simul 8:343
172. Deutsch H-P, Binder K (1992) Macromolecules 26:6214
173. Wilding N, Müller M, Binder K (1996) J Chem Phys 105:802
174. Baschnagel J, Binder K, Wittmann HP (1993) J Phys Condens Matter 5:1597
175. Paul W, Baschnagel J (1995) In: Binder K (ed) Monte Carlo and molecular dynamics simulations in polymer science. Oxford University Press, New York, p 307
176. Larson RG (1988) J Chem Phys 89:1842
177. Larson RG (1994) Macromolecules 27:4198
178. Dotera T, Hatano A (1986) J Chem Phys 105:8413
179. Zhi Y, Wang W (2007) Macromolecules 40:2872
180. Reister E, Müller M, Binder K (2001) Phys Rev E 64:041804
181. Hashimoto T (1993) In: Cahn RW, Haasen P, Kramer E (eds) Materials science and technology, vol 12. Wiley-VCH, Weinheim, p 251
182. Binder K, Fratzl P (2001) In: Kostorz G (ed) Phase transformations of materials. Wiley-VCH, Weinheim, p 409
183. Vacatello M, Avitabile G, Corradini P, Tuzi A (1980) J Chem Phys 73:548

184. Yoon DY, Vacatello M, Smith GD (1995) In: Binder K (ed) Monte Carlo and molecular dynamics simulations in polymer science. Oxford University Press, New York, p 433
185. Vacatello M (1997) Macromol Theory Simul 6:613
186. Dünweg B, Kremer K (1993) J Chem Phys 99:6983
187. Dünweg B, Stevens M, Kremer K (1995) In: Binder K (ed) Monte Carlo and molecular dynamics simulations in polymer science. Oxford University Press, New York, p 125
188. Murat M, Kremer K (1988) J Chem Phys 108:4340
189. Asakura S, Oosawa F (1954) J Chem Phys 22:1255
190. Asakura S, Oosawa F (1958) J Polym Sci 33:183
191. Vrij A (1976) Pure Appl Chem 48:471
192. Poon WCK (2002) J Phys Condens Matter 14:R859
193. Binder K, Horbach J, Vink RLC, De Virgiliis A (2008) Soft Matter 4:1555
194. Ilett SM, Orrock A, Poon WCK, Pusey PN (1995) Phys Rev E 51:1344
195. Hennequin Y, Aarts DGAL, Indekeu J, Lekkerkerker HNW, Bonn D (2008) Phys Rev Lett 100:178305
196. Aarts DGAL, Schmidt M, Lekkerkerker HNW (2004) Science 304:847
197. Aarts DGAL, Lekkerkerker HNW (2004) J Phys Condens Matter 16:S4231
198. Derks D, Aarts DGAL, Bonn D, Lekkerkerker HNW, Imhof A (2006) Phys Rev Lett 97:038301
199. Bolhuis PG, Louis AA, Hansen JP, Meyer EJ (2001) J Chem Phys 114:4296
200. Bolhuis PG, Louis AA, Hansen JP (2002) Phys Rev Lett 89:128302
201. Bolhuis PG, Louis AA (2002) Macromolecules 35:1860
202. Louis AA (2002) J Phys Condens Matter 14:9187
203. Rotenberg R, Dzubiella J, Louis AA, Hansen JP (2004) Mol Phys 102:1
204. Zausch J, Virnau P, Binder K, Horbach J, Vink RL (2009) J Chem Phys 130:064906
205. Harmandaris VA, Reith D, Van der Vegt NFA, Kremer K (2007) Macromol Chem Phys 208:2109
206. . Krushev S (2002) PhD thesis, University of Mainz
207. Gray GG, Gubbins KE (1984) Theory of molecular fluids, vol 1: fundamentals. Clarendon, Oxford
208. Müller EA, Gelb LG (2003) Ind Eng Chem Res 42:4123
209. Potoff JJ, Panagiotopoulos AZ (2000) J Chem Phys 112:6411
210. Maitland GC, Rigby M, Smith EB, Wakeham WA (1981) Intermolecular forces their origian and determination. Clarendon, Oxford
211. Fender BEF, Halsey GD Jr (1962) J Chem Phys 36:1881
212. Smith FT (1972) Phys Rev A 5:1708
213. Kong CL (1973) J Chem Phys 59:2464
214. Potoff JJ, Errington JR, Panagiotopoulos AZ (1999) Mol Phys 97:1073
215. Marcelli G, Sadus RJ (1999) J Phys Chem 111:1533
216. Raabe G, Sadus RJ (2003) J Chem Phys 119:6691
217. Yoshida H (1990) Phys Lett A 150:262
218. Rapaport DC (2005) The art of molcecular dynamics simulation, 2nd edn. Cambridge University Press, Cambridge
219. Hockney R, Eastwood J (1981) Comupter simulation using particles. McGraw-Hill, New York
220. Hill TL (1963) Thermodynamics of small systems. Benjamin, New York
221. Gross DHE (2001) Microcanonical thermodynamics. World Scientific, Singapore
222. Berendsen HJC, Postma JPM, van Gunsteren WF, DiNola A, Haak JR (1984) J Chem Phys 81:3684
223. Nose S (1991) Progr Theor Phys Suppl 103:1
224. Hoover WG (1986) Phys Rev A 31:1695
225. Hoogerbrugge PJ, Koelman JMVA (1992) Europhys Lett 19:155
226. Espanol P, Warren P (1995) Europhys Lett 30:191

227. Soddemann T, Dünweg B, Kremer K (2003) Phys Rev E 68:046702
228. Das SK, Horbach J, Binder K (2003) J Chem Phys 119:1547
229. Das SK, Horbach J, Binder K, Fisher ME, Sengers JV (2006) J Chem Phys 125:024506
230. Zervopoulou E, Mavrantzas VG, Theodorou DN (2001) J Chem Phys 115:2860
231. Tuckerman M, Berne BJ, Martyna GJ (1992) J Chem Phys 97:1990
232. Martyna GJ, Tuckerman ME, Tobias DJ, Klein ML (1996) Mol Phys 87:1117
233. Binder K, Horbach J, Kob W, Paul W, Varnik F (2004) J Phys Condens Matter 16:S429
234. Yelash L, Virnau P, Paul W, Binder K, Müller M (2008) Phys Rev E 78:031801
235. Rowlinson JS, Widom B (1982) Molecular theory of capillarity. Clarendon, Oxford
236. Alejandre J, Tildesley DJ, Chapela GA (1975) J Chem Phys 102:4374
237. Werner A, Schmid F, Müller M, Binder K (1997) J Chem Phys 107:8175
238. Werner A, Schmid F, Müller M, Binder K (1999) Phys Rev E 59:728
239. Milchev A, Binder K (2001) J Chem Phys 115:983
240. Vink RLC, Horbach J, Binder K (2005) J Chem Phys 122:134905
241. Sariban A, Binder K (1991) Macromolecules 24:578
242. Bartke J, Hentschke R (2007) Phys Rev E 75:061503
243. MacDowell LG, Virnau P, Müller M, Binder K (2004) J Chem Phys 120:5293
244. MacDowell LG, Shen VK, Errington JR (2006) J Chem Phys 125:034705
245. Schrader M, Virnau P, Binder K (2009) Phys Rev E 79:061104
246. Andersen HC (1980) J Chem Phys 72:2384
247. Widom B (1963) J Chem Phys 39:2808
248. Möller D, Fischer J (1990) Mol Phys 69:463
249. Lotfi A, Vrabec J, Fischer J (1992) Mol Phys 76:1319
250. Stoll J, Vrabec J, Hasse H, Fischer J (2001) Fluid Phase Equilib 179:339
251. Metropolis N, Rosenbluth AW, Rosenbluth MN, Teller AH, Teller E (1953) J Chem Phys 21:1087
252. Nath SK, Escobedo FA, de Pablo JJ (1998) J Chem Phys 108:9905
253. Panagiotistopoulos AZ (1987) Mol Phys 61:813
254. Sadus RJ (1996) Mol Phys 87:979
255. Vorholz J, Harismiadis VI, Rumpf B, Panagiotopoulos AZ, Maurer G (2000) Fluid Phase Equilib 170:203
256. Zhang L, Siepmann JI (2005) J Chem Phys B 109:2911
257. Houndonougho Y, Jin H, Rajagpphalan B, Wong K, Kuczera K, Subramaniam B, Laird B (2006) J Phys Chem B 110:13195
258. Ferrenberg AM, Swendsen RH (1988) Phys Rev Lett 61:2635
259. Ferrenberg AM, Swendsen RH (1989) Phys Rev Lett 63:1195
260. Torrie GM, Valleau JP (1977) J Comp Phys 23:187
261. Virnau P, Müller M (2004) J Chem Phys 120:10925
262. Berg BA, Neuhaus T (1991) Phys Lett B 267:249
263. Wilding NB (1995) Phys Rev E 52:602
264. Wilding NB (1997) J Phys Condens Matter 9:585
265. Potoff JJ, Errington JR, Panagiotopoulos AZ (1999) Mol Phys 97:1073
266. Vink RLC, Horbach J, Binder K (2005) Phys Rev E 71:011401
267. Lenart PJ, Panagiotopoulos AZ (2006) Ind Eng Chem Res 45:6929
268. Vink RLC, Horbach J (2004) J Chem Phys 121:3253
269. Addison CJ, Hansen JP, Louis AA (2005) Chem Phys Chem 6:1760
270. Cheng H, de Fernandez MEP, Zollweg JA, Streett WB (1989) J Chem Eng Data 34:319
271. Sebastian HM, Sinnick JJ, Liu H-M, Chao KC (1980) J Chem Eng Data 25:138
272. Brunner E, Hültenschmidt W, Schlichthärle G (1987) J Chem Thermodyn 19:273
273. Brunner E (1988) J Chem Thermodyn 20:273
274. Brunner E (1988) J Chem Thermodyn 20:1397
275. Scott RL, van Konynenburg PH (1970) Discuss Faraday Soc 49:87
276. Flory PJ (1956) Proc R Soc London A 234:60

277. Odijk T (1986) Macromolecules 19:2313
278. Semenov AN, Khokhlov AR (1986) Sov Phys Usp 31:988
279. Baumgärtner A (1986) J Chem Phys 84:1905
280. Kolinsky A, Skolnick J, Yaris R (1986) Macromolecules 19:2560
281. Wilson MR, Allen MP (1993) Mol Phys 80:277
282. Dijkstra M, Frenkel D (1995) Phys Rev E 51:5891
283. Strey HH, Parsegian VA, Podgornik R (1999) Phys Rev E 59:999
284. Vega C, McBride C, MacDowell LG (2001) J Chem Phys 115:4203
285. McBride C, Vega C, MacDowell LG (2001) Phys Rev E 64:011703
286. Vega C, McBride C, MacDowell LG (2002) Phys Chem Chem Phys 4:853
287. McBride C, Vega C (2002) J Chem Phys 117:10370
288. Khokhlov AR (1979) Polym Sci USSR 21:2185
289. Grosberg AY, Khokhlov AR (1981) Adv Polym Sci 41:53
290. Bellemans A, De Vos E (1973) J Polym Sci Polym Symp 42:1195
291. Okamoto H (1976) J Chem Phys 64:2686
292. Okamoto H (1983) J Chem Phys 79:3976
293. Okamoto H (1985) J Chem Phys 83:2587
294. Meier DJ (1969) J Polym Sci C 26:81
295. de Gennes PG (1978) In: Liebert J (ed) Solid state physics, suppl 14. Academic, New York, p 1
296. Leibler L, Orland H, Wheeler JC (1983) J Chem Phys 79:3550
297. Noolandi J, Hong KM (1983) Macromolecules 16:1443
298. Halperin A (1987) Macromolecules 20:2943
299. Bug ALR, Cates ME, Safran SA, Witten TA (1987) J Chem Phys 87:1824
300. Marques CM, Joanny JF, Leibler L (1988) Macromolecules 21:1051
301. Nagarajan R, Ganesh K (1989) J Chem Phys 90:5843
302. Ligoure C (1991) Macromolecules 24:2968
303. Viduna D, Milchev A, Binder K (1998) Macromol Theory Simul 7:649

Adv Polym Sci (2011) 238: 389–418
DOI: 10.1007/12_2010_94
© Springer-Verlag Berlin Heidelberg 2010
Published online: 31 July 2010

Modeling of Polymer Phase Equilibria Using Equations of State

Gabriele Sadowski

Abstract The most promising approach for the calculation of polymer phase equilibria today is the use of equations of state that are based on perturbation theories. These theories consider an appropriate reference system to describe the repulsive interactions of the molecules, whereas van der Waals attractions or the formation of hydrogen bonds are considered as perturbations of that reference system. Moreover, the chain-like structure of polymer molecules is explicitly taken into account. This work presents the basic ideas of these kinds of models. It will be shown that they (in particular SAFT and PC-SAFT) are able to describe and even to predict the phase behavior of polymer systems as functions of pressure, temperature, polymer concentration, polymer molecular weight, and polydispersity as well as – in case of copolymers – copolymer composition.

Keywords Copolymers · Equation of state · Modeling · Polymers · Solubility · Sorption · Thermodynamics

Contents

1 Introduction 392
2 Equations of State 392
3 Estimation of Model Parameters 399
4 Modeling of Homopolymer Systems 401
5 Extension to Copolymers 405
6 Accounting for the Influence of Polydispersity 408
7 Summary 414
References 416

G. Sadowski
TU Dortmund, Department of Biochemical and Chemical Engineering, Laboratory for Thermodynamics, Emil-Figge-Strasse 70, 44227 Dortmund, Germany
e-mail: g.sadowski@bci.tu-dortmund.de

Symbols

A	Helmholtz energy
a	Parameter of the van der Waals equation
$B_{\alpha\beta}$	Fraction of bonds between segments α and β within a copolymer
b	Parameter of the van der Waals equation
d	Temperature-dependent segment diameter
g	Radial distribution function
$g(d^+)$	Value of the radial distribution function at contact
k	Boltzmann constant
M	Molecular weight
$M_{2p,j}$	Molecular weight of pseudocomponent j
M_n	Number average of molecular weight
M_w	Weight average of molecular weight
M_z	z-Average of molecular weight
$\overline{M^k}$	kth moment of the molecular weight distribution
m	Segment number
$\bar{m}$	Average segment number
N	Number of molecules
$N*$	Number of association sites per molecule or monomer unit
n_i	Mole number
k_{ij}	Binary interaction parameter
p	Pressure
R	Ideal gas constant
T	Temperature
V	Volume
v	Molar volume
v_{00}	Segment volume (parameter of SAFT)
x_i	Mole fraction of component i (solvent or polymer)
$x_{2p,j}$	Mole fraction of pseudocomponent j within polymer
$W(M)$	Continuous molecular weight distribution
w_i	Weight fraction of component i
$w_{2p,j}$	Weight fraction of pseudocomponent j in polymer
z	Compressibility factor
z_α, z_β	Fraction of segments α or β in a copolymer

Abbreviations

HDPE	High-density polyethylene
L	Liquid
LL	Liquid–liquid

LDPE	Low-density polyethylene
MA	Methylacrylate
MWD	Molecular weight distribution
PA	Propylacrylate
PC-SAFT	Perturbed Chain Statistical-Associating-Fluid Theory
PR	Peng–Robinson
PHCT	Perturbed Hard-Chain Theory
PHSC	Perturbed Hard-Sphere-Chain Theory
PSCT	Perturbed Soft-Chain Theory
SAFT	Statistical-Associating-Fluid Theory
SAFT-VR	SAFT with Variable Range
SRK	Soave–Redlich–Kwong

Greek Letters

α, β	Segment type
ε	Dispersion energy parameter
ε_{AA}	Association-energy parameter
η	Reduced density
κ_{AA}	Association volume parameter
ρ	Number density (molecules per volume)
σ	Temperature-independent segment diameter
φ_i	Fugacity coefficient of component i in the mixture
$\varphi_{2p,j}$	Fugacity coefficient of polymer pseudocomponent j in the mixture

Superscripts

assoc	Contribution due to association
chain	Contribution due to chain formation
disp	Dispersion (van der Waals attraction)
disp	Dispersion contribution according to the PC-SAFT model
hc	Hard-chain contribution
hs	Hard-sphere contribution
id	Ideal gas
pert	Perturbation
ref	Reference
res	Residual
I,II	Phases I and II

1 Introduction

Mutual solubility of polymers and volatile organic substances are of importance for many applications in polymer chemistry and polymer engineering. Polymerizations, which should be performed in homogeneous phase, require the complete miscibility of monomer, polymer, solvent (liquid or supercritical) and other additives. Subsequently, the extraction of the polymer product from the reaction mixture requires a phase split (into two liquid phases or into a vapor and a liquid phase) to obtain a polymer product of high purity on one side and the remaining monomer on the other side. In this context, the devolatilization of polymers is of particular interest. Another example is the use of polymer membranes for the separation of two volatile organic compounds. Here, besides the knowledge of diffusivity, the solubility (sorption) of the different components in the polymer membrane is also an important prerequisite for an efficient process.

However, experimental data on polymer solubility are often scarce. Considerable experimental effort is generally required for determining these properties of polymer systems. Thermodynamics can provide powerful and robust tools for modeling of experimental data and even for prediction of the thermodynamic behavior.

2 Equations of State

Equations of state are traditionally equations that give the pressure p as a function of temperature, molar mixture volume, and composition $p(T, v, x_i)$. A well-known example is the van der Waals equation of state [1], which reads as:

$$p = \frac{RT}{v - b} - \frac{a}{v^2} \tag{1}$$

This equation contains two parameters a and b that are related to the interaction energy of the molecules and to the size of the molecules, respectively. Therefore, they are called pure-component parameters and are usually determined by fitting to experimental liquid-density and vapor-pressure data. Applying equations of states to mixtures is, in most cases, done by applying a one-fluid theory. This means that the parameters of a virtual "mixture molecule" are obtained by so-called mixing rules from the model parameters of the pure components, e.g., by:

$$a = \sum_i \sum_j x_i \, x_j \, a_i \, a_j \, (1 - k_{ij})$$

$$b = \sum_i x_i \, b_i \tag{2}$$

k_{ij} in (2) is a binary parameter that corrects for deviations from the mixing rule of the interaction energy parameter and needs to be fitted to binary data.

Modeling of Polymer Phase Equilibria Using Equations of State

The van der Waals equation of state, as well as related expressions such as the models of Soave–Redlich–Kwong (SRK) [2] or Peng–Robinson (PR) [3], are based on the following assumptions:

- The molecules are spherical. Introducing the acentric factor into SRK and PR improved the modeling for nonspherical molecules. Because the critical properties as well as the vapor pressure are needed to determine the acentric factor of a component, this is only applicable to volatile components.
- There exist no specific interactions (e.g., polar interactions, hydrogen bonding) between the molecules, which leads to a statistical distribution of the molecules in the mixture.

Due to these assumptions (in particular the first one), these models cannot reasonably be applied to polymer systems. Therefore, over the last 30–40 years different approaches have been developed that explicitly account for the chain-like structure of polymer molecules as well as for specific interactions.

One early considered approach was to extend Flory–Huggins-like lattice models by introducing empty lattice sites (holes) so that the number of holes in the lattice is a measure of the density of the system. Density changes in the system are realized via a variation of the hole number. Equations of state based on this idea are, for example, the Lattice-Fluid Theory from Sanchez and Lacombe [4] and the Mean-Field Lattice-Gas theory from Kleintjens and Koningsveld [5].

Another approach for obtaining an equation of state is based on the partition function of a system derived from statistical mechanics. One of these models is the Perturbed Hard-Chain Theory (PHCT) proposed by Beret and Prausnitz [6]. It was subsequently extended and modified by Cotterman et al. [7] and Morris et al. [8] as the Perturbed Soft-Chain Theory (PSCT).

An alternative is the application of so-called perturbation theories (e.g., Barker and Henderson [9], Weeks et al. [10]). The main assumption here is that the residual (the difference from an ideal gas state) part of the Helmholtz energy of a system A^{res} (and thereby also the system pressure) can be written as the sum of different contributions, whereas the main contributions are covered by the Helmholtz energy of a chosen reference system A^{ref}. Contributions to the Helmholtz energy that are not covered by the reference system are considered as perturbations and are described by A^{pert}:

$$A = A^{\text{id}} + A^{\text{res}} = A^{\text{id}} + A^{\text{ref}} + A^{\text{pert}} \tag{3}$$

$$p = p^{\text{id}} + p^{\text{res}} = p^{\text{id}} + p^{\text{ref}} + p^{\text{pert}} \tag{4}$$

An appropriate reference system (at least for small solvent molecules) is the hard-sphere (hs) system. In a hard-sphere system, the molecules are assumed to be spheres of a fixed diameter and do not have any attractive interactions. Such a reference system covers the repulsive interactions of the molecules, which are considered to mainly contribute to its thermodynamic properties. Moreover,

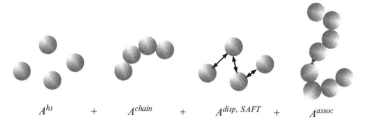

Fig. 1 Helmholtz energy contributions of SAFT

analytical expressions are available for $A^{\text{ref}} = A^{\text{hs}}$ and $p^{\text{ref}} = p^{\text{hs}}$ for hard-sphere systems (e.g., Carnahan and Starling [11]).

Deviations of real molecules from the reference system may occur, e.g., due to attractive interactions (dispersion), formation of hydrogen bonds (association), or the nonspherical shape of the molecules (which can be understood as the formation of chains from spherical segments). These contributions are usually assumed to be independent of each other and are accounted for by different perturbation terms. Depending on the kind of considered perturbation and on the expression used for its description, different models have been developed. One of the first models derived from that idea was the Statistical-Associating-Fluid Theory (SAFT) (Chapman et al. [12, 13]; Huang and Radosz [14, 15]).

In SAFT, a chain-like molecule (solvent molecule or polymer) is assumed to be a chain of m identical spherical segments. Starting from a reference system of m hard spheres (A^{hs}), this model considers three perturbation contributions: chain formation (A^{chain}), attractive interactions of the (nonbonded) segments (A^{disp}), and association via a certain number of association sites (A^{assoc}) (Fig. 1):

$$A^{\text{res}} = \underbrace{m\,A^{\text{hs}} + A^{\text{chain}}}_{A^{\text{hc}}} + m A^{\text{disp, SAFT}} + A^{\text{assoc}} \qquad (5)$$

The Carnahan–Starling formulation is used for A^{hs}; the segment–segment dispersion A^{disp} is described using a fourth-order power series with respect to reversed temperature (Chen and Kreglewski [16]); and the contribution of chain formation and the association term are based on the work of Wertheim [17].

Subsequently, various perturbation theories were developed that are also based on (5) but differ in the specific expressions used for the different types of perturbations. Examples are the Perturbed Hard-Sphere-Chain Theory (PHSC) ([18, 19]), SAFT-VR [20], and models proposed by Chang and Sandler [21], Hino and Prausnitz [22], and Blas and Vega [23].

A widely used model of this kind is the Perturbed Chain SAFT (PC-SAFT) model [24–26], which was particularly developed to improve the modeling of chain-like molecules, e.g., polymers. As the main improvement, PC-SAFT considers the hard chain as a reference system, which is of course much more appropriate for polymers and other chain-like molecules than the hard-sphere

Modeling of Polymer Phase Equilibria Using Equations of State

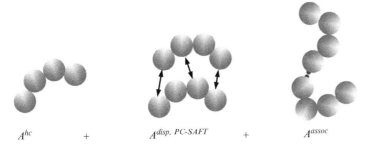

Fig. 2 Helmholtz energy contributions of PC-SAFT

system (Fig. 2). The hard-chain system is obtained as in SAFT as the sum of the hard-sphere and chain-formation contributions A^{hs} and A^{chain}. Since the dispersion term $A^{disp,PC\text{-}SAFT}$ now describes the attraction of chain molecules rather than that of nonbonded segments, it appears to be a function of chain length m:

$$A^{res} = A^{hc} + A^{disp,PC-SAFT}(m) + A^{assoc} \quad (6)$$

Modeling with PC-SAFT (as well as with original SAFT and related models) requires three pure-component parameters for a nonassociating molecule. As for the above-mentioned van der Waals equation, these parameters have a physical meaning. The first parameter of PC-SAFT is the segment diameter σ (SAFT uses the segment volume v_{00}), which corresponds to the van der Waals parameter b. Parameter ε is the energy related to the interaction of two segments, corresponding to the van der Waals parameter a. The third parameter considers the deviation from the spherical shape of the molecules: the segment number m. In the case of a polymer molecule, the latter is proportional to the molecular weight. To describe a binary system, again an additional binary parameter (k_{ij}) is used to correct for deviations of the geometric mean of the energy parameter.

According to (6), different expressions are used to describe the various contributions within the PC-SAFT model. The hard-chain contribution reads as:

$$\frac{A^{hc}}{NkT} = \bar{m}\frac{A^{hs}}{NkT} - \sum_i x_i(m_i - 1) \ln g_{ii}(d_{ii}^+) \quad (7)$$

$$\frac{A^{hs}}{NkT} = \frac{1}{\zeta_0}\left[\frac{3\zeta_1\zeta_2}{(1-\zeta_3)} + \frac{\zeta_2^3}{\zeta_3(1-\zeta_3)^2} + \left(\frac{\zeta_2^3}{\zeta_3^2} - \zeta_0\right)\ln(1-\zeta_3)\right] \quad (8)$$

$$\bar{m} = \sum_i x_i m_i \quad (9)$$

$$\zeta_n = \frac{\pi}{6}\rho\sum_i x_i m_i d_i^n; \quad \zeta_3 = \eta \quad (10)$$

Here, g is the radial distribution function, which is a function of the distance of two segments. In the case of chain formation, this distance is identical to the segment diameter d. Thus, $g(d^+)$ is the value of the radial distribution function at contact. It can be estimated as [27, 28]:

$$\ln g_{ij}\left(d_{ij}^+\right) = \frac{1}{1-\zeta_3} + \left(\frac{d_i d_j}{d_i + d_j}\right)\frac{3\zeta_2}{(1-\zeta_3)^2} + \left(\frac{d_i d_j}{d_i + d_j}\right)^2 \frac{2\zeta_2^2}{(1-\zeta_3)^3} \tag{11}$$

The (temperature-dependent) segment diameter d is calculated from the temperature-independent one according to:

$$d_i = \sigma_i\left[1 - 0.12\exp\left(\frac{-3\varepsilon_i}{kT}\right)\right] \tag{12}$$

The expressions for $g_{ij}(d_{ij}{}^+)$ and A^{hs} used in original SAFT as well as in PC-SAFT are based on the work of Boublik [27] and Mansoori et al. [28], who derived them for mixtures of hard spheres. For pure substances, these expressions become identical to the simpler one proposed by Carnahan and Starling [11]. Kouskoumvekaki et al. [29] also applied these simplified expressions to mixtures (simplified PC-SAFT) and obtained, in most cases, almost similar modeling results. Because for pure substances the expressions for $g_{ij}(d_{ij}{}^+)$ and A^{hs} become identical for PC-SAFT and simplified PC-SAFT, the pure-component parameters for PC-SAFT and simplified PC-SAFT are identical.

The contribution due to attractive interactions of segments of different chains $A^{\mathrm{disp,PC\text{-}SAFT}}$ is formulated as a power series with respect to reversed temperature. Considering only the first two terms of this series and accounting for their dependence on density and segment number, $A^{\mathrm{disp,PC\text{-}SAFT}}$ reads as:

$$\frac{A^{\mathrm{disp,PC\text{-}SAFT}}}{kTN} = \frac{A_1}{kTN} + \frac{A_2}{kTN} \tag{13}$$

$$\frac{A_1}{kTN} = -2\pi\rho I_1(\eta,\bar{m})\sum_i\sum_j x_i x_j m_i m_j \left(\frac{\varepsilon_{ij}}{kT}\right)\sigma_{ij}{}^3 \tag{14}$$

$$\frac{A_2}{kTN} = -\pi\rho\bar{m}\left(1 + Z^{\mathrm{hc}} + \rho\frac{\partial Z^{\mathrm{hc}}}{\partial\rho}\right)^{-1} I_2(\eta,\bar{m})\sum_i\sum_j x_i x_j m_i m_j \left(\frac{\varepsilon_{ij}}{kT}\right)^2 \sigma_{ij}{}^3 \tag{15}$$

With:

$$I_1(\eta,\bar{m}) = \sum_{i=0}^{6} a_i(\bar{m})\eta^i \quad \text{and} \quad I_2(\eta,\bar{m}) = \sum_{i=0}^{6} b_i(\bar{m})\eta^i \tag{16}$$

Modeling of Polymer Phase Equilibria Using Equations of State

and:

$$a_i = a_{0i} + \frac{\bar{m} - 1}{\bar{m}} a_{1i} + \frac{\bar{m} - 1}{\bar{m}} \frac{\bar{m} - 2}{\bar{m}} a_{2i} \quad (\text{analogously for } b_i) \tag{17}$$

$$\sigma_{ij} = \frac{1}{2}(\sigma_i + \sigma_j); \quad \varepsilon_{ij} = \sqrt{\varepsilon_i \varepsilon_j}(1 - k_{ij}) \tag{18}$$

$a_{0i}, a_{1i}, a_{2i}, b_{0i}, b_{1i}$ and b_{2i} in (17) are model constants and are given in Table 1. k_{ij} in (18) is again a binary parameter that corrects for deviation from the geometric mean mixing rule and has to be determined by fitting to binary experimental data.

The Helmholtz energy change due to the formation of hydrogen bonds (association) is captured by A^{assoc}. A detailed description as well as the corresponding expressions can be found, e.g., in [12]. The main assumption is that association can be described by a short-range but very strong association potential (Fig. 3). This leads to two more parameters required for the modeling of an associating molecule: the association volume κ_{AA} (corresponds to the potential width r_{AA}) and the association strength ε_{AA} (the potential depth).

For application, one needs to define the number N^* of so-called association sites, meaning the number of donor and acceptor sites respectively, by which a molecule is able to form hydrogen bonds. Different association sites may have different association parameters (in most cases they are assumed to be identical). Details can be again found in [12] and [14].

Moreover, Helmholtz energy expressions are available that can account for interactions due to dipole moments [30–32], quadrupole moments [33–35], or even charges [36–38] of the molecules. They have already been successfully applied in combination with SAFT or PC-SAFT but will not be considered within this work.

Given the expression for the Helmholtz energy, other thermodynamic properties needed for phase-equilibrium calculations can be derived by applying textbook thermodynamics. Thus, system pressure and the chemical potentials of the mixture components can be obtained by applying the following relations:

$$p = -\left(\frac{\partial A}{\partial V}\right)_{T,n_i} \tag{19}$$

and

$$RT \ln \varphi_i = \left(\frac{\partial A^{\text{res}}}{\partial n_i}\right)_{T,V,n_{j \neq i}} - RT \ln z \quad \text{with} \quad z = \frac{p V}{NkT} \tag{20}$$

Phase-equilibrium calculations are finally performed using the classical phase-equilibrium conditions:

Table 1 Model constants for the PC-SAFT equation of state

i	a_{0i}	a_{1i}	a_{2i}	b_{0i}	b_{1i}	b_{2i}
0	0.9105631445	−0.3084016918	−0.0906148351	0.7240946941	−0.5755498075	0.0976883116
1	0.6361281449	0.1860531159	0.4527842806	2.238279186	0.6995095521	−0.2557574982
2	2.686134789	−2.503004726	0.5962700728	−4.002584949	3.892567339	−9.155856153
3	−26.54736249	21.41979363	−1.724182913	−21.00357682	−17.21547165	20.64207597
4	97.75920878	−65.25588533	−4.130211253	26.85564136	192.6722645	−38.80443005
5	−159.5915409	83.31868048	13.77663187	206.5513384	−161.8264617	93.62677408
6	91.29777408	−33.74692293	−8.672847037	−355.6023561	−165.2076935	−29.66690559

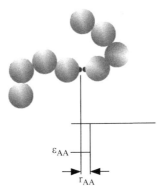

Fig. 3 Association potential as used in SAFT and PC-SAFT

$$x_i^I \varphi_i^I = x_i^{II} \varphi_i^{II} \tag{21}$$

which have to be fulfilled for all components *i* of the mixture.

3 Estimation of Model Parameters

Among the three pure-component parameters for nonassociating components there are two parameters that are related to the size of the molecule: the segment diameter σ and the segment number m. The third parameter, the energy parameter ε, decribes the attractive interactions between two molecules. For volatile components, these parameters are determined by simultaneously fitting to physical properties, which can on the one hand be caluated by an equation of state and are on the other hand related to the size and the interactions of the molecules. Such properties are, e.g., liquid-density data (related to molecule size) and vapor pressures (related to the intermolecular interactions). These parameters have already been determined for a huge number of relevant solvents and can be found in extensive parameter tables, e. g., in [15] (SAFT) and [24, 39] (PC-SAFT).

However, polymers exhibit neither a measurable vapor pressure nor any other property that can be directly related to the energy parameter. However, determining all three parameters by fitting only to liquid-density data mostly does not yield meaningful energy parameters that are suitable for binary calculations.

Considering the example of polycarbonate: in the literature only density data for a polymer of unknown molecular weight are available [40]. Assuming the molecular weight (M_w) to be 100,000 g/mol, and fitting the three parameters for the SAFT model to these data, one obtains $m = 4043.5$, $\varepsilon/k = 387.83$ K and $v_{00} = 17.114$ cm³/mol [41]. Since the segment number of polymers is proportional to their molecular weight, the ratio m/M is usually fitted instead of the absolute segment number, which led to $m/M = 0.04043$ for polycarbonate.

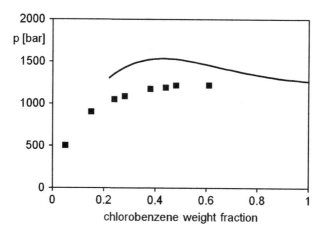

Fig. 4 Vapor–liquid equilibrium of the polycarbonate/chlorobenzene system at 140°C. *Symbols* represent experimental data; *lines* show predictions with SAFT [41]. Polycarbonate parameters were fitted to density data only

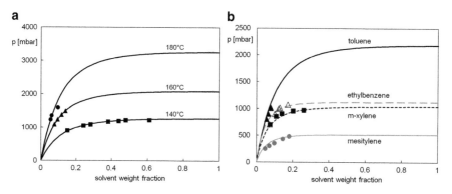

Fig. 5 Vapor–liquid equilibrium in polycarbonate/solvent systems. (**a**) Polycarbonate/ chlorobenzene at various temperatures. (**b**) Polycarbonate with various solvents at 140°C. *Symbols* represent experimental data. *Lines* show SAFT calculations [41]

Using these parameters for predicting the vapor–liquid equilibrium in the polycarbonate/chlorobenzene system ($k_{ij} = 0$) at 140°C leads to results shown in Fig. 4. It is obvious that the results are very unsatisfactory; a description of the experimental data also fails for any other value of the binary parameter k_{ij}. Therefore, the polycarbonate parameters m (uncertain since the molecular weight for the density data was unknown) and ε/k were refitted to the binary data in Fig. 5a. Using these new parameters $m/M = 0.0080$, $\varepsilon/k = 256.97$ K and $v_{00} = 17.114$ cm^3/mol, the experimental data can now be described very well.

Although the polymer parameters were also fitted to binary data, they still have the character of the pure-component parameters. This is confirmed by calculations of other polymer/solvent systems, which are illustrated in Fig. 5b. Using the same polycarbonate parameters as determined for the chlorobenzene system, the

Modeling of Polymer Phase Equilibria Using Equations of State

Table 2 PC-SAFT pure-component parameters of polymers

Polymer	m/M (mol/g)	σ (Å)	ε/k (K)	N^*	ε_{AA}/k (K)	κ_{AA}	References
LDPE	0.0263	4.0127	249.5				[25]
HDPE	0.0263	4.0127	252.0				[25]
Poly(propylene)	0.02305	4.10	217.0				[25]
Poly(1-butene)	0.014	4.20	230.0				[25]
Poly(isobutene)	0.02350	4.10	265.5				[25]
Polystyrene	0.0190	4.1071	267.0				[25]
Poly(methyl acrylate)	0.0309	3.50	243.0				[42]
Poly(methyl methacrylate)	0.0262	3.60	245.0				[42]
Poly(ethyl acrylate)	0.0271	3.65	229.0				[42]
Poly(propyl acrylate)	0.0262	3.80	225.0				[42]
Poly(butyl acrylate)	0.0259	3.59	224.0				[42]
Poly(butyl methacrylate)	0.0268	3.75	233.8				[42]
Poly(vinyl acetate)	0.03211	3.3972	204.65				[26]
Poly(acrylic acid)	0.016	4.20	249.5	2	2,035	0.33584	[43]
Poly(methacrylic acid)	0.024	3.70	249.5	2	2,610	0.07189	[43]
Poly(dimethyl siloxane)	0.0346	3.382	165.0				[44]

N^* number of association sites per monomer unit

Table 3 PC-SAFT pure-component parameters of solvents used in this chapter

Solvent	m/M (mol/g)	σ (Å)	ε/k (K)	References
Ethene	0.05679	3.4450	176.47	[24]
Ethane	0.05344	3.5206	191.42	[24]
Propylene	0.04657	3.5356	207.19	[24]
Propane	0.04540	3.6184	208.11	[24]
1-Butene	0.04075	3.6431	222.00	[24]
Butane	0.04011	3.7086	222.88	[24]
Pentane	0.03728	3.7729	231.2	[24]
Carbon dioxide	0.04710	2.7852	169.21	[24]
Methylmethacrylate	0.03060	3.6238	265.69	[45]
Toluene	0.03055	3.7169	285.69	[24]

experimental data for toluene, m-xylene, ethylbenzene, and mesitylene can also be described using very small values for the binary parameters k_{ij}.

This approach was proven also for other equations of state, like for example PC-SAFT. Fitting polymer parameters to liquid densities of the polymer and to one binary polymer/solvent system is an established method for determining polymer parameters. Table 2 summarizes the so-determined PC-SAFT parameters for a series of polymers. PC-SAFT parameters for solvents used in this work can be found in Table 3.

4 Modeling of Homopolymer Systems

As an example for modeling of a vapor–liquid equilibrium in a polymer/solvent system, Fig. 6 shows the results for the polyethylene/toluene binary mixture. The experimental data shown were determined for two different (relatively low)

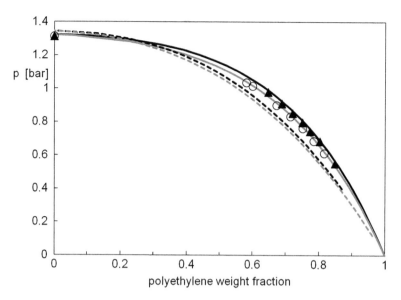

Fig. 6 Vapor–liquid equilibrium of toluene/polyethylene for different molecular weights of the polymer (*triangles* 6,220 g/mol, *circles* 1,710 g/mol). *Symbols* represent experimental data [46]. *Solid lines* show predictions ($k_{ij} = 0$) with PC-SAFT and *dashed lines* with SAFT [25]

molecular weights of the polymer, which are compared to predictions using the SAFT and PC-SAFT models, respectively. The binary parameter k_{ij} is zero in both cases. Thus, the lines are not fitted to the shown binary data but were calculated using pure-component information only. This clearly shows that SAFT as well as PC-SAFT are able to predict the vapor–liquid equilibrium in this system and even correctly consider the molecular weight dependence. Moreover, it can be seen that PC-SAFT performs slightly better than SAFT.

Whereas in the vapor–liquid equilibrium calculation at low pressures, the model is only used to correct for the nonideality of the liquid phase, the results for liquid–liquid equilibrium calculations and even whether a demixing is calculated or not, is completely determined by the model used. Therefore, liquid–liquid calculations are always much more challenging for any model. Figure 7 demonstrates the ability of PC-SAFT to model also liquid–liquid demixing in polymer systems with high accuracy. The calculations are performed for the system polypropylene/n-pentane at three different temperatures. Although only temperature-independent pure-component and binary parameters were used, the experimental data can be described very well and the model even correctly captures the temperature dependence of the miscibility gap.

As already seen in Fig. 6, polymer phase equilibria do depend on the molecular weight of the polymer. This is even more pronounced for liquid–liquid equilibria, where the polymer distributes to the two liquid phases (in the case of vapor–liquid equilibria, the polymer is only present in the liquid phase whereas the vapor contains only the pure solvent).This fact needs of course also to be accounted for

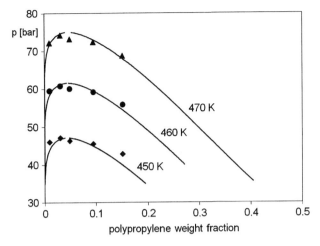

Fig. 7 Liquid–liquid equilibrium of polypropylene/n-pentane at three temperatures (polypropylene: M_w = 50.4 kg/mol, M_w/M_n = 2.2). Comparison of experimental cloud points (*symbols*) [47] with PC-SAFT calculations (*lines*) [25]. k_{ij} = 0.0137. The polymer was assumed to be monodisperse with M_w being the molecular weight

in the modeling. As polymers of different molecular weight consist of the same type of monomers (segments) and differ in the number of monomers (corresponds to segment number) only, the only parameter that needs to be changed for polymers of different molecular weights is the segment number. The latter is already given as function of molecular weight (see Table 2), which makes it easy to account for different molecular weights. Figure 8 presents solubility curves for polyethylene (LDPE) in ethene as a function of the polymer molecular weight. The same set of parameters for polyethylene (Table 2) was used to describe the whole range of molecular weights using PC-SAFT. As can be seen, the model captures the influence of the molecular weight very well and is thus able to predict the phase behavior as function of the polymer length.

Similarly to Fig. 5b for vapor–liquid equilibria, Fig. 9 illustrates the ability of PC-SAFT to model liquid–liquid equilibria of LDPE dissolved in a variety of solvents. The amount of polymer is about 5 wt% for all cases. Using PC-SAFT, the experimental cloud points can be described with high accuracy. Although for each system only one binary temperature-independent parameter is used, the model even captures the changing slope of the cloud-point curves, from a negative slope for ethene and ethane to the positive slope for C_3 and C_4 solvents.

The modeling can, of course, also be extended to systems containing more than two components. In this case, it is assumed that the phase behavior in ternary or higher systems is still dominated by binary interactions. This means that only binary interactions are accounted for and the modeling is completely based on the pure-component and binary parameters determined before for the pure substances and binary subsystems, respectively.

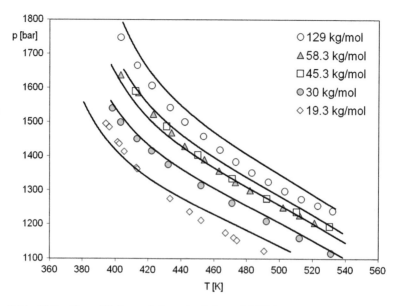

Fig. 8 Liquid–liquid equilibrium of the polyethylene (LDPE)/ethene system as function of polymer molecular weight. Polymer weight fraction is about 5 wt%. *Symbols* represent experimental cloud points (Latz and Buback, 2002, personal communication), *lines* show predictions with PC-SAFT. All polymers were assumed to be monodisperse

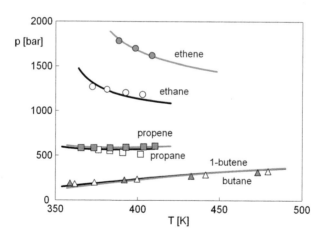

Fig. 9 Cloud-point data of different polyethylene (LDPE)/solvent systems. *Symbols* represent experimental data [48, 49]. *Lines* show PC-SAFT calculations [26]

Thus, calculations for ternary systems are usually pure predictions (unless binary parameters were determined by fitting to ternary data, which is sometimes done).

Figure 10 shows the modeling results for the ternary system poly(methylmethacrylate) (PMMA)/methylmethacrylate (MMA) /carbon dioxide [45]. At pressures

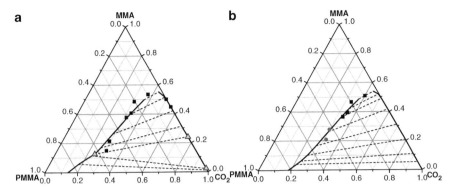

Fig. 10 Phase behavior of the system PMMA (M_w = 18 kg/mol)/MMA/CO$_2$ at 65°C [45]. *Squares* and *circles* represent experimental cloud-point data. The *solid lines* (phase boundary) and *dashed lines* (tie lines) are calculated with PC-SAFT. *Triangles* represent the VLLE region. (**a**) p = 100 bar, (**b**) p = 150 bar

of 100 bar this system exhibits a three-phase region that can be suppressed by increasing the system pressure up to 160 bar. Moreover, this behavior strongly depends on temperature and the molecular weight of the polymer. Using only pure-component and binary parameters for the three subsystems the phase behavior and even the presence or absence of the three-phase region can be described by PC-SAFT in excellent agreement with the experimental data.

5 Extension to Copolymers

The original versions of the above-mentioned perturbation theories (e.g., SAFT or PC-SAFT) consider a molecule as a chain of identical segments. Extensions of these models allow for taking into account different types of segments and can therefore describe copolymer systems [50–52]. The extension of PC-SAFT to copolymers is referred to as copolymer PC-SAFT [53].

Now, the monomer segments are allowed to differ in size as well as in attractive or electrostatic interactions (such as dispersion energy, association, polarity) (Fig. 11). Thus, each of the different monomer units is described by its own parameter set (for nonassociating and nonpolar monomers: m, σ, ε).

The relative amount of monomer units is usually given by the experimental polymer characterization. The relative amount of segments α and β is described by segment fractions z_α and z_β ($z_\alpha + z_\beta = 1$) within the copolymer, which are defined as:

$$z_\alpha = \frac{m_\alpha}{m} \quad \text{(analogously for } \beta\text{)} \tag{22}$$

Fig. 11 Copolymer chain with different segment types

whereas m_α is the number of α segments in the copolymer (according to Table 2, each polymer unit could consist of more than one segment). In addition, bond fractions $B_{\alpha\beta}$ are used to describe, at least to certain extent, the arrangement of the segments in the polymer chain. Here, a bond fraction $B_{\alpha\beta}$ is the fraction of all bonds within the copolymer, which is a bond between a segment α and a segment β. The bond fractions are estimated on the basis of the molecular structure of the copolymer. In the case of block copolymers, the values of the respective bond fractions are clearly defined. Out of $(m-1)$ bonds, there is only one bond of type $\alpha-\beta$, i.e., $B_{\alpha\beta} = \frac{1}{(m-1)}$. The bond fraction of type $\alpha-\alpha$ is simply given by $B_{\alpha\alpha} = \frac{z_\alpha(m-2)}{(m-1)}$ (the same for $B_{\beta\beta}$) [52]. In the case of an alternating copolymer, there exist only one bond type $\alpha-\beta$, i.e., $B_{\alpha\beta} = 1$.

The bond fractions in a random copolymer are usually not exactly known and can only be estimated. One possibility is to assume that if segment type α is in the majority, there exist no bonds between two β-segments. Hence, the bond fractions for such a copolymer are [53]:

$$B_{\beta\beta} = 0; \quad B_{\alpha\beta} = \frac{2z_\alpha m}{(m-1)}; \quad B_{\alpha\alpha} = 1 - B_{\alpha\beta} \qquad (23)$$

Taking this into account, the expression for the hard-chain contribution can be extended to copolymer systems [52]:

$$\frac{A^{hc}}{NkT} = \bar{m}\frac{A^{hs}}{NkT} - \sum_i x_i(m_i - 1)\sum_\alpha \sum_\beta B_{i\alpha i\beta} \ln g_{i\alpha i\beta}\left(d^+_{i\alpha i\beta}\right) \qquad (24)$$

whereas the segment number m_i of a copolymer is determined by:

$$m_i = \sum_\alpha m_{i\alpha} = M_i \sum_\alpha \left(\frac{m}{M}\right)_{i\alpha} w_{i\alpha} \qquad (25)$$

with $w_{i\alpha}$ being the weight fraction of α-segments within the copolymer. All other expressions [(8)–(18)] are used as for the homopolymer system considering the copolymer/solvent system consisting of three types of segments (the solvent segments and (at least) two different types of polymer segments). The concentration $x_{i\alpha}$ of copolymer segments of type α is given by the product of the copolymer concentration x_i and the segment fraction $z_{\alpha i}$:

$$x_{i\alpha} = x_i \, z_{i\alpha} \qquad (26)$$

In copolymer PC-SAFT, the required segment parameters for a copolymer unit are taken from the corresponding homopolymers (Table 2). The description of a copolymer/solvent system requires three binary parameters: two for the interactions of the solvent with the respective monomer segments α and β, and a third for the interactions between the unlike monomer segments. Whereas the first two parameters can be determined from homopolymer/solvent systems, the third parameter is the only one that has to be determined from copolymer/solvent data.

Figure 12 shows the results for the modeling of the solubility of the copolymer poly(ethylene-*co*-1-butene) in propane. The pure-component parameters for poly(ethylene) (HDPE), poly(1-butene), and propane as well as the binary parameters for HDPE/propane and poly(1-butene)/propane were used as determined for the homopolymer systems.

The interaction parameter for the ethylene segment/butene segment interaction was fitted to one point of the cloud-point curve of the copolymer containing 35% butene monomers. Using this approach, the solubility can be predicted over the whole range of copolymer compositions.

Results of similar quality can also be obtained for copolymers in which the segments are not as similar as in the previous example. Figure 13 shows the solubility of ethylene-*co*-alkylacrylate copolymers in ethane [42]. For the poly(ethylene-*co*-propylacrylate) copolymer, the solubility increases with an increasing amount of propylacrylate monomers in the copolymer backbone (Fig. 13b). In contrast, the poly(ethylene-*co*-methylacrylate) solubility first increases (cloud-point pressure decreases) due to favorable methylacrylate–ethene interactions but,

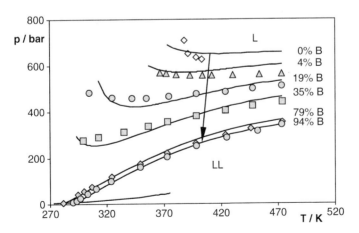

Fig. 12 Cloud-point pressures in the system poly(ethylene-*co*-1-butene)/propane for different copolymer compositions (*B* is mole percent butene in the backbone; 0% B = LDPE). Polymer weight fraction is about 0.05 wt%. *Symbols* represent experimental data [54]. *Lines* show PC-SAFT calculations [53]

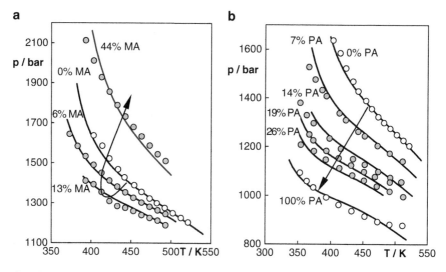

Fig. 13 Cloud-point pressures for poly(ethylene-*co*-alkylacrylate)/ethene systems at different copolymer compositions (mole percent of acrylate monomer in the backbone is indicated). (**a**) Solubility of poly(ethylene-*co*-ethylacrylate) (0% MA = LDPE). (**b**) Solubility of poly(ethylene-*co*-propylacrylate) [0% PA = LDPE; 100% PA = poly(propylacrylate)]. Polymer content is about 5 wt%. *Symbols* represent experimental data and *lines* show PC-SAFT calculations [42]

after reaching a maximum (pressure minimum), solubility again decreases because of the strong interactions between the methylacrylate monomers (Fig. 13a). The homopolymer poly(methylacrylate) can not be dissolved in ethene, even at pressures up to 3,000 bar. Although the two systems show this qualitatively different behavior, the copolymer version of PC-SAFT is able to model the solubility in both cases in almost quantitative agreement with the experimental data.

6 Accounting for the Influence of Polydispersity

So far, all polymers in this chapter were assumed to be and were modeled as being monodisperse. However, polymers always have a certain polydispersity, which is usually described as the ratio of the weight-average (M_w) and number average (M_n) of the polymer molecular weight distribution. As can be seen from Fig. 6, the influence of molecular weight on the vapor–liquid equilibrium of a polymer/solvent system is rather small. The reason is that the polymer, as soon as it has a molecular weight above a certain level, does not show a measurable vapor pressure and thus does not participate in the partitioning between the liquid and the vapor phase. For long-enough polymers this holds independently of the molecular weight, thus polydispersity does not influence vapor–liquid equilibria to a notable amount and usually does not need to be considered in the modeling.

This changes dramatically when considering liquid–liquid equilibria. Smaller polymers are more soluble than longer ones (see, e.g., Fig. 8). This also means that the phase boundary of polymer systems is strongly influenced by the molecular weight distribution of the polymer. Figure 14 shows the cloud-point curve of a polydisperse LDPE (M_n = 43 kg/mol, M_w = 118 kg/mol, M_z = 231 kg/mol) in ethene in comparison with PC-SAFT results obtained by monodisperse calculations using either M_n, M_w, or M_z as the polymer molecular weight. Although PC-SAFT is in general able to describe the phase behavior of that system (see Fig. 8), none of the calculations is able to describe the experimental data in Fig. 14.

Obviously, the polydispersity of the polymer also needs to be considered in the phase-equilibrium calculations. Assuming a system containing a solvent 1 and a polydisperse polymer 2, the phase-equilibrium conditions have to be applied to the solvent as well as to every polymer species. Equation (21) becomes:

$$x_1^I \varphi_1^I = x_1^{II} \varphi_1^{II} \quad \text{for the solvent} \tag{27}$$

$$x_2^I x_{2p,j}^I \varphi_{2j}^I = x_2^{II} x_{2p,j}^{II} \varphi_{2j}^{II} \quad \text{for each polymer species} \tag{28}$$

with x_2 being the mole fraction of the polymer ($x_2 = 1 - x_1$), and $x_{2p,j}$ meaning the mole fraction of polymer species j within the solvent-free polymer. The latter have to fulfill the normalization condition:

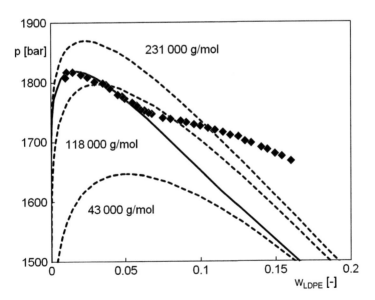

Fig. 14 Phase equilibrium in the system ethene/polyethylene (LDPE; M_n = 43,000 g/mol, M_w = 118,000 g/mol, M_z = 231,000 g/mol). *Symbols* represent experimental data [55]. *Lines* show calculations using the SAFT model. *Dashed lines* show monodisperse calculations using M_n, M_w, and M_z, respectively. *Solid line* shows a calculation using two pseudocomponents as given in Table 4

$$\sum_j x_{2p,j} = 1 \tag{29}$$

which is valid in each of the phases. Rearranging (28), summing up over all polymer components, and using (29) one obtains:

$$1 = \frac{x_2^I}{x_2^{II}} \sum_j x_{2p,j}^I \frac{\varphi_{2j}^I}{\varphi_{2j}^{II}} \tag{30}$$

For given temperature, pressure, and polymer distribution in one phase ($x_{2p,j}^I$), equations (27) and (30) can be used to determine the unknowns x_1^I and x_1^{II}. Depending on the expressions for the fugacity coefficients φ_1^{II} and $\varphi_{2p,j}^{II}$, additional unknowns, e.g., $\bar{m}_2^{II} = \sum x_{2p,j}^{II} m_{2p,j}$ will have to be determined. The $m_{2p,j}$ are the segment numbers of the various polymer species (which are of course the same in the two phases). Additional equations can easily be obtained by multiplying (28), e.g., with $m_{2p,j}$, again rearranging, summing up over all polymer species, and using (29) to eliminate the unknown molecular weight distribution in the second phase:

$$\bar{m}_2^{II} = \frac{x_2^I}{x_2^{II}} \sum_{j=1} x_{2p,j}^I m_{2p,j} \frac{\varphi_{2j}^I}{\varphi_{2j}^{II}} \tag{31}$$

Applying the approach given by (29)–(31), rather than the classical approach of considering only (27) and (28), has the advantage that the number of equations to be solved numerically is independent of the number of considered polymer species.

After solving (27), (30), and (31), the missing concentrations of the single polymer species (molecular weight distribution) in the second phase can easily be obtained from rearranging (28):

$$x_{2p,j}^{II} = \frac{x_2^I}{x_2^{II}} x_{2p,j}^I \frac{\varphi_{2j}^I}{\varphi_{2j}^{II}} \tag{32}$$

A very elegant version of this approach is the so-called continuous thermodynamics [56–59]. It can be considered as a reformulation of the classical thermodynamic relationships that allows for using continuous molecular weight distribution functions $W(M)$ rather than the mole fractions of discrete polymer components. Using this approach, e.g., (30) becomes:

$$1 = \frac{x_2^I}{x_2^{II}} \int_M W(M) \frac{\varphi_2^I(M)}{\varphi_2^{II}(M)} \, dM \tag{33}$$

For certain combinations of analytical molecular weight distributions (e.g., Schulz–Flory distribution) and fugacity-coefficient expressions, the integral in (33) can be solved analytically and no (time-consuming) summation is required.

An application of this approach to describe polymer fractionations is described in detail by Enders [60].

Direct use of (28) to (32) requires choosing a certain number of polymer components that can be used within the calculations. One could, e.g., think of dividing the polymer molecular weight distribution into (not necessarily equally spaced) intervals and choosing the polymer (pseudo)components such that each of them represents one of these intervals (Fig. 15). Depending on the shape of the molecular weight distribution, this approach requires about 5–20 polymer components for representing the phase boundary of the polydisperse system.

The number of pseudocomponents can be remarkably decreased by choosing them in a such a way that they represent relevant moments $\overline{M^k}$ of the molecular weight distribution. These moments are related to the molecular weight averages M_n, M_w and, M_z by:

$$M_n = \overline{M^1}; \qquad \bar{M}_w = \frac{\overline{M^2}}{\overline{M^1}}; \qquad \bar{M}_z = \frac{\overline{M^3}}{\overline{M^2}}; \qquad \overline{M^k} = \sum_j x_{2p,j} M_j^k \qquad (34)$$

Taking into account that each pseudocomponent is characterized by two properties (its mole fraction $x_{2p,j}$ and its molecular weight $M_{2p,j}$), n pseudocomponents can reproduce $2n$ moments of a polymer distribution. This means, that, using only two pseudocomponents, one can exactly reproduce M_n, M_w, and M_z. It also means that the mole fractions of the two components have to add to unity ($\overline{M^0}$). Using three pseudocomponents, even two additional moments of the molecular weight distribution can be covered.

Table 4 gives two sets of pseudocomponents (two and three components) determined for the poly(ethylene) used in Fig. 14, which cover the given number of moments.

The pseudocomponents differ only in molecular weight. Thus, one can assume that all parameters except the segment number stay the same for all polymer species. Since the segment number is directly proportional to the molecular weight (see Table 2), it can easily be determined for each pseudocomponent.

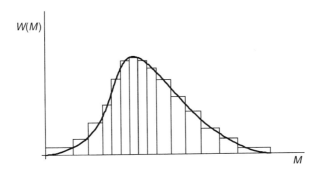

Fig. 15 Molecular weight distribution $W(M)$ and its representation by pseudocomponents

Table 4 Pseudocomponents for poly(ethylene) considered in Fig. 14

$\overline{M^k}$	k = {0, 1, 2, 3}		k = {−1, 0, 1, 2, 3, 4}	
j	$M_{2p,j}$ (g/mol)	$x_{2p,j}$	$M_{2p,j}$ (g/mol)	$x_{2p,j}$
1	2.8614×10^4	0.93970	2.7676×10^3	0.54529
2	2.6717×10^5	0.06030	8.3769×10^4	0.44588
3			4.6883×10^5	0.00883

M^k moment of molecular weight distribution, $M_{2p,j}$ molecular weight of the pseudocomponent, $x_{2p,j}$ mole fraction of the pseudocomponent in the solvent-free system

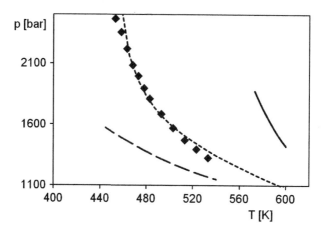

Fig. 16 Solubility of poly($E_{96.2}$-co-$AA_{3.8}$) in ethene. Polymer properties can be found in Table 5. Polymer concentration in the mixture is about 3 wt%. *Symbols* represent experimental data [61] compared with results form PC-SAFT calculation [43]. *Dashed line* shows monodisperse calculation using M_n; *solid line* shows monodisperse calculation using M_w; and *dotted line* shows calculation using two pseudocomponents given in Table 6

Figure 14 contains the results obtained with PC-SAFT when using the set of two pseudocomponents left-hand side from Table 4, in comparison with the experimental data and with the monodisperse calculations. As can be seen, the results of the calculations can be improved remarkably when accounting for the polydispersity by using only two pseudocomponents.

As a second example, the phase equilibrium in the system poly(ethylene-*co*-acrylic acid)/ethene is illustrated in Fig. 16. The polymer has a polydispersity index M_w/M_n of about 3.3. The solubility of this copolymer was first modeled assuming the polymer to be monodisperse. Neither modeling using M_n nor using M_w as the molecular weight of the polymer could satisfactorily describe the experimental data (Fig. 16). In a second step, the polydispersity of the copolymer was accounted for by using two pseudocomponents (see Table 6). As also shown in Fig. 16, without changing or refitting any parameters, the modeling is now in almost perfect agreement with the experimental data. This again demonstrates the importance of accounting for polydispersity in the modeling as well as in the interpretation of the experimental data [43].

Modeling of Polymer Phase Equilibria Using Equations of State

Table 5 Properties of poly(ethylene-*co*-acrylic acid) copolymer samples

Copolymer	Acrylic acid content (mol%)	M_n (g/mol)	M_w (g/mol)	M_w/M_n
Poly($E_{96.2}$-*co*-$AA_{3.8}$)	3.8	19,090	63,250	3.3
Poly($E_{97.6}$-*eo*-$AA_{2.4}$)	2.4	22,713	258,214	11.4
Poly($E_{96.9}$-*co*-$AA_{3.1}$)	3.1	19,930	235,146	11.8
Poly($E_{96.3}$-*co*-$AA_{3.7}$)	3.7	23,443	227,427	9.7
Poly($E_{95.4}$-*co*-$AA_{4.6}$)	4.6	23,730	205,024	8.6

Table 6 Pseudocomponents for the copolymers poly(ethylene-*co*-acrylic acid)

Copolymer	Pseudocomponents j	$w_{2p,j}$	$M_{2p,j}$ (g/mol)
Poly($E_{96.2}$-*co*-$AA_{3.8}$)	1	0.5	10,400
	2	0.5	116,100
Poly($E_{97.6}$-*co*-$AA_{2.4}$)	1	0.673	15,433
	2	0.327	757,509
Poly($E_{96.9}$-*co*-$AA_{3.1}$)	1	0.673	13,554
	2	0.327	692,543
Poly($E_{96.3}$-*co*-$AA_{3.7}$)	1	0.669	158,74
	2	0.331	655,259
Poly($E_{95.4}$-*co*-$AA_{4.6}$)	1	0.666	16,023
	2	0.334	582,165

$M_{2p,j}$ molecular weight of the pseudocomponent, $w_{2p,j}$ weight fraction of the pseudocomponent in the solvent-free system

Using the same pure-component and binary parameters, the phase boundary of other poly(ethylene-*co*-acrylic acid) samples varying in copolymer composition, molecular weight, and with even higher polydispersity can also be described successfully, as demonstrated in Fig. 17.

In some cases, e.g., for polymer fractionations, not only the phase boundary, but also the molecular weight distribution in the coexisting phases is of interest.

In this case, it is useful to use the whole molecular weight distribution (see [60]) or a higher number of pseudocomponents for the modeling. As an example, Fig. 18 shows the molecular weight distributions of polystyrene in the two coexisting phases observed with cyclohexane/carbon dioxide solvent mixture [63]. An initially bimodal mixture of two polystyrene samples (40 kg/mol and 160 kg/mol) was mixed with cyclohexane and carbon dioxide at 170°C and different pressures to generate two liquid phases.

From the experimental data shown in Fig. 18, it becomes obvious that the shorter polymer species preferably dissolve in the polymer-lean phase whereas the species of higher molecular weight accumulate in the polymer-rich phase. The separation effect is best at low pressures where almost no longer polymers are found in the polymer-lean phase. The selectivity decreases with increasing pressures due to increasing solubility, also of the longer polymer species. Modeling this system using about 30 pseudocomponents (right-hand side of Fig. 18) leads to the same conclusion and shows an excellent agreement with the experimental findings [64].

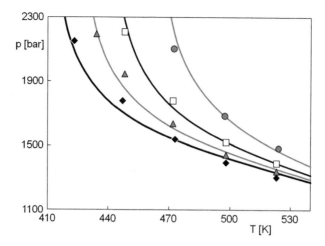

Fig. 17 Solubilities of poly(ethylene-*co*-acrylic acid) copolymers with different acrylic acid contents in ethene. The polymer concentration is 5 wt%. The properties of the copolymers are given in Table 5. *Lines* show predictions using PC-SAFT and the pseudocomponents given in Table 6 [43]. *Symbols* represent experimental data [62]: *Circles* poly($E_{95.4}$-*co*-$AA_{4.6}$), *squares* poly($E_{96.3}$-*co*-$AA_{3.7}$), *triangles* poly($E_{96.9}$-*co*-$AA_{3.1}$), *diamonds* poly($E_{97.6}$-*co*-$AA_{2.4}$)

7 Summary

State-of-the-art equations of state like SAFT and PC-SAFT are based on a sound physical background and are able to explicitly account for various molecular properties such as the nonspherical shape, the ability to form hydrogen bonds or the polarity. These models usually require three (for nonassociating molecules) or five pure-component parameters (for associating molecules) that have a physical meaning. Whereas for solvents these parameters are usually determined from fitting to vapor pressures and liquid densities, a different approach is applied to polymers. Because vapor pressures are not accessible for these molecules, polymer parameters are fitted to the experimental data of a polymer/solvent binary system. It could be shown that although fitted to binary data, these parameters have a physical meaning and do characterize the considered polymer molecule. Using these parameters, the phase equilibrium of homopolymer systems can be described as a function of temperature, pressure, polymer concentration, molecular weight, and even the polymer molecular weight distribution.

The extension to copolymer systems is straightforward. The parameters for the different polymer units as well as the binary parameters for their interaction with the solvent can be used as determined for the respective homopolymer systems. Using only one additional binary parameter that describes the interactions of the unlike monomer units in the copolymer system, the phase behavior of copolymer systems can be described and even predicted over a wide range of copolymer compositions.

Modeling of Polymer Phase Equilibria Using Equations of State 415

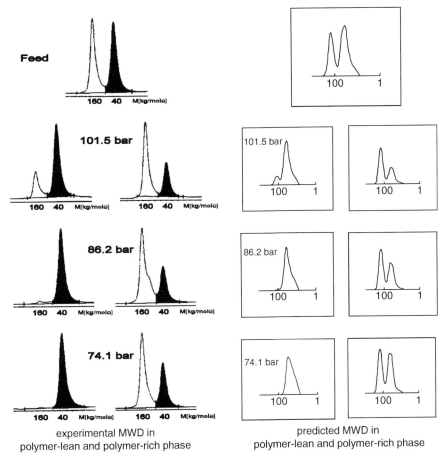

Fig. 18 Molecular weight distributions of polystyrene (mixture of two almost monodisperse samples of 40 kg/mol and 160 kg/mol) in the coexisting phases of a polystyrene/cyclohexane/carbon dioxide mixture at 170°C and varying pressures [63]. The two *left-hand columns* give the GPC analysis of the polymer-lean and polymer-rich phases. The *two right-hand columns* show the molecular weight distributions as calculated using the SAFT model using 30 pseudocomponents [64]

Polymers are often polydisperse with respect to molecular weight. Whereas this is of minor importance for the solvent sorption in polymers (vapor–liquid equilibrium), this fact usually remarkably influences the polymer solubility (liquid–liquid equilibrium). Therefore, polydispersity needs to be accounted for in interpretation and modeling of experimental data. This can be done by applying continuous thermodynamics as well as by choosing a representative set of pseudocomponents. It was shown that a meaningful estimation of the phase boundary is possible when using only two or three pseudocomponents as soon as they reflect the important moments (M_n, M_w, M_z) of the molecular weight distribution.

Finally, one can say that the thermodynamics of polymer systems is (of course) the same as for small molecules because all thermodynamic relationships stay the same – but is also completely different because the models and approaches used have to explicitly account for the fact that polymers are chain molecules and are usually themselves mixtures.

References

1. van der Waals JD (1873) Over de Continuiteit van den gas- en vloeistoftoestand. Sijthoff, Leiden
2. Soave G (1972) Equilibrium constants from a modified Redlich–Kwong equation of state. Chem Eng Sci 271:197–1203
3. Peng D-Y, Robinson DP (1976) A new two-constant equation of state. Ind Eng Chem Fundam 15:59–64
4. Sanchez IC, Lacombe RH (1976) An elementary molecular theory of classical fluids: pure fluids. J Phys Chem 80:2352–2362
5. Kleintjens LA, Koningsveld R (1980) Liquid–liquid phase separation in multicomponent polymer systems XIX. Mean-field lattice-gas treatment of the system n-alkane/linear polyethylene. Colloid Polymer Sci 258:711–718
6. Beret S, Prausnitz JM (1975) Perturbed Hard-Chain Theory: an equation of state for fluids containing small or large molecules. AIChE J 21:1123–1132
7. Cotterman RL, Schwarz BJ, Prausnitz JM (1986) Molecular thermodynamics for fluids at low and high densities. Part 1. Pure fluids containing small or large molecules. AIChE J 32:1787–1798
8. Morris WO, Vimalchand P, Donohue MD (1987) The Perturbed-Soft-Chain Theory: an equation of state based on the Lennard–Jones potential. Fluid Phase Equilib 32:103–115
9. Barker JA, Henderson D (1967) Perturbation theory and equation of state for fluid II. A successful theory of liquids. J Chem Phys 47:4714–4721
10. Weeks JD, Chandler D, Anderson HC (1971) Role of repulsive forces in determining the equilibrium structure of simple liquids. J Chem Phys 54:5237–5247
11. Carnahan NF, Starling KE (1969) Equation of state for nonattracting rigid spheres. J Chem Phys 51:635–636
12. Chapman WG, Gubbins KE, Jackson G, Radosz M (1989) SAFT: Equation-of-State Model for associating fluids. Fluid Phase Equilib 52:31–38
13. Chapman WG, Gubbins KE, Jackson G, Radosz M (1990) New reference equation of state for associating liquids. Ind Eng Chem Res 29:1709–1721
14. Huang SH, Radosz M (1990) Equation of state for small, large, polydisperse, and associating molecules. Ind Eng Chem Res 29:2284–2294
15. Huang SH, Radosz M (1991) Equation of state for small, large, polydisperse, and associating molecules: extension to fluid mixtures. Ind Eng Chem Res 30:1994–2005
16. Chen SS, Kreglewski A (1977) Applications of the augmented van der Waals theory of fluids I. Pure fluids. Ber Bunsen-Ges Phys Chem 81:1048–1052
17. Wertheim MS (1987) Thermodynamic perturbation theory of polymerization. J Chem Phys 87:7323–7331
18. Song Y, Lambert SM, Prausnitz JM (1994) Equation of state for mixtures of hard sphere chains including copolymers. Macromolecules 27:441–448
19. Song Y, Lambert SM, Prausnitz JM (1994) A perturbed hard-sphere-chain equation of state for normal fluids and polymers. Ind Eng Chem Res 33:1047–1057

20. Gil-Villegas A, Galindo A, Whitehead PJ, Mills SJ, Jackson G, Burgess AN (1997) Statistical associating fluid theory for chain molecules with attractive potentials of variable range. J Chem Phys 106:4168–4186
21. Chang J, Sandler SI (1994) A completely analytic perturbation theory for the square-well fluid of variable well width. Mol Phys 81:745–765
22. Hino T, Prausnitz JM (1997) A perturbed hard-sphere-chain equation of state for normal fluids and polymers using the square-well potential of variable width. Fluid Phase Equilib 138:105–130
23. Blas FJ, Vega LF (1997) Thermodynamic behaviour of homonuclear Lennard–Jones chains with association sites from simulation and theory. Mol Phys 92:135–150
24. Gross J, Sadowski G (2001) Perturbed-Chain SAFT: an equation of state based on a perturbation theory for chain molecules. Ind Eng Chem Res 40:1244–1260
25. Gross J, Sadowski G (2002) Modeling polymer systems using the Perturbed-Chain SAFT equation of state. Ind Eng Chem Res 41:1084–1093
26. Tumakaka F, Gross J, Sadowski G (2002) Modeling of polymer phase equilibria using Perturbed-Chain SAFT. Fluid Phase Equilib 194–197:541–551
27. Boublik T (1970) Hard-sphere equation of state. J Chem Phys 53:471–472
28. Mansoori GA, Carnahan NF, Starling KE, Leland TW Jr (1971) Equilibrium thermodynamic properties of hard spheres. J Chem Phys 54:1523–1525
29. Kouskoumvekaki I, von Solms N, Michelsen ML, Kontogeorgis G (2004) Application of the Perturbed Chain SAFT equation of state to complex polymer systems using simplified mixing rules. Fluid Phase Equilib 215:71–78
30. Jog PK, Chapman WG (1999) Application of Wertheim's thermodynamic perturbation theory to dipolar hard sphere chains. Mol Phys 97:307–319
31. Karakatsani EK, Spyriouni T, Economou IG (2005) Extended statistical associating fluid theory (SAFT) equations of state for dipolar fluids. AIChE J 51:2328–2342
32. Gross J, Vrabec J (2006) An equation of state contribution for polar components: dipolar molecules. AIChE J 52:1194–1201
33. Karakatsani EK, Economou IG (2006) Perturbed-chain statistical associating fluid theory extended to dipolar and quadrupolar molecular fluids. J Phys Chem B 110:9252–9261
34. Gross J (2005) An equation of state contribution for polar components: quadrupolar molecules. AIChE J 51:2556–2568
35. Vrabec J, Gross J (2008) A molecular based approach to dipolar und quadrupolar fluids: vapor-liquid equilibria simulation and an equation of state contribution for dipole-quadrupole interactions. J Phys Chem B 112:51–60
36. Cameretti L, Sadowski G, Mollerup J (2005) Modelling of aqueous electrolyte solutions with PC-SAFT. Ind Eng Chem Res 44:3355–3362
37. Held C, Cameretti L, Sadowski G (2008) Modeling aqueous electrolyte solutions. Part 1: fully dissociated electrolytes. Fluid Phase Equilib 270:87–96
38. Behzad B, Patel BH, Galindo A, Ghotbi C (2005) Modeling electrolyte solutions with the SAFT-VR equation using Yukawa potentials and the mean-spherical approximation. Fluid Phase Equilib 236:241–255
39. Tihic A, Kontogeorgis GM, von Solms N, Michelsen ML (2006) Application of the simplified perturbed-chain SAFT equation of state using an extended parameter table. Fluid Phase Equilib 248:29–43
40. Danner RP, High MS (1993) Handbook of polymer solution thermodynamics. DIPPR, AIChE/Wiley, New York
41. Sadowski G, Mokrushina L, Arlt W (1997) Finite and infinite dilution activity coefficients in polycarbonate systems. Fluid Phase Equilib 139:391–403
42. Becker F, Buback M, Latz H, Sadowski G, Tumakaka F (2004) Cloud-point curves of ethylene–(meth)acrylate copolymers in fluid ethene up to high pressures and temperatures – experimental study and PC-SAFT modeling. Fluid Phase Equilib 215:263–282

418   G. Sadowski

43. Kleiner M, Tumakaka F, Sadowski G, Latz H, Buback M (2006) Phase equilibria in polydisperse and associating copolymer solutions: poly(ethene-co-(meth)acrylic acid)–monomer mixtures. Fluid Phase Equilib 241:113–123

44. Krueger K-M, Pfohl O, Dohrn R, Sadowski G (2006) Phase equilibria and diffusion coefficients in the poly(dimethylsiloxane) + n-pentane system. Fluid Phase Equilib 241:138–146

45. Goernert M, Sadowski G (2008) Phase-equilibrium measurement and modeling of the PMMA/MMA/carbon dioxide ternary system. J Supercrit Fluids 46:218–225

46. Wohlfarth C (1994) Vapour-liquid equilibrium data of binary polymer solutions; physical sciences data 44. Elsevier, Amsterdam

47. Martin TM, Lateef AA, Roberts CB (1999) Measurements and modeling of cloud point behavior for polypropylene/n-pentane and polypropylene/n-pentane/carbon dioxide mixtures at high pressures. Fluid Phase Equilib 154:241–259

48. Hasch BM (1994) Hydrogen bonding and polarity in ethylene copolymer-solvent mixtures: experiment and modeling. Ph.D. Thesis, Johns Hopkins University, Baltimore, MD

49. Dietzsch H (1999) Hochdruck-Copolymerisation von Ethen und (Meth)Acrylsäureestern: Entmischungsverhalten der Systeme Ethen/Cosolvens/Poly(Ethen-co-Acrylsäureester) – Kinetik der Ethen-Methylmethacrylat-Copolymerisation. Ph.D. Thesis, Georg August-Universität zu Göttingen, Germany

50. Song Y, Hino T, Lambert SM, Prausnitz JM (1996) Liquid-liquid equilibria for polymer solutions and blends, including copolymers. Fluid Phase Equilib 117:69–76

51. Banaszak M, Chen CK, Radosz M (1996) Copolymer SAFT equation of state. Thermodynamic perturbation theory extended to heterobonded chains. Macromolecules 29:6481–6486

52. Shukla KP, Chapman WG (1997) SAFT equation of state for fluid mixtures of hard chain copolymers. Mol Phys 91:1075–1081

53. Gross J, Spuhl O, Tumakaka F, Sadowski G (2003) Modeling copolymer systems using the Perturbed-Chain SAFT equation of state. Ind Eng Chem Res 42:1266–1274

54. Chen S-J, Banaszak M, Radosz M (1995) Phase behavior of poly(ethylene-1-butene) in subcritical and supercritical propane: ethyl branches reduce segment energy and enhance miscibility. Macromolecules 28:1812–1817

55. de Loos TW, Poot W, Diepen GAM (1983) Fluid phase equilibriums in the system polyethylene + ethylene. 1. Systems of linear polyethylene + ethylene at high pressure. Macromolecules 16:111–117

56. Kehlen H, Rätzsch MT (1980) Continuous thermodynamics of multicomponent mixtures. In: Proceedings of the 6th international conference on thermodynamics, Merseburg, Germany, pp 41–51

57. Gualtieri JA, Kincaid JM, Morrison G (1982) Phase equilibria in polydisperse fluids. J Chem Phys 77:521–535

58. Rätzsch MT, Kehlen H (1983) Continuous thermodynamics of complex mixtures. Fluid Phase Equilib 14:225–234

59. Cotterman RL, Prausnitz JM (1985) Flash calculations for continuous or semicontinuous mixtures using an equation of state. Ind Eng Chem Proc Des Dev 24:434–443

60. Enders S (2010) Theory of random copolymer fractionation in columns. Adv Polym Sci. doi: 10.1007/12_2010_92

61. Buback M, Latz H (2003) Cloud-point pressure curves of ethene/poly(ethylene-co-meth acrylic acid) mixtures. Macromol Chem Phys 204:638–645

62. Beyer C, Oellrich LR (2002) Cosolvent studies bath the system ethylene/poly(ethylene-co-acrylic acid): Effects of solvent, density, polarity, hydrogen bonding, and copolymer composition. Helvetica Chimica Acta 85:659–670

63. Bungert B, Sadowski G, Arlt W (1997) Supercritical antisolvent fractionation: measurements in the systems monodisperse and bidisperse polystyrene-cyclohexane-carbon dioxide. Fluid Phase Equilib 139:349–359

64. Behme S, Sadowski G, Arlt W (1999) Modeling of the separation of polydisperse polymer systems by compressed gases. Fluid Phase Equilib 158–160:869–877

Index

Alkanes, in CO_2 366
  –in dipolar solvents 370
Ammonia 338, 370
Ammonium poly(acrylate) 131
Aqueous solutions 67

Baker–Williams fractionation 271, 288, 309
Blend solutions 57
Block copolymers 18, 46, 375
tert-Butyl acetate (TBA) 32

Cell theory 4, 95
Cellulose 54
  –swelling in water 43
CH/PS/PVME 60
Chloroform, vapor pressure 36
Coarse-grained models 341, 329
Compatibilizers 180
Cononsolvency 55, 59
Continuous polymer fractionation (CPF) 275, 291, 316
Continuous spin fractionation (CSF) 293
Continuous thermodynamics 271
Copolymers 389
  –fractionation 271
  –linear random 38
Counterion condensation 67
  –Manning 94, 101
Critical micelle concentration (CMC) 375
Cross-fractionation 305
Cyclohexane/poly(vinyl methyl ether) (CH/PVME) 28

Debye–Hückel theory 95
Dextran 45
Dextran–ethyl carbonate 288
Differential membrane osmometry 81
Dissipative particle dynamics (DPD) thermostat 358
Drop analysis 188

Electromotive force 81
Emulsifying agents 180
Epoxide resins 288
Equation of state 389, 392
Equilibrium dialysis (EQDIA) 85
Ethene/polyethylene 409
Ethylene-*co*-alkylacrylate 407
Ethylene–α-olefin 288
Ethylene vinyl acetate 296
Excess Gibbs energy 67
Explosive decompression failure (XDF) 138

Flame ionization detector (FID) 25
Flory–Huggins theory 4
Fractionation in column 271
Freezing point depression 80

Gas sorption 137, 140
Gas–polymer interactions 138
Gel deswelling 81
Gibbs energy 5, 75
  –polyelectrolytes, aqueous solutions 91

420 Index

Glass transitions 137
   –high pressures 169

HDPE/propane 407
High impact polystyrenes (HIPS) 180
High pressure 137
Hole theory 4
Homopolymers 401
   –branched 16
   –linear 5, 27
   –nonlinear 37
HS–GC 24
Hydrofluorocarbons (HFCs) 139
Hydrogen sulfide 338, 370

Ideal mixture 81
Interfacial partitioning 180
Interfacial tension 180
   –binary polymer blends 189
Inverse gas chromatography (IGC) 24, 26
Isopiestic experiments 81
Isotropic–nematic transition 371

Kuhn–Mark–Houwink relation 13, 28

Lattice theory 278
   –fluid 4
LCST 33
LDPE 403
Liouville's theorem 356
Liquid–liquid equilibria (LLE) 271, 276, 284
Lorentz–Berthelot combining rule 370

Mean field lattice gas model 4
Medium density polyethylene (MDPE) 150
Membrane osmometry 25, 81
Methyl acetate 38
Methyl ethyl ketone (MEK)/PDMS 27
Methylacrylate–ethene 407
Methylmethacrylate (MMA) 404
Metropolis algorithm 362
Micelle formation 375
Mixed solvents 1, 21
Modeling 1, 389
Molecular dynamics (MD) 329, 333
Monte Carlo 329, 362
   –step (MCS) per monomer 364

NMMO/H_2O/Solucell 400 54
Nylon 189

Osmometry 24, 25
Osmotic coefficient 67, 99

PDMS 191
Peng–Robinson (PR) 393
PEO 165, 190
Perturbation theories 4, 393
Perturbed chain statistical associating fluid
   theory (PC-SAFT) 332, 394
Perturbed hard-chain theory (PHCT) 4, 393
Phase diagrams 1, 329
PMA(Az) 165
PMMA/MMA/carbon dioxide 404
PMMA/PnBMA 204
Poisson–Boltzmann equation 95
Poly(2-acrylamido-2-methyl-1-propane
   sulfonic acid) (HPAMS) 77
Poly(acrylates) 84
Poly(acrylic acid) 77, 238
Poly(allylaminohydrochloride) (PAAm) 78
Poly(butadiene-*co*-styrene) 169
Poly(1-butene) 401, 407
Poly(1-butene)/propane 407
Poly(*n*-butyl methacrylate) (PnBMA) 204
Poly(diallyldimethyl ammonium chloride)
   (PDADMAC) 78
Poly(2,6-dimethyl-1,4-phenylene oxide) 181
Poly(dimethyl siloxanes) (PDMS) 190, 191,
   229, 401
Poly(ethyl acrylate) 401
Poly(ethyl ethylene) 188, 216
Poly(ethylene glycols) 190
Poly(ethylene oxide) (PEO) 165, 190
Poly(ethylene sulfonates) 85
Poly(ethylene terephthalate) 189
Poly(ethylene-*co*-1-butene) 407
Poly(ethylene-*co*-acrylic acid) 275, 413
Poly(ethylene-*co*-acrylic acid),
   pseudocomponents 413
Poly(ethylene-*co*-alkylacrylate)/ethene 408
Poly(ethylene-*co*-methylacrylate) 407
Poly(ethylene-*co*-propylacrylate) 407
Poly(ethyleneimine) (PEI) 78
Poly(hexyl methyl siloxane) 235
Poly(methacrylic acid) 77
Poly(2-(methacryloyloxy) ethyl trimethyl
   ammonium chloride) (PMETAC) 78

Index 421

Poly(methylacrylate) 408
Poly(methylmethacrylate) (PMMA) 16, 39, 191, 404
Poly(phosphoric acid) (HPP) 77
Poly(propylacrylate) 408
Poly(propylene oxide) (PPO) 190
Poly(sodium ethylene sulfate) (NaPES) 82
Poly(sodium methacrylate) 128
Poly(sodium styrene sulfonate) (NaPSS) 82
Poly(sodium vinyl sulfate) (NAPVAS) 84
Poly(styrene carboxylic acid) 77
Poly(styrene sulfonic acid) 77
Poly(styrene-*ran*-methyl methacrylate) 38
Poly(trimethylammonium methyl methacrylate) (PTMAC) 78
Poly(vinyl acetate) 204
Poly(vinyl amine) (PVAm) 78
Poly(vinyl benzoic acid) 77
Poly(vinyl ethylene) (PVE) 216
Poly(vinyl methyl ether) (PVME) 17, 28, 40, 48, 255
Poly(vinyl sulfonic acid) (HPVS) 77
Poly(vinyl sulfuric acid) (HPVAS) 77
Poly(vinylbenzene trimethylammonium chloride) (PVBTMAC) 78
Poly(vinylidene fluoride) (PVDF) 150, 162
Poly(2-vinylpyridine) (P2VP) 226
1,2-Polybutadiene 27, 34
1,4-Polybutadiene 27, 34, 44
Polycarbonate 399
Polydispersity 74, 408
Polyelectrolytes 67
Polyethylene (PE) 189
  –united atom model 335
Polyethylene/toluene 401
Polyethylene oxide (PEO) 31, 36
Polymer blends 1, 18
Polymer incompatibility 50
Polymer interfaces 180
Polymer solutions 1
Polymer–polymer interfaces 196
Polypropylene (PP) 193
Polypropylene/*n*-pentane 402
Polysaccharides 17
Polystyrene (PS) 16, 48, 139, 180, 192
Polyvinyl acetate (PVA) 169
Pressure decay 142
PS/PBD/styrene 208
PS/PMMA 192
PS-*b*-poly(acrylic acid) 238
PS-*b*-poly(ethyl ethylene) 216

PS-*b*-poly(hexyl methyl siloxane)-*b*-PS 235
Pullulan 45
PVC 226
PVME/PS 48

Salt effects 67
Sanchez–Lacombe theory 4, 35
Scaling law 95
Scanning transitiometry 144
Segment molar Gibbs energy 5
Self-assembling 137
Self-consistent mean field (SCMF) theory 206
Soave–Redlich–Kwong (SRK) 393
Sodium carboxymethylcellulose (NaCMC) 77, 84
Sodium dextran sulfate (NaDS) 77
Sodium poly(acrylate) 130
Sodium poly(styrene sulfonate) 91
Solubility 137, 389
Solutions, unmixing 329
Sorption 389
SSF/SPF 300
Statistical-associating-fluid theory (SAFT) 4, 332, 394
Stepwise fractionation 299
Stiff polymers 371
Styrene/butadiene diblock (SB) 47
Styrene–acrylic acid 288
Styrene–acrylonitrile 288
Styrene–butadiene 288
Styrene–2-methoxyethyl methacrylate 288
Styrene–methyl methacrylate 288
Successive precipitation fractionation (SPF) 285
Successive solution fractionation (SSF) 285
Supercritical carbon dioxide 150
Supercritical fluids (SCFs) 140
Surface light scattering 184

Temperature rising elution fractionation (TREF) 275
Temperature-programmed column fractionation (TPCF) 275
Ternary mixtures 1, 21, 53
Thermodynamic perturbation theory (TPT) 332
Thermodynamics 1, 67, 389
THF/PS/PVME 60

TL/PDMS 27
Toluene 30
TPT1-MSA 336
Transitiometry 137, 144

UCST 33
UNIFAC model 4
UNIQUAC model 4

Vapor pressure osmometry 25, 81
Vapor–liquid equilibrium 67
   –aqueous polyelectrolyte solutions 80
Velocity–Verlet algorithm 356
Verlet algorithm 356
Vibrating-wire (VW)–pressure-volume-
   temperature (pVT) technique 140
Vibrating-wire technique/sensor 137, 141
Vinyl chloride–vinyl acetate 288